AF358644

Advances in Marine Engineering: Geological Environment and Hazards II

Contents

About the Editors

Xiaolei Liu

Dr. Xiaolei Liu is Professor and Doctoral supervisor at Ocean University of China (OUC). He graduated in environmental engineering with a Ph.D. from OUC in 2014 when he started working on marine engineering geological disasters and environmental problems at the College of Environmental Science and Engineering of OUC. He is currently the director of the Institute of Marine Engineering Geology and the Environment at OUC and the deputy director of Shandong Provincial Key Laboratory of Marine Environment and Geological Engineering. Dr. Liu is the PI of more than 10 national and provincial scientific research projects in marine engineering geology and has been consulted on many submarine disaster problems for development of offshore wind power and underwater oil and gas throughout China. He has gained 7 Ministerial and Provincial-Level Science and Technology Awards, published more than 120 academic papers, and authorized more than 60 invention patents. Dr. Liu is the chairman of Committee on Marine Geo-disaster and Geo-environment (TC-4) in International Consortium on Geo-disaster Reduction (ICGdR), the deputy director of Marine Engineering Geo-disaster Prevention and Control Branch of Chinese Society for Rock Mechanics and Engineering (CSRME), and the editorial board member of several international journals.

Xingsen Guo

Dr. Xingsen Guo is a Research Fellow at University College London, focusing on marine engineering geo-disasters and geo-environmental issues. He earned his PhD from Dalian University of Technology (DUT) in 2021. Dr. Guo has published 68 peer-reviewed articles in leading journals, including *Acta Geotechnica*, *Coastal Engineering*, *Computers and Geotechnics*, *Engineering Geology*, *Landslides*, *Ocean Engineering*, and *Physics of Fluids*. His work has been widely recognized, with two ESI hot papers, seven ESI highly cited papers, and a cover article. He also serves on the editorial boards of several journals, such as *Marine Georesources & Geotechnology* and *Frontiers in Earth Science*, and and serves as the Secretary-General of the Marine Geo-disaster and Geo-environment Committee (TC-4) in the International Consortium on Geo-disaster Reduction (ICGdR). His accolades include being named among the World's Top 2% Scientists (2024), the Geoenvironmental Disasters Best Paper Award (2024), the BGA (British Geotechnical Association) Poster Award (2024), the JMSE Travel Award (2023), the ICGdR Outstanding Young Scientist Award (2023), the 2nd National Postdoctoral Innovation and Entrepreneurship Competition Silver Award (2023), the BGA Fund Award (2022), and several Excellent Doctoral Dissertation Awards (ICGdR, the Chinese Society for Rock Mechanics & Engineering, Liaoning Province, and DUT) between 2021 and 2023.

Thorsten Stoesser

Dr. Thorsten Stoesser is the leader of the Fluid Mechanics Research Group in the Department of Civil, Environmental, and Geomatic Engineering at University College London. His research interest is in developing advanced computational fluid dynamics (CFD) tools and their application to solve environmental fluid mechanics problems. Thorsten has published over 100 peer-reviewed journal papers on developing, testing, and applying advanced CFD methods to predict the hydrodynamics and transport processes in rivers, estuaries, and coastal waters, the fluid–structure interaction of marine turbines, and the nearfield dynamics of jets and plumes. For his research, Prof. Stoesser twice received the American Society of Civil Engineers (ASCE) Karl Emil Hilgard Hydraulic Prize

(2012 and 2016); in 2015, he won the International Association of Hydro-Environmental Research (IAHR) Harold Shoemaker Award; and in 2016, he won the Institution of Civil Engineers' George Stephenson Medal. He has received over GBP 5 M in funding from industry, government institutions, and research councils, including the DFG (Germany), NSF (USA), and EPSRC (UK).

Journal of
Marine Science and Engineering

Editorial

Advances in Marine Engineering: Geological Environment and Hazards II

Xingsen Guo [1,2,3,4,5], Xiaolei Liu [1,6,*] and Thorsten Stoesser [2]

1   Shandong Provincial Key Laboratory of Marine Environment and Geological Engineering, Ocean University of China, Qingdao 266100, China; xingsen.guo@ucl.ac.uk
2   Department of Civil, Environmental and Geomatic Engineering, University College London, London WC1E 6BT, UK; t.stoesser@ucl.ac.uk
3   State Key Laboratory of Geohazard Prevention and Geoenvironment Protection, Chengdu University of Technology, Chengdu 610059, China
4   State Key Laboratory of Coastal and Offshore Engineering, Dalian University of Technology, Dalian 116024, China
5   Shandong Provincial Key Laboratory of Ocean Engineering, Ocean University of China, Qingdao 266100, China
6   Laboratory for Marine Geology, Qingdao Marine Science and Technology Center, Qingdao 266237, China
*   Correspondence: xiaolei@ouc.edu.cn

Citation: Guo, X.; Liu, X.; Stoesser, T. Advances in Marine Engineering: Geological Environment and Hazards II. *J. Mar. Sci. Eng.* **2024**, *12*, 1253. https://doi.org/10.3390/jmse12081253

Received: 25 May 2024
Revised: 3 June 2024
Accepted: 8 July 2024
Published: 25 July 2024

1. Introduction

In October 2021, the editorial office invited Prof. Xiaolei Liu, from the Ocean University of China, Prof. Thorsten Stoesser, from University College London, and Dr. Xingsen Guo, from University College London, to serve as guest editors for the special issue titled *"Advances in Marine Engineering: Geological Environment and Hazards"* [1]. Their task was to collect high-quality papers in the field of marine geological environments and hazards, including marine geological environments, marine geological hazards, marine engineering geology, marine hydrodynamics, marine fluid mechanics, and marine geotechnical engineering. As of October 2022, this Special Issue comprised one review paper, one editorial paper, and thirteen research papers, all of which have been compiled into a single journal issue [2]. The success of this Special Issue led to an invitation from the editorial office to extend it until April 2024, with a subsequent Special Issue titled *"Advances in Marine Engineering: Geological Environment and Hazards II"*.

As of April 2024, the new Special Issue includes one review paper [3] and fifteen research papers [4–18] that cover different aspects related to the subject. These papers showcase the latest advancements in research, introducing state-of-the-art concepts, sophisticated methodologies, and valuable data. Their collective contributions are poised to significantly advance the development of marine geological environments and hazards. A comprehensive synthesis of the key findings and noteworthy contributions from each paper within this Special Issue is presented in the following section.

2. Papers Details

Chen et al. [3] conducted a comprehensive review of seabed response induced by nonlinear internal waves, providing an overview of the theories, models, and limited observations that have contributed to our current understanding. They highlight that the pressure disturbance generated by nonlinear internal waves results from the combined effects of interface displacement and near-bottom acceleration. Recent observations in the South China Sea have revealed pressure magnitudes of up to 4 kPa, representing the largest known disturbance caused by nonlinear internal waves. During the shoaling and breaking of these waves, intense pore-pressure variations occur in approximately the top 1 m of the weakly conductive seabed, resulting in transient liquefaction and the appearance of pebbles on the local seabed. The review emphasizes the importance of in situ observations to validate theoretical knowledge and enhance our ability to model the multiscale interaction process between the seabed and internal waves in the future.

Alhaddad et al. [4] assessed the efficacy of an empirical evaluation method aimed at determining the trajectory of failure development during breaching. This involved investigating whether the breach would progressively enlarge (destabilizing breaching) or diminish (stabilizing breaching). The researchers conducted extensive large-scale experiments, observing both stabilizing and destabilizing failure modes. They point out significant inaccuracies in existing methods used for evaluating failure modes, revealing an average absolute percentage error of up to 92% in practical applications. Consequently, Alhaddad et al. recommend the adoption of more sophisticated three-dimensional numerical simulation methods to precisely predict stable and unstable modes during underwater slope failure. The outcomes of this study are important to improving the accuracy of safety assessments for underwater infrastructure and flood protection structures.

Jiang et al. [5] conducted a comprehensive investigation into the monotonic and cyclic performance of composite bucket foundations using centrifuge modeling. They subjected the foundations to monotonic loads to simulate extreme wave conditions and cyclic loads to mimic long-term serviceability conditions. Through well-designed experiments, the researchers examined the effects of different loading types, soil strength, and load eccentricities. Their findings elucidated the failure patterns of ocean structures by analyzing parameters such as rotation, displacement, and pore pressure. They revealed that the variation in failure mechanisms is largely determined by soil resistance. Additionally, the researchers discovered an intriguing phenomenon: non-symmetric loading does not impact bucket foundations as severely as symmetric loading. The outcomes of this study are important for controlling the deformation of underwater infrastructure by harnessing deep-soil resistance.

Kan et al. [6] assessed the impact of the physical–mechanical properties of seafloor sediments on sound speed and formulated prediction equations for sediment sound speed based on either single or dual physical–mechanical parameters on the East China Sea shelf. The researchers highlighted that the determination coefficients of the dual-parameter prediction equations for sediment sound speed all exceeded 0.90, surpassing those of the single-parameter equations. This indicates superior predictive performance. The prediction equations developed in this study serve as valuable additions to marine geoacoustic models, significantly enhancing our understanding of the acoustic properties of seafloor sediments on the East China Sea shelf. They hold substantial significance for obtaining accurate acoustic property information in this region.

Li et al. [7] employed a combination of in situ techniques, sediment sampling, and laboratory measurements to gather data on sediment acoustic properties (such as sound speed and attenuation) and physical properties (including particle composition, density, porosity, and mean grain size) in the northwestern South China Sea. Their investigation revealed notable differences between laboratory and in situ measurements of acoustic properties, particularly for shallow-water coarse-grained sediments and deep-sea sediments, with laboratory measurements generally yielding higher values. Building on this observation, the researchers established relationships between measured attenuation and physical properties, as well as between sound speed and mean grain size, which differed from previous empirical equations. Furthermore, their study unveiled significant variations in sediment acoustic and physical properties in the downslope direction, with more gradual variations observed in the along-slope direction.

Li et al. [8] delved into the effects of high temperatures on the microstructure and mechanical characteristics of high-strength fiber composite materials (HSFCM), crucial for understanding material behavior in harsh environments. Through thorough experimentation and analysis, the study revealed the significant impacts of high temperatures on the mechanical properties of HSFCM. Specifically, the material exhibited decreased strength and hardness at elevated temperatures, attributed to alterations in its microstructure. Notably, high temperatures induced softening of the matrix material and failure at fiber interfaces, leading to a decline in overall performance. Moreover, the study investigated the mechanism by which high temperatures affect the microstructure of HSFCM, uncovering distortions in internal lattice structures and grain growth as contributors to structural instability, thereby influencing mechanical properties. This understanding provides valuable guidance for future research into material behavior under extreme conditions.

Meng et al. [9] utilized the horizontal-to-vertical spectral ratio (HVSR) method to evaluate silt liquefaction hazards within the Yellow River Delta and explored its applicability in seismic engineering. Their study revealed a fundamental frequency ranging from 0.8 to 9.8 Hz, with corresponding amplification values ranging from 1.8 to 3.5. Soil characteristics exerted a significant influence on liquefaction susceptibility, with higher vulnerability index values indicating increased potential. Particularly noteworthy were the elevated values observed in the southwestern beaches of the Yellow River Delta, suggesting heightened liquefaction susceptibility in that area. Consequently, it is important to consider the liquefaction potential of sediment deposits in engineering endeavors and assessments of foundation stability. The HVSR method provides a rapid means of assessing liquefaction potential and holds promise for applications in seismic engineering and geology.

Niu et al. [10] introduced novel prediction methods for evaluating the hydrodynamic performance of box-typed free surface breakwater under actual marine wave conditions. They developed an improved viscous numerical wave tank incorporating a mass source wave maker, which demonstrated high accuracy in simulating wave behaviors using computational fluid dynamics (CFD) technology. Their study extensively delineated the wave force characteristics of the box-typed free surface breakwater and identified key influencing factors. These findings are invaluable for enhancing the design and behavior analysis of marine breakwaters.

Sun et al. [11] formulated a predictive model for assessing borehole instability in horizontal wells situated within hydrate-bearing clayey-silt formations. Their research emphasized the optimal azimuthal arrangement of the borehole (60–120°) to mitigate drilling risks. In their paper, they suggested that hydrate dissociation could increase collapse pressure and compromise borehole stability. These insights hold significant implications for drilling operations in hydrate-bearing clayey-silt sediments within the northern South China Sea region.

Sun et al. [12] developed a deep learning model to accurately assess the dynamic changes in the physical properties of seafloor sediments, such as density, water content, and porosity. They utilized both the empirical formula and the deep learning model to map the spatial distribution and temporal variations of these properties in the hydrate test area of the South China Sea over a period of 12 days. Their analysis demonstrated that the deep learning model provided a better fit and more precise predictions compared to the empirical formula. Based on in situ observations within the hydrate zone, they identified a four-layer structure in the sediment's physical properties and explored the potential impact of hydrate decomposition. This study contributes to the monitoring and early warning of changes in seafloor sediment properties, supporting the safety of marine engineering projects.

Wang et al. [13], through systematic experimentation, investigated the bonding strength, shear performance, and tensile strength of HIRA-type materials in various bonding scenarios. Subsequently, they meticulously tracked and observed the development of the material's bonding capability over extended periods. Their observations revealed the resilience of HIRA-type materials in anchoring solids over different time scales, as evidenced by the fluctuation patterns in bonding strength over time. Moreover, through in-depth analysis of the experimental data, they identified patterns indicating rapid changes or persistent increases in material performance at specific time points. These findings are crucial for enhancing our understanding of the material's long-term reliability. The comprehensive experimental data generated by this study serve as essential resources for future maritime engineering structure design endeavors aimed at enhancing stability and safety.

Wang et al. [14] introduced a methodology aimed at evaluating the susceptibility of wave-induced seabed liquefaction, utilizing deterministic analysis and empirical Bayesian Kriging (EBK). This approach integrates the safety factor, determined by combining cyclic stress ratio and cyclic resistance ratio, as key evaluation parameters. It employs EBK interpolation within ArcGIS to delineate the susceptibility of wave-induced seabed liquefaction in a localized study area. In their study, Wang et al. implemented this method in the Chengdao area, calculating safety factors. They subsequently generated wave-induced seabed liquefaction susceptibility maps under 5-year, 10-year, and 25-year wave conditions. These maps facilitated discussions on the engineering geological conditions of the Chengdao area. This methodology offers a quantitative means of assessing liquefaction susceptibility in the study area, providing valuable insights for the selection and maintenance of marine engineering facilities. As such, it holds considerable engineering significance.

Xie et al. [15] proposed a convenient simulation method for the runout simulation of submarine landslides based on the Eulerian analysis technology of Abaqus/Explicit, providing a large deformation calculation method in an explicit finite element scheme. A flume experiment and a simple submarine landslide case were used to validate the proposed model. The simulation results reported in this paper demonstrate good consistency with those of the flume experiment and other widely validated numerical methods. Focusing on the potential mass movements of submarine landslides in the Shenhu Sea area on the northern slope of the South China Sea, the runout process was analyzed under different combinations of soil parameter cases by using the convenient method proposed. The shear strain softening and rate-dependency effects are highly involved in the runout process. The simulated landslide's failure mode is consistent with the geophysical interpretation of existing landslide characteristics.

Xu et al. [16] conducted numerical simulations to confirm the high-speed movement of submarine turbidity currents. They validated the underwater movement dynamics of turbidity currents through laboratory flume experiments and analyzed the propagation speed and amplitude of excitation waves induced by their movement. The aim was to identify the controlling factors and formulate expressions for the propagation speed of these excitation waves. Through extensive numerical simulations across various field scales, the researchers monitored the propagation speed of excitation waves and the resulting changes in surface elevation. Their findings revealed that the propagation speed of excitation waves greatly exceeds the speed of the turbidity current itself, with its magnitude solely dictated by water depth. This research offers a fresh perspective on the rapid movement of turbidity currents in submarine canyons, enhancing our comprehension of sediment resuspension and transport in underwater environments.

Yu et al. [17] investigated the vortex-induced vibration (VIV) characteristics of a riser model tensioned by a buoyancy can, which experiences low-frequency vortex-induced motion. They conducted a series of model tests involving the combined riser model and buoyancy can under uniform flow, regular wave, and wave–current conditions. The researchers employed various methods, such as mode superposition, Euler angle conversion, band-pass filtering, and signal processing techniques, such as fast Fourier and wavelet transforms, to analyze the testing data. By exploring the dynamic interactions between the real structure's environmental loads, dynamic boundary conditions, and riser VIV under different environmental scenarios, the study revealed crucial insights. The authors underscored the importance of considering the dynamic boundary of the platform or vessel motion in VIV fatigue assessments during riser design. This highlights the necessity for a comprehensive understanding of environmental influences on riser performance and structural integrity.

Yu et al. [18] presented a novel testing methodology for investigating soil–structure interaction in marine engineering, aiming to offer insights into the safety assessment of marine structures within the Yellow River Delta. The researchers conducted cyclic shear tests on the steel–silt interface under constant normal load conditions, systematically examining the influence of normal stress, shear amplitude, roughness, and water content on key parameters, such as the interface shear strength, shear stiffness, and damping ratio. Their findings underscored the substantial impact of normal stress and shear amplitude on the shear strength, stiffness, and damping ratio at the interface. Moreover, roughness and water content emerged as pivotal factors shaping the variation of these parameters with the number of cycles. Importantly, the outcomes of this study are poised to provide valuable technical guidance for understanding other forms of soil–structure interactions in marine engineering applications.

Author Contributions: Conceptualization, X.G., X.L. and T.S.; investigation, X.G.; resources, X.L. and T.S.; writing—original draft preparation, X.G.; writing—review and editing, X.L. and T.S. All authors have read and agreed to the published version of the manuscript.

Funding: Financial support for this work is provided by the National Natural Science Foundation of China (42277138), the Fundamental Research Funds for the Central Universities (202441003), the Shandong Province National-Level Leading Talent Supporting Project (2022GJJLJRC-15), the Opening Fund of State Key Laboratory of Geohazard Prevention and Geoenvironment Protection at Chengdu University of Technology (SKLGP2023K001), the Opening Fund of the State Key Laboratory of Coastal and Offshore Engineering at Dalian University of Technology (LP2310), the Shandong Provincial Key Laboratory of Ocean Engineering with a grant from the Ocean University of China (kloe202301), and the Shandong Provincial Key Laboratory of Marine Environment and Geological Engineering with a grant from the Ocean University of China (MEGE2024001 and MEGE2024002).

Acknowledgments: We express our sincere gratitude to all the authors, the academic editors, and the reviewers for their invaluable contributions to this work. Additionally, we extend our thanks to the Editorial Board for their support, with special appreciation to the contact editors for their assistance in managing the daily procedures.

Conflicts of Interest: The authors declare no conflicts of interest.

References

1. Guo, X.; Liu, X.; Stoesser, T. Advances in Marine Engineering: Geological Environment and Hazards. *J. Mar. Sci. Eng.* **2023**, *11*, 475. [CrossRef]
2. Liu, X.; Stoesser, T.; Guo, X. *Advances in Marine Engineering: Geological Environment and Hazards*; MDPI Books: Basel, Switzerland, 2023.
3. Chen, T.; Li, Z.; Nai, H.; Liu, H.; Shan, H.; Jia, Y. Seabed Dynamic Responses Induced by Nonlinear Internal Waves: New Insights and Future Directions. *J. Mar. Sci. Eng.* **2023**, *11*, 395. [CrossRef]
4. Alhaddad, S.; Weij, D.; van Rhee, C.; Keetels, G. Stabilizing and Destabilizing Breaching Flow Slides. *J. Mar. Sci.* **2023**, *11*, 560. [CrossRef]
5. Jiang, M.; Lu, Z.; Cai, Z.; Xu, G. Centrifuge Modelling of Composite Bucket Foundation Breakwater in Clay under Monotonic and Cyclic Loads. *J. Mar. Sci.* **2024**, *12*, 469. [CrossRef]
6. Kan, G.; Lu, J.; Meng, X.; Wang, J.; Zhang, L.; Li, G.; Chen, M. Prediction Model of the Sound Speed of Seafloor Sediments on the Continental Shelf of the East China Sea Based on Empirical Equations. *J. Mar. Sci.* **2023**, *12*, 27. [CrossRef]
7. Li, G.; Wang, J.; Meng, X.; Hua, Q.; Kan, G.; Liu, C. Seafloor Sediment Acoustic Properties on the Continental Slope in the Northwestern South China Sea. *J. Mar. Sci.* **2024**, *12*, 545. [CrossRef]
8. Li, Y.; Meng, Q.; Zhang, Y.; Peng, H.; Liu, T. Effect of High Temperature on Mechanical Properties and Microstructure of HSFCM. *J. Mar. Sci.* **2023**, *11*, 721. [CrossRef]
9. Meng, Q.; Li, Y.; Wang, W.; Chen, Y.; Wang, S. A case study assessing the liquefaction hazards of silt sediments based on the horizontal-to-vertical spectral ratio method. *J. Mar. Sci.* **2023**, *11*, 104. [CrossRef]
10. Niu, G.; Chen, Y.; Lv, J.; Zhang, J.; Fan, N. Determination of Formulae for the Hydrodynamic Performance of a Fixed Box-Type Free Surface Breakwater in the Intermediate Water. *J. Mar. Sci.* **2023**, *11*, 1812. [CrossRef]
11. Sun, X.; Hu, Q.; Li, Y.; Chen, M.; Zhang, Y. Stability characteristics of horizontal wells in the exploitation of hydrate-bearing clayey-silt sediments. *J. Mar. Sci.* **2022**, *10*, 1935. [CrossRef]
12. Sun, Z.; Fan, Z.; Zhu, C.; Li, K.; Sun, Z.; Song, X.; Xue, L.; Liu, H.; Jia, Y. Study on the Relationship between Resistivity and the Physical Properties of Seafloor Sediments Based on the Deep Neural Learning Algorithm. *J. Mar. Sci.* **2023**, *11*, 937. [CrossRef]
13. Wang, K.; Meng, Q.; Zhang, Y.; Peng, H.; Liu, T. Experimental Study on the Bonding Performance of HIRA-Type Material Anchor Solids Considering Time Variation. *J. Mar. Sci.* **2023**, *11*, 798. [CrossRef]
14. Wang, Y.; Guo, X.; Liu, J.; Hou, F.; Zhang, H.; Gao, H.; Liu, X. A Methodology for Susceptibility Assessment of Wave-Induced Seabed Liquefaction in Silt-Dominated Nearshore Environments. *J. Mar. Sci. Eng.* **2024**, *12*, 785. [CrossRef]

15. Xie, Q.; Xu, Q.; Xiu, Z.; Liu, L.; Du, X.; Yang, J.; Liu, H. Convenient Method for Large-Deformation Finite-Element Simulation of Submarine Landslides Considering Shear Softening and Rate Correlation Effects. *J. Mar. Sci.* **2024**, *12*, 81. [CrossRef]
16. Xu, G.; Sun, S.; Ren, Y.; Li, M.; Chen, Z. Propagation Velocity of Excitation Waves Caused by Turbidity Currents. *J. Mar. Sci. Eng.* **2024**, *12*, 132. [CrossRef]
17. Yu, C.; Zhang, S.; Zhang, C. Coupling Effects of a Top-Hinged Buoyancy Can on the Vortex-Induced Vibration of a Riser Model in Currents and Waves. *J. Mar. Sci. Eng.* **2024**, *12*, 751. [CrossRef]
18. Yu, P.; Dong, J.; Guan, Y.; Wang, Q.; Jia, S.; Xu, M.; Liu, H.; Yang, Q. Experimental Investigation on the Cyclic Shear Mechanical Characteristics and Dynamic Response of a Steel–Silt Interface in the Yellow River Delta. *J. Mar. Sci.* **2023**, *11*, 223. [CrossRef]

Short Biography of the Authors

Xingsen Guo is a research fellow in the Department of Civil, Environmental, and Geomatic Engineering at University College London (UCL) and a member of the Fluid Mechanics Research Group at UCL since 2020. He obtained his PhD at the Dalian University of Technology (DUT) in 2021. His research mainly concerns marine engineering geology (e.g., surficial sediment and hydrodynamic environments) and geotechnical engineering (e.g., pipeline and pile), marine geological hazards (e.g., seabed stability and submarine landslides), and computational fluid dynamics in turbidity currents and deep-sea mining plumes. He has published over 60 peer-reviewed journal papers (e.g., *Acta Geotechnica, Coastal Engineering, Computers and Geotechnics, Engineering Geology, Landslides, Ocean Engineering,* and *Physics of Fluids*) and has been recognized with an ESI hot paper, five ESI highly cited papers, and a cover paper. He also serves as a valuable editorial board member for several journals (e.g., *Marine Georesources & Geotechnology, FDMP-Fluid Dynamics & Materials Processing,* and *Frontiers in Earth Science*). He has organized four Special Issues in the *Journal of Marine Science and Engineering* and *Frontiers in Marine Science* and served as the Secretary-General of the Marine Geo-disaster and Geo-environment Committee in the ICGdR and the Deputy Secretary-General of the Coastal and Offshore Engineering Disaster and Environmental Protection Committee in the Seismological Society of China. For his research, he was honored with the Geoenvironmental Disasters Best Paper Award in 2024, the BGA (British Geotechnical Association) Poster Award in 2024, the ICGdR Outstanding Young Scientist Award in 2023, the Liaoning Province Excellent Doctoral Dissertation Award in 2023, the 2nd National Postdoctoral Innovation and Entrepreneurship Competition Silver Award in 2023, the JMSE Travel Award in 2023, the ICGdR Excellent Doctoral Dissertation Award in 2022, the Excellent Doctoral Dissertation Award of the Chinese Society for Rock Mechanics & Engineering (CSRME) in 2021, the DUT Excellent Doctoral Dissertation Award in 2022, the BGA Fund Award in 2022, and the Liu Huixian Earthquake Engineering Scholarship Award awarded by the Huixian Earthquake Engineering Foundation (China) and the US–China Earthquake Engineering Foundation (USA) in 2020.

Xiaolei Liu works as a professor and doctoral supervisor at the Ocean University of China (OUC). He is the director of the Institute of Marine Engineering Geology and the Environment at OUC and the deputy director of the Shandong Provincial Key Laboratory of Marine Environment and Geological Engineering. Additionally, he is the chairman of the Committee on Marine Geo-disaster and Geo-environment (TC-4) in the International Consortium on Geo-disaster Reduction (ICGdR). His research interests are focused on marine engineering, the geological environment, and related disasters, including the phase transition mechanism of submarine sediment, the evolution mechanism of fluidized sediment movement, and probe techniques for observing seabed interface layers. He has obtained the first prize of the Science Research Famous Achievement Award in Higher Institution, the first prize of the Qingdao City Science and Technology Progress Award, and the special prize of the Technological Invention Award from the China Association of Oceanic Engineering. He has undertaken more than 10 national scientific research projects and published over 120 peer-reviewed journal papers in mainstream journals within the international marine engineering geology field such as *Engineering Geology, Journal of Geophysical Research: Oceans, Landslides, Marine Geology, Ocean Engineering,* etc. At the same time, he has published three monographs and authorized more than 60 invention patents from China, the United States, Japan and Europe.

Thorsten Stoesser is the leader of the Fluid Mechanics Research Group in the Department of Civil, Environmental, and Geomatic Engineering at University College London. His research interest is in developing advanced computational fluid dynamics (CFD) tools and their application to solve environmental fluid mechanics problems. Thorsten has published over 100 peer-reviewed journal papers on developing, testing, and applying advanced CFD methods to predict the hydrodynamics and transport processes in rivers, estuaries and coastal waters, the fluid–structure interaction of marine turbines, and the nearfield dynamics of jets and plumes. For his research, Prof. Stoesser received twice the American Society of Civil Engineers (ASCE) Karl Emil Hilgard Hydraulic Prize (2012 and 2016); in 2015, he won the International Association of Hydro-Environmental Research (IAHR) Harold Shoemaker Award; and in 2016, he won the Institution of Civil Engineers' George Stephenson Medal. He has received over GBP 5 M in funding from industry, government institutions, and research councils, including the DFG (Germany), NSF (USA), and EPSRC (UK).

Journal of
*Marine Science
and Engineering*

Article

A Methodology for Susceptibility Assessment of Wave-Induced Seabed Liquefaction in Silt-Dominated Nearshore Environments

Yueying Wang [1], Xingsen Guo [1,2,*], Jinkun Liu [3], Fang Hou [3], Hong Zhang [4], Han Gao [1] and Xiaolei Liu [1,5,*]

[1] Shandong Provincial Key Laboratory of Marine Environment and Geological Engineering, Ocean University of China, Qingdao 266100, China; wyy98@stu.ouc.edu.cn (Y.W.); gaohan9009@stu.ouc.edu.cn (H.G.)

[2] Department of Civil, Environmental, and Geomatic Engineering, University College London, London WC1E 6BT, UK

[3] Sinopec Petroleum Engineering Corporation, Dongying 257026, China; speljk@163.com (J.L.); slecchf@126.com (F.H.)

[4] College of Engineering, Ocean University of China, Qingdao 266100, China; zhanghong9645@ouc.edu.cn

[5] Laboratory for Marine Geology, Qingdao Marine Science and Technology Center, Qingdao 266237, China

* Correspondence: xingsen.guo@ucl.ac.uk (X.G.); xiaolei@ouc.edu.cn (X.L.)

Abstract: Wave-induced seabed liquefaction significantly jeopardizes the stability of marine structures and the safety of human life. Susceptibility assessment is key to enabling spatial predictions and establishing a solid foundation for effective risk analysis and management. However, the current research encounters various challenges, involving an incomplete evaluation system, poor applicability of methods, and insufficient databases. These issues collectively hinder the accuracy of susceptibility assessments, undermining their utility in engineering projects. To address these challenges, a susceptibility assessment method with the safety factor was developed as the key assessment parameter, allowing for a comprehensive susceptibility assessment across the silt-dominated nearshore environment using Empirical Bayesian Kriging (EBK). The safety factor is determined by combining the cyclic stress ratio (CSR) and the cyclic resistance ratio (CRR), which characterize wave loadings and sediment properties in the study area, respectively. This method was applied in the Chengdao region of the Yellow River Estuary, China, a typical silt-dominated nearshore environment where wave-induced liquefaction events have been reported as being responsible for multiple oil platform and pipeline accidents. By collecting the regional wave and seabed sediment data from cores spanning from 1998 to 2017, the safety factors were calculated, and a zonal map depicting the susceptibility assessment of wave-induced seabed liquefaction was created. This study can serve as a valuable reference for the construction and maintenance of marine engineering in liquefaction-prone areas.

Keywords: wave-induced seabed liquefaction; susceptibility assessment; safety factor; cyclic stress ratio; cyclic resistance ratio; Empirical Bayesian Kriging

Citation: Wang, Y.; Guo, X.; Liu, J.; Hou, F.; Zhang, H.; Gao, H.; Liu, X. A Methodology for Susceptibility Assessment of Wave-Induced Seabed Liquefaction in Silt-Dominated Nearshore Environments. *J. Mar. Sci. Eng.* **2024**, *12*, 785. https://doi.org/10.3390/jmse12050785

Academic Editor: Jianhong Ye

Received: 9 April 2024
Revised: 24 April 2024
Accepted: 27 April 2024
Published: 8 May 2024

1. Introduction

Seabed liquefaction is a phenomenon associated with sediment instability. Under the influence of external forces (e.g., wave action, earthquakes, and human activities), the internal stress state of the sediment changes. Specifically, when the excess pore pressure is equivalent to the effective stress of the overlying sediment, the state of the sediment changes from solid to fluid, and the sediment becomes liquefied [1–3]. Wave-induced seabed liquefaction is caused by an external load cyclically applied to the seabed, generating differential loading on the seafloor through pressure waves that induce a series of cyclic shear stresses in the underlying sediment. If the induced shear stress exceeds the strength, it may lead to substantial deformation or liquefaction failure [1]. This sequence of this event can lead to various engineering safety problems and compromise the stability of marine structures [4,5]. It poses significant risks to human lives and property, as illustrated in Figure 1.

The damage of walls in Magoodhoo Island, Maldives	The oscillation and settling of the platform in the Bay of Campeche, Mexico	The floatation of submarine pipeline in the Shatt Al Arab Delta
(a)	(b)	(c)

Figure 1. Effects of seabed liquefaction on coastal environments. Because of seabed liquefaction, differential settlement of coastal areas causes cracks in nearshore houses and walls (**a**); sediment strength is reduced, causing destabilization of offshore platforms (**b**); and the seabed soil around the trench turned into a dense liquid that flooded the trench and caused the pipeline to fail (**c**) (images from Di Fiore et al. [6], Chávez et al. [7], and Damgaard et al. [8]).

To predict and evaluate the occurrence probability and severity of seabed liquefaction effectively, numerous researchers have started applying risk assessment and management techniques to this field [9–12]. These studies contribute to the quantitative analysis of the spatial extent of disasters, geological disaster mechanisms, and triggering factors [13,14]. The insights derived from these analyses provide recommendations for pre-disaster planning, disaster response, and post-disaster recovery efforts, ultimately mitigating losses. This field has garnered significant attention from governments, industries, and the scientific research community [15]. Nevertheless, the complex nature of wave-induced seabed liquefaction mechanisms makes it challenging to identify assessment factors. This difficulty hinders the advancement of risk assessment and management in this specific context. Among these challenges, the susceptibility assessment of seabed liquefaction emerges as a pivotal aspect in the risk assessment of seabed liquefaction. It directly impacts the accuracy of risk assessment and requires the advancement of reliable theories and practical methodologies. Susceptibility assessment of wave-induced seabed liquefaction primarily involves the creation of a disaster inventory, selection of conditioning factors, and identification of assessment methods [16]. Because of the complexity of wave-induced seabed liquefaction mechanisms, the choice of different monitoring tools for liquefaction under different sediment conditions directly affects the establishment of the disaster inventory and the selection of condition factors, and ultimately determines the research methods [17]. Therefore, priority should be given to addressing the accuracy of susceptibility assessment methods. The existing research methods lack the capability to adequately evaluate susceptibility across various sediment types, highlighting the necessity for the development of new methods to enhance evaluation accuracy [17,18].

Currently, susceptibility assessment methods for seabed liquefaction predominantly revolve around statistical approaches and deterministic analysis. The statistical approach enables both qualitative and quantitative evaluations of geological disasters by employing statistical principles to describe relationships among conditioning factors [19]. Consequently, this method, including bivariate statistical analysis, multivariate statistical analysis, and machine learning (such as neural network models and maximum entropy models), finds extensive application in the susceptibility analysis of marine geological disasters [16]. However, the statistical approach primarily focuses on the probability of disasters and allocates less consideration to the mechanism of disasters [19]. In contrast, the deterministic approach focuses on the mechanics of disaster occurrence. It has gained significant attention from researchers for its ability to quantitatively assess susceptibility to wave-induced seabed liquefaction [16]. A 3-D assessment methodology for seabed liquefaction has been proposed, based on cyclic triaxial tests, liquefaction potential evaluation criteria, and the nearshore spectral windwave (NSW) model. Nonetheless, this method did not take into account the effects of pore water pressure dissipation that could take place during a storm period, rendering it unsuitable for completing the susceptibility assessment of shallow sediment liquefaction [20]. Di Fiore et al. explored wave-induced seabed liquefaction through cyclic triaxial tests and evaluated the liquefaction potential of the Magoodhoo coralline island and coastal regions [6]. However, their calculation method for CSR (the cyclic stress ratio) was only applicable to wave heights ranging from 0.2 to 3.0 m. Furthermore, the evaluation factor was solely wave height, resulting in an oversimplified assessment model that compromises result accuracy [6]. In summary, the current evaluation method lacks a comprehensive consideration of evaluation factors for generating zoning maps based on evaluation results, and it falls short in assessing different sediment types. Therefore, it is crucial to introduce a more precise methodology to overcome these challenges.

In this study, a method for the susceptibility assessment of wave-induced seabed liquefaction is introduced, with a particular focus on safety factors, which have been widely used in susceptibility assessment for various natural processes. This method characterizes wave loading through CSR and employs CRR (the cyclic resistance ratio) to represent sediment properties to calculate safety factors. The calculation of CSR follows the method proposed by Ishihara et al. [21], while CRR analysis integrates considerations of sediment characteristics. Utilizing this approach in the Chengdao region involved calculating CSR values for diverse wave recurrence periods (5-year, 10-year, and 25-year). Concurrently, sediment properties determined CRR at varying depths, assessed through cyclic triaxial torsion shear tests at sampling points. Establishing the smallest safety factor among these points facilitated the identification of the wave-induced seabed liquefaction susceptibility assessment. The susceptibility zoning map was generated using Empirical Bayesian Kriging (EBK) in the Arc Geographic Information System 10.7 (ArcGIS 10.7). This method is applicable to seabed areas with silt-dominated environments and can maximize the accuracy of small-scale susceptibility assessments of wave-induced seabed liquefaction, even when data are limited.

2. Methodology

2.1. Susceptibility Assessment Model of Wave-Induced Seabed Liquefaction

Currently, various methods exist for the susceptibility assessment of wave-induced seabed liquefaction, commonly categorized as a statistical approach or deterministic analysis [16]. Deterministic analysis, as discussed earlier, allows for a more comprehensive exploration of the mechanism behind wave-induced seabed liquefaction, aligning closely with its characteristics and thus finding extensive application [16]. The safety factor employed in this study falls within a deterministic analysis, offering quantitative insights into liquefaction susceptibility. This method was initially proposed by Seed et al. in 1971 [22]. Its versatility allows for application across study areas with diverse sediment types and various scenarios, promising a more precise susceptibility of wave-induced seabed liquefaction [23]. This method employs CSR to signify the extent of external load impact on sediment liquefaction and uses CRR to indicate sediment resistance properties [22,24]. The safety factor can be determined by comparing CSR and CRR at varying depths according to Equation (1):

$$FS = \frac{CRR}{CSR} \tag{1}$$

where FS is the safety factor; CSR is the cyclic stress ratio; and CRR is the cyclic resistance ratio. The seabed liquefaction susceptibility of each area is further classified according to the characteristics of the study area and the probability of liquefaction, combined with the safety factor [25].

It is usually considered that the liquefaction will not occur in the region when FS > 1 and that the liquefaction will occur in the region when FS < 1. The interrelation between the safety factor and reliability had been explored, proposing that the safety factor can be specified based on reliability levels, and the allocation of the safety factor should be based on having the same reliability associated with each failure mode [26]. The influence of geotechnical parameter variability on the liquefaction potential of tailing dams has been explored [27]. The safety factor for different liquefaction probabilities has been compared, highlighting the use of probabilistic methods for failure probability and the safety factor against liquefaction analysis [27]. Therefore, the division intervals of safety factors need to be determined based on the actual conditions of the research area, considering the characteristics of disaster events, and based on different liquefaction probabilities.

2.2. Cyclic Stress Ratio

CSR represents the ratio of the amplitude of shear stress to the effective confining stress, which is used to quantify the deformability or strength of soils under cyclic loading conditions [22]. According to linear wave theory, a wave event can be described by fundamental wave characteristics [28]. Thus, various authors have suggested simplifying the CSR calculation based on this theory [21]. When analyzing wave conditions, the following criteria must be considered: (1) to reduce computational complexity, the irregular loads, induced by actual waves on the seabed surface, are simplified to uniform loads and (2) because this method is designed for shallow sea areas, it assumes that all waves are linear [23,29,30].

Regarding the characteristics of waves, they can be considered to consist of an infinite number of wave trains with the same amplitude and wavelength. The passage of such an array of waves over the ocean creates harmonic pressure waves on the seafloor, increasing the pressure under the crest and reducing it under the trough, as shown in Figure 2 [21]. Considering the rectangular coordinate system, the sea surface elevation of a wave field η may be represented by Ishihara and Yamazaki's Equation (2) [21]:

$$\eta(x,t) = \frac{H}{2}\cos(kx - \omega t) \tag{2}$$

where H is the wave height, $k = \left(\frac{2\pi}{L}\right)$ is the wave number, L is the wave length, $\omega = \left(\frac{2\pi}{T}\right)$ is the wave frequence, T is the wave period, x is the spatial coordinate in the x direction, and t is the time coordinate. For linear waves, k and ω are related by the dispersion relationship according to Equation (3) [1]:

$$\omega^2 = gk\tanh kd \tag{3}$$

where g is the gravitational acceleration. Equation (3) also can be rewritten as Equation (4):

$$L = \frac{gT^2}{2\pi}\tan h\left(\frac{2\pi d}{L}\right) \tag{4}$$

where L is the wave length, T is the wave period, and d is the water depth.

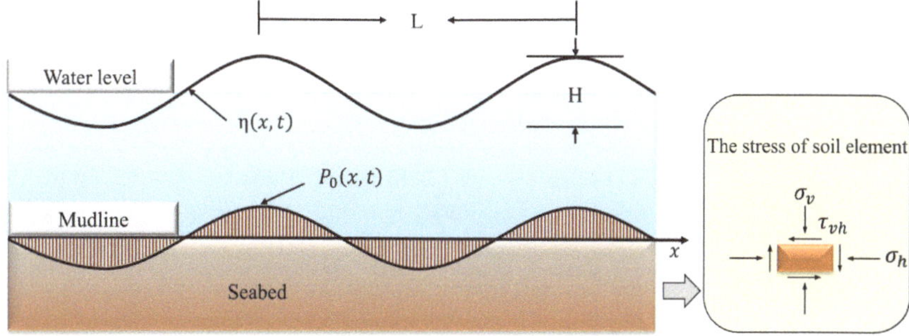

Figure 2. Wave-induced cyclic pressures and stresses.

Regarding the characteristics of wave-induced cyclic stress, the seabed deposit is assumed to consist of a homogeneous elastic material extending to an infinite depth. Boussinesq's classical solution for the two-dimensional plane strain problem can be employed to calculate the stresses acting on it [31]. The stress induced in the seabed is, therefore, analyzed by applying a sinusoidally changing load on the horizontal surface from minus to plus infinity according to Equation (5):

$$p(x) = p_0 \cos\left(\frac{2\pi}{L}x - \frac{2\pi}{T}t\right) \tag{5}$$

where L is the wave length, x is the spatial coordinate in the x direction, T is the wave period, and t is the time coordinate. According to the small amplitude wave, the amplitude of the pressure fluctuation exerted on the sea bottom, p_0, by the travelling wave is given by Horikawa's Equation (6) [32]:

$$p_0 = \frac{\rho_w g H}{2\cos h(2\pi d/L)} \tag{6}$$

where ρ_w is the density of seawater, g is the gravitational acceleration, H is the wave height, d is the water depth, and L is the wave length. Considering the uniform distribution and infinite depth of the sediments, the amplitude of maximum shear stress can be determined by analyzing the relationship between the horizontal stress, vertical stress, and shear stress of the harmonic pressure wave. The analysis utilizes the classical Boussinesq solution for two-dimensional plane strain problems according to Equation (7) [21]:

$$\tau_{vh} = 2\pi z p_0 \frac{1}{L}\left(e^{-2\pi z/L}\right) \tag{7}$$

where τ_{vh} is the shear stress, p_0 is the amplitude of the pressure fluctuation exerted on the sea bottom, L is the wave length, and z is the sampling depth. Since the magnitude of σ_v' is expressed by $\sigma_v' = \rho' g z$, CSR is given by Equation (8) [21]:

$$\text{CSR} = \frac{(\tau_{vh})_{max}}{\sigma_v'} = \frac{2\pi p_0}{\rho' g L}\exp\left(-\frac{2\pi z}{L}\right) \tag{8}$$

where τ_{vh} is the shear stress, σ_v' is the vertical effective stress, p_0 is the amplitude of the pressure fluctuation exerted on the sea bottom, ρ' is the submerged unit mass of soils in the seabed, g is the gravitational acceleration, L is the wave length, and z is the sampling depth. According to the formula, comprehensive data on fundamental wave characteristics like wave height, wavelength, wave period, and water depth is essential. This information not only allows for the calculation of CSR but also facilitates the simulation of CSR under extreme wave conditions.

2.3. Cyclic Resistance Ratio

CRR is a parameter that measures the resistance of sediments to liquefaction. It represents the ratio between the average cyclic shear stress and the vertical effective stress of sediments [22]. Different sediment types require specific calculation methods, and the operational ease varies among various test methods. The field test methods of CRR under seismic liquefaction were summarized by referring to the research results of Youd et al., as shown in Table 1 [24]. The primary field test includes SPT (Standard Penetration Test), CPT (Cone Penetration Test), measurement of V_S (Shear Wave Velocity), and BPT. SPT is a test conducted during a test boring in the field to measure the approximate soil resistance to penetration of a split-spoon sampler at various depths below the ground surface [33]. This test was intended to measure the number of blows required to drive a standard sampling tube into the ground to a certain depth [34]. As for CPT, an elongated metal probe (typically conical in shape) is pushed into the ground, and the resistance and soil displacement during the penetration process are measured. It is performed by pushing an instrumented probe with a specific diameter into the earth at a constant speed while simultaneously measuring the cone resistance and sleeve friction resistance [35]. The V_S test is a geoengineering test used to determine the shear wave velocity in soil or rock. A shear wave is a wave that propagates through shear deformation in a material, and its speed is related to the shear modulus of the material. Shear wave velocity tests are often used to study the elastic properties and mechanical behavior of soil or rock [36]. When conducting liquefaction assessments, SASWs are useful for V_S profile surveys of liquefaction sites [37]. Becker penetration resistance is defined as the number of blows required to drive the casing through an increment of 300 mm [24]. The formula for calculating CRR corresponding to the above method is mentioned in Table 2. It is essential to choose the appropriate method based on the actual conditions of CRR calculation, significantly enhancing the accuracy of CRR and FS and consequently improving the precision of susceptibility assessment results.

Table 1. Comparison of advantages and disadvantages of various field tests for CRR.

Feature	Test Type			
	SPT [38]	CPT [39]	V_S [37]	BPT [24]
Types of stress–strain behavior influencing test	Partially drained, large strain	Drained, large strain	Small strain	Partially drained, large strain
Quality control and repeatability	Poor to good	Very good	Good	Poor
Soil types in which the test is recommended	No gravel	No gravel	All	Gravel

Table 2. Summary of methods for calculating CRR.

Source	Formula	Testing Method
[38]	$CRR = \exp\left[\dfrac{(N_{1,60}*(1+\theta_1*\text{FC})-\theta_6*\ln(M_w)-\theta_3*\ln\left(\frac{\sigma'_v}{P_a}\right)+\theta_4*\text{FC}+\sigma_\varepsilon*\Phi^{-1}(P_L)}{\theta_6} \right]$	SPT
[39]	$CRR = \exp\left\{ \dfrac{\left[q_{c,1}^{1.045}+q_{c,1}\left(0.110*R_f\right)+\left(0.001*R_f\right)-0.848*\ln(M_w)-0.002*\ln(\sigma'_v)-20.923+1.632*\Phi^{-1}(P_L)\right]}{7.177} \right\}$	CPT
[37]	$CRR = \exp\left\{ \dfrac{\left[(0.0073*V_{s1})^{2.8011}-2.6168*\ln(M_w)-0.0099*\ln(\sigma'_v)+0.0028*\text{FC}+0.4809*\Phi^{-1}(P_L)\right]}{1.946} \right\}$	SASW

Note: $N_{1,60}$ is the standard penetration test blowcount value corrected for overburden, energy, equipment, and procedural factors; θ_i is the set of unknown model coefficients; FC is the fines content; M_w is the earthquake moment magnitude; σ'_v is the vertical effective stress; P_a is the atmospheric pressure (1 atm); Φ is the standard cumulative normal distribution; P_L is the probability of triggering of liquefaction; $q_{c,1}$ is the normalized cone tip resistance; R_f is the friction ratio; V_{s1} is the effective stress normalized shear-wave velocity.

The above method is suitable for the calculation of CRR in the terrestrial environment. However, the monitoring difficulty in the marine environment is greater than that in the terrestrial environment, and the in situ test is more difficult and costly [40,41]. Therefore, the laboratory test has attracted more attention in the field of submarine geological disaster susceptibility assessment. The cyclic triaxial torsion shear test is a widely used sediment testing method suitable for saturated sand and saturated silt in nearshore areas [31]. It offers a streamlined computation process with fewer evaluation factors, facilitating a rapid assessment of sediment resistance to liquefaction. This test reflects the cyclic loading characteristics of the sediment, a characteristic that is one of the determining factors in the susceptibility assessment of wave-induced seabed liquefaction [31]. Introduced by Ishihara et al., this method selects the dynamic stress ratio when the double amplitude deviator strain reaches 5% as CRR for the sediments at the given point [21]. Based on this standard, they conducted a large number of cyclic triaxial tests, summarized the relevant rules, and proposed the calculation according to the relative density of sediments and the coefficient of earth pressure at rest, as shown in Equation (9) [21]:

$$\text{CRR} = 0.0019 * D_r * \frac{1 + 2K_0}{3} \tag{9}$$

where D_r is the relative density and K_0 is the coefficient of earth pressure at rest. However, this method is applicable to sandy sediments, but not to silt, so some scholars have found another universal rule, which indicates that the cyclic stress ratio at 100 load cycles is determined to be closest to this criterion, as illustrated in Equation (10) [21,23]. Since there exists an exponential relationship between the number of cyclic loadings and the cyclic dynamic stress in the cyclic triaxial test, as depicted in Equation (11), the CRR can be calculated by fitting this exponential relationship with convincing data obtained through the tests [23].

$$\text{CRR} = \left(\frac{\tau_L}{\sigma_v'}\right)_{100} = \left(\frac{\sigma_d}{2\sigma_C}\right)_{100} = \frac{3\sigma_d}{2(\sigma_1 + \sigma_3)} \tag{10}$$

where τ_L is the horizontal shear stress, σ_v' is the vertical effective stress, σ_d is the cyclic dynamic stress, σ_C is the average consolidation pressures, σ_1 is the axial consolidation pressures, and σ_3 is the lateral consolidation pressures.

$$\frac{\tau_L}{\sigma_v'} = a * \ln N + b \tag{11}$$

where τ_L is the horizontal shear stress, σ_v' is the vertical effective stress, N is the number of cyclic loadings, and a and b are the fitting coefficients.

2.4. Geospatial Interpolation

The previous process allows us to calculate the safety factor for a specific location. However, to complete a susceptibility assessment of marine geological disasters for the entire study area, spatial interpolation tools within the Arc Geographic Information System 10.7 (ArcGIS 10.7) need to be employed. These tools maximize the utilization of existing data points to comprehensively delineate the susceptibility assessment of liquefaction across the entire area [9]. To reconstruct a continuous attribute distribution over the entire study area, the spatial interpolation method utilizes a finite set of sampling points $S = \{(x_i, f_i), i = 1, 2, 3, \ldots, n\}$ to establish an interpolation function $(f : x \to x)$ that estimates attribute values at any given data point within the study area [10]. Spatial interpolation can be categorized into geostatistical interpolation and deterministic interpolation. A widely used geostatistical interpolation method is EBK, which is valued for its simplicity and accurate results [42,43]. This method accounts for introduced error by estimating the underlying semivariance function, whereas other kriging methods calculate the semivariance function from a known data location and apply this single semivariance function to make predictions at an unknown location [9,44]. By not considering the uncertainty in the estimation of the semivariance function, other kriging methods underestimate the standard error of the prediction, while the EBK is more accurate [44]. Additionally, EBK requires minimal interactive modeling, excels at predicting unstable data, and is particularly well-suited for smaller datasets [9].

2.5. Susceptibility Assessment System of Wave-Induced Seabed Liquefaction

Based on the above discussion of the susceptibility assessment and spatial interpolation methods, the specific steps for partitioning the susceptibility assessment of wave-induced seabed liquefaction are shown in Figure 3.

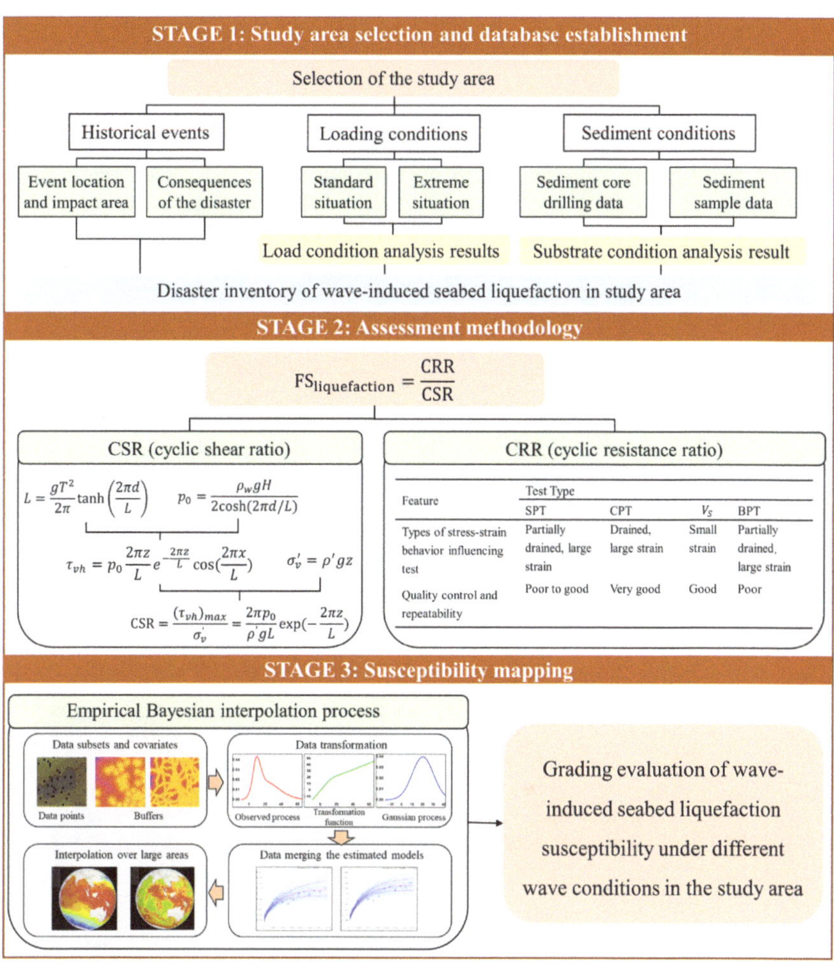

In the figure, STAGE 2 shows:

$$\text{FS}_{\text{liquefaction}} = \frac{\text{CRR}}{\text{CSR}}$$

$$L = \frac{gT^2}{2\pi}\tanh\left(\frac{2\pi d}{L}\right) \qquad p_0 = \frac{\rho_w gH}{2\cosh(2\pi d/L)}$$

$$\tau_{vh} = p_0\frac{2\pi z}{L}e^{-\frac{2\pi z}{L}}\cos\left(\frac{2\pi x}{L}\right) \qquad \sigma_v' = \rho' gz$$

$$\text{CSR} = \frac{(\tau_{vh})_{max}}{\sigma_v'} = \frac{2\pi p_0}{\rho' gL}\exp\left(-\frac{2\pi z}{L}\right)$$

Feature	Test Type			
	SPT	CPT	V_s	BPT
Types of stress-strain behavior influencing test	Partially drained, large strain	Drained, large strain	Small strain	Partially drained, large strain
Quality control and repeatability	Poor to good	Very good	Good	Poor

Figure 3. Process for the susceptibility assessment of wave-induced seabed liquefaction (modified from [14,22,43,44]).

This study proceeds through the following steps:

(1) Identify the study area through a comprehensive review of historical data. Summarize the fundamental wave characteristics within this region and establish a comprehensive disaster inventory.

(2) Calculate the CSR of all measurement points using Equation (8), considering the measured water depth and wave parameters of different wave recurrence periods. Utilize the cyclic triaxial torsion shear test data from sampling points to calculate CRR at the known depth of each sampling point. Select the minimum CRR value for each measurement point as the CRR value for susceptibility assessment. Calculate the safety factor of wave-induced seabed liquefaction at the known depth of each sampling point under the different wave conditions by applying Equation (1). Select the minimum safety factor value for each point under varying wave conditions.

(3) Analyze and process the data through ArcGIS 10.7, incorporating interpolation methods to generate a liquefaction susceptibility zoning map for the study area. Subsequently, conduct an in-depth analysis of the zoning results.

3. A Case Study in a Silt-Dominated Nearshore Environment

3.1. Study Area

Numerous studies indicate that nearshore silty seabed are typical areas prone to liquefaction, posing a significant threat to the stability of nearshore marine engineering [7,45–47]. In this context, the Chengdao Island area, where silty sediments are widely distributed, has been selected as the study area. The Chengdao region is situated to the north of the Yellow River Estuary in China, exhibiting a southwest-to-northeast topographical gradient, as depicted in Figure 4. This area primarily consists of two subdeltas [47]. The first subdelta was formed when the Yellow River was redirected to the Shenxiangou flow path from 1953 to 1964, while the second subdelta was established when the river was diverted to the Diaokou flow path from 1964 to 1976 [48]. The underwater Yellow River Estuary spans an area of approximately 3000 km². A distinctive feature of this region is its remarkably gentle gradient, typically less than 0.4°, and an exceptionally low slope drop ranging from 1% to 5% [48]. This makes the area highly susceptible to storm surge hazards [1]. Between 1949 and 2005, more than 150 storm surge events occurred in the region, with an average of two to three events per year, of which 36 storm surges caused geological disasters, most of which occurred in the summer and fall [49].

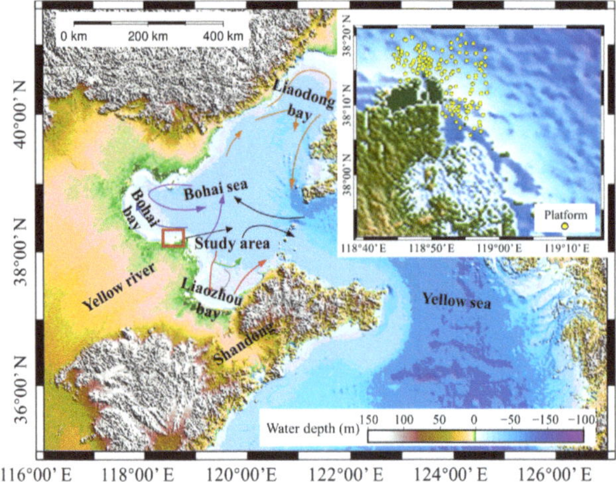

Figure 4. Overview of the Chengdao region and the study zone. The large image shows the location of the Chengdao region and the study zone, with arrows indicating the direction of ocean currents in the study area; the small image in the upper right corner represents the study zone, where yellow indicates the distribution of offshore platforms in existing publicly available data (images modified from Zhang et al. [47] and Wang et al. [50]).

The Chengdao region holds significant engineering importance as it encompasses the Chengdao oil field, which is a crucial oil-producing area. Numerous engineering facilities, including oil platforms, submarine pipelines, and breakwaters, have been established in this region [51–53]. However, the marine engineering environment in the Chengdao region is notably complex, with frequent occurrences of seabed liquefaction leading to numerous destabilization accidents involving offshore engineering facilities. For instance, in November 2003, a seabed deformation sliding event occurred near the oil production platform CB12B, which led to the rupture and interruption of two submarine cables. In May 2009, the CB25A-CB25B subsea crude oil transmission pipeline fractured near the endpoint of CB25A [52]. The investigation data indicated that the accident site was approximately 50–60 m from the CB25A platform, 80 m from the intersection of the Center II-CB1A subsea pipeline, and 200 m from the Center II-CB1A submarine cable. Furthermore, a capsizing incident occurred on the Shengli Operation No. 3 maintenance platform near the CB22C well group platform of Chengdu Island Oilfield in 2010, putting 36 people in danger [54]. Unfortunately, two individuals lost their lives [7]. The Shengli Operation No. 3 platform and the adjacent seabed were adversely affected by intense wind and waves induced by a typhoon. This led to the liquefaction of the soft substrate beneath the platform, and as the liquefaction zone gradually expanded, it triggered instability, ultimately causing the capsizing of the platform [7].

3.2. Cyclic Stress Ratio under Various Wave Return Periods

The Chengdao region is influenced by prevailing monsoon winds, experiencing dominant north winds in the winter and south winds in the summer [52]. Wave patterns in this area are primarily controlled by surface winds, characterized by short wind zones, rapid wave growth, high wave heights, short wave periods, and limited attenuation distances. These characteristics are associated with the semi-enclosed nature of the Chengdao region, which is connected solely to the Yellow Sea through the narrow Bohai Strait. Furthermore, the engineering geological conditions of the underwater delta impact wave propagation, limiting the distance waves can travel [55]. Under normal sea conditions, wave heights typically do not exceed 1.5 m. However, during extreme sea conditions, wave heights can surpass 5.8 m, and the measured maximum current velocity can reach 1.5 m/s [1]. It is estimated that the extreme water level may increase by 1.8–3.2 m per year, with an average increase of 2.2 m in the Chengdao region [46]. These extreme conditions are primarily generated by weather phenomena such as typhoons, cyclones, and cold waves [56]. Observational and numerical simulation data provide a breakdown of wave information for different wind conditions in the Chengdao region, as shown in Table 3.

Table 3. Corresponding wave elements under different wave recurrence periods in the Chengdao region (data from [46]).

Water Depth	5-Year Return Period			10-Year Return Period			25-Year Return Period		
(m)	$H_{1/10}$ (m)	L (m)	T (s)	$H_{1/10}$ (m)	L (m)	T (s)	$H_{1/10}$ (m)	L (m)	T (s)
3	2.0	42.0	8.0	2.0	42.6	8.1	2.0	44.2	8.4
4	2.6	48.5	8.0	2.6	49.2	8.1	2.6	50.9	8.4
5	3.4	53.9	8.0	3.4	54.5	8.1	3.4	56.6	8.4
6	4.0	58.4	8.0	4.0	58.9	8.1	4.0	61.2	8.4
7	4.3	61.4	8.0	4.6	62.3	8.1	4.6	64.9	8.4
8	4.4	64.9	8.0	4.6	65.8	8.1	4.9	68.7	8.4
9	4.5	65.4	8.0	4.7	68.6	8.1	5.0	71.6	8.4
10	4.6	70.9	8.0	4.8	71.1	8.1	5.1	74.2	8.4
14	5.0	79.9	8.0	5.2	79.2	8.1	5.6	82.8	8.4

Note: $H_{1/10}$ is the effective wave height.

According to linear wave theory, an extreme wave event can be considered to consist of many different waves with characteristics, and the wave components can be characterized by wave period, wavelength, and wave height [31]. The peak wave stress in the study area is calculated under the wave conditions of a 5-year, 10-year, and 25-year return period to analyze the distribution of cyclic stress within the study area across varying wave conditions and sampling depths. Considering that the maximum liquefaction depth in the Chengdao region is less than 15 m, CSRs for three different wave conditions at sampling depths of 5 m, 10 m, and 15 m were further calculated [13]. The interpolation method was utilized to generate a zonal map of the peak wave pressures under different wave conditions ($\rho_w = 1025$ kg/m^3 is the density of seawater and $g = 9.8$ m/s^2 is the gravitational acceleration), as shown in Figure 5.

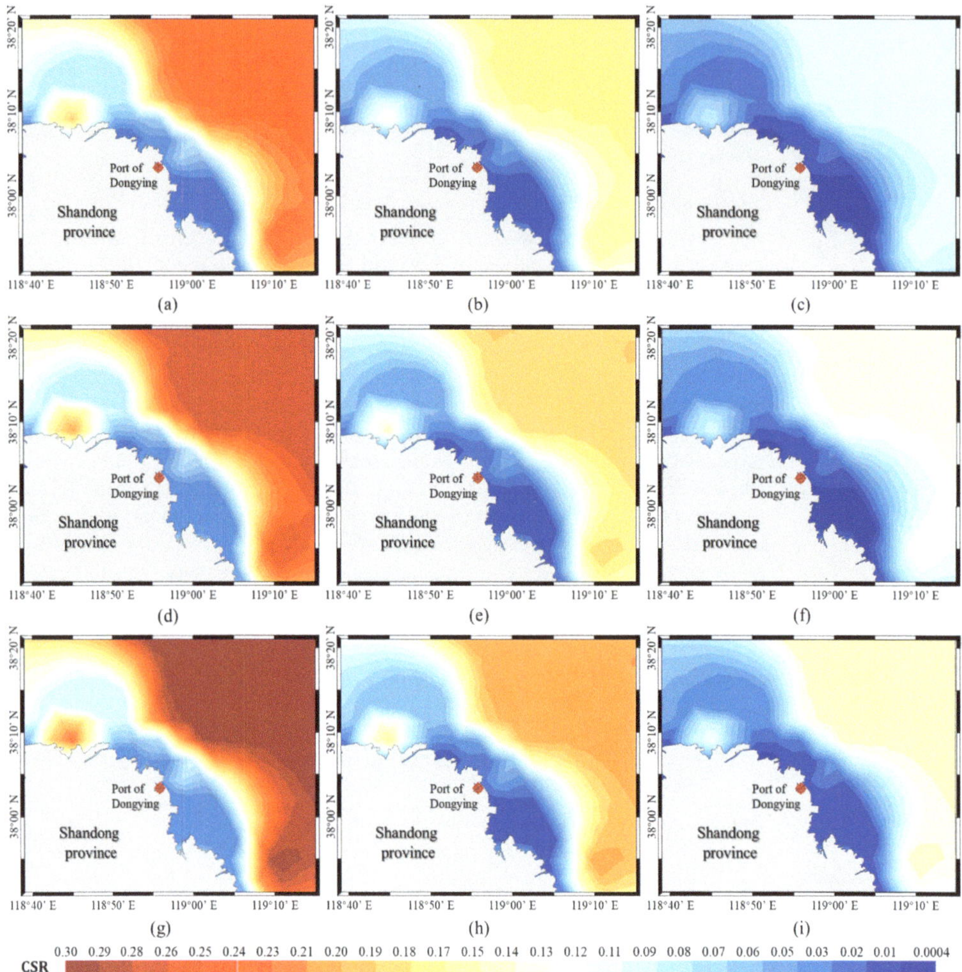

Figure 5. CSR of the study area (the small image in Figure 4) under different wave conditions at different sampling depths: (**a**) CSR of the wave condition in a 5-year return period at 5 m depth; (**b**) CSR of the wave condition in a 5-year return period at 10 m depth; (**c**) CSR of the wave condition in a 5-year return period at 15 m depth; (**d**) CSR of the wave condition in a 10-year return period at 5 m depth; (**e**) CSR of the wave condition in a 10-year return period at 10 m depth; (**f**) CSR of the wave condition in a 10-year return period at 15 m depth; (**g**) CSR of the wave condition in a 25-year return period at 5 m depth; (**h**) CSR of the wave condition in a 25-year return period at 10 m depth; and (**i**) CSR of the wave condition in a 25-year return period at 15 m depth.

Overall, a discernible spatial regularity characterizes the distribution of CSR, with contours closely aligned with the coastline. The lowest CSR value is observed under the wave conditions of a 5-year return period, situated at a depth of 15 m in the southeastern offshore area of the study region, with a value of 0.0004. Conversely, the highest CSR value is found under the wave conditions of a 25-year return period, located at a depth of 5 m in the northeastern offshore area, with a value of 0.305. A deeper analysis of the CSR distribution at different depths but the same wave conditions reveals a general increasing trend from nearshore to offshore areas. However, a distinct spatial pattern emerges in the northern nearshore part of the Chengdao region, characterized by a concentration of higher CSR values where numerous offshore platforms are located. Additionally, CSR displays a negative correlation with sampling depth, ranging from 0.03 to 0.31 at 5 m depth, 0.01 to 0.23 at 10 m depth, and 0.0004 to 0.16 at 15 m depth. This indicates that deeper sediment experiences a reduced impact from wave loading. By analyzing the distribution of CSR at the same depth but under different wave conditions, it can be found that higher wave conditions lead to an increased CSR, which, in turn, amplifies the likelihood of seabed liquefaction. The precision of the interpolation results is primarily attributed to the limited distribution of interpolation points. However, regarding the interpolation of CSR, the known data points are mostly distributed in nearshore areas, with fewer points in offshore areas. As a result, the accuracy of interpolation results in offshore areas will be lower than in nearshore areas.

3.3. Cyclic Resistance Ratio of Sediments

The sediment types in the Chengdao region can be categorized into four types, including clay, silty clay, silt, and silty sand, as shown in Figure 6 [52]. In the Chengdao oil field beach and shallow marine stratigraphy, silt is the main sediment type [51]. However, the nearshore area experiences strong hydrodynamic forces and sediment redeposition, leading to a more intricate sediment composition [49]. This results in an alternation between coarser and finer sediment types, with silt predominating locally. Moreover, the distribution of silt in the far shore area slightly surpasses that of clay [46,57].

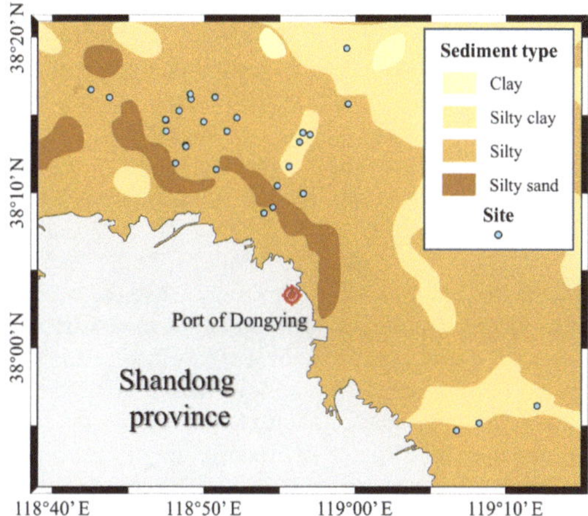

Figure 6. Distribution of sediment types in the study area. Blue dots represent sample points (data from Liu et al. [51] and Liu et al. [52]).

The sediment strength in the Chengdao region exhibits a notable pattern, which can be observed in Figure 7, depicting the distribution of undrained shear strength in depth ranges of 0–5 m, 5–10 m, and 10–15 m. The exact calculation results are presented in Appendix A. Because of more rapid sedimentation, the shallow sediments usually do not reach complete consolidation, so their undrained shear strengths are lower, and the range of variation is smaller [57]. On the contrary, the undrained shear strength of the deeper sediments roughly shows a gradual increase along the delta and outward [57]. This phenomenon results from a gradual reduction in sedimentation rates and a progressive thinning of sediment particles. Furthermore, regional trends in the sediment water content, the pore ratio, and the liquid–plastic limit follow a similar pattern, with these parameters gradually increasing with water depth and distance from the Diaokou inlet [48]. These characteristics collectively create conditions characterized by low consolidation, a high water content, a high pore ratio, and low strength, which increases the risk of geological disasters such as sliding, thixotropy, and liquefaction.

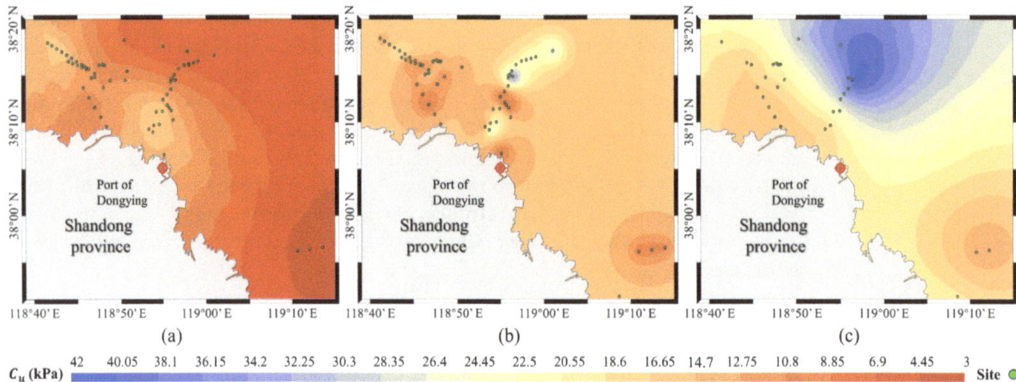

Figure 7. Distribution of undrained shear strength in the Chengdao region at different sampling depths of 0–5 m (**a**), 5–10 m (**b**), and 10–15 m (**c**).

To calculate CRR, this study used the cyclic triaxial shear test data. The method for this experiment was the consolidation of undrained shear, with an isotropic consolidation process (i.e., axial consolidation pressure (σ_1) equaled lateral consolidation pressure (σ_3)). After consolidation was completed, the confining pressure and axial pressure remained unchanged, while a sinusoidal axial load with a frequency of 1 Hz was applied. This caused cyclic shear stress on a 45° inclined plane within the soil sample, simulating the shear forces induced by wave loading within the seabed. The same set of soil samples was subjected to at least three different amplitudes of dynamic loading to obtain a relationship curve between the number of cycles and the dynamic stress ratio. Data from 29 groups of cyclic triaxial tests conducted between 1998 and 2017 with sediment samples in the study area were collected and organized. The relationship between the dynamic stress ratio and the number of cycles at various depths was fitted to calculate the cyclic resistance ratio (CRR) for each sampling point. The exact calculation results are provided in Appendix B. The minimum CRR value at each point was selected as the final CRR. Figure 8 only presents the relationship between the dynamic stress ratio and the number of cycles for a portion of different sampling points, along with the method for determining CRR. However, because of the nonuniform distribution of 29 sampling points, it was necessary to utilize Empirical Bayesian Interpolation to extrapolate the CRR distribution across the continuous surface of the study area, as illustrated in Figure 9.

CRR mainly reflects the ability of sediments in the study area to resist liquefaction. In general, variations in CRR within the study area are relatively minor and closely linked to sediment type. The northwestern region predominantly features silt sediments, with most of this area exhibiting CRR values exceeding 0.27, indicating greater resistance to liquefaction. Furthermore, CRR correlates with undrained shear strength, which is notably lower in the southeastern region at different depths, resulting in a CRR within the range of 0.2 to 0.22. Unlike the distribution of CSR, CRR does not follow a coastline-based distribution pattern. Instead, they display one high CRR center and three low CRR centers arranged in a concentric circle formation. These interpolation results are influenced by the distribution of sampling points and the selection of the interpolation method.

A comprehensive analysis of CSR, CRR, and safety factors at various sampling depths at the same location reveals interesting patterns. Under the same wave condition, CSR and CRR exhibit a gradual decrease as sampling depth increases. The reduction in CSR (Figure 5) is associated with the attenuation of stress as waves propagate downward through the sediment layers. Furthermore, the reduction in the CRR trend could be attributed to the robust hydrodynamic forces in the nearshore area, which generate intense shear stresses, inhibiting the settling of sediment particles toward the seabed and consequently resulting in a decline in CRR with greater depth. However, as sampling depth increases, the magnitude of reduction in CSR surpasses that of CRR, resulting in an increase in the safety factor and a decreased susceptibility to wave-induced seabed liquefaction. This implies that the depth of sediment may significantly influence the susceptibility to wave-induced seabed liquefaction, leading to a reduced susceptibility to disasters.

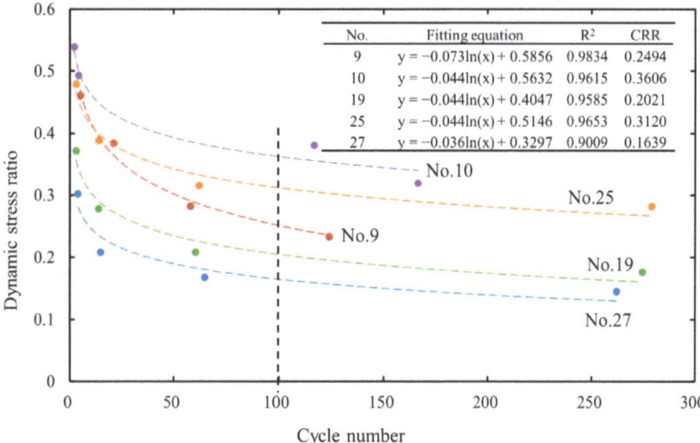

Figure 8. The relationship between the dynamic stress ratio of the selected sampling points and the number of cycles. The dotted line represents the dynamic stress ratio corresponding to the cycle number of 100, which is considered the CRR of the sampling point.

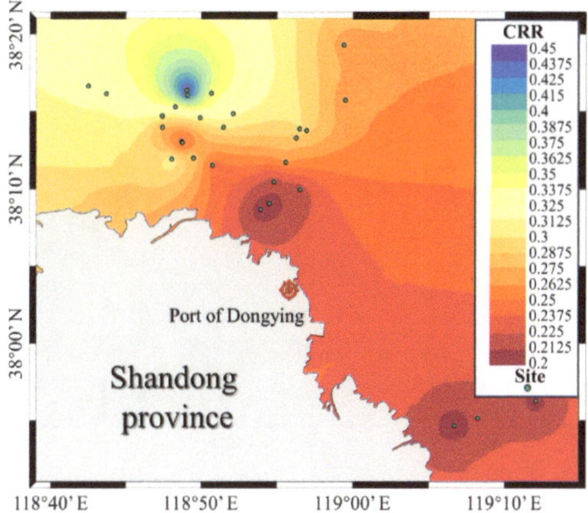

Figure 9. The interpolated distribution represents the minimum values of known CRRs for the sampling points within the 15 m depth range.

3.4. Susceptibility Assessment

The safety factor for wave-induced seabed liquefaction at various known depths of sampling points was calculated following the previously outlined procedure. These calculations were conducted under different wave conditions corresponding to 5-year, 10-year, and 25-year return periods. We considered the smallest safety factor among those calculated at different known depths as the safety factor for wave-induced seabed liquefaction at that specific point, as detailed in Table 4. Subsequently, we utilized EBK in ArcGIS 10.7 to generate a distribution map of the safety factor of wave-induced seabed liquefaction, as shown in Figure 10.

Table 4. CSRs, CRRs, and the safety factor minima calculated for 29 core sampling points.

No.	Latitude (N)	Longitude (E)	Sampling Depth (m)	CSR-5y	CSR-10y	CSR-25y	CRR	FS-5y	FS-10y	FS-25y
1	38°09′52″	118°56′33″	12.5	0.0965	0.1022	0.1161	0.1836	1.9023	1.7960	1.5809
2	38°11′26″	118°50′48″	3	0.0951	0.1031	0.1069	0.1948	2.0477	1.8904	1.8220
3	38°16′32″	118°42′33″	5	0.1148	0.1213	0.1337	0.3379	2.9431	2.7853	2.5262
4	38°16′02″	118°43′45″	10	0.0641	0.0684	0.0772	0.3031	4.7245	4.4323	3.9275
5	38°15′12″	118°48′18″	15	0.0211	0.0218	0.0238	0.3151	14.9494	14.4820	13.2178
6	38°13′47″	118°56′31″	3	0.2702	0.2962	0.3299	0.3605	1.3339	1.2171	1.0926
7	37°56′13″	119°12′04″	9	0.1860	0.2087	0.2327	0.1792	0.9631	0.8586	0.7699
8	38°10′22″	118°54′49″	4.2	0.1544	0.1685	0.1799	0.1697	1.0989	1.0066	0.9433
9	38°15′38″	118°59′30″	8	0.2001	0.2241	0.2491	0.2494	1.2466	1.1128	1.0015
10	38°14′35″	118°47′27″	2.3	0.0645	0.0642	0.0668	0.3606	5.5879	5.6146	5.3979
11	38°14′45″	118°52′09″	5	0.1148	0.1213	0.1337	0.3314	2.8872	2.7325	2.4783
12	38°13′53″	118°51′31″	5	0.1018	0.1078	0.1193	0.3250	3.1923	3.0155	2.7243
13	38°12′58″	118°48′44″	12	0.0114	0.0118	0.0129	0.1121	9.8112	9.5096	8.6919
14	38°15′56″	118°49′06″	4	0.0985	0.1043	0.1155	0.6891	6.9947	6.6040	5.9662
15	38°16′16″	118°49′03″	7	0.0737	0.0784	0.0878	0.7281	9.8829	9.2914	8.2944
16	37°55′07″	119°08′13″	9	0.1740	0.1921	0.2187	0.2288	1.3151	1.1907	1.0461
17	38°13′53″	118°47′28″	2	0.0423	0.0429	0.0447	0.2204	5.2083	5.1376	4.9361
18	38°11′49″	118°48′05″	2	0.1196	0.1263	0.1387	0.4844	4.0513	3.8358	3.4931
19	38°12′54″	118°48′46″	3.1	0.0366	0.0372	0.0380	0.2020	5.5149	5.4297	5.3092
20	38°11′38″	118°55′36″	6.3	0.2101	0.2323	0.2619	0.3528	1.6788	1.5184	1.3475
21	38°13′40″	118°56′58″	14	0.1292	0.1459	0.1657	0.3082	2.3848	2.1123	1.8604
22	38°16′04″	118°50′43″	5	0.1148	0.1213	0.1337	0.3299	2.8741	2.7201	2.4670
23	38°14′37″	118°47′25″	5	0.0457	0.0477	0.0491	0.5014	10.9623	10.5121	10.2190
24	38°14′29″	118°49′57″	10	0.0364	0.0373	0.0400	0.3103	8.5222	8.3192	7.7590
25	38°13′11″	118°56′17″	3.6	0.2580	0.2829	0.3166	0.3120	1.2032	1.1027	0.9853
26	37°54′38″	119°06′41″	4	0.1794	0.1987	0.2185	0.1669	0.9300	0.8397	0.7635
27	38°08′35″	118°53′56″	8	0.0328	0.0336	0.0359	0.1639	4.9898	4.8789	4.5715
28	38°08′59″	118°54′31″	10	0.0259	0.0266	0.0286	0.1884	7.2663	7.0817	6.5748
29	38°19′12″	118°59′24″	5	0.2489	0.2778	0.3054	0.2455	0.9860	0.8835	0.8038

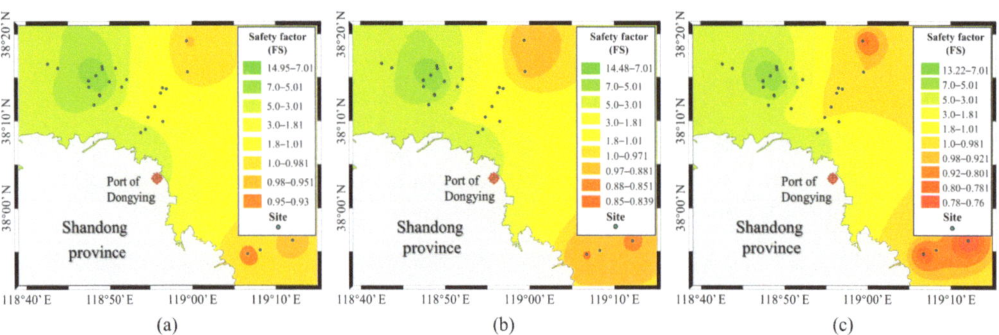

Figure 10. Grading evaluation of wave-induced seabed liquefaction susceptibility in the Chengdao region under wave conditions in a 5-year return period (**a**), a 10-year return period (**b**), and a 25-year return period (**c**), respectively.

Figure 10 reveals the susceptibility assessment result of wave-induced seabed liquefaction in the Chengdao region. The zoning of susceptibility refers to previous research by Chang et al. [48]. The safety factor and liquefaction probability of sediments against liquefaction were calculated under different wave conditions, taking into account the variability in wave and soil parameters. Based on the Chengdao region, a division of the safety factor was proposed. If the probability is controlled within 5%, the safety factor must be greater than 1.8, i.e., the area is a non-liquefaction zone when FS > 1.8 and a liquefaction susceptibility zone when FS < 1.8 [48]. Broadly, the safety factor in the western part of the study area surpasses that in the eastern part, showing substantial spatial variation. The extent of the susceptibility zone expands with more severe wave conditions. Under the wave condition in a 5-year return period, the safety factor in the study area ranges from 0.93

to 14.95, with the liquefaction susceptibility zone encompassing 64.46% of the total study area. Under the wave condition in a 10-year return period, the safety factor ranges from 0.839 to 14.48, covering 65.07% of the study area. Under the wave condition in a 25-year return period, the safety factor varies from 0.76 to 13.22, with the susceptibility zone encompassing 68.89% of the study area. The distribution of the safety factor shows a concentric circle structure centered on the sampling point, with a high safety factor center occurring in the northwestern area and a low safety factor center in the southeastern area. The high safety factor center was located in the northeast of the study area, and its range gradually expanded and developed significantly with an enhancement in wind and wave action, while the range of the high safety factor center in the northwest gradually decreased and the range of the low safety factor center in the southeast gradually expanded (Figure 10). Notably, the safety factor in the western part of the study area under the 25-year return period wave conditions ranges between 0.98 and 1.0. There is an area extending southwestward along the sampling points, and its range changes most significantly, from 18.73% for the wave condition in a 5-year return period to 43.11% for the wave condition in a 25-year return period. Combining the distribution of sampling points and susceptibility areas, it is observed that most sampling points are located in high-safety areas. However, there are still some sampling points with lower safety factors. This is of concern because some important submarine engineering facilities are in these zones. For example, submarine cables from KD48 to KD34C, near borehole K1-1, are buried in an area with the lowest safety factor. This site is particularly susceptible to wave-induced seabed liquefaction under different wave conditions.

The method can realize the preliminary susceptibility assessment of wave-induced seabed liquefaction, which provides a reference value for the construction of marine engineering and maintenance of facilities, but there remains room for its enhancement. Firstly, the accuracy of susceptibility assessments in areas with limited or sparse data points is compromised because of their spatial distribution and the lack of data. Secondly, the current method does not fully account for the influence of sampling depth changes on the safety factor, making it unable to provide a comprehensive assessment of the three-dimensional spatial susceptibility to wave-induced seabed liquefaction. Consequently, the evaluation results may deviate from the actual conditions. To enhance the accuracy of this method, future research can explore wave-induced seabed liquefaction principles through extensive indoor experiments, take into account the impact of sampling depth on susceptibility more comprehensively, and potentially employ machine learning-based models to achieve a more precise evaluation. These improvements could lead to more reliable results in evaluating susceptibility and better support marine engineering construction and facility maintenance decisions.

4. Conclusions

This study modifies the formula for calculating the safety factor, which makes the method suitable for studying the susceptibility assessment of seafloor sediment liquefaction under wave loads. This method employs the safety factor as a primary evaluation criterion and combines an analysis of the cyclic stress ratio (CSR) and the cyclic resistance ratio (CRR) to provide a comprehensive assessment. The Chengdao region in China, characterized by extensive silt distribution, was selected as the study area. To create a regional evaluation zoning map for the susceptibility of wave-induced seabed liquefaction, Empirical Bayesian Kriging (EBK) in the Arc Geographic Information System 10.7 (ArcGIS 10.7) was utilized. The following conclusions can be drawn from this study:

(1)   A method for the susceptibility assessment of wave-induced seabed liquefaction was modified using the safety factor. This safety factor was determined by employing CSR to account for the wave characteristics in the study area and CRR to reflect the sediment properties. This method was then applied to the Chengdao region, where a substantial amount of data was collected to evaluate the susceptibility of wave-induced seabed liquefaction. The EBK method was further utilized to create zoning

maps for the susceptibility assessment of wave-induced seabed liquefaction in the Chengdao region under varying wave conditions.

(2) CSR primarily characterizes the wave conditions in the study area and is a crucial factor for the susceptibility assessment of wave-induced seabed liquefaction. In this study, CSR values are calculated using basic wave elements based on linear wave theory. To gain a deeper insight into the impact of waves on the susceptibility assessment of wave-induced seabed liquefaction and enhance the accuracy of the evaluation, we believe that multiple extreme wave conditions should be calculated for CSR.

(3) CRR primarily represents the sediment characteristics of the study area, and its calculation method should be tailored to the sediment types in the study area. This study presents several testing methods, such as SPT (Standard Penetration Test), CPT (Cone Penetration Test), measurement of V_S (shear wave velocity), and BPT, with a specific focus on the cyclic triaxial test, which is convenient for rapid computation, and the data are easily obtainable, making it applicable to various types of sediments.

(4) The safety factor serves as an effective indicator of wave-induced seabed liquefaction susceptibility assessment in the study area. The safety factor integrates the wave characteristics and sediment properties of the study area. Combined with spatial interpolation methods, it allows for a continuous assessment of wave-induced seabed liquefaction susceptibility across the study area. The specific categorization of susceptibility levels also needs to consider the characteristics of the study area and liquefaction probabilities. Through the calculation of safety factors, it was observed that CSR is more sensitive to changes in sampling depth compared with CRR, exerting a greater impact on the safety factor. Therefore, in future studies, it is crucial to consider the sampling depth as a key reference factor for a more comprehensive evaluation of wave-induced seabed liquefaction susceptibility in three-dimensional space.

Overall, the methodology of this study provides a tool for the susceptibility assessment of wave-induced seabed liquefaction, setting the stage for subsequent risk assessment of such disasters. This method can be directly applied to seabed areas in the study region where the sediment type is silt, providing a spatial reference for stability assessment and site selection of marine engineering facilities within the study area. When this method is applied to other types of sediment in the study area, the formula of the cyclic resistance ratio needs to be adjusted according to the characteristics of the sediment. However, we also need to recognize that there is still room for improvement in this method, especially in the amount and distribution of data, and more in-depth studies are needed to enhance the accuracy and reliability of the assessment.

Author Contributions: Conceptualization, Y.W., X.G. and X.L.; methodology, Y.W. and X.G.; software, Y.W.; formal analysis, J.L. and F.H.; investigation, J.L. and F.H.; resources, J.L. and F.H.; data curation, J.L. and F.H.; writing—original draft preparation, Y.W. and H.G.; writing—review and editing, Y.W., X.G., H.Z. and X.L.; visualization, H.G.; funding acquisition, X.G., H.Z. and X.L. All authors have read and agreed to the published version of the manuscript.

Funding: This research was financially supported by the Fundamental Research Funds for the Central Universities (202441003), the National Natural Science Foundation of China (42207181), the Shandong Province National-Level Leading Talent Supporting Project (2022GJJLJRC-15), the Opening Fund of the State Key Laboratory of Geohazard Prevention and Geoenvironment Protection at Chengdu University of Technology (SKLGP2023K001), and the Shandong Provincial Key Laboratory of Ocean Engineering with grant at Ocean University of China (kloe202301).

Institutional Review Board Statement: Not applicable.

Informed Consent Statement: Not applicable.

Data Availability Statement: All data, models, and code generated or used during this study appear in this submitted article.

J. Mar. Sci. Eng. **2024**, 12, 785

Acknowledgments: We are indebted to the National Resource Sharing Service Platform—National Marine Science Data Center for furnishing the 2018 annual wave element data for the Chengdao area. We acknowledge with gratitude the provision of cyclic triaxial test data from the sampling points, the comprehensive physical and mechanical property data of sediment in the Chengdao area by the Sinopec Petroleum Engineering Corporation, conducted in collaboration with Ocean University of China, the First Institute of Oceanography, Ministry of Natural Resources, Qingdao Huanhai Marine Engineering Survey Institute, and Sinopec Petroleum Engineering Corporation. Additionally, we would like to extend our appreciation to Mengjiao Liu for her invaluable assistance in the design and production of the figures.

Conflicts of Interest: Jinkun Liu and Fang Hou were employed by the Sinopec Petroleum Engineering Corporation. The remaining authors declare that the research was conducted in the absence of any commercial or financial relationships that could be construed as a potential conflict of interest.

Abbreviations and Symbols

C_u	undrained shear strength
D_r	relative density
K_0	coefficient of earth pressure at rest
M_w	earthquake moment magnitude
$N_{1,60}$	standard penetration test blowcount value corrected for overburden, energy, equipment, and procedural factors
P_L	probability of triggering of liquefaction
P_a	atmospheric pressure (1 atm)
R_f	friction ratio
V_s	shear wave velocity
V_{s1}	effective stress normalized shear-wave velocity
p_0	amplitude of the pressure fluctuation exerted on the sea bottom
$q_{c,1}$	normalized cone tip resistance
θ_i	set of unknown model coefficients
ρ'	submerged unit mass of soils in the seabed
ρ_w	density of seawater
σ_1	axial consolidation pressures
σ_3	lateral consolidation pressures
σ_h	horizontal normal stress
σ_C	average consolidation pressures
σ_d	cyclic dynamic stress
σ_v'	vertical effective stress
τ_L	horizontal shear stress
τ_{vh}	shear stress
ArcGIS	Arc Geographical Information System
ASCE	American Society of Civil Engineers
BPT	Becker Penetration Test
CPT	Cone Penetration Test
CRR	cyclic resistance ratio
CSR	cyclic stress ratio
EBK	Empirical Bayesian Kriging
FC	fines content
FS	safety factor
NSW	nearshore spectral windwave
SASW	Spectral Analysis of Surface Waves Method
SPT	Standard Penetration Test
z	sampling depth
$H_{1/10}$	effective wave height
Φ	standard cumulative normal distribution
H	wave height
L	wave length
N	number of cyclic loadings
T	wave period

d	water depth
g	gravitational acceleration
k	wave number
t	time coordinate
ω	wave frequency

Appendix A. Data on Undrained Shear Strength (C_u)

Table A1. The mean undrained shear strength of sediments in the range of 0–5 m.

No.	Longitude (E)	Latitude (N)	C_u (kPa)	No.	Longitude (E)	Latitude (N)	C_u (kPa)	No.	Longitude (E)	Latitude (N)	C_u (kPa)
1	38°11′37.527″	118°55′36.164″	8.98	24	38°11′37.527″	118°55′36.164″	13.79	46	38°13′31.550″	118°56′28.545″	8.50
2	38°13′58.479″	118°47′50.427″	31.00	25	38°13′40.223″	118°56′58.128″	10.64	47	38°14′51.026″	118°57′20.382″	7.50
3	38°14′35.813″	118°51′24.383″	9.50	26	38°15′37.895″	118°59′29.534″	4.89	48	38°15′28.451″	118°58′59.854″	7.83
4	38°13′38.735″	118°51′10.270″	22.50	27	38°8′35.324″	118°53′55.815″	6.67	49	38°15′21.194″	118°58′28.706″	9.67
5	38°15′11.221″	118°48′55.427″	6.25	28	38°8′58.727″	118°54′31.065″	9.67	50	38°15′13.898″	118°57′57.407″	9.00
6	38°15′14.378″	118°48′47.141″	7.33	29	38°12′51.577″	118°46′33.851″	16.13	51	38°14′47.305″	118°47′12.758″	23.00
7	38°15′17.532″	118°48′38.830″	7.67	30	38°11′51.329″	118°47′11.960″	10.57	52	38°15′06.877″	118°46′40.886″	23.00
8	38°15′11.792″	118°48′18.430″	17.50	31	38°10′51.110″	118°47′50.093″	7.093	53	38°14′55.867″	118°46′55.874″	20.33
9	38°13′52.956″	118°47′27.762″	10.00	32	38°9′50.856″	118°48′28.209″	8.79	54	38°15′15.444″	118°46′23.098″	23.25
10	38°13′38.289″	118°49′32.524″	8.00	33	38°8′51.122″	118°49′05.992″	10.83	55	38°15′35.003″	118°45′50.350″	16.00
11	38°15′22.392″	118°45′24.160″	6.50	34	38°10′31.501″	118°56′31.520″	13.50	56	38°15′54.592″	118°45′17.536″	20.33
12	38°15′10.623″	118°45′54.851″	6.50	35	38°10′48.020″	118°56′17.790″	12.33	57	38°16′14.163″	118°44′44.746″	22.25
13	38°14′58.852″	118°46′25.540″	5.50	36	38°11′04.538″	118°56′04.057″	18.67	58	38°16′33.734″	118°44′11.951″	24.67
14	38°14′47.079″	118°46′56.226″	4.50	37	38°5′56.911″	118°55′41.783″	17.33	59	38°16′53.283″	118°43′39.146″	18.00
15	38°14′20.103″	118°47′48.698″	5.00	38	37°56′24.973″	119°13′28.386″	3.00	60	38°17′12.862″	118°43′06.349″	12.00
16	38°10′26.026″	118°55′31.334″	25.00	39	37°56′23.958″	119°13′29.568″	4.50	61	38°17′29.534″	118°42′38.384″	11.00
17	38°9′05.299″	118°55′10.320″	16.50	40	37°56′13.302″	119°12′04.466″	4.25	62	38°12′53.977″	118°48′46.351″	8.75
18	38°8′17.413″	118°54′23.899″	32.5	41	37°56′01.755″	119°10′40.617″	4.33	63	38°17′47.306″	118°51′05.372″	4.83
19	38°12′42.606″	118°56′28.897″	14.75	42	38°9′33.488″	118°56′42.908″	10.00	64	38°17′10.451″	118°55′38.071″	3.45
20	38°13′47.422″	118°56′30.587″	4.87	43	38°15′33.877″	118°44′55.901″	22.49	65	38°16′38.159″	118°58′19.233″	3.46
21	38°12′10.039″	118°56′02.593″	17.18	44	38°14′13.013″	118°56′51.961″	9.00	66	38°16′14.429″	119°1′19.094″	4.82
22	38°10′22.429″	118°54′48.940″	31.17	45	38°13′52.281″	118°56′40.254″	8.67	67	38°15′14.770″	118°58′28.480″	4.15
23	38°13′38.989″	118°56′56.266″	7.36								

Table A2. The mean undrained shear strength of sediments in the range of 5–10 m.

No.	Longitude (E)	Latitude (N)	C_u (kPa)	No.	Longitude (E)	Latitude (N)	C_u (kPa)	No.	Longitude (E)	Latitude (N)	C_u (kPa)
1	38°11′37.527″	118°55′36.164″	5.24	22	38°10′22.429″	118°54′48.940″	13.61	43	38°15′33.877″	118°44′55.901″	23.60
2	38°13′58.479″	118°47′50.427″	22.80	23	38°13′38.989″	118°56′56.266″	36.38	44	38°14′13.013″	118°56′51.961″	26.00
3	38°14′35.813″	118°51′24.383″	13.00	24	38°11′37.527″	118°55′36.164″	9.380	45	38°13′52.281″	118°56′40.254″	23.13
4	38°15′11.661″	118°49′07.406″	17.00	25	38°15′37.895″	118°59′29.534″	22.70	46	38°13′31.550″	118°56′28.545″	26.00
5	38°15′11.221″	118°48′55.427″	8.73	26	38°8′35.324″	118°53′55.815″	9.67	47	38°14′51.026″	118°57′20.382″	21.38
6	38°10′26.026″	118°55′31.334″	25.00	27	38°8′58.727″	118°54′31.065″	18.00	48	38°15′28.451″	118°58′59.854″	26.00
7	38°15′14.378″	118°48′47.141″	13.00	28	38°12′51.577″	118°46′33.851″	11.80	49	38°15′21.194″	118°58′28.706″	23.88
8	38°15′17.532″	118°48′38.830″	12.75	29	38°11′51.329″	118°47′11.960″	9.62	50	38°15′13.898″	118°57′57.407″	21.50
9	38°15′11.792″	118°48′18.430″	16.00	30	38°10′51.110″	118°47′50.093″	5.44	51	38°14′47.305″	118°47′12.758″	8.33
10	38°13′52.956″	118°47′27.762″	13.67	31	38°9′50.856″	118°48′28.209″	12.81	52	38°14′55.867″	118°46′55.874″	14.40
11	38°13′38.289″	118°49′32.524″	10.25	32	38°8′51.122″	118°49′05.992″	14.40	53	38°15′15.444″	118°46′23.098″	14.75
12	38°15′22.392″	118°45′24.160″	9.50	33	38°10′31.501″	118°56′31.520″	6.50	54	38°15′35.003″	118°45′50.350″	16.75
13	38°15′10.623″	118°45′54.851″	7.50	34	38°10′48.020″	118°56′17.790″	4.50	55	38°15′54.592″	118°45′17.536″	19.00
14	38°14′58.852″	118°46′25.540″	11.67	35	38°11′04.538″	118°56′04.057″	6.00	56	38°16′14.163″	118°44′44.746″	15.00
15	38°14′47.079″	118°46′56.226″	10.17	36	38°5′56.911″	118°55′41.783″	6.00	57	38°16′33.734″	118°44′11.951″	17.75
16	38°14′20.103″	118°47′48.698″	8.50	37	37°56′24.973″	119°13′28.386″	9.00	58	38°16′53.283″	118°43′39.146″	16.00
17	38°9′05.299″	118°55′10.320″	23.50	38	37°56′23.958″	119°13′29.568″	10.50	59	38°17′12.862″	118°43′06.349″	12.00
18	38°8′17.413″	118°54′23.899″	35.00	39	37°56′13.302″	119°12′04.466″	8.50	60	38°17′29.534″	118°42′38.384″	14.67
19	38°12′42.606″	118°56′28.897″	24.50	40	37°56′01.755″	119°10′40.617″	8.50	61	38°12′53.977″	118°48′46.351″	10.00
20	38°13′47.422″	118°56′30.587″	32.58	41	37°51′30.354″	119°8′32.770″	14.00	62	38°16′14.429″	119°1′19.094″	35.78
21	38°12′10.039″	118°56′02.593″	10.77	42	38°9′33.488″	118°56′42.908″	15.50				

Table A3. The mean undrained shear strength of sediments in the range of 10–15 m.

No.	Longitude (E)	Latitude (N)	C_u (kPa)	No.	Longitude (E)	Latitude (N)	C_u (kPa)	No.	Longitude (E)	Latitude (N)	C_u (kPa)
1	38°11′37.527″	118°55′36.164″	32.54	10	38°15′22.392″	118°45′24.160″	13.25	19	38°11′51.329″	118°47′11.960″	11.65
2	38°15′11.661″	118°49′07.406″	20.00	11	38°15′10.623″	118°45′54.851″	25.00	20	38°10′51.110″	118°47′50.093″	13.22
3	38°15′11.221″	118°48′55.427″	13.00	12	38°10′26.026″	118°55′31.334″	43.00	21	38°9′50.856″	118°48′28.209″	16.17
4	38°15′14.378″	118°48′47.141″	25.00	13	38°12′42.606″	118°56′28.897″	12.15	22	38°8′51.122″	118°49′05.992″	11.00
5	38°15′17.532″	118°48′38.830″	22.50	14	38°13′47.422″	118°56′30.587″	16.70	23	37°56′13.302″	119°12′04.466″	11.00
6	38°15′11.792″	118°48′18.430″	24.50	15	38°10′22.429″	118°54′48.940″	12.00	24	37°56′01.755″	119°10′40.617″	25.67
7	38°13′52.956″	118°47′27.762″	30.00	16	38°8′35.324″	118°53′55.815″	20.00	25	37°51′30.354″	119°8′32.770″	22.00
8	38°13′38.289″	118°49′32.524″	19.00	17	38°8′58.727″	118°54′31.065″	14.87	26	38°17′29.534″	118°42′38.384″	17.50
9	38°11′37.527″	118°55′36.164″	14.00	18	38°12′51.577″	118°46′33.851″	15.60	27	38°12′53.977″	118°48′46.351″	40.00

Table A4. Sediment cyclic triaxial torsion shear test data in the research area and CRR.

No.	Longitude (E)	Latitude (N)	Lateral Consolidation Pressures (kPa) ($\sigma3'$)	Sampling Depth	Cycle Number	Dynamic Stress Ratio	Fitting Formula	R^2	CRR
1	38°09′52.4517″	118°56′32.8744″	125	12.5	8 13 24 68	0.2261 0.2294 0.2120 0.1898	$y = -0.019\ln(x) + 0.2711$	0.925	0.1836

Appendix B. Data on CRR

No.	Longitude (E)	Latitude (N)	Lateral Consolidation Pressures (kPa) ($\sigma3'$)	Sampling Depth	Cycle Number	Dynamic Stress Ratio	Fitting Formula	R^2	CRR
2	38°11′25.8671″	118°50′47.5194″	30	3	1 50 1495	0.2617 0.2104 0.1540	$y = -0.015\ln(x) + 0.2639$	0.9952	0.1948
3	38°16′32.3239″	118°42′32.5437″	50	5	11 56 205	0.4450 0.3560 0.3080	$y = -0.047\ln(x) + 0.5543$	0.9888	0.3379
4	38°16′02.2446″	118°43′45.4653″	100	10	30 76 1366	0.3320 0.3190 0.2340	$y = -0.027\ln(x) + 0.4274$	0.9873	0.3031
5	38°15′11.7918″	118°48′18.4296″	150	15	1 12 1283	0.3240 0.3186 0.3066	$y = -0.002\ln(x) + 0.3243$	0.9983	0.3151
6	38°13′47.4217″	118°56′30.5870″	30	3	6 25 56	0.5877 0.4942 0.3916	$y = -0.082\ln(x) + 0.7381$	0.9609	0.3605
7	37°56′13.3016″	119°12′04.4655″	90	9	1 48 1434	0.3005 0.2041 0.1035	$y = -0.027\ln(x) + 0.3035$	0.9975	0.1792
8	38°10′22.4291″	118°54′48.9404″	42	4.2	7 13 31	0.5548 0.4721 0.3323	$y = -0.145\ln(x) + 0.8374$	0.9913	0.1697
9	38°15′37.8945″	118°59′29.5341″	80	8	5 21 58 124	0.4600 0.3830 0.2810 0.2330	$y = -0.073\ln(x) + 0.5856$	0.9834	0.2494

J. Mar. Sci. Eng. **2024**, 12, 785

Table A4. *Cont.*

No.	Longitude (E)	Latitude (N)	Lateral Consolidation Pressures (kPa) ($\sigma 3'$)	Sampling Depth	Cycle Number	Dynamic Stress Ratio	Fitting Formula	R^2	CRR
10	38°14′35.2878″	118°47′26.9503″	20	2.3	2 4 117 166	0.5379 0.4920 0.3797 0.3186	$y = -0.044\ln(x) + 0.5632$	0.9615	0.3606
11	38°14′44.8627″	118°52′08.9572″	50	5	232 254 2143	0.3100 0.3000 0.2400	$y = -0.03\ln(x) + 0.4696$	0.9907	0.3314
12	38°13′52.7815″	118°51′30.6558″	50	5	162 782 1665	0.3430 0.4130 0.4350	$y = 0.0402\ln(x) + 0.1399$	0.9915	0.3250
13	38°12′58.4168″	118°48′44.4721″	120	12	2 10 200	0.2780 0.2310 0.0750	$y = -0.045\ln(x) + 0.3193$	0.9835	0.1121
14	38°15′56.3888″	118°49′06.2177″	40	4	19 102 598 2589 8372	0.7793 0.7142 0.5692 0.4523 0.4265	$y = -0.063\ln(x) + 0.9792$	0.9774	0.6891
15	38°16′15.7953″	118°49′03.2053″	70	7	23 165 1042 2997 9471	0.8483 0.6870 0.5178 0.4892 0.4597	$y = -0.067\ln(x) + 1.0366$	0.9555	0.7281
16	37°55′06.6434″	119°08′13.1352″	90	9	15 16 94	0.3154 0.3002 0.2338	$y = -0.042\ln(x) + 0.4222$	0.9794	0.2288
17	38°13′52.9560″	118°47′27.7619″	20	2	3 17 77 306	0.2896 0.2473 0.2193 0.2086	$y = -0.018\ln(x) + 0.3033$	0.9569	0.2204

Table A4. Cont.

No.	Longitude (E)	Latitude (N)	Lateral Consolidation Pressures (kPa) ($\sigma3'$)	Sampling Depth	Cycle Number	Dynamic Stress Ratio	Fitting Formula	R^2	CRR
18	38°11′49.3949″	118°48′04.5013″	20	2	339 1495 7508	0.4025 0.1801 0.1199	$y = -0.09\ln(x) + 0.8989$	0.8863	0.4844
19	38°12′53.9775″	118°48′46.3513″	30	3	3 14 60 274	0.3712 0.2775 0.2080 0.1757	$y = -0.044\ln(x) + 0.4047$	0.9585	0.2021
20	38°11′37.5267″	118°55′36.1643″	63	6.3	8 35 95	0.5097 0.4208 0.3537	$y = -0.063\ln(x) + 0.6429$	0.9994	0.3528
21	38°13′40.2232″	118°56′58.1280″	140	14	7 47 90	0.3976 0.3406 0.3066	$y = -0.035\ln(x) + 0.4694$	0.985	0.3082
22	38°16′03.9100″	118°50′42.5115″	50	5	232 254 2143	0.3100 0.2991 0.2400	$y = -0.03\ln(x) + 0.4681$	0.9882	0.3299
23	38°14′36.9329″	118°47′24.9800″	50	5	10 25 63	0.7246 0.6138 0.5531	$y = -0.095\ln(x) + 0.9389$	0.968	0.5014
24	38°13′40.2232″	118°56′58.1280″	100	10	11 31 64	0.3870 0.3580 0.3200	$y = -0.036\ln(x) + 0.4761$	0.9574	0.3103
25	38°16′03.9100″	118°50′42.5115″	36	3.6	3 14 62 279	0.4780 0.3880 0.3140 0.2810	$y = -0.044\ln(x) + 0.5146$	0.9653	0.3120
26	37°54′37.9495″	119°06′41.0577″	40	4	7 13 31	0.5528 0.4710 0.3300	$y = -0.145\ln(x) + 0.8346$	0.9907	0.1669

Table A4. *Cont.*

No.	Longitude (E)	Latitude (N)	Lateral Consolidation Pressures (kPa) ($\sigma3'$)	Sampling Depth	Cycle Number	Dynamic Stress Ratio	Fitting Formula	R^2	CRR
27	38°08′35.3239″	118°53′55.8153″	80	8	4 15 65 262	0.3012 0.2079 0.1674 0.1451	$y = -0.036\ln(x) + 0.3297$	0.9009	0.1639
28	38°08′58.7269″	118°54′31.0652″	100	10	4 16 66 274	0.3000 0.2267 0.1930 0.1740	$y = -0.029\ln(x) + 0.3219$	0.9225	0.1884
29	38°19′12.0000″	118°59′24.0000″	50	5	1 5 123	0.4007 0.3008 0.2504	$y = -0.029\ln(x) + 0.379$	0.8479	0.2455

References

1. Rahman, M.S.; Jaber, W.Y. A Simplified Drained Analysis for Wave-Induced Liquefaction in Ocean Floor Sands. *Soils Found.* **1986**, *26*, 57–68. [CrossRef] [PubMed]
2. Guo, X.; Fan, N.; Liu, Y.; Wang, Z.; Xie, X.; Jia, Y. Deep seabed mining: Frontiers in engineering geology and environment. *Int. J. Coal. Sci. Technol.* **2023**, *10*, 23. [CrossRef]
3. Wang, H.; Liu, H. Evaluation of storm wave-induced silt seabed instability and geo-hazards: A case study in the Yellow River delta. *Appl. Ocean Res.* **2016**, *58*, 135–145. [CrossRef]
4. Guo, X.; Fan, N.; Zheng, D.; Fu, C.; Wu, H.; Zhang, Y.; Song, X.; Nian, T. Predicting impact forces on pipelines from deep-sea fluidized slides: A comprehensive review of key factors. *Int. J. Min. Sci. Technol.* **2024**, *34*, 211–225. [CrossRef]
5. Guo, X.; Liu, X.; Zheng, T.; Zhang, H.; Lu, Y.; Li, T. A mass transfer-based LES modelling methodology for analyzing the movement of submarine sediment flows with extensive shear behavior. *Coast. Eng.* **2024**, *191*, 104531. [CrossRef]
6. Di Fiore, V.; Punzo, M.; Cavuoto, G.; Galli, P.; Mazzola, S.; Pelosi, N.; Tarallo, D. Geophysical approach to study the potential ocean wave-induced liquefaction: An example at Magoodhoo Island (Faafu Atoll, Maldives, Indian Ocean). *Mar. Geophys. Res.* **2020**, *41*, 9. [CrossRef]
7. Chávez, V.; Mendoza, E.; Silva, R.; Meneses, A.; Pérez, D.; Clavero, M.; Benedicto, I.; Losada, M. Failure of seabeds with a high mud content: An experimental study. *Coast. Eng. Proc.* **2014**, *1*, sediment.47. [CrossRef]
8. Damgaard, J.; Sumer, B.; Teh, T.; Palmer, A.; Foray, P.; Osorio, D. Guidelines for Pipeline On-Bottom Stability on Liquefied Noncohesive Seabeds. *J. Waterw. Port Coast. Ocean Eng.* **2006**, *132*, 300–309. [CrossRef]
9. Pilz, J.; Spock, G. Why do we need and how should we implement Bayesian kriging methods. *Stoch. Environ. Res. Risk Assess.* **2008**, *22*, 621–632. [CrossRef]
10. Fabijanczyk, P.; Zawadzki, J.; Magiera, T. Magnetometric assessment of soil contamination in problematic area using empirical Bayesian and indicator kriging: A case study in Upper Silesia, Poland. *Geoderma* **2017**, *308*, 69–77. [CrossRef]
11. Wang, Z.; Zheng, D.; Guo, X.; Gu, Z.; Shen, Y.; Nian, T. Investigation of offshore landslides impact on bucket foundations using a coupled SPH–FEM method. *Geoenviron. Disasters* **2024**, *11*, 2. [CrossRef]
12. Goeldner-Gianella, L.; Grancher, D.; Robertsen, Ø.; Anselme, B.; Brunstein, D.; Lavigne, F. Perception of the risk of tsunami in a context of high-level risk assessment and management: The case of the fjord Lyngen in Norway. *Geoenviron. Disasters* **2017**, *4*, 7. [CrossRef]
13. Du, X.; Sun, Y.; Song, Y.; Xiu, Z. Wave-induced liquefaction hazard assessment and liquefaction depth distribution: A case study in the Yellow River Estuary, China. *IOP Conf. Ser. Earth Environ. Sci.* **2020**, *569*, 012011. [CrossRef]
14. Fell, R.; Corominas, J.; Bonnard, C.; Cascini, L.; Leroi, E.; Savage, W.Z. Guidelines for landslide susceptibility, hazard and risk zoning for land use planning. *Eng. Geol.* **2008**, *102*, 85–98. [CrossRef]
15. Kvalstad, T.J. What is the Current "Best Practice" in Offshore Geohazard Investigations? A State-of-the-Art Review. In Proceedings of the Offshore Technology Conference, Houston, TX, USA, 30 April–3 May 2007. [CrossRef]
16. Liu, X.; Wang, Y.; Zhang, H.; Guo, X. Susceptibility of typical marine geological disasters: An overview. *Geoenviron. Disasters* **2023**, *10*, 10. [CrossRef]
17. Zhang, X.; Lei, L.; Xu, C. Large-scale landslide inventory and their mobility in Lvliang City, Shanxi Province, China. *Nat. Hazards Res.* **2022**, *2*, 111–120. [CrossRef]
18. Gamboa, D.; Omira, R.; Terrinha, P. A database of submarine landslides offshore West and Southwest Iberia. *Sci. Data* **2021**, *8*, 185. [CrossRef] [PubMed]
19. Locat, J.; Lee, H.J. Submarine landslides: Advances and challenges. *Can. Geotech. J.* **2011**, *39*, 193–212. [CrossRef]
20. Chang, C.; Chien, L.; Chang, Y. 3-D liquefaction potential analysis of seabed at nearshore area. *J. Mar. Sci. Technol.* **2004**, *12*, 2. [CrossRef]
21. Ishihara, K.; Yamazaki, A. Analysis of wave-induced liquefaction in seabed deposits of sand. *Soils Found.* **1984**, *24*, 85–100. [CrossRef]
22. Seed, H.B.; Idriss, I.M. Simplified Procedure for Evaluating Soil Liquefaction Potential. *J. Soil Mech. Found. Div.* **1971**, *97*, 1249–1273. [CrossRef]
23. Nataraja, M.S.; Gill, H.S. Ocean wave-induced liquefaction analysis. *J. Geotech. Eng.* **1983**, *109*, 17881. [CrossRef]
24. Youd, T.L.; Idriss, I.M.; Andrus, R.D.; Arango, I.; Castro, G.; Christian, J.T.; Dobry, R.; Finn, W.D.L.; Harder, L.F.; Hynes, M.E.; et al. Liquefaction resistance of soils: Summary report from the 1996 NCEER and 1998 NCEER/NSF workshops on evaluation of liquefaction resistance of soils. *J. Geotech. Geoenviron. Eng.* **2001**, *127*, 817–833. [CrossRef]
25. Vongchavalitkul, S. Probabilistic Safety Factor of Soil Liquefaction. *Appl. Mech. Mater.* **2012**, *217*, 2414–2418. [CrossRef]
26. Elishakoff, I. Interrelation between Safety Factors and Reliability. NASA/CR—2001-211309. 2001. Available online: https://ntrs.nasa.gov/api/citations/20020011027/downloads/20020011027.pdf (accessed on 1 November 2001).
27. Espósito, T.; Assis, A.; Giovannini, M. Influence of the Variability of Geotechnical Parameters on the Liquefaction Potential of Tailing Dams. *Int. J. Surf. Min. Reclam. Environ.* **2002**, *16*, 304–313. [CrossRef]
28. Madsen, O.S. Wave-induced pore pressures and effective stresses in a porous bed. *Geotechnique* **1978**, *28*, 377–393. [CrossRef]
29. Demars, K.R. Transient Stresses Induced in Sandbed by Wave Loading. *J. Geotech. Eng.* **1983**, *109*, 591–602. [CrossRef]
30. Zen, K.; Yamazaki, H. Mechanism of Wave-Induced Liquefaction and Densification in Seabed. *Soils Found.* **1990**, *30*, 90–104. [CrossRef] [PubMed]

31. Seed, H.B.; Rahman, M.S. Wave-induced pore pressure in relation to ocean floor stability of cohesionless soils. *Mar. Georesour. Geotechnol.* **1978**, *3*, 123–150. [CrossRef]
32. Horikawa, K. *Coastal Engineering: An Introduction to Ocean Engineering*; University of Tokyo Press: Tokyo, Japan, 1978.
33. Ameratunga, J.; Sivakugan, N.; Das, B.M. Standard Penetration Test. In *Correlations of Soil and Rock Properties in Geotechnical Engineering*; Ameratunga, J., Sivakugan, N., Eds.; Springer: New Delhi, India, 2016; pp. 87–113.
34. Bolton Seed, H.; Tokimatsu, K.; Harder, L.F.; Chung, R.M. Influence of SPT Procedures in Soil Liquefaction Resistance Evaluations. *J. Geotech. Eng.* **1985**, *111*, 1425–1445. [CrossRef]
35. Olsen, R. Normalization and Prediction of Geotechnical Properties Using the Cone Penetrometer Test (CPT). 1994. 326p. Available online: https://apps.dtic.mil/sti/citations/ADA285193 (accessed on 1 August 1994).
36. Tokimatsu, K.; Uchida, A. Correlation Between Liquefaction Resistance and Shear Wave Velocity. *Soils Found.* **1990**, *30*, 33–42. [CrossRef] [PubMed]
37. Kayen, R.; Moss, R.; Thompson, E.; Seed, R.; Cetin, K.; Kiureghian, A.; Tanaka, Y.; Tokimatsu, K. Shear-Wave Velocity–Based Probabilistic and Deterministic Assessment of Seismic Soil Liquefaction Potential. *J. Geotech. Geoenviron. Eng.* **2013**, *139*, 407–419. [CrossRef]
38. Cetin, K.O.; Seed, R.B.; Kayen, R.E.; Moss, R.E.S.; Bilge, H.T.; Ilgac, M.; Chowdhury, K. SPT-based probabilistic and deterministic assessment of seismic soil liquefaction triggering hazard. *Soil Dyn. Earthq. Eng.* **2018**, *115*, 698–709. [CrossRef]
39. Moss, R.E.S.; Seed, R.B.; Kayen, R.E.; Stewart, J.P.; Kiureghian, A.D.; Cetin, K.O. CPT-Based Probabilistic and Deterministic Assessment of In Situ Seismic Soil Liquefaction Potential. *J. Geotech. Geoenviron. Eng.* **2006**, *132*, 1032–1051. [CrossRef]
40. Lee, H.J. The Role of Laboratory Testing in the Determination of Deep-Sea Sediment Engineering Properties. In *Deep-Sea Sediments*; Springer: Boston, MA, USA, 1974; pp. 111–127. [CrossRef]
41. Iwasaki, T.; Arakawa, T.; Tokida, K. Simplified procedures for assessing soil liquefaction during earthquakes. *Int. J. Soil Dyn. Earthq. Eng.* **1984**, *3*, 49–58. [CrossRef]
42. Silva, M.P.N.; Guedes, C.C.; Vander, F.M.; Mascarenhas, R.D.O.; Salvador, F.A. Evaluating geostatistical methods along with semi-destructive analysis for forensic provenancing organic-rich soils in humid subtropical climate. *Forensic Sci. Int.* **2022**, *341*, 111508. [CrossRef] [PubMed]
43. Gribov, A.; Krivoruchko, K. Empirical Bayesian kriging implementation and usage. *Sci. Total Environ.* **2020**, *722*, 137290. [CrossRef] [PubMed]
44. Krivoruchko, K.; Gribov, A. Evaluation of empirical Bayesian kriging. *Spat. Stat.* **2019**, *32*, 100368. [CrossRef]
45. Jia, Y.; Zhang, L.; Zheng, J.; Liu, X.; Jeng, D.; Shan, H. Effects of wave-induced seabed liquefaction on sediment re-suspension in the Yellow River Delta. *Ocean Eng.* **2014**, *89*, 146–156. [CrossRef]
46. Liu, X.; Zhang, H.; Zheng, J.; Guo, L.; Jia, Y.; Bian, C.; Li, M.; Ma, L.; Zhang, S. Critical role of wave–seabed interactions in the extensive erosion of Yellow River estuarine sediments. *Mar. Geol.* **2020**, *426*, 106208. [CrossRef]
47. Zhang, H.; Lu, Y.; Liu, X.; Li, X.; Wang, Z.; Ji, C.; Zhang, C.; Wang, Z.; Jing, S.; Jia, Y. Morphology and origin of liquefaction-related sediment failures on the Yellow River subaqueous delta. *Mar. Pet. Geol.* **2023**, *153*, 106262. [CrossRef]
48. Chang, F. Study on Mechanism of Wave-Induced Submarine landslide at the Yellow River Estuary. Ph.D. Thesis, Ocean University of China, Qingdao, China, 2009.
49. Ma, B. Liquefaction of Seabed Silt under Storm Waves. Master's Thesis, First Institute of Oceanography, MNR, Qingdao, China, 2015. (In Chinese)
50. Wang, Z.; Du, X.; Sun, Y.; Song, Y.; Dong, L.; Zhou, Q.; Jiang, W. Risk zonation of submarine geological hazards in the Chengdao area of the Yellow River subaqueous delta. *Front. Mar. Sci.* **2023**, *10*, 1285437. [CrossRef]
51. Liu, X.; Lu, Y.; Yu, H.; Ma, L.; Li, X.; Li, W.; Zhang, H.; Bian, C. In-situ observation of storm-induced wave-supported fluid mud occurrence in the subaqueous Yellow River delta. *J. Geophys. Res. Oceans.* **2022**, *127*, e2021JC018190. [CrossRef]
52. Liu, Z.; Chen, Q.; Zheng, C.; Han, Z.; Cai, B.; Liu, Y. Oil spill modeling of Chengdao oilfield in the Chinese Bohai Sea. *Ocean Eng.* **2022**, *255*, 111422. [CrossRef]
53. Guo, X.; Liu, X.; Li, M.; Lu, Y. Lateral force on buried pipelines caused by seabed slides using a CFD method with a shear interface weakening model. *Ocean Eng.* **2023**, *280*, 114663. [CrossRef]
54. Chu, Z.X.; Sun, X.G.; Zhai, S.K.; Xu, K.H. Changing pattern of accretion/erosion of the modern Yellow River (Huanghe) subaerial delta, China: Based on remote sensing images. *Mar. Geol.* **2006**, *227*, 13–30. [CrossRef]
55. Zhang, H.; Liu, X.; Jia, Y.; Du, Q.; Sun, Y.; Yin, P.; Shan, H. Rapid consolidation characteristics of Yellow River-derived sediment: Geotechnical characterization and its implications for the deltaic geomorphic evolution. *Eng. Geol.* **2020**, *270*, 105578. [CrossRef]
56. Cao, L.H.; Zhai, K.; Pu, J.J.; Hou, Z.M.; Gao, W. Study on regional distribution pattern of subaqueous shore slope sediment movement in Chengdao sea area. *Geol. Environ. Sci.* **2013**, *29*, 49–54. (In Chinese)
57. Zhang, W.; Liang, R.; Mou, X. Sea Bottom Sediment Characteristics and Engineering Geologic Properties in the Chengdao Oil Field Sea Area. *Adv. Mar. Sci.* **2005**, *3*, 305–312. (In Chinese)

Journal of
Marine Science and Engineering

Article

Coupling Effects of a Top-Hinged Buoyancy Can on the Vortex-Induced Vibration of a Riser Model in Currents and Waves

Chi Yu [1], Sheng Zhang [2] and Cheng Zhang [2,*]

[1] Guangdong Energy Group Science and Technology Research Institute Co., Ltd., Guangzhou 510630, China
[2] School of Marine Science and Engineering, South China University of Technology, Guangzhou 511442, China
*   Correspondence: zhangcheng@scut.edu.cn

Abstract: In order to investigate the effects of the top-end dynamic boundary of risers caused by floater motions on their vortex-induced vibration (VIV) characteristics, a combined model comprising a buoyancy can with a relatively simple structural form and a riser is taken as the research object in the present study. The aspect ratios of the buoyancy can and the riser model are 5.37 and 250, respectively. A set of experimental devices is designed to support the VIV test of the riser with a dynamic boundary stimulating the vortex-induced motion (VIM) of the buoyancy can under different uniform flow and regular wave conditions. Several data processing methods are applied in the model test, i.e., mode superposition, Euler angle conversion, band pass filter, fast Fourier transform, and wavelet transform. Based on the testing results, the effect of low-frequency VIM on the high-frequency VIV of the riser is discussed in relation to a single current, a single wave, and a combined wave and current. It is found that the coupling effect of VIM on the riser VIV presents certain orthogonal features at low current velocities. The effect of the cross-flow VIM component on VIV is far more prominent than that of its counterpart, the in-line VIM, with increasing flow velocity. The VIM in the combined wave–current condition significantly enhances the modulation of vibration amplitude and frequency, resulting in larger fluctuation peaks of vibration response and further increasing the risk of VIV fatigue.

Keywords: vortex-induced vibration; dynamic boundary; vortex-induced motion; riser; buoyancy can

Citation: Yu, C.; Zhang, S.; Zhang, C. Coupling Effects of a Top-Hinged Buoyancy Can on the Vortex-Induced Vibration of a Riser Model in Currents and Waves. *J. Mar. Sci. Eng.* **2024**, *12*, 751. https://doi.org/10.3390/jmse12050751

Academic Editor: Fuping Gao

Received: 23 March 2024
Revised: 17 April 2024
Accepted: 28 April 2024
Published: 30 April 2024

1. Introduction

Marine risers in offshore oil and gas production systems provide the most critical connection between floating platforms and underwater equipment [1]. The service lives of marine risers are very long in most scenarios, and the fatigue damage is, thus, a significant aspect of their designs. Vortex-induced vibration (VIV), generated through the interactions between the platform, the current, and the riser, is a primary factor that causes fatigue damage to risers. VIV is a typical fluid–structure interaction whose response characteristics are related to both flow states and structural attributes [2,3]. In recent decades, the VIV of marine risers and pipelines immersed in ocean currents has attracted wide concern from researchers and engineers in offshore oil and gas engineering [4–6].

The steady ocean current is often regarded as the dominant factor affecting the VIV of risers and has been investigated in previous studies [7–10]. Steady shear flow is commonly used to characterize the spatial variation in ocean currents. Zhou et al. [11] examined the VIV of a deep-sea riser under bidirectional steady shear flow in a model test. Based on the finite element approach, Jiang et al. [12] examined the two-dimensional VIV and multi-mode responses of fluid-conveying risers in steady shear flow. Besides the steady ocean current, the VIV of a riser system is affected by other complex factors, e.g., floater motions. The surrounding equivalent flow velocity and riser tension are influenced by the combined effect of the floater motion and various marine environmental loads, which undergo time-varying characteristics [13]. The real-time relative flow velocity and the tension of risers are the critical parameters for VIV. By considering the coupling effect of the overall response caused by the top floater and various marine environmental loads on riser VIV, the study of the unsteady fluid–structure interaction of the riser can contribute to the accurate and systematic understanding of VIV.

In engineering, the flow velocity of ocean currents usually varies on a temporal scale. One of the most common examples is oscillating flow, or the combination of steady and oscillatory flows. Oscillations can be introduced by floating bodies, vessels, or platforms subjected to surface waves. Recently, the riser VIV in oscillatory flow, or subjected to oscillatory forces, has received a lot of attention. Jung et al. [14] conducted an experimental study on the VIV of low-tension submarine cables, in which the flow field was categorized as either uniform flow or a combination of uniform and oscillatory flow. It was observed that there is a periodic enhancement and weakening in the vibration amplitude and frequency under the combined flow, which may contribute to more severe fatigue damage to the structure. By carrying out a dynamic response experiment on a steel catenary riser (SCR), Grant et al. [15] discovered intermittent VIV that was solely excited by the top platform's motion. Wang et al. [16] conducted a systematic experiment on this phenomenon, in which the relative oscillatory currents were achieved by driving the periodic riser motion through a forced motion device. In the article, the relationship between the maximum reduced velocity, the KC number ($KC = U_m \cdot T_f / D$ with U_m and T_f being velocity amplitude and period of oscillatory flow and D being riser diameter), and the riser VIV was also analyzed. Based on the above, Wang et al. [17] conducted an experimental study on the VIV characteristics of an SCR induced by the motion of a floating body on the water surface. The results indicate that in low-KC-number conditions, the time-varying VIV characteristics are highly dependent on the KC number of different pipe positions. Taheri et al. [18] focused on the interaction between oscillatory flow and cylindrical structures in oblique directions and identified different flow regimes affected by structural vibration through numerical simulation. In contrast to the VIV studies focusing on first-mode-dominated flexible pipes in oscillatory flow, Ren et al. [19] investigated the features of a higher-mode-dominated VIV through a model test in the ocean basin. In addition, Neshamar et al. [20] and Deng et al. [21] elaborated on the vibration trajectories of a flexible cylinder in oscillatory flow, which present certain similarities to those of an elastically mounted rigid cylinder under the same conditions.

Some researchers evaluated the effects of a floating body on the riser VIV by exerting excitation directly on the top of the riser. The dynamics of a riser simultaneously affected by the platform sway and VIV were numerically studied by Chen et al. [22]. The displacement level of the riser was found to be amplified several times when platform sway was included in the model. The lateral excitation on the riser top has been used to simulate the sway motion of a platform and has been confirmed to have great effects on VIV; the related results and conclusions are expressed in [23,24]. In terms of the heave excitation on the riser top, Li et al. [25] investigated the VIV of a catenary riser conveying fluid using a semi-empirical model. The effects of double-degree-of-freedom (2DOF) excitations on riser VIV have also been considered. Zhang et al. [13,26] performed a series of model tests on the unsteady VIV of a riser model in uniform flow, affected by single-degree-of-freedom (1DOF) and 2DOF harmonic top-end excitations stimulated by servo motors. They found that the cross-flow (CF) component of the dynamic boundary induces oscillatory characteristics in in-line (IL) VIV, while the IL dynamic boundary component promotes the counterpart of CF VIV. The heave and sway excitations were numerically imposed on the top of a riser by Zhang et al. [27], who found that the combined heave and swing motions result in an increase in CF and IL VIV amplitudes. In order to take the important platform motion into account while evaluating the VIV of a free-hanging riser, Qu et al. [28] modified the wake oscillator by introducing relative oscillatory flow velocity. In addition, Wang et al. [29], Duan et al. [30], and many other researchers have also focused on the motion of floating bodies in VIV studies and suggested that this factor should be considered in VIV response and fatigue evaluation.

Although researchers have gradually identified through various research works that the oscillatory features associated with the dynamic boundaries of floating bodies significantly affect the riser VIV, more specific studies, especially with experimental components, are still at the preliminary stage and introduce many simplifications. For example, the complex platform motions on the riser top are simplified into simple harmonic motions, which leads to a disconnection between the dynamic boundary and the environmental loads. It is difficult to accurately reflect the coupling effect of the riser's dynamic boundary caused by floater motion. Meanwhile, the differences and associations between the CF and IL VIVs affected by multiple-degree-of-freedom dynamic boundaries need to be further clarified. The specific relationships of VIV amplitudes and frequencies with the parameters of dynamic boundaries are also notable issues to be addressed.

In view of the research challenges in VIVs affected by dynamic boundaries, the present study experimentally examines the coupling effects of a top-hinged buoyancy can on the VIV of a riser model in currents and waves. A buoyancy can with a relatively simple circular cylinder structure is selected as the top floating body to generate a top-end dynamic boundary with high recognizability in the current or wave environment. A series of uniform flow, regular wave, and wave–current conditions are designed. The methods of mode superposition, Euler angle conversion, band pass filter, and signal processing, such as fast Fourier and wavelet transforms, are applied to process the testing data. The VIV responses of a riser model, coupled with the vortex-induced motion (VIM) of buoyancy can, are investigated by conducting the model test. By inducing the dynamic boundary of the real structure under different environmental conditions, the relationships between environmental loads, the dynamic boundary, and the riser VIV are discussed. The summarized conclusions for the fluid–structure interaction of VIV are helpful for further understanding complex VIV. The other parts of the paper are organized as follows: The test details are described in Section 2, followed by the experimental data processing in Section 3. In Section 4, the coupling effect of VIM on VIV under uniform currents is discussed, taking the VIVs of three current velocities and two vibration directions into account. Based on this, the influence of waves on the VIV is considered in Section 5, which includes a comparison of the isolated current, the isolated wave, and the combined wave–current. Finally, Section 6 presents the conclusions.

2. Experimental Set-Up

2.1. Test Devices

In this paper, an experiment on the VIV of a riser is conducted at a water basin that is 50 m long, 30 m wide, and 10 m deep. The simulation of uniform flow is formed through the relative movement between a towing carriage and still water, while waves are generated using a wave maker in some cases. The performance of the rocker-flap wave maker covers a wave period of 0.5–4.0 s and produces a maximum wave height of 0.4 m.

The testing devices from our previous study [13] are introduced here as Figure 1. The total testing structure is arranged vertically, with the top end connected to the crane. The bottom end of the riser model is hinged with the lower towing truss. The towing truss height and distance between the left and right steel beams are 6.5 m and 3 m, respectively. The lower part of the riser is connected to the truss by a hinge, and the upper part is hinged to the buoyancy can.

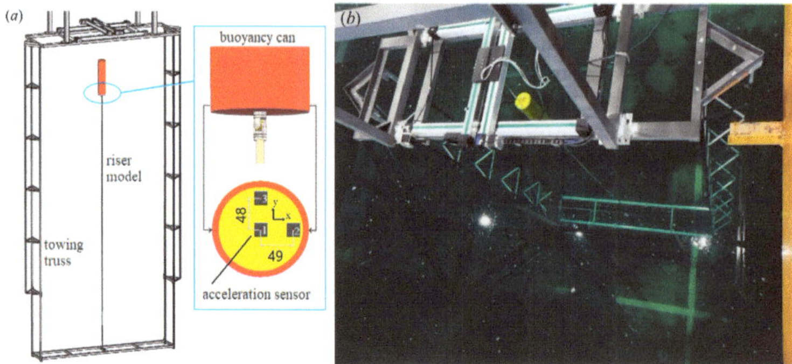

Figure 1. Diagrams of the VIV experimental devices: (**a**) sketch map and (**b**) experimental photo [13].

2.2. Test Model

The present riser model is composed of different composite materials, which include copper pipes to provide stiffness, PTFE pipes, waterproof glue, and other auxiliary materials. More details can be found in [26]. The model is shown in Figure 2a.

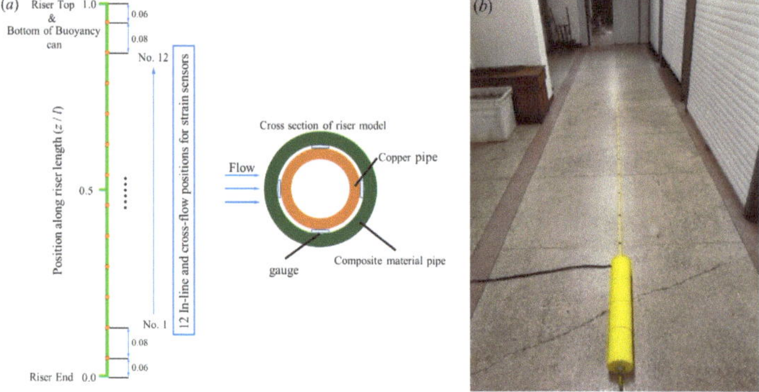

Figure 2. Riser and buoyancy can models: (**a**) strain gauge arrangement and riser cross-section; (**b**) picture of the combined model.

In this experiment, strain gauges are applied to measure the strain data along the model. Then, the model displacements of the riser are obtained using the mode superposition method. The arrangement of the strain gauges is displayed in Figure 2b; its design principle refers to [13]. The scale parameters of the composite riser model are described in Table 1. The first six orders of natural frequencies of the riser model tensioned by buoyancy can are listed in Table 2.

Table 1. Riser model parameters.

Aspect Ratio	Length (m)	Outer Diameter (m)	Inner Diameter (m)	Mass Ratio	EI (N·m²)	EA (N)
250	5	0.02	0.008	2.33	42.62	1.470×10^6

Table 2. Natural frequencies of the tensioned riser model.

Order Number	1	2	3	4
Frequency (Hz)	0.993	2.344	4.639	7.666

A regular cylinder made of fiberglass is used as the buoyancy can. It is rigid, waterproof, and divided into four cabins. The lower part of the buoyancy can is hinged on the riser by a cardan joint. Since the aspect ratios of the buoyancy cans in actual free-standing hybrid riser systems are mostly between 4 and 6 [31,32], the experimental model scale was also selected to be in this range. The parameters are listed in Table 3, and a picture of the model can be seen in Figure 2a. The natural frequencies of the riser-hinged buoyancy can in the cross-flow and in-line directions are very close, specifically 0.219 Hz. There are three acceleration sensors arranged inside the buoyancy can. The specific arrangement is exhibited in Figure 1a, where the numerical unit is a millimeter. The 6-DOF motion measurement is obtained through data processing, and the method is described in the next section.

Table 3. Buoyancy can parameters.

Aspect Ratio	Length (m)	Diameter (m)	Displacement (kg)	Mass Iratio
5.37	0.805	0.15	14.23	0.34

2.3. Testing Conditions

The nonlinear top-end dynamic boundary of the riser is the VIM of the buoyancy can. In this experiment, the VIV is investigated in different conditions, involving single uniform flow, single wave, and combined wave–current, respectively. The top of the buoyancy can is nearly 10 mm below the water's surface under the still water condition. The drifting distance of the buoyancy can significantly increase in large-velocity currents, causing the wires of the measuring instrument to be tightened and further affecting the experiment. Therefore, the flow velocity of the current is kept within the range of 0.1–0.44 m/s, with an interval of 0.02 m/s for different cases. The reduced velocities of the riser model and buoyancy can, represented by U_r and U_{rc}, respectively, are defined as

$$U_r = \frac{U}{f_{n1}D}, \ U_{rc} = \frac{U}{f_{nc}D_c} \tag{1}$$

where U is the steady flow velocity; D and D_c are the outer diameters of the riser model and buoyancy can, respectively; f_{n1} is the first-order natural frequency of the tensioned riser model; and f_{nc} is the natural frequency of the hinged buoyancy can. The ranges of U_r and U_{rc} are 4.89–21.51 and 3.04–13.39, respectively. The Reynolds numbers corresponding to the riser model and buoyancy can, i.e., Re and Re_c, respectively, are defined as

$$Re = \frac{UD}{\nu}, \ Re_c = \frac{UD_c}{\nu} \tag{2}$$

where ν is the kinematic viscosity of water. The ranges of Re and Re_c are 2000–8800 and 15,000–66,000, respectively. The wave period range is 1.5–2.5 s, with an interval of 0.5 s. The range of the wave height is 0.1–0.3 m, with an interval of 0.05 m. All of the test conditions are summarized in Table 4. The experimental errors of the vibration strains range from 5% to 20% after multiple replications, which could be caused by the instability of the generated equivalent flow and waves.

Table 4. Wave and flow parameters in different test conditions.

Item	Wave Height (m)	Period (s)	Flow Velocity (m/s)
CM-01-01~CM-01-18	0	0	0.1~0.44
WM-01-01~WM-01-05	0.2	1.5~2.5	0
WM-02-01~WM-02-04	0.1~0.3	2	0
WM-03-01~WM-03-18	0.2	2	0.1~0.44

3. Data Processing

The strain data are directly measured from the experiment. The strain data of a certain model section need to be preliminary processed to separate the VIV strain from the total strain. Then, the strain data need to be converted into time-history displacements. The methods for both the preliminary processing of the strain and the transformation from strain to displacement are very similar to those presented by Zhang et al. [26]; therefore, only a brief summary is presented in the following.

Based on the small deformation assumption, the relationship between the riser model's strain and displacement in matrix form is described by

$$P_{t\times n} = \varepsilon_{t\times m} C_{m\times n}^T \left(C_{n\times m} C_{m\times n}^T \right)_{n\times n}^{-1} / R \tag{3}$$

$$Y_{t\times m} = P_{t\times n} \Phi_{n\times m} \tag{4}$$

where $\varepsilon_{t\times m}$ and $Y_{t\times m}$ represent the strains and displacements in matrix form, m and t are the total number of measure points and the time, $P_{t\times n}$ is a modal weight matrix with n being the modal order, $C_{n\times m}$ is a matrix composed of a modal shape function after a second derivative to length coordinate, $\Phi_{n\times m}$ is the modal shape matrix, and R is the riser radius. The modal weight matrix is computed using Equation (3) based on the strain data and then processed into a displacement matrix through Equation (4).

In addition to the mode superposition method, other approaches are applied to process the experimental data, e.g., band pass filter, Fourier transform, and wavelet transform, and have been validated by Zhang et al. [13,26].

Three acceleration sensors are installed inside the buoyancy can to measure its three-directional accelerations. The acceleration in the local coordinate system, which is moving and rotating with the buoyancy can, can be measured by the sensors fixed on the bottom of the buoyancy can. Then, the motion of the buoyancy can is obtained after Euler angle conversion and integration for the local acceleration. The specific processing method is introduced as follows:

The Euler angle is assumed as $\Theta = [\phi, \theta, \varphi]^T$, where roll, pitch, and yaw are arranged in sequence. The relationship between the acceleration vector a in the local coordinate system and the acceleration vector A in the global coordinate system is described by

$$A = R(\Theta)a \tag{5}$$

where $R(\Theta)$ is the Euler angle conversion matrix:

$$R(\Theta) = \begin{bmatrix} \cos\varphi\cos\theta & -\sin\varphi\cos\phi + \cos\varphi\sin\theta\sin\phi & \sin\varphi\sin\phi + \cos\varphi\sin\theta\cos\phi \\ \sin\varphi\cos\theta & \cos\varphi\cos\phi + \sin\varphi\sin\theta\sin\phi & -\cos\varphi\sin\phi + \sin\varphi\sin\theta\cos\phi \\ -\sin\theta & \cos\theta\sin\phi & \cos\theta\cos\phi \end{bmatrix} \tag{6}$$

The angles of ϕ, θ, and φ can be obtained through

$$\dot{\phi}(t) = \frac{1}{l_{21}} \int_0^t (a_{2z} - a_{1z})d\tau, \ \phi(t) = \int_0^t \dot{\phi}(t)d\tau \tag{7}$$

$$\dot{\theta}(t) = \frac{1}{l_{31}} \int_0^t (a_{3z} - a_{1z})d\tau, \ \theta(t) = \int_0^t \dot{\theta}(t)d\tau \tag{8}$$

$$\dot{\varphi}(t) = \frac{1}{l_{21}} \int_0^t (a_{2x} - a_{1x})d\tau \tag{9}$$

where a_{1z} and a_{1x} are the center point accelerations in the z and x directions, a_{2z} and a_{2x} are the starboard point accelerations in the z and x directions, a_{3z} is the bow point acceleration in the z direction, and l_{21} and l_{31} are the distances between two accelerators. The three-directional accelerations in the global coordinate system are then obtained through

Equations (5) and (7)–(9). Then, the corresponding displacement results are calculated using a high-precision integral.

4. Coupling effects of VIM on VIV under Uniform Flow

When the buoyancy can and riser are immersed in flow, different fluid–structure interaction phenomena can be captured, i.e., low-frequency VIM and high-frequency VIV. The effect of the low-frequency 2DOF VIM of the buoyancy can on the high-frequency VIV of the riser is discussed in terms of low flow velocity. A comparative analysis is first carried out on the buoyancy can's VIM features, such as amplitude and frequency. On this basis, the role of the VIM frequency in the riser's VIV is qualitatively discussed.

4.1. VIM Characteristics of the Buoyancy Can

The effect of the planar motion of buoyancy on VIV is investigated in this paper. The experimental results of the buoyancy can are filtered to eliminate the high-frequency motions of pitch and roll and obtain the planar VIM. Figure 3 shows the VIM results of the buoyancy can in the cross-flow and in-line directions at three flow velocities, i.e., 0.14 m/s, 0.26 m/s, and 0.44 m/s, respectively. The results include time-history motion, FFT spectral density curves, and motion trajectories in two directions. The system identification method is applied to improve the recognition of the motion trajectory [33].

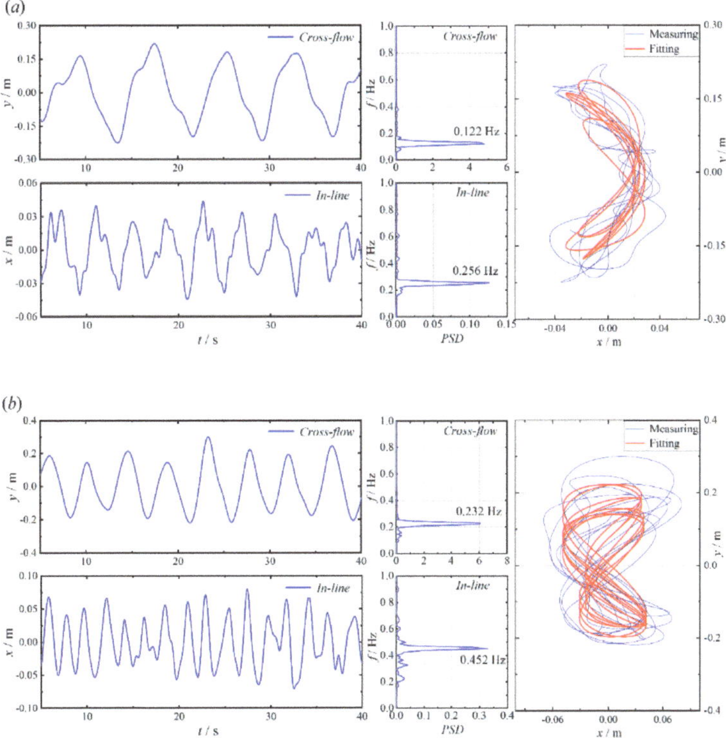

Figure 3. *Cont.*

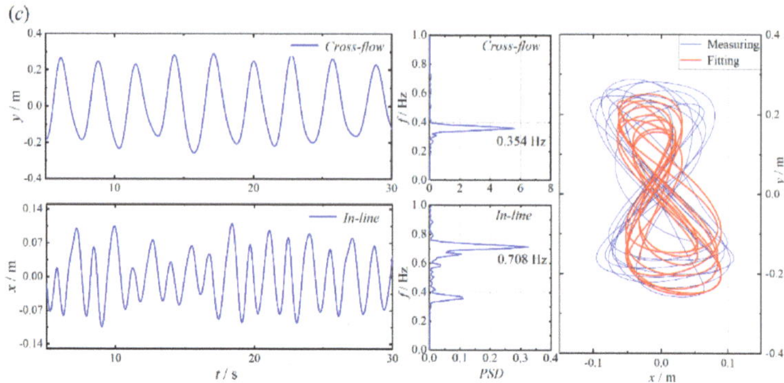

Figure 3. Time−history displacements, FFT frequencies, and motion trajectories of the buoyancy can at different velocities: (**a**) U = 0.14 m/s, (**b**) U = 0.26 m/s, and (**c**) U = 0.44 m/s.

It is observed that the cross-flow motion of the buoyancy can is obvious even at the low flow velocity condition of U = 0.14 m/s, and its main frequency is 0.122 Hz. The in-line motion is relatively small, but its frequency is approximately two times higher than that of its cross-flow counterpart, which follows the typical fluid–structure interaction law. There is a phase difference of $\pi/2$ between the cross-flow and in-line motions, which leads to the trajectory changing to a crescent shape. At the medium flow velocity condition of U = 0.26 m/s, the amplitudes and frequencies of motions in the two directions present a growing trend when compared with the case at low flow velocity. The in-line and cross-flow motion frequencies are 0.452 Hz and 0.232 Hz, respectively, which maintain a near-twice relationship. Their motion trajectories appear in an "8" shape. When U = 0.44 m/s, the amplitudes and frequencies of the in-line and cross-flow motions continue to increase, and the motion frequencies in the two directions reach 0.354 Hz and 0.708 Hz, respectively. The motion trajectories evolve into a relatively standard "8" shape. Kang et al. [33,34] observed an "8"-shaped motion trajectory at medium to high flow velocities in a VIM experiment of a line-tethered buoyant can, which is similar to the results in this paper.

The above analysis shows that most of the motion trajectories of the buoyancy can under medium to high flow velocities are in an "8" shape. The maximum amplitude range of the buoyancy can's cross-flow motion is 0.1 m–0.3 m, and the motion frequency ranges from 0.1 Hz to 0.35 Hz. The frequency of the in-line motion is double that of the cross-flow. The parameters selected in the present experiment are generally reliable and helpful for studying the impact of VIM boundaries on VIV.

The root mean square (RMS) amplitude and frequency of the buoyancy can displacement at different reduced velocities are shown in Figure 4 to further analyze the amplitude and frequency features of the VIM. As a comparison, the experimental VIM results of Kang et al. [33,34] for a buoyancy can are also presented. In their study, a thin line with a length of nearly 3 m is connected with the buoyancy can [33,34]. The buoyancy can in the present study is hinged with a 5 m riser model. The two restriction methods cause differences in the response of the buoyancy can.

The amplitude and frequency results obtained in this study show a similar trend to those in the literature, indicating the reliability of the experiment. There are also some differences for some values, which are caused by the different testing parameters. The VIM amplitude first grows with increasing U_r, and local peaks appear at U_{rc} = 5.5 and 4.3 for cross-flow and in-line VIMs, respectively. After an interval of decreasing amplitude, ending at U_{rc} = 6.7, the cross-flow VIM presents continuous growth with a rising U_{rc}. The amplitude of the typical VIV for a cylinder usually presents a falling trend with increasing reduced velocity, defined as the lower branch by Williamson and Govardhan [35]. However, once the mass ratio of the cylinder is below a certain critical value, the vibration amplitude will not decrease at larger, reduced velocities; instead, it will continuously increase [35].

This statement may help to explain the VIM amplitude feature of the buoyancy can due to the similarity of the structure shape and the common point in fluid–structure interaction. A mass ratio of 0.34 was measured for the buoyancy can, which is small enough for many fluid–structure interaction scenarios; thus, the motion presents a higher amplitude at larger U_{rc}. It is found that the present results are close to those of Kang et al. in terms of the VIM frequency [33,34], and there is no obvious lock-in phenomenon. The planar motions of the buoyancy can are coupled with roll and pitch. When the resonance of cross-flow or in-line motion occurs, the roll and pitch may absorb the energy and disturb the resonance, which makes it hard to present a lock-in phenomenon. The cross-flow motion approaches the natural frequency of the buoyancy can in the vicinity of $U_r = 7.3$–7.9, which is larger than $U_r = 5.5$, corresponding to the amplitude peak that may be caused by the effect of added mass. The variation in f_y/f_{ny} versus U_{rc} is close to the frequency line of $St = 0.16$, which features the vortex shedding frequency. The frequency ratio of the cross-flow and in-line motions is near 2, inducing the "8"-shaped motion trajectory [36].

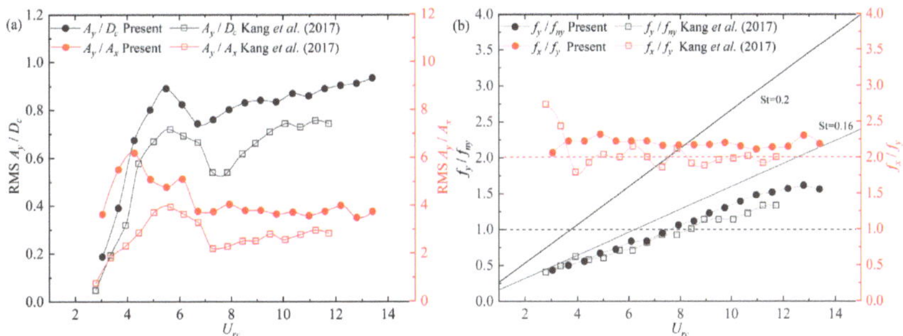

Figure 4. VIM of the buoyancy can: (**a**) RMS amplitude and (**b**) primary frequency [33,34].

4.2. VIV Responses Coupled with VIM

The VIV amplitudes and frequencies of the present riser model with the hinged buoyancy can under steady current were discussed in our previous study [26], which can provide a basic reference. The relative velocity between the riser and its equivalent flow oscillates due to the VIM of the buoyancy can. A coupled analysis of the motions of the two structures is conducted here; its purpose is to reveal the oscillation effects of VIM on VIV. A flow condition of $U = 0.26$ m/s is first selected to examine the specific effects, and the results are illustrated in Figure 5. The left subfigure represents the VIM trajectory of the buoyancy can, as well as the cross-flow and in-line motion periods, where D_c is the diameter of the buoyancy can. The right subfigures exhibit time-history vibration displacements, temporary frequencies, and displacements.

When $U = 0.26$ m/s, the motion trajectory of the buoyancy can presents an asymmetric "8" shape. According to the phase difference analysis of the VIV trajectory by Jauvtis and Williamson [37], the phase difference between the cross-flow and in-line VIM is approximately within the range 0–$\pi/4$. Due to the instability of the overall riser vibration under this condition, there is an obvious amplitude modulation phenomenon in the in-line and cross-flow vibrations. It is observed that the overall amplitude modulation periods of the vibration in the two directions are similar, where the value of 4.4 s is close to the cross-flow VIM period. The cross-flow vibration frequency is time-varying near the main frequency of 1.5 Hz, which is smaller than the second-order natural frequency of 2.344 Hz. The natural frequency of the model is obtained from the decaying test in static water, which cannot accurately consider the added mass related to structural acceleration and thus introduces a difference in the frequency. The maximum value of the fluctuation is close to the second-order natural frequency; thus, there are combined first- and second-order features manifested in the vibration mode. The in-line vibration varies intermittently and contains high-order harmonic frequencies. The main frequency is close to the second-order natural frequency, so the in-line mode is in the transitional state from first- to second-order, but the total amplitude is much lower than that of the cross-flow vibration. At this current velocity, consistently periodic amplitude modulations, time-varying frequency fluctuations, and high-order harmonic phenomena in the riser VIV are induced by the cross-flow VIM. Conversely, the effect of in-line flow VIM is not significant.

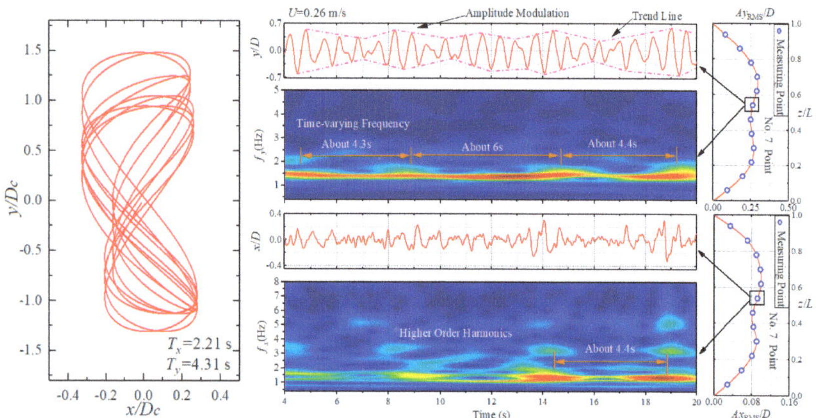

Figure 5. Experimental results of buoyancy can and riser involving time−history displacement, time-varying frequency, and RMS vibration amplitude at a flow velocity of 0.26 m/s.

We presented the results of conditions for other current velocities, i.e., $U = 0.14$ m/s and $U = 0.44$ m/s, in Figures 16 and 17 of Zhang et al. [13]. The results are summarized again in Figure 6 here for comparison and discussion. It is observed that the oscillation period of riser displacement related to the VIM of buoyancy tends to reduce with an increase in flow velocity. The stability of the overall riser vibration is better in the high-velocity condition, while the in-line and cross-flow amplitude modulation is more obvious in the low-velocity state. When $U = 0.14$ m/s, the cross-flow vibration mode is in the first order. As the flow velocity increases, there is a trend of transition from the first to the second order for the cross-flow vibration. The in-line VIM affects the cross-flow VIV more at lower flow velocities, while the cross-flow VIM corresponds to the in-line VIV, which is called an orthogonal coupling effect. As the U rises to 0.44 m/s, the primary and harmonic frequency components, coupled through in-line and cross-flow vibrations, are captured, providing the two-direction coupled feature of VIV. The cross-flow and in-line VIVs are in the second- and third-order states, respectively. In fact, the highest mode detected in this study is the third order, which occurs for in-line vibration at $U = 0.44$ m/s.

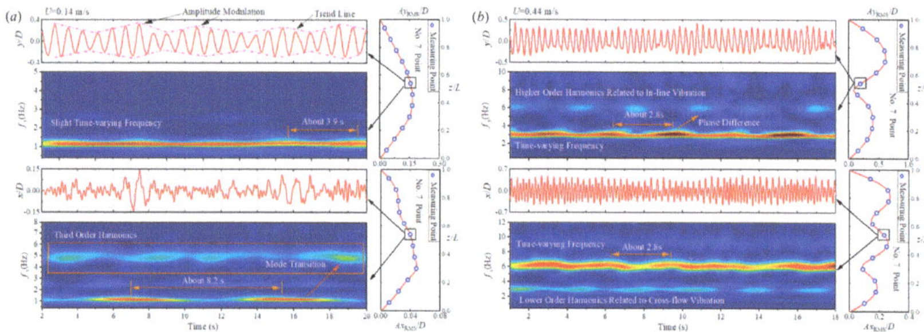

Figure 6. Experimental results of the riser involving time−history displacement, time−varying frequency, and RMS vibration amplitude at (**a**) $U = 0.14$ m/s and (**b**) $U = 0.44$ m/s [13].

The results under the three flow velocities are illustrated in Figure 7 to observe the waveform features of the riser vibration affected by VIM. It is found that the cross-flow vibration presents stronger regularity than the in-line vibration under the three flow velocities. When affected by the VIM, the cross-flow vibration appears as an amplitude variation in the first standing wave at $U = 0.14$ m/s. When $U = 0.26$ m/s, the VIM induces the cross-flow vibration to repeatedly transform between a first-order standing wave and a second-order traveling wave. At a higher flow velocity of 0.44 m/s, the VIV presents a slight amplitude variation in the waveform, even though the VIM is relatively strong. The stability of the in-line vibration waveform is poorer at $U = 0.14$ m/s and $U = 0.26$ m/s. It is observed that the vibration order and amplitude change violently over time, mainly due to the conversion of low-order to high-order waveforms. The periodic features of VIV become prominent at $U = 0.44$ m/s, mainly appearing as a periodic transition between the third-order standing wave in the low-amplitude region and the mixed waveforms of the standing and traveling waves in the high-amplitude region, caused by the periodic velocity oscillation motivated by VIM.

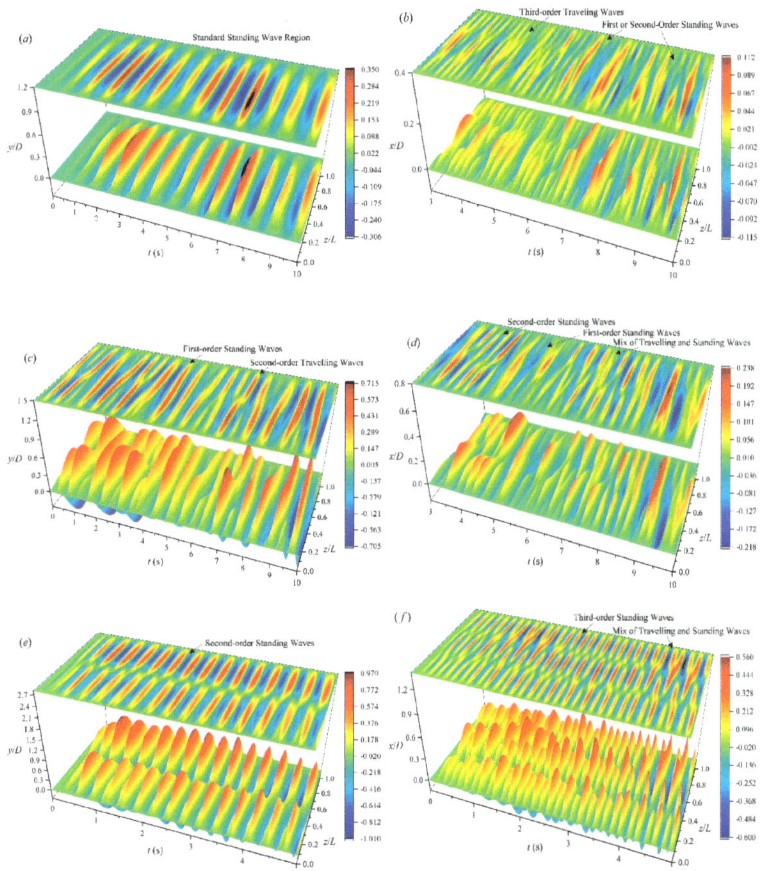

Figure 7. Vibration waveforms at different flow velocities and directions: (**a**) $U = 0.14$ m/s at cross−flow direction; (**b**) $U = 0.14$ m/s at in−line direction; (**c**) $U = 0.26$ m/s at cross−flow direction; (**d**) $U = 0.26$ m/s at in−line direction; (**e**) $U = 0.44$ m/s at cross−flow direction; and (**f**) $U = 0.44$ m/s at in−line direction.

5. Coupling Effects of Wave Load

Waves are one of the most common marine environmental loads acting on floating bodies [38,39]. In the experiment, the whole model is towed in waves, producing the current−wave load. Regular waves are designed in this experiment to investigate the mechanism more clearly based on the testing results. The parameters of wave height and period are set to 0.2 m and 2 s, respectively. The flow velocity at 0.3 m/s is selected as an example for analysis. The boundary effect of VIM on VIV under the combined wave–current load is discussed here.

The VIM and VIV results of the combined model at different conditions, i.e., isolated current, isolated wave, and combined wave and current, respectively, are depicted in Figure 8. The left-side subfigures show the planar motion trajectories and the motion periods of the buoyancy can; the middle three subfigures present the time-history displacements of the same riser position as Figure 5; and the right-side subfigures show the RMS amplitudes along the riser.

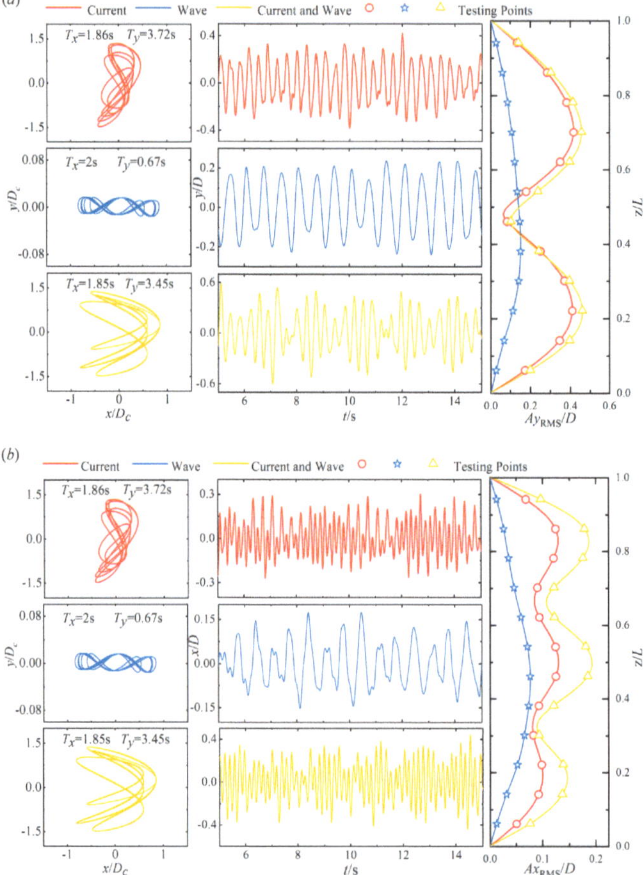

Figure 8. Experimental results under the isolated current, isolated wave, and combined wave and current: (**a**) cross−flow direction and (**b**) in−line direction.

When the buoyancy can is immersed in the isolated current or the combined wave and current, it is found to move with a certain "8"-shaped trajectory. When it is only subjected to wave action, the motion trajectory is a flat, woven shape that is very close to a horizontal line. By comparing the amplitude and frequency characteristics of VIM at isolated current and combined wave–current conditions, it is found that the in-line motion amplitude at the wave–current condition is significantly greater than that at the isolated current condition. The cross-flow motion periods are 3.45 s and 3.72 s at the combined wave–current and isolated current conditions, respectively. This indicates that the waves mainly increase the in-line motion amplitude of the VIM while also reducing the cross-flow VIM period.

Due to the significant influence of the wave on the VIM of the buoyancy can, the coupling between VIM and VIV varies. In the cross-flow vibration condition, the influence of a 1DOF dynamic VIV boundary exhibits an orthogonal effect. When the in-line and cross-flow boundaries coexist with a small difference between them, the former is the dominant effect factor. Under the isolated current, the in-line VIM is much smaller than that of the cross-flow, and the modulation period of the riser VIV amplitude matches that of the cross-flow VIM period. At the combined wave and current state, the wave increases the in-line VIM significantly, which causes a periodic amplitude modulation in the riser VIV, while the period is very similar to that of its in-line counterpart. Similar variation occurs in the in-line vibration, and the wave alters the effect of VIM on VIV. It is found that a first-order riser VIV is excited under the isolated wave when observing the riser RMS amplitude further. The combined wave and current increase the riser VIV amplitude, which is obvious in the in-line VIV.

The vibration frequency is further analyzed. The time–frequency wavelet results under different conditions, i.e., the isolated current, the isolated wave, and the combined wave and current, respectively, are illustrated in Figure 9. Under the isolated current condition, the time-varying fluctuation effect of VIM on the main frequency of the cross-flow VIV (2.61 Hz) is weak. The VIV vibration frequency under isolated waves hardly changes over time. A frequency fluctuation in the cross-flow direction is observed when the two conditions are combined. The fluctuation period is close to the in-line VIM period, while the peak fluctuation exceeds 3 Hz. Although a significant temporal variation feature is exhibited in the in-line VIV under the isolated current condition, the wave characteristics undergo obvious changes. Under the combined wave and current conditions, the fluctuation period becomes shorter and closer to that of the in-line VIM period, while the maximum fluctuation exceeds 6 Hz. The above phenomenon indicates that the oscillation effect of VIM is enhanced by the wave, which leads to higher-order frequencies of the riser VIV.

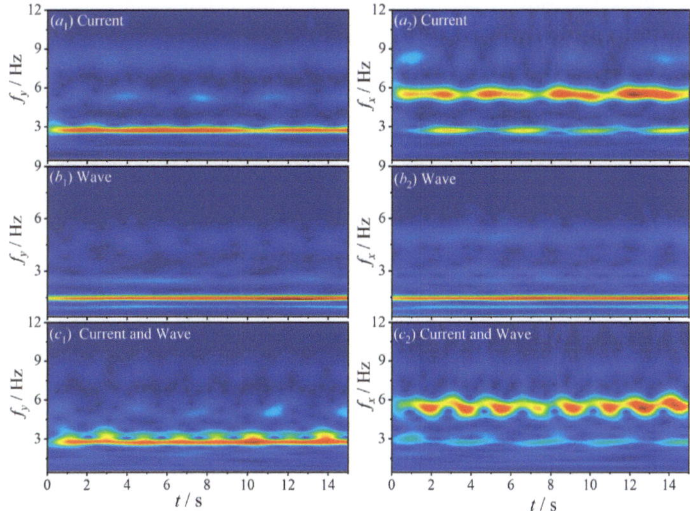

Figure 9. Time-varying frequencies of measuring point No. 7: (**a₁**) isolated current, (**b₁**) isolated wave, and (**c₁**) combined wave and current for cross−flow vibration, while (**a₂**–**c₂**) correspond to in−line vibraton.

Finally, the riser vibration waveform is analyzed, and the relevant results are shown in Figure 10. Under the isolated current and combined wave–current conditions, the second-order standing wave is the dominant wave of cross-flow vibration, and a locally weak traveling wave also emerges. Under the isolated wave, the vibration is mainly a first-order standing wave. There is a waveform transition with a period of 2 s, and the wave has relatively little influence on the cross-flow vibration waveform. For the in-line vibration, the vibration waveform under the isolated wave switches between first-order and second-order, and the overall switching period is close to 2 s. Under the combined wave–current and isolated current conditions, both the third-order combined waveform and the third-order standing wave switch periodically, but the switching period of each is different. There is little influence of the wave on the riser vibration waveforms, especially the cross-flow vibration. Moreover, the wave does not change the in-line vibration waveform. However, the switching period of the combined waveform and standing wave varies between the cross-flow VIM period and that of the in-line VIM.

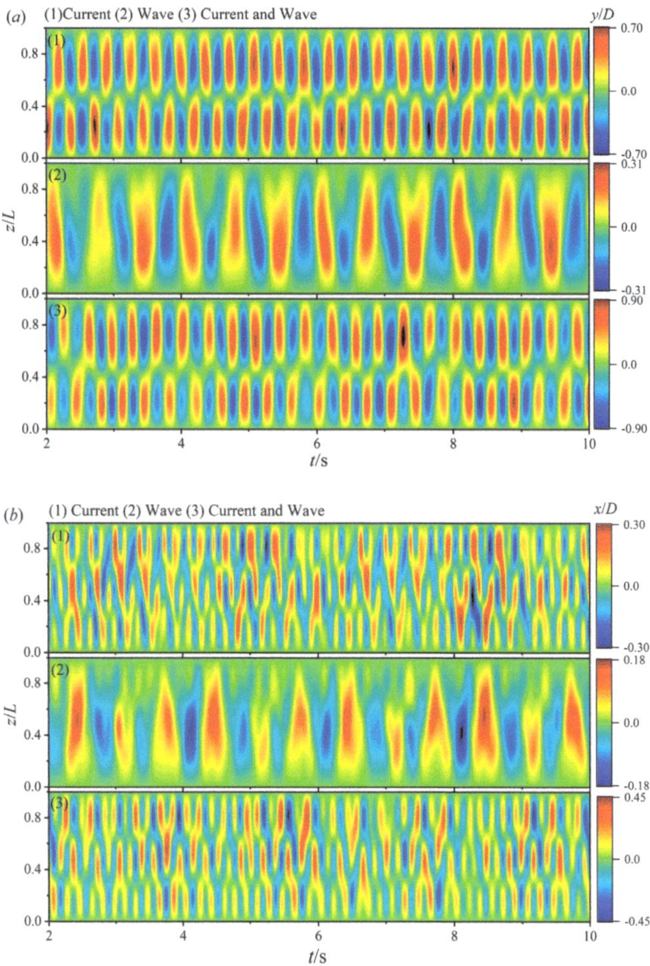

Figure 10. Riser vibrating waveform results under isolated current, isolated wave, and combined wave and current in (**a**) cross−flow and (**b**) in−line directions.

6. Conclusions

In this paper, the VIV characteristics of a riser model, hinged with a buoyancy can on top, are experimentally investigated under uniform current and wave conditions. Firstly, the VIM features are analyzed and compared with the literature for validation. The coupling effects of the buoyancy can's VIM on the riser VIV at different flow velocities are discussed, while the wave is also included in some cases. The following conclusions are drawn:

In terms of the VIM results, the amplitude and frequency results are close to those in the literature, which indicates the reliability of the measurements and data processing in the VIM experiment. The different parameters between the present VIM experiment and those in the literature led to the differences in local values to some extent, which are assumed to be in a reasonable range. Meanwhile, the test set-up and methods were also validated by Zhang et al. (2020, 2022) [26,27], implying that the design of the experiment in the present study satisfies the basic requirements.

The VIM features of the buoyancy can are connected with a riser model, involving the amplitudes, frequencies, and trajectories, which present high similarity to a scenario in which a buoyancy can is moored by a line. This indicates that the small-amplitude and high-frequency VIV have little impact on the large-amplitude and low-frequency VIM of the floater. The VIM may induce low-frequency oscillations in the riser VIV, including periodic modulation of the vibration amplitude, time-varying frequency fluctuations, periodic reproduction of high-order or low-order harmonics, and a switch between the standing wave and the traveling wave. At low current velocity conditions, a certain orthogonal relationship is observed for VIM and VIV. With the increase in flow velocity, this orthogonal effect is no longer significant. The effect of cross-flow VIM on VIV is far more prominent than that of its in-line counterpart.

Waves can significantly amplify the in-line VIM amplitude and allow its in-line component to play a dominant role in the VIV coupling effect. This promotes the oscillation characteristics of VIV, similar to the in-line VIM period. The combined wave–current VIM enhances the amplitude modulation and the overall RMS amplitude of the riser VIV. The time-varying frequency characteristics of VIV are also more obvious in the wave–current condition, resulting in larger fluctuation peaks of frequency. This may further increase the risk of VIV fatigue. The time-scale VIV waveform at the combined wave and current condition is almost consistent with that excited by the isolated current.

Overall, considering the effect of buoyancy can VIM on amplifying the VIV response of the riser model, the dynamic boundary of the platform or the vessel motion should be considered in VIV fatigue assessments when designing risers. In subsequent work, irregular motions of the floating body with more degrees of freedom are expected to be coupled to the top of the risers. This is an area that still needs further investigation. Moreover, further CFD numerical investigations on riser VIV features, especially flow evolutions and hydrodynamic forces, will be helpful for revealing the coupling mechanism of dynamic boundaries with VIV.

Author Contributions: Conceptualization, C.Y. and C.Z.; methodology, C.Z.; software, S.Z.; validation, S.Z. and C.Y.; formal analysis, C.Z.; investigation, C.Z. and C.Y.; resources, C.Y.; data curation, S.Z.; writing—original draft preparation, S.Z.; writing—review and editing, C.Z. and C.Y.; visualization, C.Y.; supervision, C.Z.; project administration, C.Z.; funding acquisition, C.Y. All authors have read and agreed to the published version of the manuscript.

Funding: This research was funded by the Key-Area Research and Development Program of Guangdong Province [grant number 2022B0101100001], the National Natural Science Foundation of China [grant number 52201310], and the Fundamental Research Funds for the Central Universities [grant number D2230600].

Data Availability Statement: We agree to share our research data, which can be obtained by emailing the corresponding author at zhangcheng@scut.edu.cn.

Acknowledgments: We really appreciate Zhuang Kang, who is working at Harbin Engineering University, for providing us with some important experimental equipment during the research.

Conflicts of Interest: Author Chi Yu was employed by the company Guangdong Energy Group Science and Technology Research Institute Co., Ltd. The remaining authors declare that the research was conducted in the absence of any commercial or financial relationships that could be construed as a potential conflict of interest.

References

1. Liu, G.; Li, H.; Qiu, Z.; Leng, D.; Li, Z.; Li, W. A mini review of recent progress on vortex-induced vibrations of marine risers. *Ocean Eng.* **2020**, *195*, 106704. [CrossRef]
2. Zheng, R.; Wang, C.; He, W.; Zhang, Z.; Ma, K.; Ren, M. The Experimental Study of Dynamic Response of Marine Riser under Coupling Effect of Multiparameter. *J. Mar. Sci. Eng.* **2023**, *11*, 1787. [CrossRef]
3. Cai, Q.; Li, Z.; Chan, R.W.; Luo, H.; Duan, G.; Huang, B.; Wu, H. Study on the Vibration Characteristics of Marine Riser Based on Flume Experiment and Numerical Simulation. *J. Mar. Sci. Eng.* **2023**, *11*, 1316. [CrossRef]
4. Chaplin, J.R.; Bearman, P.W.; Huarte, F.H.; Pattenden, R.J. Laboratory measurements of vortex-induced vibrations of a vertical tension riser in a stepped current. *J. Fluids Struct.* **2005**, *21*, 3–24. [CrossRef]
5. Trim, A.D.; Braaten, H.; Lie, H.; Tognarelli, M.A. Experimental investigation of vortex-induced vibration of long marine risers. *J. Fluids Struct.* **2005**, *21*, 335–361. [CrossRef]
6. Hong, K.S.; Shah, U.H. Vortex-induced vibrations and control of marine risers: A review. *Ocean Eng.* **2018**, *152*, 300–315. [CrossRef]
7. Fang, S.M.; Niedzwecki, J.M.; Fu, S.; Li, R.; Yang, J. VIV response of a flexible cylinder with varied coverage by buoyancy elements and helical strakes. *Mar. Struct.* **2014**, *39*, 70–89. [CrossRef]
8. Wu, J.; Lie, H.; Larsen, C.M.; Liapis, S.; Baarholm, R. Vortex-induced vibration of a flexible cylinder: Interaction of the in-line and cross-flow responses. *J. Fluids Struct.* **2016**, *63*, 238–258. [CrossRef]
9. Aswathy, M.S.; Sarkar, S. Effect of stochastic parametric noise on vortex induced vibrations. *Int. J. Mech. Sci.* **2019**, *153*, 103–118. [CrossRef]
10. Zou, X.; Xie, B.; Zang, Z.; Chen, E.; Hou, J. Vortex-Induced Vibration and Fatigue Damage Assessment for a Submarine Pipeline on a Sand Wave Seabed. *J. Mar. Sci. Eng.* **2023**, *11*, 2031. [CrossRef]
11. Zhou, W.; Duan, M.; Chen, R.; Wang, S.; Li, H. Test Study on Vortex-Induced Vibration of Deep-Sea Riser under Bidirectional Shear Flow. *J. Mar. Sci. Eng.* **2022**, *10*, 1689. [CrossRef]
12. Jiang, T.; Liu, Z.; Dai, H.; Wang, L.; He, F. Nonplanar multi-modal vibrations of fluid-conveying risers under shear cross flows. *Appl. Ocean Res.* **2019**, *88*, 187–209. [CrossRef]
13. Zhang, C.; Kang, Z.; Ni, W.; Xie, Z. Coupling effects of top-end dynamic boundary with a planar trajectory of "8" on three-dimensional VIV characteristics. *Mar. Struct.* **2022**, *86*, 103304. [CrossRef]
14. Jung, D.H.; Park, H.I.; Koterayama, W. A numerical and experimental study on dynamics of a towed low-tension cable. In Proceedings of the ISOPE International Ocean and Polar Engineering Conference, Kitakyushu, Japan, 26–31 May 2002; ISOPE: Mountain View, CA, USA, 2002; p. ISOPE-I.
15. Grant, R.G.; Litton, R.W.; Mamidipudi, P. Highly compliant rigid (HCR) riser model tests and analysis. In Proceedings of the Offshore Technology Conference, Houston, TX, USA, 3–6 May 1999; OTC: Tokyo, Japan, 1999; p. OTC-10973.
16. Wang, J.; Fu, S.; Baarholm, R.; Wu, J.; Larsen, C.M. Fatigue damage induced by vortex-induced vibrations in oscillatory flow. *Mar. Struct.* **2015**, *40*, 73–91. [CrossRef]
17. Wang, J.; Fu, S.; Larsen, C.M.; Baarholm, R.; Wu, J.; Lie, H. Dominant parameters for vortex-induced vibration of a steel catenary riser under vessel motion. *Ocean Eng.* **2017**, *136*, 260–271. [CrossRef]
18. Taheri, E.; Zhao, M.; Wu, H. Numerical investigation of the vibration of a circular cylinder in oscillatory flow in oblique directions. *J. Mar. Sci. Eng.* **2022**, *10*, 767. [CrossRef]
19. Ren, H.; Zhang, M.; Cheng, J.; Cao, P.; Xu, Y.; Fu, S.; Liu, C. Experimental investigation on vortex-induced vibration of a flexible pipe under higher mode in an oscillatory flow. *J. Mar. Sci. Eng.* **2020**, *8*, 408. [CrossRef]
20. Neshamar, O.E.; O'Donoghue, T. Flow-induced vibration of a cantilevered cylinder in oscillatory flow at high KC. *J. Fluids Struct.* **2022**, *109*, 103476. [CrossRef]
21. Deng, D.; Zhao, W.; Wan, D. Numerical study of vortex-induced vibration of a flexible cylinder with large aspect ratios in oscillatory flows. *Ocean Eng.* **2021**, *238*, 109730. [CrossRef]
22. Chen, W.; Li, M.; Zheng, Z.; Guo, S.; Gan, K. Impacts of top-end vessel sway on vortex-induced vibration of the submarine riser for a floating platform in deep water. *Ocean Eng.* **2015**, *99*, 1–8. [CrossRef]
23. He, F.; Dai, H.; Huang, Z.; Wang, L. Nonlinear dynamics of a fluid-conveying pipe under the combined action of cross-flow and top-end excitations. *Appl. Ocean Res.* **2017**, *62*, 199–209. [CrossRef]
24. Yuan, Y.; Xue, H.; Tang, W. Nonlinear dynamic response analysis of marine risers under non-uniform combined unsteady flows. *Ocean Eng.* **2020**, *213*, 107687. [CrossRef]
25. Li, X.; Yuan, Y.; Duan, Z.; Xue, H.; Tang, W. Numerical investigation on vortex-induced vibration of catenary riser conveying fluid under top-end heave excitation. *Ships Offshore Struct.* **2024**, 1–18. [CrossRef]
26. Zhang, C.; Kang, Z.; Stoesser, T.; Xie, Z.; Massie, L. Experimental investigation on the VIV of a slender body under the combination of uniform flow and top-end surge. *Ocean Eng.* **2020**, *216*, 108094. [CrossRef]
27. Zhang, B.; Chai, Y.; Li, F.; Chen, Y. Three-dimensional nonlinear vortex-induced vibrations of top-tension risers considering platform motion. *Ocean Eng.* **2022**, *263*, 112393. [CrossRef]
28. Qu, Y.; Fu, S.; Xu, Y.; Huang, J. Application of a modified wake oscillator model to vortex-induced vibration of a free-hanging riser subjected to vessel motion. *Ocean Eng.* **2022**, *253*, 111165. [CrossRef]

29. Wang, J.; Jaiman, R.K.; Adaikalaraj, P.F.B.; Shen, L.; Tan, S.B.; Wang, W. Vortex-induced vibration of a free-hanging riser under irregular vessel motion. In Proceedings of the International Conference on Offshore Mechanics and Arctic Engineering, Busan, Korea, 18–24 June 2016; American Society of Mechanical Engineers: New York, NY, USA, 2016; Volume 49934, p. V002T08A040.
30. Duan, J.; Zhou, J.; Wang, X.; You, Y. Vortex-induced vibration of a flexible fluid-conveying riser due to vessel motion. *Int. J. Mech. Sci.* **2022**, *223*, 107288. [CrossRef]
31. Kang, Z.; Li, H.; Sun, L.P.; Sha, Y.; Cao, J. Overall design and analysis of free-standing vertical risers. *Ship Ocean Eng.* **2011**, *40*, 154–159.
32. Chen, H.F.; Xu, S.P.; Guo, H.Y. Parametric study of global response behavior of deepwater free standing hybrid risers. *J. Ship Mech.* **2011**, *15*, 996–1004.
33. Kang, Z.; Ni, W.; Ma, G.; Xu, X. A model test investigation on vortex-induced motions of a buoyancy can. *Mar. Struct.* **2017**, *53*, 86–104. [CrossRef]
34. Kang, Z.; Ni, W.; Zhang, L.; Ma, G. An experimental study on vortex induced motion of a tethered cylinder in uniform flow. *Ocean Eng.* **2017**, *142*, 259–267. [CrossRef]
35. Williamson, C.H.K.; Govardhan, R. A brief review of recent results in vortex-induced vibrations. *J. Wind Eng. Ind. Aerodyn.* **2008**, *96*, 713–735. [CrossRef]
36. Liu, D.; Ai, S.; Sun, L.; Soares, C.G. Vortex-induced vibrations of catenary risers in varied flow angles. *Int. J. Mech. Sci.* **2024**, *269*, 109086. [CrossRef]
37. Jauvtis, N.A.; Williamson, C.H.K. The effect of two degrees of freedom on vortex-induced vibration at low mass and damping. *J. Fluid Mech.* **2004**, *509*, 23–62. [CrossRef]
38. Guo, X.; Fan, N.; Zheng, D.; Fu, C.; Wu, H.; Zhang, Y.; Song, X.; Nian, T. Predicting impact forces on pipelines from deep-sea fluidized slides: A comprehensive review of key factors. *Int. J. Min. Sci. Technol.* **2024**, *34*, 211–225. [CrossRef]
39. Guo, X.; Liu, X.; Zheng, T.; Zhang, H.; Lu, Y.; Li, T. A mass transfer-based LES modelling methodology for analyzing the movement of submarine sediment flows with extensive shear behavior. *Coast. Eng.* **2024**, *191*, 104531. [CrossRef]

Journal of
Marine Science and Engineering

Article

Seafloor Sediment Acoustic Properties on the Continental Slope in the Northwestern South China Sea

Guanbao Li [1,2,*], Jingqiang Wang [1,2], Xiangmei Meng [1,2], Qingfeng Hua [1,2], Guangming Kan [1,2] and Chenguang Liu [1,2]

[1] Key Laboratory of Marine Geology and Metallogency, First Institute of Oceanography, MNR, Qingdao 266061, China; wangjqfio@fio.org.cn (J.W.); mxmeng@fio.org.cn (X.M.); hqf@fio.org.cn (Q.H.); kgming135@fio.org.cn (G.K.); lcg@fio.org.cn (C.L.)

[2] Laboratory for Marine Geology, Qingdao Marine Science and Technology Center, Qingdao 266237, China

[*] Correspondence: gbli@fio.org.cn

Abstract: The acoustic properties of seafloor sediments on continental slopes play a crucial role in underwater acoustic propagation, communication, and detection. To investigate the acoustic characteristics and spatial distribution patterns of sediments on the continental slope, a geoacoustic experiment was conducted in the northwestern South China Sea. The experiment covered two sections: one crossing the shelf and slope in the downslope direction, and the other near the shelf break in the along-slope direction. In situ techniques, sediment sampling, and laboratory measurements were used to acquire data on sediment acoustic properties (such as sound speed and attenuation) and physical properties (including particle composition, density, porosity, and mean grain size). The experimental findings revealed several key points: (1) Acoustic properties of shallow water coarse-grained sediments and deep-sea sediments were higher when measured in the laboratory compared to in situ measurements. (2) Relationships between measured attenuation and physical properties, as well as between sound speed and mean grain size, showed deviations from previous empirical equations. (3) Sediment acoustic and physical properties exhibited significant variations in the downslope direction, while showing gradual variations in the along-slope direction. These variations can be attributed to sedimentary environmental factors such as material sources, hydrodynamic conditions, and water depth.

Keywords: sound speed; attenuation; sediment; continental slope; continental shelf; in situ measurement; South China Sea

Citation: Li, G.; Wang, J.; Meng, X.; Hua, Q.; Kan, G.; Liu, C. Seafloor Sediment Acoustic Properties on the Continental Slope in the Northwestern South China Sea. *J. Mar. Sci. Eng.* **2024**, *12*, 545. https://doi.org/10.3390/jmse12040545

Academic Editor: Dimitris Sakellariou

Received: 28 February 2024
Revised: 20 March 2024
Accepted: 21 March 2024
Published: 25 March 2024

1. Introduction

The acoustic properties of sediments, including sound speed, attenuation, and density, play a crucial role in the transmission and reflection of sound waves in seafloor sediments and at the sediment–water interface [1]. Understanding these properties is essential for underwater navigation, communication, and detection using acoustic techniques [2]. These properties are influenced by sediment composition, texture, and depositional processes, making them valuable in marine sedimentary environments, marine engineering, and oceanographic studies [3,4]. Acoustic properties of sediments vary spatially, depending on sediment distribution patterns shaped by environmental conditions during formation [5,6]. Hamilton's 1980 study in the North Pacific Ocean categorized sediment acoustic and physical properties into three physiographic provinces: continental terrace, abyssal plain, and abyssal hill [6]. Among these provinces, continental terrace, including continental shelf and slope, exhibit diverse sediment types and complex acoustic property variations, thus large impacts on the sound propagation.

Over the past two decades, studies have focused on the frequency-dependence of sediment acoustic properties, especially for sandy sediments [7–12], with data collection primarily in shallow or coastal areas. A few investigations and studies were carried out on sediment acoustic property distribution in continental slope regions. Wang et al. [13] investigated acoustic properties of surface sediments in the South China Sea, finding higher sound speeds on shelves compared to slopes and troughs. Kim et al. [4] studied the sound speed distribution of surficial sediments in the Ulleung Basin, east of the Korean Peninsula, and identified geoacoustic provinces reflecting sediment textures and properties consistent with marine geological features in shelf and slope environments. Tian et al. [14] analyzed sound speed data from sediment samples at 270 stations in the northern South China Sea, identifying two geoacoustic provinces consistent with water depth, and they found that sound speed ratios were less than 1 in the relatively deeper slope region and more than 1 in shallower shelf region, and the latter was further divided into two sub-provinces based on sediment type. Studies mentioned above are all based on sediment acoustic data obtained by sampling and laboratory measurement methods, although some have attempted to make in situ corrections using seafloor temperatures and water pressures [4]. However, comparison between in situ and laboratory measurements in the west Pacific Ocean deeper than 5000 m had showed the limitations in fully correcting laboratory sound speed into the in situ one in the deep waters due to structural perturbations and state changes of sediment sample [15]. This hinders accurate recognition of spatial variability of acoustic properties based solely on laboratory measurement data. Furthermore, most previous studies, including that of Hamilton [6], lacked analysis of spatial distribution of acoustic attenuation.

In recent years, in situ sediment acoustic measurement techniques were rapidly developed in China and widely used in the field survey (e.g., [15–17]), providing a technical basis to overall and accurately understand the distribution of sediment acoustic properties, not only sound speed, but also attenuation. Using a hydraulic-driven in situ sediment acoustic measurement system, Li et al. [18] observed strong down-slope variations and relative along-slope uniformity of sound speed and attenuation in the northern South China Sea, implying the peculiarity of sound property distribution on the continental slope. However, due to the sparsity of measurement stations and lacking sediment samples, detailed analysis of the sedimentary environment was not provided. To further understand sediment acoustic property variations on the continental slope and their environmental controlling factors, a sediment acoustic experiment based on in situ and laboratory acoustic measurement techniques was conducted on the northwestern South China Sea (SCS), along two sections, one parallel and the other perpendicular to the slope. The results of this experiment and some new findings are presented in the following.

2. Regional Settings

The experimental area is situated in the northwestern SCS (Figure 1), extending southeastward from the middle part of continental shelf near Hainan Island to the Xisha Trough, with water depth ranging from 90 m to 1900 m. The Xisha Trough is a northeast-trending depression located on the western continental slope of the South China Sea, serving as a transition zone between the northwest shelf and the central basin of the South China Sea.

Different current systems dominate the shelf and the slope. The SCS northwestern shelf is influenced by the Guangdong Coastal Current (GCC), which shifts direction seasonally due to monsoon climate, flowing northeastward in summer and southwestward in winter [19]. The Xisha Trough is affected by the SCS Deep Water Current (DWC), which originates from the Pacific Ocean, enters SCS through the Luzon strait, and flows southwest along the SCS northern slope into Xisha Trough [20].

Sediments in this region come from nearby large rivers (e.g., the Pearl River, and Red River) and small mountain rivers (e.g., the rivers on Hainan and Taiwan Island) [21–23]. These sediment sources, along with the flow patterns, determine the distribution of sediment types in the area. Various sediment types are found in the shelf area [23,24], arranged roughly in a northeast-trending belt. Offshore of Hainan Island, there are gravelly sands which transition to silty sands, sandy silt, and clayey silt as one moves away from the island. Fine sands are also locally present, along with a layer of muddy sediments on the inner shelf near Hainan Island. From the outer part of the shelf to the Xisha Trough, clayey silt is the dominant sediment type [24].

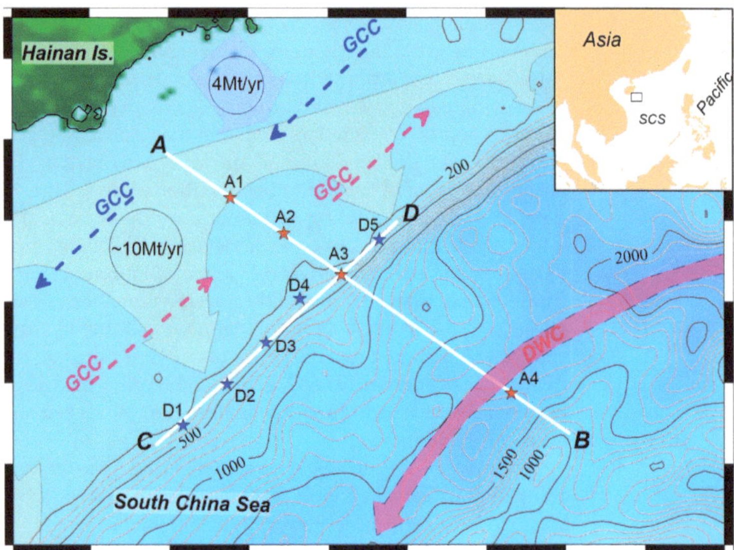

Figure 1. Location map of the two sections (solid while lines) and stations (pentagrams in red and blue, respectively) in the experiment. The upper right inlet shows the location of the study area. The dashed line with an arrow indicates the flow direction of the GCC (dark blue for winter, pink for summer), modified from Liu et al. [25]. The thick solid-brown line with arrows indicates the direction of the DWC (modified from Zheng and Yan [20]). Grey arrows depict sediment transport paths from the Red River, while light-blue arrows show transport from the Wanquan River on Hainan Island, with the arrow size indicating discharge amount (according to Zhao et al. [26]). Isobaths at 100 m intervals are represented by light-grey solid lines, with bathymetry data sourced from GEBCO [27].

3. Materials and Methods

The goal of the experiment is to study the distribution and variation pattern of sediment acoustic properties on the continental slope and its adjacent shelf, so two near-perpendicular sections in cross-slope and along-slope directions, respectively, were established as depicted in Figure 1. Section AB, approximately perpendicular to the isobath direction, comprises four stations: A1 and A2 in the shelf area (depth ~100 m), A3 out of the shelf break (depth 244.4 m), and A4 in the Xisha Trough center (depth 1864.8 m). In situ measurements were conducted at all these stations using the ballast in situ acoustic measurement system (BISAMS), with sediment samples collected simultaneously using a short gravity corer attached on BISAMS for acoustic and physical property measurement in laboratory. Section CD, roughly parallel to the isobath direction, includes stations D1–D5 and intersects section AB at station A3. These six stations are all near the shelf break, with water depths mostly between 350 m and 400 m, except for D4 (depth ~300 m). In situ measurements were not performed at D1–D5 due to the bad weather conditions, and only sediment samples were collected using a gravity corer.

The BISAMS used for in situ measurements on section AB has been previously described (e.g., [15,17]). This system is capable of measuring in situ sound speed and attenuation at a depth of 80 cm below the seafloor, using a 30 kHz central frequency and a 10 MHz sampling frequency. The BISAMS consists of four acoustic transducers, one transmitter and three receivers, which penetrate the seafloor under the pressure of the ballast weight. The transducers transmit and receive pulse waveform signals which passes through the sediment. The sound speed and attenuation can be calculated using the travel time difference and amplitude difference of the signals received by three channels (Figure 2), respectively, according to the formula suggested by Wang et al. [17]. During each measurement at a station, the waveform signals were transmitted and received six times, and the averaging algorithm was used to eliminate the noise. Furthermore, the average value of three receiving channels is used to reduce measurement error. The BISAMS demonstrates stable performance and high measurement accuracy, and has been utilized in seafloor acoustic measurement expeditions in the East China Sea [17], the South China Sea [28], and the west Pacific Ocean [15].

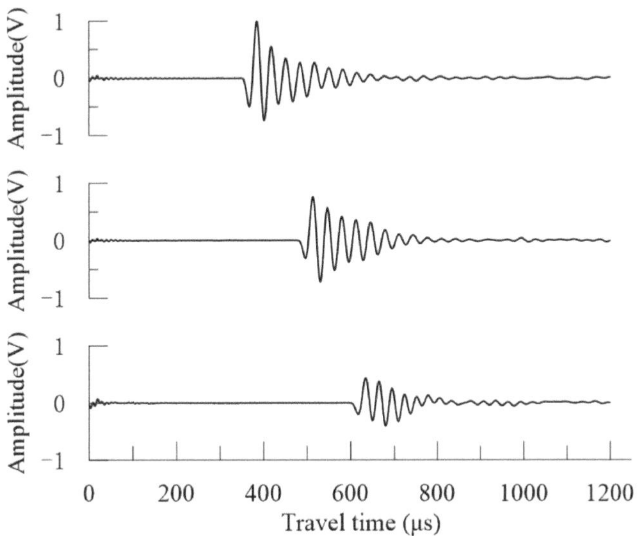

Figure 2. Pulse waveform signals acquired using three receiving channels of BISAMS.

The ratio of sediment sound speed to the seawater sound speed under the same temperature and pressure conditions is defined as the sound speed ratio (SSR), which is experimentally and theoretically proved to remain constant [29–31]. During the experiment, a self-contained sound speed profiler (SVP, Valeport Ltd., Devon, UK) was attached to the BISAMS, and the seawater sound speed was acquired simultaneous at each in situ measurement. Then, the SSR is calculated as the ratio of sediment sound speed measured using the BISAMS to near-seafloor seawater sound speed measured using the SVP.

The BISAMS is equipped with a 75 mm diameter short gravity corer, so a sediment sample with length equivalent to the penetration depth can be collected synchronously during the situ measurement of section AB. For sections CD, the sediment sampling was conducted only, using a 110 mm diameter, 3 m long gravity corer. All sediment samples were measured in laboratory for acoustic and physical property.

In the laboratory, sound speed and acoustic attenuation were measured with an acoustic measurement system, as detailed by Wang et al. [32], which comprises a digital signal generator and a waveform recorder, a power amplifier and a preamplifier, a couple of transmitting and receiving planar transducer with central frequency of 100 kHz, and an acoustic measuring platform. The sample was segmented into 20–30 cm lengths and placed on the platform. Acoustic signals at 100 kHz frequency were emitted and received by transducers at the sample ends. Sound speed was determined using the time-of-flight (TOF) method based on signal travel time and sample length. Acoustic attenuation coefficients were calculated by comparing signal amplitudes and propagation path lengths between full-length and segmented samples using the coaxial gap attenuation measurement method [15,33]. For each segment sample, the sound speed and attenuation were measured at least three times, and the laboratory sound speed and attenuation values were obtained by averaging segment measurements. Sample temperature was recorded, and the sound speed in seawater at that temperature (with salinity of 35‰) was calculated using Mackenzie's formula [15]. The ratio of laboratory sediment sound speed to seawater sound speed was defined as the laboratory sound speed ratio. Sound attenuation coefficient measurements, whether in situ or in the laboratory, were divided by frequency to determine the attenuation factor [3,6].

Following the acoustic measurement, the samples underwent physical property measurements to determine parameters such as particle composition, mean grain size (MGZ), bulk density, water content, and porosity. Laboratory procedures to determine the physical properties of the sediment samples followed the specifications for oceanographic survey GB/T 12763.8-2007 [34] and the standard for geotechnical testing method GB/T 50123-1999 [35], belonging to National Standard of the People's Republic of China. The wet bulk density is the weight of the mineral solids and porewater per unit volume and was measured using a cutting-ring method using a steel ring sampler (diameter 6 cm, height 2 cm) which was pushed into the sediment sample. The sampler with sediment core in it was weighed to determine the wet bulk density. The grain specific gravity was measured on dried samples with a pycnometer (volume 50 mL). Porosity is the ratio in percent of the volume of voids to the total volume of the sediment mass. Porosity was calculated for each sample using measurements of water content, wet bulk density and grain specific gravity. The calculation equation of porosity was,

$$e = \frac{\rho(1 + w)}{\rho_s} - 1 \tag{1}$$

$$n = \frac{e}{1 + e} \tag{2}$$

where e is the void ratio; ρ is wet bulk density; w is water content; ρ_s is grain specific gravity; n is porosity.

A sieving method was used to determine the composition, mean grain size, sand content, clay content, and sorting coefficient. The apertures of sieve include 2.0 mm, 1.0 mm, 0.5 mm, 0.25 mm, 0.1 mm, and 0.075 mm. The sediments were then classified using Shepard's ternary diagram [36] based on the obtained sand and clay contents. The mean grain size was determined by plotting the particle distribution curve and identifying the values corresponding to d16, d50, and d84.

4. Results

4.1. Sediment Types

Table 1 presents the particle composition and sediment types at all stations, with corresponding Shepard ternary diagram in Figure 3. Sections AB and CD exhibit distinct sediment distribution patterns. Four stations on section AB display varying sediment types, with finer sediments in shallow waters. Station A1 consists of silty sand with a high sand content of 63.8%, while stations A2 and A3 transition to higher silt content and lower clay content towards the outer shelf, categorized as silt and sandy silt, respectively. Station A4 is characterized by silt particles with a clay content of 33.4%, classified as clayey silt. In contrast, stations D1 to D5 on section CD have lower sand content, predominantly composed of silt and clay. Silt content ranges from 39.9% to 65.7%, and clay content is slightly lower, ranging from 32.9% to 54.2%. Stations D1, D4, and D5 have higher silt content, classified as clayey silt, while D2 and D3 have higher clay content, classified as silty clay. Despite the proximity of station A3 to D4 and D5 on section CD, its particle composition differs significantly. Conversely, station A4, located in deep water, shares similarities in sediment type and particle composition with the shallower water stations on section CD.

Table 1. Sediment type, particle component, and physical properties for each stations.

Station	Sediment Type	Sand Content (%)	Silt Content (%)	Clay Content (%)	Density (kg/m³)	Porosity (%)	Mean Grain Size (φ)
A1	Silty sand	63.8	29.7	6.5	1876	48.6	3.994
A2	Silt	5.2	81.3	13.5	1574	66.4	6.018
A3	Sandy silt	18.0	71.2	10.8	1634	63.3	5.494
A4	Clayey silt	5.7	60.9	33.4	1406	76.7	7.231
D1	Clayey silt	1.7	56.0	42.3	1602	66.6	7.820
D2	Silty clay	2.8	48.0	49.2	1584	64.9	7.953
D3	Silty clay	5.9	39.9	54.2	1521	69.9	8.064
D4	Clayey silt	2.4	60.8	36.8	1599	64.9	7.108
D5	Clayey silt	1.4	65.7	32.9	1713	59.2	7.139

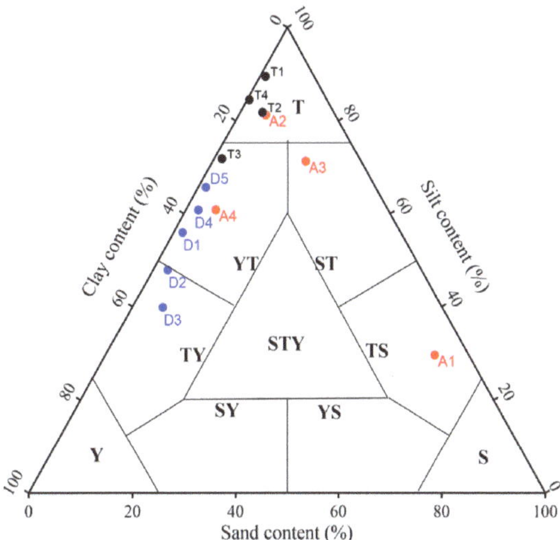

Figure 3. Shepard ternary diagram of sediment types in the experimental area. Red and blue dots represent particle composition at stations on sections AB and CD, respectively. The diagram also includes the percentage of sediment particle fractions at four stations in the West Pacific Ocean (black dots, data from Wang et al. [15]).

4.2. Acoustic Properties and Corresponding Physical Properties on Section AB

Tables 1 and 2 displays the physical and acoustic properties, respectively, and Figure 4 illustrates the variation of both acoustic and physical properties along the section AB. The acoustic and physical properties at each station exhibit significant variations in the down-slope direction, corresponding to changes in sediment type. Generally, as water depth increases from the shelf to the trough, acoustic properties (both in situ and in laboratory) and density decrease, while porosity and mean grain size increase. In situ acoustic speed ranges from 1446–1594 m/s, attenuation coefficient ranges from 0.89–7.63 dB/m, and SSR varies from 0.970–1.038. The laboratory acoustic speed ranges from 1473–1613 m/s, attenuation coefficient ranges from 14.91–47.28 dB/m, and SSR varies from 0.978–1.072. Physical properties show density ranging from 1406–1876 kg/m³, porosity from 48.6–76.7%, and MGZ from 3.994–7.231 φ. As shown in the figure and tables, lab acoustic properties are consistently higher than in situ ones at all stations, with the most significant difference observed in attenuation, particularly at station A1. Station A1 stands out as having a SSR greater than 1, indicating higher sediment sound speed compared to seafloor seawater sound speed. For stations from the shelf edge to the trough, SSR values less than 1 imply that sediment sound speed is lower than seafloor seawater sound speed.

Table 2. Acoustic properties measured in situ and in laboratory for each stations. Acoustic properties include sound speed (*V*), sound speed ratio (SSR), attenuation coefficient (α), and attenuation factor (*k*); the subscript "insitu" and "lab" represent data acquisition using the in situ or laboratory measurement method, respectively.

Station	Depth (m)	V_{insitu} (m/s)	SSR_{insitu}	α_{insitu} (dB/m)	k_{insitu} (dB/m/kHz)	V_{lab} (m/s)	SSR_{lab}	α_{lab} (dB/m)	k_{lab} (dB/m/kHz)
A1	92	1594	1.038	7.63	0.254	1613	1.072	47.28	0.426
A2	143	1499	0.986	0.89	0.030	1486	0.988	18.84	0.170
A3	244	1499	0.994	1.48	0.049	1504	0.998	14.91	0.134
A4	1865	1446	0.970	3.96	0.132	1473	0.979	17.17	0.155
D1	385	—	—	—	—	1462	0.976	38.81	0.343
D2	379	—	—	—	—	1480	0.987	21.51	0.190
D3	356	—	—	—	—	1465	0.981	41.17	0.364
D4	298	—	—	—	—	1479	0.986	30.37	0.269
D5	376	—	—	—	—	1499	0.995	20.38	0.180

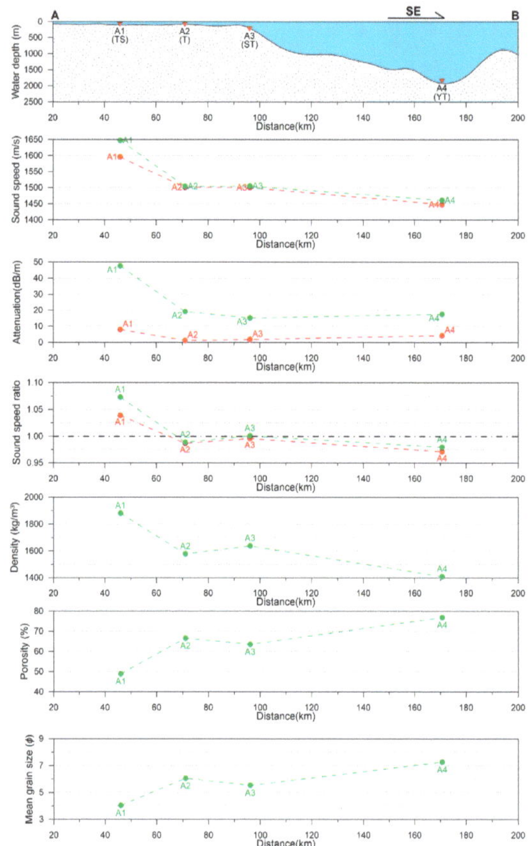

Figure 4. Variation of sediment acoustic and physical properties along section AB (NW-SE direction). Red dots represent acoustic properties measured using the in situ technique, and green dots represent acoustic and physical properties of sediment samples measured in the laboratory. The black dash-dotted line (SSR = 1) indicates that the sediment sound speed is equal to the overlying seawater sound speed.

4.3. Acoustic Properties and Corresponding Physical Properties on Section CD

Table 2 displays the acoustic properties of each station in section CD, while Table 1 presents the corresponding physical properties. Figure 5 illustrates the variations in acoustic and physical properties along the section, including station A3 at the intersection with section AB. The properties along this section exhibit distinct differences from those of section AB. From southwest to northeast, there is a consistent trend of increasing sound speed, SSR, and density, along with a gradual decrease in porosity and MGZ, but the variability of these parameters, except attenuation, is notably reduced compared to section AB. The laboratory sound speed ranged from 1462 to 1504 m/s, attenuation from 14.91 to 41.17 dB/m, and SSR from 0.976 to 0.998. In terms of physical properties, density ranged from 1521 to 1713 kg/m^3, porosity from 59.2% to 69.9%, and MGZ from 5.494 to 8.064 ϕ. Attenuation coefficient exhibited significant fluctuations along the section, ranging from 14.91 to 41.17 dB/m, with no distinct correlation observed with water depth or sediment type. Similar to station A3, SSR values of D1~D5 are all below 1, indicating that sediment sound speed near the shelf break is generally lower than seafloor seawater sound speed.

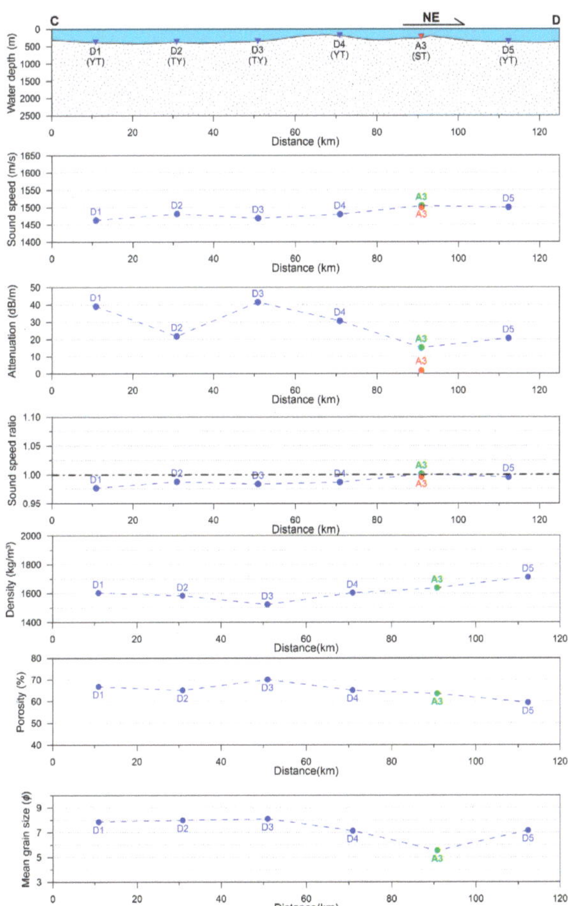

Figure 5. Variation of sediment acoustic and physical properties along section CD (SW-NE direction). Red dots represent acoustic properties measured using the in situ technique, and green and blue dots represent acoustic and physical properties of sediment samples measured in the laboratory. The black dash-dotted line (SSR = 1) indicates that the sediment sound speed is equal to the overlying seawater sound speed.

5. Discussion

5.1. Differences between In Situ and Laboratory Acoustic Properties Measurements

In the field work, sediment acoustics properties are typically obtained through in situ or laboratory techniques [3]. It is essential to compare the results from these two methods, considering the potential perturbation during sampling and measuring processes for the laboratory acoustic properties measurements. Richardson and Briggs [37] observed close agreement between laboratory and in situ measurements for gas-rich muddy sediments in Eckernförde Bay, Baltic Sea, and hard-packed sandy sediments in the northeastern Gulf of Mexico, despite differences in measurement frequencies (400 kHz in the laboratory and 58 kHz in situ). In contrast, Gorgas et al. [38] observed a little higher sound speed and greater attenuation measured in laboratory than in situ, in heterogeneous soft seabed sediments, Eel River shelf, and California.

In our experiment, in situ acoustic measurements were obtained at four stations in section AB, along with laboratory measurement using synchronously collected sediment samples. The comparison revealed varying degrees of differences between in situ and laboratory acoustic properties, with laboratory measurements generally higher for sound speed, sound speed ratio, and attenuation coefficient (Figure 4 and Table 2). This is consistent with the higher prediction of the laboratory acoustic–physical-property regressions than the in situ ones, particularly for sandy sediments [3]. Considering the frequency difference between in situ and laboratory measurement, the dispersion effect might be attributed to the higher laboratory acoustic properties [38], especially for sandy sediments, although the mechanism of acoustic attenuation dispersion remains debated [12,39–41]. Besides, sediment disturbance during sampling and transferring to the laboratory may also contribute to discrepancies between in situ and laboratory measurements. Li et al. [42] found shipboard measurements to align more closely with in situ measurements compared to laboratory results, particularly for fine-grained sediments, possibly due to perturbations and pore water loss or redistribution during the sample encapsulation, handling, and storage after their recovery.

In the case of deep-sea fine-grained sediments, the situation is different. Wang et al. [15] conducted in situ and laboratory sediment acoustic measurements from four stations (T1–T4 in Figure 3) in the western Pacific Ocean, with water depths exceeding 5000 m. The comparison also revealed discrepancies between the laboratory and in situ data, with higher sound speed ratios and attenuation coefficients in the laboratory results. These differences could not be fully explained by the frequencies dispersion based on poroelastic theoretical model, and the authors attributed them to sediment sample state changes caused by the deep-sea sample collection process. The water column pressure in deep sea might be another thinkable factor to explain this discrepancy. The weight of overlying water compacts seafloor sediments in their in situ state, thus increase their density and sound speed. Laboratory experiment had confirmed the sediment sound speed increase with increasing ambient pressure, but the sound speed ratio is almost invariable when the pressure change [31], meaning the water column pressure is negligible to affect the in situ and laboratory SSR discrepancy.

The findings in this experiment align well with existing knowledge. Station A1 has sand-dominated sediment, and shows higher laboratory acoustic properties, with the most significant deviation from its in situ ones. Stations A2 and A3 contain fine-grained sediment from shallower waters, with laboratory measurements closely matching in situ values. Station A4, despite also having fine-grained sediment, shows a larger discrepancy between laboratory and in situ measurements due to significant changes from deep-sea to laboratory conditions. These results highlight the need for caution when using laboratory acoustic measurements for both deep-sea and offshore coarse-grained sediments.

5.2. Relationship between Acoustic and Physical Properties

The variation trends of acoustic and physical properties along two sections (see Figures 4 and 5) demonstrates a strong correlation between them. Previous researches had established empirical equations linking these properties, enabling the prediction of acoustic properties from physical properties and vice versa [3,17,37,43,44]. While early equations were derived from laboratory data, recent studies have introduced empirical equations based on in situ measurements, which generally align in terms of trends, with some discrepancies existing [3,17].

To assess these equations for prediction of sound speed and attenuation based on physical properties, two classical and two recent empirical equations are chosen to compare with the measured data in this experiment. These equations are: (1) ISSAMS empirical equations, established by Richardson and Briggs [37] based on in situ measured data at 38 kHz or 58 kHz using in situ sediment acoustic measurements system (ISSAMS); (2) BISAMS empirical equations, established by Wang et al. [17] based on in situ measured data at 30 kHz using BISAMS; (3) R&B empirical equations, established by Richardson and Briggs [37] based on laboratory measured data at 400 kHz; (4) SCS empirical equations, established by Li et al. [44] based on laboratory measured data at 100 kHz. For ISSAMS, BISAMS, and R&B equations, the relationships of SSR and attenuation factors with density, porosity, and MGZ are included, while for SCS equations, only the relationships of SSR with density, porosity, and MGZ are discussed.

As presented in Figure 6a–f, both laboratory and in situ results in this experiment show good agreement with the SSR vs. density and SSR vs. porosity empirical equations. In section AB, in situ values fall between ISSAMS and BISAMS curves, while laboratory measurements fall between SCS and R&B curves. In section CD, laboratory values are slightly below SCS and R&B curves, while falling between the two in situ curves. In contrast, agreement between the measurements and empirical SSR vs. MGZ curves is low. Most in situ and laboratory values are below SSR vs. MGZ curves, indicating the limitation of acoustic prediction based on MGZ.

It is noteworthy that D1–D5 have MGZ values similar to A4, but closer density and porosity values to A2 and A3, which correlate better with SSR, also proving the poor prediction based on MGZ. In fact, the multiplicity of porosity vs. MGZ relationship had previously been noticed, as shown in Figure 7 in Buckingham [8], which he attributed to the difference of grain smoothness (Δ, rms roughness) in various sediment. Using different Δ to fit the measured porosity vs. MGZ relationship on the two sections, as shown in Figure 7, the good agreement of theoretical curves to measured data seems to confirm the assertions of Buckingham. As a sediment characteristic, MGZ only provides limited insight into the sedimentary environment, and other factors, such as sorting and psephicity, also influence sediment arrangement and therefore acoustic and physical characteristics. Consequently, predicting acoustic properties based solely on particle composition or MGZ is associated with higher uncertainty.

As far as the correlation between attenuation factor and physical properties, a general trend is that the attenuation factor decreases with porosity and MGZ and increases with density (Figure 6d–f). However, this correlation is weaker compared to that between SSR and physical properties. For in situ attenuation, the relationship between attenuation factor and density or porosity is close to the BISAMS curve for A2 to A3, and to the ISSAMS curve for A1 and A4 (Figure 6d,e). Stations A1–A3 show a close relationship between the attenuation factor and MGZ following the BISAMS curve, while station A4 roughly follows the ISSAMS curve (Figure 6f). For the relationship between laboratory attenuation and physical properties, all stations align more closely with the ISSAMS curves rather than the laboratory R&B curves. These results highlight the high uncertainty in predicting attenuation based on physical properties, as suggested by Jackson and Richardson [3].

To different degree the measured data depart from the previous equations, as shown in Figure 6a–f. Even among these equations differences also exist, not only between the in situ and laboratory ones, but also those based on same type of measurement technique. Jackson and Richardson [3] attribute it to difference between the acoustic measurement technology and the physical property measurement technology. Wang et al. [17] suggest the physical properties measurement in laboratory might contribute more to such discrepancy. During the measurement by different authors, the disturbances in the sample collection and measurement process varied to different degree, causing the measured physical properties depart various degree to the true value. As a result, the acoustical properties prediction based on empirical equations can only provide the preliminary estimation when the measured data are not available.

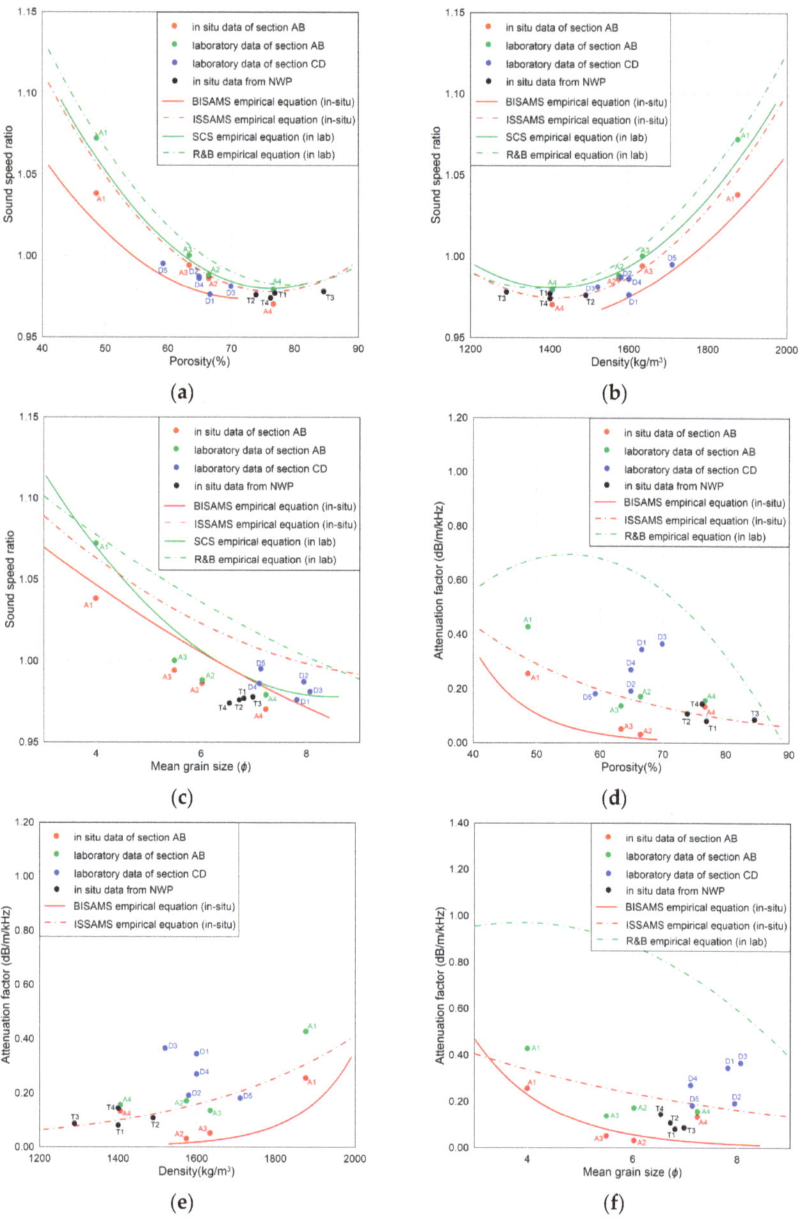

Figure 6. Relationship between measured acoustic properties and physical properties and their comparison with existing empirical relationship curves: (**a**) SSR vs. porosity, (**b**) SSR vs. density, (**c**) SSR vs. MGZ, (**d**) attenuation factor vs. porosity, (**e**) attenuation factor vs. density, and (**f**) attenuation factor vs. MGZ.

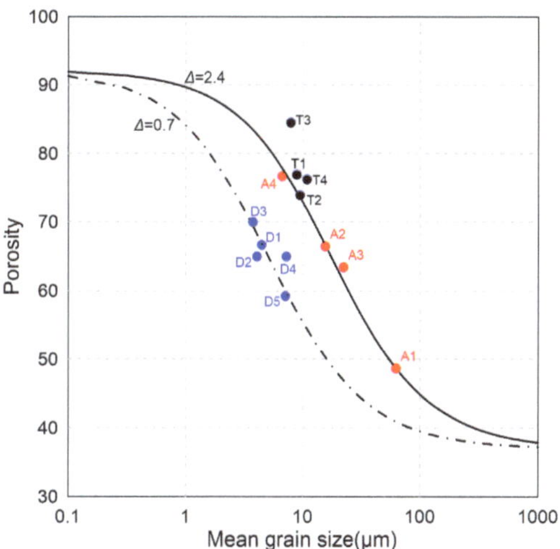

Figure 7. Relationship between measured porosity and MGZ (dots in different colors), with comparison to theoretical curves (solid and dotted lines) calculated from Equation (17) of Buckingham [45]. Note the difference of Δ (rms roughness) taken for the two curves.

5.3. Sedimentary Environment Controls on Sediment Acoustic Properties

Sediment acoustic properties are inherent characteristics of sediments, closely linked to their composition and structure, which are influenced by the environmental conditions during their formation [3,4,6]. In the study area, sediment acoustic properties and physical properties exhibit significant down-slope changes and gradual along-slope changes. These variations are associated with regional sediment distribution patterns and the sedimentary environment.

The study area is situated in the continental terrace in the north-western part of the South China Sea. Continental terraces serve as crucial pathways for terrigenous materials between continents and oceans, playing a significant role in the Earth's source-sink system [46]. Rivers transport a substantial amount of terrigenous materials to the shelf and slope, where they are deposited, or further carried to the open ocean by ocean dynamics [47]. In the South China Sea, coastal currents and deep currents, associated with the East Asian monsoon, the Kuroshio current, and Western Pacific deep waters passing through the Luzon Strait are prominent factors to mold the sediment distribution pattern [48–52]. Studies had identified multiple sources of sediment in different regions of this area. Sediments on the shelf and slope primarily originate from the Red River and small rivers on Hainan Island [22]. In the trough, sediment transport is influenced by gravity currents and materials from Taiwan Island through bottom currents in the northern South China Sea [23]. On the shelf, sediments exhibit a zonal distribution parallel to the shoreline, with coarser sediments closer to land and finer sediments outside [24], resulting in the gradual reduction of the acoustic properties away from the Hainan Island. The distribution pattern along the terraces remains stable, with fine-grained sediments from the Red River transported over long distances and decreasing in clay content towards the northeast due to the blockage by Hainan Island [26]. Conversely, short-distance transport from Hainan Island increases, leading to an increase in relatively coarse-grained material and a decrease in clay content towards the northeast [22], potentially contributing to the increase in acoustic properties in that direction.

The deep-sea deposits at station A4 exhibit unique characteristics compared to other stations. While their particle composition is similar to stations D1–D5 on the outer shelf, the physical and acoustic properties of A4 differ significantly (Tables 1 and 2). In contrast, stations in the deep western Pacific Ocean show similar acoustic and physical properties to

A4 (Figure 3), despite differences in particle composition. This highlights the influence of deep sea environment on the sediment acoustic properties.

6. Conclusions

Acoustic and physical properties of the seabed in the northwestern South China Sea were studied through in situ and laboratory measurements on the shelf and slope along two sections: down-slope and along-slope. The analysis of the results revealed the following conclusions:

(1) Acoustic properties of seafloor sediments exhibit systematic changes with increasing offshore distance and water depth from the shelf to the slope, showing significant decreases in sound speed and attenuation in the down-slope direction. In the along-slope direction, especially near the shelf break, acoustic speed and attenuation are relatively consistent or change minimally compared to the down-slope direction.

(2) Discrepancies exist between acoustic properties obtained from in situ and laboratory measurements, particularly for deep-sea and offshore coarse-grained sediments. Caution should be exercised when using laboratory measurements for these sediment types, with preference given to in situ measurements.

(3) Comparison of measured acoustic properties with predictions from empirical equations for acoustic-physical properties indicates that predicted sound speed using density and porosity aligns more closely with measured results, while predicted sound attenuation shows inconsistencies. Predictions based on mean grain size result in deviations in both sound speed and attenuation, highlighting limitations of the empirical prediction method.

(4) Changes in acoustic properties of seafloor sediments on the shelf and slope are influenced by material sources, hydrodynamic conditions, and water depth. Regional variations in acoustic properties are more prominent on the shelf due to source influence and shallow-water hydrodynamics, while properties on the slope and in deep water are primarily controlled by water depth.

Author Contributions: Conceptualization, G.L.; methodology, J.W. and X.M.; validation, G.L., G.K., and C.L.; formal analysis, G.L. and J.W.; investigation, G.L. and Q.H.; data curation, G.K.; writing—original draft preparation, G.L. and J.W.; writing—review and editing, G.L. and G.K.; visualization, Q.H. and C.L.; supervision, C.L.; project administration, G.L.; funding acquisition, G.L. All authors have read and agreed to the published version of the manuscript.

Funding: This research was funded by National Natural Science Foundation of China under contract No. 42076082, Laoshan Laboratory under contract No. LSKJ202204802, Shandong Provincial Natural Science Foundation under contract No. DKXZZ202206.

Institutional Review Board Statement: Not applicable.

Informed Consent Statement: Not applicable.

Data Availability Statement: Data are contained within the article.

Acknowledgments: The authors thank the crew of R/V Shiyan 1 for their assistance in field data acquisition and samples collection.

Conflicts of Interest: The authors declare no conflict of interest.

References

1. Katsnelson, B.; Petnikov, V.; Lynch, J. *Fundamentals of Shallow Water Acoustics*; Springer: New York, NY, USA, 2012; pp. 55–62.
2. Frisk, G.V. *Ocean and Seabed Acoustics: A Theory of Wave Propagation*; P T R Prentice-Hall: Upper Saddle River, NJ, USA, 1994; pp. 1–16.
3. Jackson, D.R.; Richardson, M.D. *High-Frequency Seafloor Acoustics*; Springer: New York, NY, USA, 2007; pp. 134–170.
4. Kim, S.R.; Lee, G.S.; Kim, D.C.; Bae, S.H.; Kim, S.P. Physical properties and geoacoustic provinces of surficial sediments in the southwestern part of the Ulleung Basin in the East Sea. *Quat. Int.* **2017**, *459*, 35–44. [CrossRef]
5. Hamilton, E.L. Sound velocity and related properties of marine sediments, North Pacific. *J. Geophys. Res.* **1970**, *75*, 4423–4446. [CrossRef]

6. Hamilton, E.L. Geoacoustic modelling of the seafloor. *J. Acoust. Soc. Am.* **1980**, *68*, 1313–1340. [CrossRef]
7. Williams, K.L.; Jackson, D.R.; Thorsos, E.I.; Tang, D.; Schock, S.G. Comparison of sound speed and attenuation measured in a sandy sediment to predictions based on the Biot theory of porous media. *IEEE J. Ocean. Eng.* **2002**, *27*, 413–428. [CrossRef]
8. Buckingham, M.J. Compressional and shear wave properties of marine sediments: Comparisons between theory and data. *J. Acoust. Soc. Am.* **2005**, *117*, 137–152. [CrossRef] [PubMed]
9. Zhou, J.; Zhang, X.; Knobles, D.P. Low-frequency geoacoustic model for the effective properties of sandy seabottoms. *J. Acoust. Soc. Am.* **2009**, *125*, 2847–2866. [CrossRef]
10. Hines, P.C.; Osler, J.C.; Scrutton, J.G.E.; Halloran, L.J.S. Time-of-flight measurements of acoustic wave speed in a sandy sediment at 0.6–20 kHz. *IEEE J. Ocean. Eng.* **2010**, *35*, 502–515. [CrossRef]
11. Yang, J.; Tang, D. Direct measurements of sediment sound speed and attenuation in the frequency band of 2–8 kHz at the Target and Reverberation Experiment site. *IEEE J. Ocean. Eng.* **2017**, *42*, 1102–1109. [CrossRef]
12. Chotiros, N.P. *Acoustics of the Seabed as a Poroelastic Medium*; Springer: New York, NY, USA, 2017; pp. 7–24. [CrossRef]
13. Wang, J.; Guo, C.; Hou, Z.; Fu, Y.; Yan, J. Distributions and vertical variation patterns of sound speed of surface sediments in South China Sea. *J. Asian Earth Sci.* **2014**, *89*, 46–53. [CrossRef]
14. Tian, Y.; Chen, Z.; Hou, Z.; Luo, Y.; Xu, A.; Yan, W. Geoacoustic provinces of the northern South China Sea based on sound speed as predicted from sediment grain size. *Mar. Geophy. Res.* **2019**, *40*, 571–579. [CrossRef]
15. Wang, J.; Li, G.; Kan, G.; Liu, B.; Meng, X. Experimental study on in situ measurement of acoustic characteristics of deep seabed sediments. *Chin. J. Geophy.* **2020**, *63*, 4463–4472, (In Chinese with English abstract).
16. Liu, B.; Han, T.; Kan, G.; Li, G. Correlations between the in situ acoustic properties and geotechnical parameters of sediments in the Yellow Sea, China. *J. Asian Earth Sci.* **2013**, *77*, 83–90. [CrossRef]
17. Wang, J.; Kan, G.; Li, G.; Meng, X.; Zhang, L.; Chen, M.; Liu, C.; Liu, B. Physical properties and in situ geoacoustic properties of seafloor surface sediments in the East China Sea. *Front. Mar. Sci.* **2023**, *10*, 1195651. [CrossRef]
18. Li, G.; Wang, J.; Liu, B.; Meng, X.; Kan, G.; Han, G.; Hua, Q.; Pei, Y.; Sun, L. In situ acoustic properties of fine-grained sediments on the northern continental slope of the South China Sea. *Ocean Eng.* **2020**, *218*, 108244. [CrossRef]
19. Li, K.Z.; Yin, J.Q.; Huang, L.M.; Lian, S.M.; Zhang, J.L.; Liu, C.G. Monsoon forced distribution and assemblages of appendicularians in the northwestern coastal waters of South China Sea. *Estuar. Coast. Shelf Sci.* **2010**, *89*, 145–153. [CrossRef]
20. Zheng, H.B.; Yan, P. Deep-water bottom current research in the Northern South China Sea. *Mar. Georesour. Geotechnol.* **2012**, *30*, 122–129.
21. Liu, Z.; Tuo, S.; Colin, C.; Liu, J.T.; Huang, C.-Y.; Selvaraj, K.; Chen, C.-T.A.; Zhao, Y.; Siringan, F.P.; Boulay, S.; et al. Detrital fine-grained sediment contribution from Taiwan to the northern South China Sea and its relation to regional ocean circulation. *Mar. Geol.* **2008**, *255*, 149–155. [CrossRef]
22. Liu, J.; Clift, P.D.; Yan, W.; Chen, Z.; Chen, H.; Xiang, R.; Wang, D. Modern transport and deposition of settling particles in the northern South China Sea: Sediment trap evidence adjacent to Xisha Trough. *Deep Sea Res. Part I* **2014**, *93*, 145–155. [CrossRef]
23. Xu, F.J.; Hu, B.Q.; Dou, Y.G.; Liu, X.T.; Wan, S.M.; Xu, Z.K.; Tian, X.; Liu, Z.Q.; Yin, X.B.; Li, A.C. Sediment provenance and paleoenvironmental changes in the northwestern shelf mud area of the South China Sea since the mid-Holocene. *Cont. Shelf Res.* **2017**, *144*, 21–30. [CrossRef]
24. Shi, X.F. *China Marine Offshore-Seabed Sediment*; China Ocean Press: Beijing, China, 2012; pp. 1–561. (In Chinese)
25. Liu, Z.; Colin, C.; Li, X.; Zhao, Y.; Tuo, S.; Chen, Z.; Siringan, F.P.; Liu, J.T.; Huang, C.-Y.; You, C.-F.; et al. Clay mineral distribution in surface sediments of the northeastern South China Sea and surrounding fluvial drainage basins: Source and transport. *Mar. Geol.* **2010**, *277*, 48–60. [CrossRef]
26. Zhao, R.; Chen, S.; Olariu, C.; Steel, R.; Zhang, J.; Wang, H. A model for oblique accretion on the South China Sea margin; Red River (Song Hong) sediment transport into Qiongdongnan Basin since Upper Miocene. *Mar. Geol.* **2019**, *416*, 106001. [CrossRef]
27. GEBCO Compilation Group. GEBCO 2023 Grid. Available online: https://doi.org/10.5285/f98b053b-0cbc-6c23-e053-6c86abc0af7b (accessed on 1 August 2023).
28. Wang, J.; Li, G.; Liu, B.; Kan, G.; Sun, Z.; Meng, X. Experimental study of the ballast in situ sediment acoustic measurement system in South China sea. *Mar. Georesour. Geotechnol.* **2018**, *36*, 515–521. [CrossRef]
29. Hamilton, E.L. Prediction of in situ acoustic and elastic properties of seafloor sediments. *Geophysics* **1971**, *36*, 266–284. [CrossRef]
30. Zou, D.P.; Williams, K.L.; Thorsos, E.I. Infuence of temperature on acoustic sound speed and attenuation of seafloor sand sediment. *IEEE J. Oceanic Eng.* **2015**, *40*, 969–980. [CrossRef]
31. Kan, G.; Zou, D.; Liu, B.; Wang, J.; Meng, X.; Li, G.; Pei, Y. Correction for effects of temperature and pressure on sound speed in shallow seafloor sediments. *Mar. Georesources Geotechnol.* **2019**, *37*, 1217–1226. [CrossRef]
32. Wang, J.; Li, G.; Kan, G.; Hou, Z.; Meng, X.; Liu, B.; Liu, C.; Sun, L. High frequency dependence of sound speed and attenuation in coral sand sediments. *Ocean Eng.* **2021**, *234*, 109215. [CrossRef]
33. Hou, Z.; Chen, Z.; Wang, J.; Zheng, X.; Yan, W.; Tian, Y.; Luo, Y. Acoustic characteristics of seafloor sediments in the abyssal areas of the South China Sea. *Ocean Eng.* **2018**, *156*, 93–100. [CrossRef]
34. *GB/T 12763.8-2007 (National Standards of People's Republic of China)*; Specifications for Oceanographic Survey—Part 8: Marine Geology and Geophysics Survey. General Administration of Quality Supervision, Inspection and Quarantine of the People's Republic of China and National Standardization Management Committee of China: Beijing, China, 13 August 2007.

35. *GB/T 50123-1999 (National Standards of People's Republic of China)*; Standard for Soil Test Method. General Administration of Quality Supervision and Inspection and Quarantine of the People's Republic of China: Beijing, China, 10 June 1999.

36. Shepard, F.P. Nomenclature based on sand-silt-clay ratios. *J. Sediment. Res.* **1954**, *24*, 151–158.

37. Richardson, M.D.; Briggs, K.B. In situ and laboratory geoacoustic measurements in soft mud and hard-packed sand sediments: Implications for high-frequency acoustic propagation and scattering. *Geo-Mar. Lett.* **1996**, *16*, 196–203. [CrossRef]

38. Gorgas, T.J.; Wilkens, R.H.; Fu, S.S.; Frazer, L.N.; Richardson, M.D.; Briggs, K.B.; Lee, H. In situ acoustic and laboratory ultrasonic sound speed and attenuation measured in heterogeneous soft seabed sediments: Eel river shelf, California. *Mar. Geol.* **2002**, *182*, 103–119. [CrossRef]

39. Buckingham, M.J. On pore-fluid viscosity and the wave properties of saturated granular materials including marine sediments. *J. Acoust. Soc. Am.* **2007**, *122*, 1486–1501. [CrossRef]

40. Buckingham, M.J. Wave speed and attenuation profiles in a stratified marine sediment: Geo-acoustic modeling of seabed layering using the viscous grain shearing theory. *J. Acoust. Soc. Am.* **2020**, *148*, 962–974. [CrossRef] [PubMed]

41. Chotiros, N.P.; Isakson, M.J. Comments on "Pore fluid viscosity and the wave properties of saturated granular materials including marine sediments" [J. Acoust. Soc. Am. 122, 1486–1501 2007]". *J. Acoust. Soc. Am.* **2010**, *127*, 2095–2098. [CrossRef] [PubMed]

42. Li, G.; Kan, G.; Meng, X. Effect of the condition changes on the laboratory acoustic velocity measurements of seafloor sediments. *Adv. Mar. Sci.* **2013**, *31*, 360–366, (In Chinese with English abstract).

43. Hamilton, E.L.; Bachman, R.T. Sound velocity and related properties of marine sediments. *J. Acoust. Soc. Am.* **1982**, *72*, 1891–1904. [CrossRef]

44. Li, G.B.; Hou, Z.Y.; Wang, J.Q.; Kan, G.M.; Liu, B.H. Empirical equations of p-wave velocity in the shallow and semi-deep sea sediments from the South China Sea. *J. Ocean Univ. China* **2021**, *20*, 532–538. [CrossRef]

45. Buckingham, M.J. Theory of acoustic attenuation, dispersion, and pulse propagation in unconsolidated granular materials including marine sediments. *J. Acoust. Soc. Am.* **1997**, *102*, 2579–2596. [CrossRef]

46. Liu, Z.F.; Zhao, Y.L.; Colin, C.; Stattegger, K.; Wiesner, M.G.; Huh, C.A. Source-to-sink transport processes of fluvial sediments in the South China Sea. *Earth Sci. Rev.* **2016**, *153*, 238–273. [CrossRef]

47. Liu, J.P.; Xue, Z.; Ross, K.; Wang, H.J.; Yang, Z.S.; Li, A.C.; Gao, S. Fate of sediments delivered to the sea by Asian large rivers: Long-distance transport and formation of remote along shore clinothems. *SEPM Sediment. Rec.* **2009**, *7*, 4–9. [CrossRef]

48. Fang, G.H.; Fang, W.D.; Fang, Y.; Wang, K. A survey of studies on the South China Sea upper ocean circulation. *Acta Oceanogr. Taiwan* **1998**, *37*, 1–16.

49. Caruso, M.J.; Gawarkiewicz, G.G.; Beardsley, R.C. Interannual variability of the Kuroshio intrusion in the South China Sea. *J. Oceanogr.* **2006**, *62*, 559–575. [CrossRef]

50. Qu, T.D.; Girton, J.B.; Whitehead, J.A. Deepwater overflow through Luzon Strait. *J. Geophys. Res.* **2006**, *111*, C01002. [CrossRef]

51. Zhao, W.; Zhou, C.; Tian, J.W.; Yang, Q.X.; Wang, B.; Xie, L.L.; Qu, T.D. Deep water circulation in the Luzon Strait. *J. Geophys. Res. Oceans* **2014**, *119*, 790–804. [CrossRef]

52. Liu, J.G.; Clift, P.D.; Yan, W.; Chen, Z.; Chen, H.; Xiang, R. Temporal and spatial patterns of sediment deposition in the northern South China Sea over the last 50,000 years. *Palaeogeogr. Palaeoclimatol. Palaeoecol.* **2017**, *465*, 212–224. [CrossRef]

Journal of
Marine Science and Engineering

Article

Centrifuge Modelling of Composite Bucket Foundation Breakwater in Clay under Monotonic and Cyclic Loads

Minmin Jiang [1,2], Zhao Lu [3,*], Zhengyin Cai [4] and Guangming Xu [4]

1   College of Civil Engineering, Henan University of Technology, Zhengzhou 450001, China; jiangmm@haut.edu.cn
2   Henan Key Laboratory of Grain and Oil Storage Facility & Safety, Henan University of Technology (HAUT), Zhengzhou 450001, China
3   Marine Geotechnical Research Center, HKUST Shenzhen-Hong Kong Collaborative Innovation Research Institute, Shenzhen 518048, China
4   Department of Geotechnical Engineering, Nanjing Hydraulic Research Institute, Nanjing 210024, China; zycai@nhri.cn (Z.C.); gmxu@nhri.cn (G.X.)
*   Correspondence: zhaolu@ust.hk

Abstract: This study investigates the monotonic and cyclic performance of composite bucket foundation breakwater in clay through centrifuge modeling. The application of monotonic loads simulates extreme wave conditions, and cyclic load corresponds to long-term serviceability conditions. In centrifuge tests, three typical soil strengths were tested, and two load eccentricities were simulated to check the influence of wave force height. Multiple measurements were conducted, including rotation angle, horizontal displacement, vertical settlement, and pore pressure variation. When soil strength increases in monotonic centrifuge tests, the ultimate bearing capacity of the bucket foundation experiences significant growth, and the foundation failure pattern varies. In responding to the monotonic test, the foundation's rotation center constantly moved downward during the loading process, indicating that the deeper soil would be activated to resist the horizontal loading. In contrast, the rotation center movement in the symmetric centrifuge test was opposed to the non-symmetric test because the deeper soil was required to provide resistance to balance the more severe load under the non-symmetric loading condition. It should be noted that non-symmetric loading does not impact the bucket foundation as seriously as symmetric loading. The utilization of deep-soil resistance in non-symmetric tests is beneficial in controlling deformation.

Keywords: soil strength; load eccentricity; bearing capacity; failure mode; excess pore pressure; displacement

Citation: Jiang, M.; Lu, Z.; Cai, Z.; Xu, G. Centrifuge Modelling of Composite Bucket Foundation Breakwater in Clay under Monotonic and Cyclic Loads. *J. Mar. Sci. Eng.* **2024**, *12*, 469. https://doi.org/10.3390/jmse12030469

Academic Editor: Dmitry A. Ruban

Received: 17 January 2024
Revised: 25 February 2024
Accepted: 27 February 2024
Published: 9 March 2024

1. Introduction

Breakwaters are important marine structures for protecting harbor facilities from severe storm wave loads. Caisson breakwater is one of the most widely used types of breakwater; however, it is often restricted in its application due to harsh conditions such as deep water and soft soil [1–8]. To address this issue, most recently, composite bucket foundation breakwater has become a competitive alternative [9–12]. A composite bucket foundation comprises four upside-down buckets combined by connecting walls, as illustrated in Figure 1. Compared with the traditional solid breakwater structure, the composite bucket foundation breakwater structure has a light weight, and the interaction between soil and the structure is strong due to the multiple compartments, making it suitable for application in coastal soft soil layers [12]. Meanwhile, it can be prefabricated on shore, and the installation is convenient, since the impact of filed construction can be controlled without cast-in-place work, etc. However, failure of the composite bucket foundation often occurs due to large displacements and excessive pore pressures when subjected to severe storm wave loads. Guo et al. [13–16] have conducted a comprehensive study about the structure and pipelines in the seabed. Therefore, understanding the performance of composite bucket foundation breakwater under storm wave loading conditions becomes considerably crucial for its design and construction in practice.

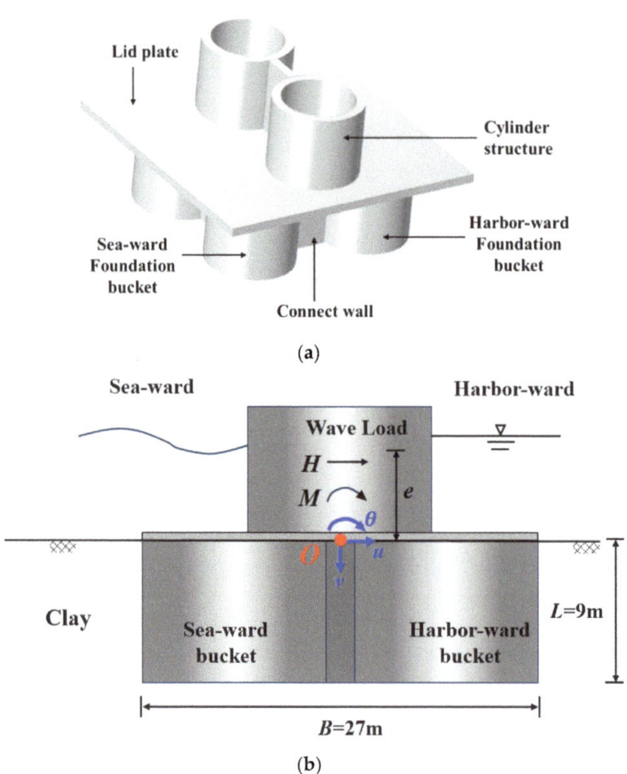

Figure 1. Composite bucket foundation breakwater. (**a**) Composite bucket foundation breakwater model. (**b**) Sketch of composite bucket foundation breakwater under storm wave load.

Over recent decades, various studies have been conducted to investigate the behavior of bucket foundations under storm wave loads. Bransby et al. [16], Achmus et al. [17], and Thieken et al. [18] have the impact of investigated bucket geometry, embedment ratio, loading conditions, and soil permeability on the bearing capacity of bucket foundations. Hong et al. [19,20], Park et al. [21,22], Wang et al. [23,24], Choo et al. [25], and Tasiopoulou et al. [26] have performed numerical and centrifuge studies to investigate the bearing capacity of bucket foundations, and equations for estimating the vertical, lateral, and tensile bearing capacities were proposed. Zhang et al. [27] have explored the cyclic failure modes of bucket foundations in silt. They indicated that the structure failed due to large displacements after long-term excitations, or foundation liquefaction in early excitations. Bransby et al. [16], Barari et al. [28], and Liu et al. [29] have studied the failure envelopes of bucket foundations, and the results indicate that failure envelopes under combined vertical, horizontal, and moment loading present an ellipsoid shape. Byrne et al. [30], Cox et al. [31], and Zhu et al. [32,33] have also studied the cyclic deformation behavior of bucket foundations. They found that a power relationship with a number of load cycles can describe vertical load-dependent displacement. Achmus et al. [17], Cox et al. [31], Wang et al. [10,24], and Grecu et al. [34] have studied the stiffness of the bucket soil system and revealed that initial stiffness is dependent on bucket geometry, the relative density of sand, and load eccentricity. Ding et al. [35] have performed a model test and numerical study to investigate the deformation mechanism and ultimate bearing capacity of composite bucket foundations for wind turbines. Kim et al. [36,37] and Hong et al. [19,20] have studied the bearing capacity of tripod bucket foundations, found that the bearing capacity of a tripod foundation was lower than that of a monopod foundation, and proposed the group efficiency factor to quantify the group effect of tripod bucket foundations.

The centrifugal model test has advantages over the 1 g model, field test, and numerical simulation method [13,16]. First, the stress state in the 1 g laboratory model test is significantly smaller than that of the engineering prototype. The engineering characteristic of soil is influenced by the stress state and has nonlinear and elastic-plastic characteristics [16]. Hence, the 1 g model test can not fully reflect the practice's stress state and response characteristics. On the other hand, field tests are resource- and time-consuming methods, so the number of tests is limited. A field test is generally conducted to check the serviceability of infrastructures and cannot investigate the extreme state. With respect to numerical simulations, the assumptions of the numerical model are generally idealized, and the parameters are simplified. The centrifugal model test provides one effective solution to the above problems. Following the similarity law, the model is placed in a high gravitational field, which can reproduce the field stress state and reflect the characteristics of geotechnical engineering. Therefore, the centrifugal test is used to investigate the characteristics of the composite bucket foundation structure in clay.

However, the majority of the above studies are limited to single-bucket foundations, while very limited attention has been paid to composite bucket foundations. In addition, the existing studies have mainly focused on wind turbine and offshore platform structures; nevertheless, the load conditions applied on breakwater differ significantly from that for those structures [35,38–40]. Moreover, the existing studies were mainly conducted by employing numerical analysis, with a focus on the stability of bucket foundations and the distribution of earth pressure against bucket walls under monotonic loading conditions [23,24,40], while experimental studies on the behaviors of bucket foundation breakwater are rather limited, especially when it is subjected to long-term cyclic loading conditions.

In this study, a series of centrifuge tests were carried out to investigate the performance of composite foundation breakwater in clay under different loading conditions. Based on a monotonic loading apparatus, a non-contact cyclic loading system was developed to apply both symmetric and non-symmetric loading. Both monotonic and cyclic loading models were conducted, simulating the extreme wave and long-term serviceability states, respectively. Three typical clay soils with different undrained strengths were used to construct the foundation soil, and two loading eccentricities were tested to check the influence of wave force height. In all tests, multiple measurements were adopted, including rotation angle, horizontal displacement, vertical settlement, and pore pressure variation. Based on the centrifuge tests, the displacement pattern at both extreme and regular service states could be analyzed.

2. Laboratory Investigations

2.1. Geotechnical Centrifuge Test Setup

The centrifuge modeling method is a powerful means to study the soil structure interaction, in that stress in the model test equals the prototype conditions [41]. The scaling laws for the geotechnical centrifuge test in this study are listed in Table 1. The tests in this study were conducted at Nanjing Hydraulic Research Institute on a 400 g-t geotechnical centrifuge, which has a radius of 5 m and a maximum acceleration of 200 g under the weight of 2000 kg. The centrifuge test was conducted at 105 g. The model container on the swing platform had an inner dimension of 685 mm × 350 mm × 475 mm (length × width × height), which corresponded to a prototype scale dimension of 72 m × 36.8 m × 47.3 m. This paper's parameters and test results are presented on a prototype scale.

Table 1. Scaling law for geotechnical centrifuge test.

Parameters	Scaling Law (Model/Prototype)
Length	$1/N$
Density	1
Force	$1/N^2$
Bending moment	$1/N^3$
Undrained shear strength	1
Moment of inertia	$1/N^4$
Flexural stiffness	$1/N^4$
Frequency	N
Time (consolidation)	$1/N^2$
Time (dynamic)	$1/N$
Acceleration	N

2.2. Modeling of Composite Bucket Foundation Breakwater

The composite bucket foundation comprised four upside-down buckets and vertical connection walls, forming five compartments. Above the foundation, a lid plate was fixed. Then, two vertical cylinder structures were erected on top of the lid plate to resist storm wave load. The structure of the composite bucket foundation breakwater is illustrated in Figure 1. Composite bucket foundation breakwater was sunk into the soil by weight and under pressure inside the compartments until the lid plate contacted the seabed.

The length of the structure, force on the structure, frequency of cyclic load, and bending moment on the structure were modeled according to the geotechnical centrifuge scaling law, as illustrated in Table 1. The model breakwater in this study was made of aluminum alloy with an elastic modulus of 70 GPa and a density of 2.7 g/cm^3. The model was designed according to the centrifuge scaling law at a scale of 1:105. The wall thickness was modelled according to the flexural rigidity law [42,43] as follows:

$$t_m = \sqrt[3]{\frac{t_p E_p}{N E_m}} \tag{1}$$

where E_p is the elastic modulus in the prototype, E_m is the elastic modulus in the centrifuge model, t_p is the thickness of the breakwater wall in the prototype, t_m is the thickness of the breakwater wall in the centrifuge model, N is the scaling number of the centrifuge test, and subscripts p and m stand for the prototype and centrifuge model, respectively. The dimensions of the breakwater were as follows: the outer diameter of the foundation bucket and cylinder structure on top of the breakwater was 12 m, the thickness of the walls was 0.3 m, the length of the foundation bucket skirt and cylinder structure were 9 m and 8.3 m, respectively, the lid plate was square with a side length of 27 m and a thickness of 0.5 m. The model composite bucket foundation breakwater is shown in Figure 1a.

2.3. Preparation of Test Soil

The offshore area in Tianjin, China, was selected as the target site, and is primarily composed of a thick layer of clay. In the centrifuge study, a clay foundation was prepared using the consolidation method. The foundation soil was constructed with clay obtained from Tianjin engineering sites, as the clay parameters are introduced in Table 2. The clay was made into a uniform slurry with a water content two times that of the liquid limit. Then, the clay was poured into a model container and consolidated under constant vertical pressure. The undrained shear strength of the clay foundation was measured by a mini cone penetrometer after consolidation; the average undrained shear strength of the clay foundation is shown in Table 3. The soil with undrained shear strength of 23 kPa represents the deep soft soil condition on the Tianjin Port breakwater project site. And the clay foundations with higher soil strengths of 37 kPa and 45 kPa (all represent typical soil strengths tested from Tianjin Port soft soil) were also simulated in the centrifugal test, to check the soil strength effect.

Table 2. Clay parameters.

The Specific Gravity of Soil G_s	Density ρ	Water Content w	Liquid Limit w_L	Plastic Limit w_P	Plasticity Index I_p
	/g·cm^{-3}	/%	/%	/%	
2.75	1.65	61.5	42	23.5	18.5

Table 3. Centrifuge test program.

Test No.	Loading Eccentricity e (m)	Soil Undrained Shear Strength S_u (kPa)	ζ_b	ζ_c	Cycle Number N
S1	4.5	23.9	-	-	-
S2	10.5	22.9	-	-	-
S3	10.5	37.4	-	-	-
S4	10.5	44.5	-	-	-
C1	4.5	22.5	1	−1	46,800
C2	4.5	22.3	1	−0.4	46,800

Note: ζ_b and ζ_c were calculated based on Equations (2) and (3).

According to the centrifuge scaling law illustrated in Table 1, the soils' densities and undrained strengths in the centrifuge model test were equal to the prototype value and time for consolidation, and the dynamic stage is $1/N^2$ and $1/N$ of the prototype value. The same pore fluid and soil were utilized in the centrifuge model study as in the prototype condition. The time-scaling law for consolidation is inconsistent with dynamic events in cyclic tests, i.e., pore pressure dissipates faster than in prototype conditions, which is known as time-scaling law conflict. However, in this study, the permeability of clay was about 1.3×10^{-7} cm/s, and the dissipation of excess pore pressure was not apparent during cyclic loads; therefore, scaling law conflict can be neglected, and pore pressure was considered to be correctly replicated in cyclic centrifuge tests [44,45].

2.4. Simulation of Storm Wave Load on Composite Bucket Foundation

In the monotonic centrifuge test, the load was applied on breakwater through an actuator, which was driven by a low-speed permanent-magnet synchronous motor. Consequently, the monotonic load was applied under a constant displacement rate. A schematic of the device is shown in Figure 2a.

In the cyclic centrifuge test, a cyclic wave load was applied through a newly developed non-contact loading device. This novel device comprised an armature iron, a load transfer unit, and two electromagnetic actuators on the sea-ward and harbor-ward sides. Armature iron was fixed beneath the top cover of the model test chamber, and a load transfer unit was installed between the two electromagnetic actuators. The cyclic load exerted by electromagnetic actuators on both sides was imposed on the armature iron, and then acted on breakwater through the load transfer unit. There was no connection between the electromagnetic actuators and breakwater; hence, the structure could be displaced freely without contact constraint. A sketch of the non-contact cyclic loading device is shown in Figure 2b. In this study, the loading system was improved, as the non-contact cyclic loading system was both symmetrically and non-symmetrically developed. It should be noted that the effect of wave pressure on the surface of the subsea foundation was not considered in this study, and this might be improved in future experiments and numerical simulations.

Clay exhibits complicated mechanical properties under cyclic loads, such as plasticity, viscosity, and hysteresis. In particular, the interaction between the breakwater structure and clay foundation is complex, which causes difficulty in applying cyclic load on the breakwater precisely. Hence, a real-time control system was developed. The cyclic load was controlled by a fuzzy increment proportional–integral–differential (abbreviated as PID) algorithm programmed by using LabVIEW (version 2016). The control signal was composed of a primary component and a feedback component. The data acquisition frequency reached 1000 Hz, conforming to the cyclic test's record requirements. A low-pass digital filter was used to eliminate the noise of test data. Schematics of the real-time control system are illustrated in Figure 2c.

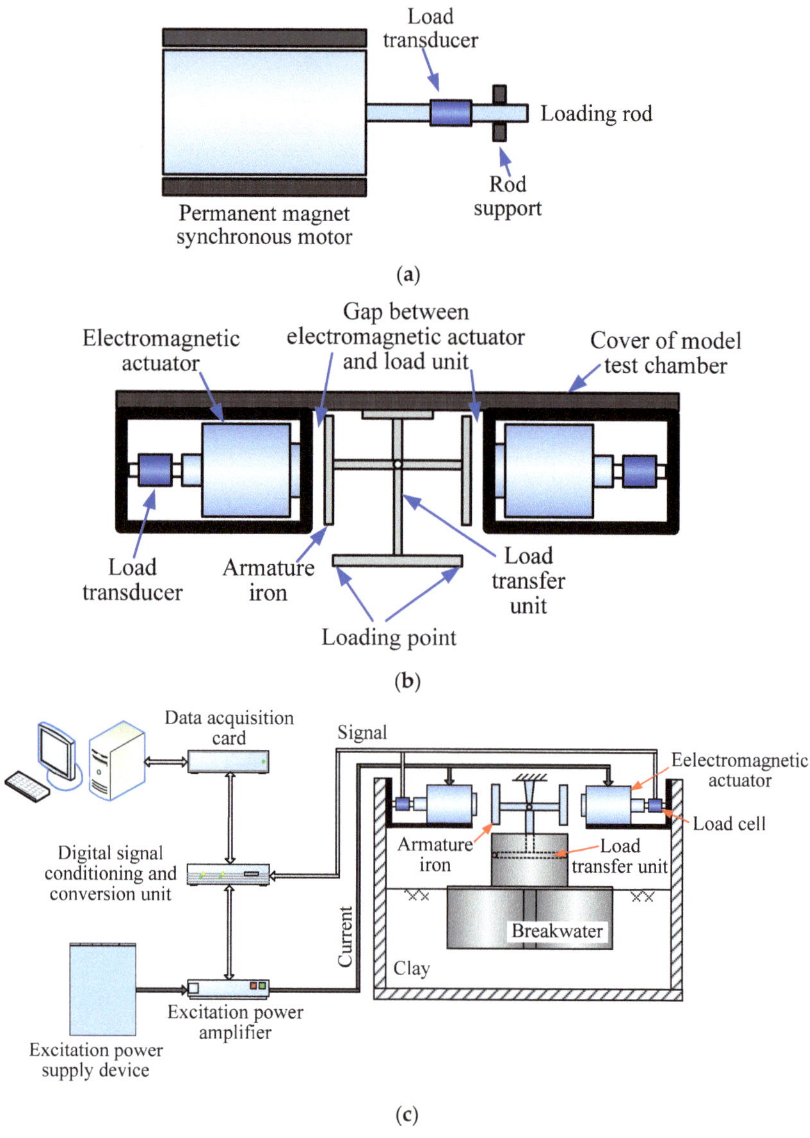

Figure 2. Load device and real-time control system in centrifuge test; (**a**) Schematic of monotonic loading device; (**b**) Schematic of non-contact cyclic loading device; (**c**) Schematic of real-time control system in cyclic test.

2.5. Test Program

A series of monotonic and cyclic tests were designed to study the behavior of composite bucket foundation breakwater under the influence of soil undrained shear strength, load eccentricity, and the characteristics of cyclic load.

Wave parameters in the Tianjin Port site are as follows: the water depth in front of the breakwater is 8.6 m, the wave height is 5.2 m, and the wave period is 8.1 s. Wave load on the breakwater structure was estimated based on Goda formulas [46,47], the design horizontal load acted by the wave crest H_d was 8.45 MN, design moment M_d is 38 MN·m, and the negative load acted by the wave trough was −3.38 MN (about −0.4 H_d). The horizontal load was 4.5 m above the seabed level, referred to as loading eccentricity e, as is illustrated in Figure 1b (equals 38 MN·m/8.45 MN).

The monotonic load was applied to the breakwater structure by a monotonic loading device at a rate of 0.02 mm/s, and the horizontal load was normalized by the design horizontal load on the Tianjin Port site. The time history of the monotonic load is illustrated in Figure 3a. Cyclic loads were applied by non-contact cyclic loading at a period of 8.1 s, which is the same as Tianjin Port's conditions. Two parameters, ζ_b and ζ_c, were employed to describe the characteristics of cyclic load:

$$\zeta_b = \frac{H_{\max}}{H_d} \tag{2}$$

$$\zeta_c = \frac{H_{\min}}{H_{\max}} \tag{3}$$

where H_d is the designed horizontal load calculated by Goda's method, and H_{max} and H_{min} are the maximum and minimum load in a single period acted by wave crest and wave trough [48]. ζ_b denotes the size of the cyclic load, and ζ_c represents the type of cyclic loading. When ζ_c is −1, the cyclic load is symmetrical; when ζ_c is −0.4, the cyclic load is a non-symmetric load corresponding to the Tianjin Port conditions. The total cycle number reached 46,800 in the cyclic centrifuge test, corresponding to 105 h of storm wave load. Schematics of the load history for the cyclic test are presented in Figure 3b.

The undrained shear strength of soil was 23.9 kPa, and load eccentricity was 4.5 m in test S1. The parameters corresponding to the Tianjin Port site situation were applied in test S1, the reference test. In tests S2, S3, and S4, the undrained shear strength of soil increased from 23 kPa to 37.4 and 45 kPa, respectively. Meanwhile, load eccentricity increased from 4.5 m to 10.5 m, and tests S2, S3, and S4 were performed to investigate the effect of the undrained shear strength of soil and the height of wave load on composite bucket foundation breakwater performance. Details of the centrifuge test program are illustrated in Table 3.

After the clay foundation was prepared, the model composite bucket foundation breakwater was jacked into seabed soil. Then, the model test chamber was placed in a forced gravitational field to simulate the clay foundation's consolidation under the breakwater's weight. Finally, the monotonic and cyclic horizontal loads were applied to the breakwater structure, until the prescribed displacement or cycle number were attained.

Schematics of the side elevation of the model setups in the monotonic and cyclic loading tests are presented in Figure 4. The displacement of the breakwater structure was measured by a high-performance laser distance sensor (abbreviated as LDS). Positive rotation and horizontal displacement are toward the harbor, and positive vertical displacement corresponds to settlement. Two pore pressure transducers (PPTs) were embedded in clay foundations outside the sea-ward bucket and harbor-ward bucket at a depth of 6 m, monitoring excess pore pressure under monotonic and cyclic loading conditions.

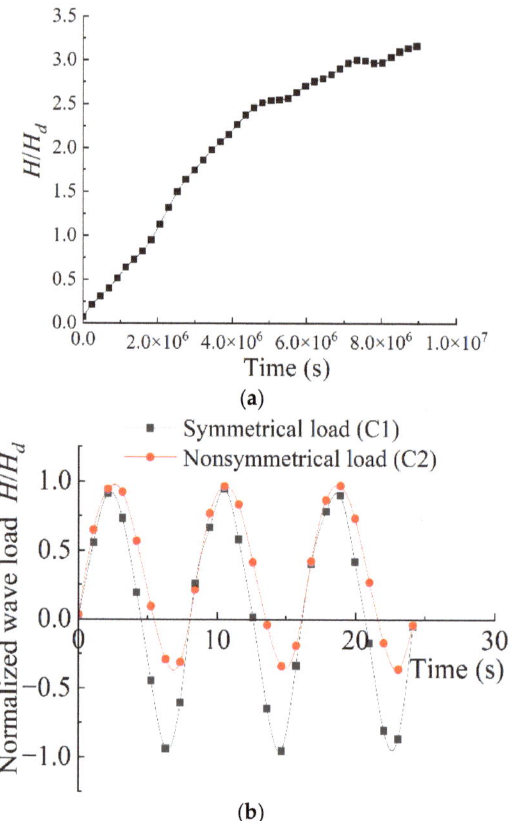

Figure 3. Time history of load in the monotonic and cyclic tests; (**a**) Monotonic load; (**b**) Cyclic load.

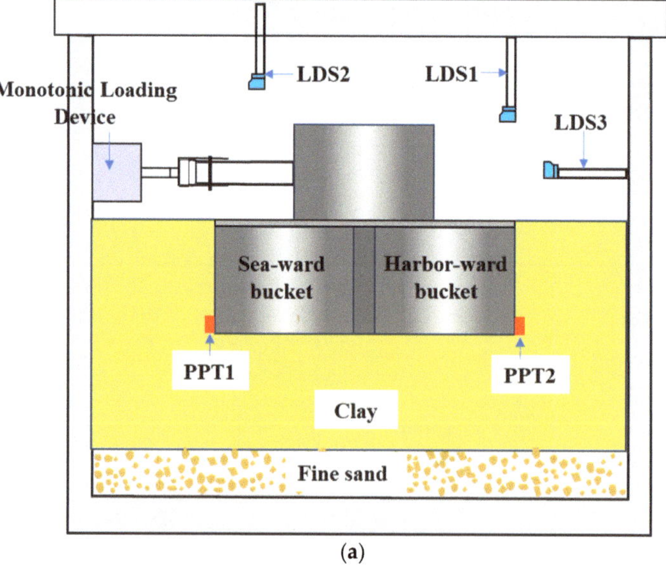

(**a**)

Figure 4. *Cont.*

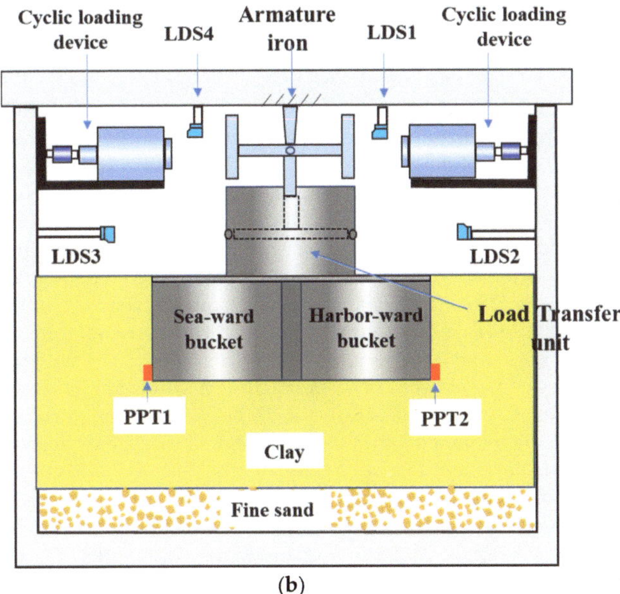

Figure 4. Sketch of experimental setup and instrumentation in centrifuge test; (**a**) Monotonic test; (**b**) Cyclic test.

3. Results

First, the repeatability of centrifuge tests was checked. Static model tests S1-1 and S1-2 were performed with identical experimental conditions, and the rotation angle responses for the two tests are presented in Figure 5. It is demonstrated that the parallel test shows good consistency in geotechnical performance. According to the locating inflection point method, the lateral capacity for the S1 model calculated by Xiao [40,49] is 2.66 H_d, which is in line with the centrifuge test result of 2.69 H_d. This demonstrates that the centrifuge experiment method is applicable in evaluating bucket foundation performance. In centrifuge models, the test condition on the bucket foundation can achieve an extreme failure state, and a more systematic study of the foundation can be conducted. For field practice models, the experiment condition is generally limited to a regular service state, which is restricted to describing the full performance of geotechnical structures.

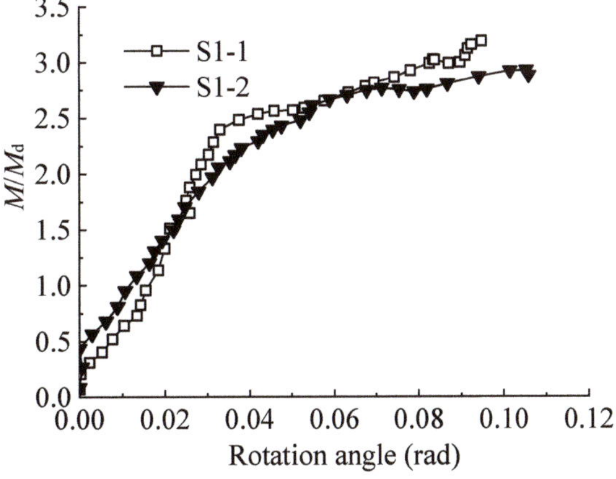

Figure 5. The repeatability check of centrifuge tests.

3.1. Behavior Breakwater under Monotonic Load

3.1.1. Displacement Properties of Breakwater

Under monotonic load toward the harbor-ward direction, the influences of soil strength and load eccentricity on the displacement of composite bucket foundation breakwater are shown in Figure 6. The load–displacement curve can be divided into two stages: the initial elastic and the plastic stages [50,51]. A distinct transition is observed in the load–rotation angle curve as the load increases significantly beyond the inflection point. The initial elastic range and plastic range are fitted with straight lines, and the value at the inflection point is defined as the bearing capacity of the breakwater [32]. The ultimate bearing capacity load is depicted in Figure 6a, denoted by H_u. When the soil undrained shear strength increases from 23 kPa to 44.5 kPa, the bearing capacity load (H_u) increases significantly from 1.31 H_d to 4.29 H_d. When the soil undrained shear strength remains 23 kPa, the ultimate bearing capacity load H_u decreases from 2.69 H_d to 1.31 H_d with the rise in loading eccentricity.

The influence of soil strength and load eccentricity on the horizontal load–horizontal displacement relationship of the bucket foundation is presented in Figure 6b. When the horizontal load exceeds the critical value (ultimate bearing capacity load H_u), horizontal displacement increases significantly. For tests with a low undrained shear strength of soils (i.e., S_u is about 23 kPa in tests S1 and S2), the transition points are found in the horizontal load–vertical displacement curve, beyond which vertical displacement increased significantly. In discussing medium soil strengths (37.4 kPa in test S3), vertical displacement increases roughly linearly with horizontal load. In contrast, for the test with a high undrained shear strength of soils (44.5 kPa in test S4), vertical displacement decreases with horizontal load after reaching the critical value.

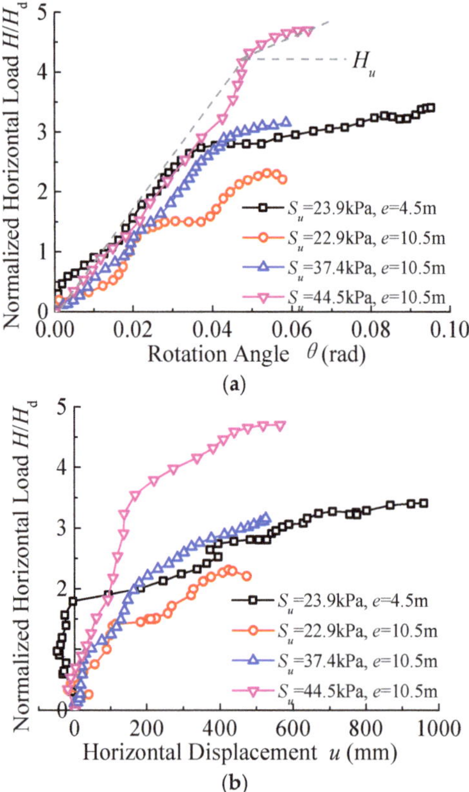

Figure 6. *Cont.*

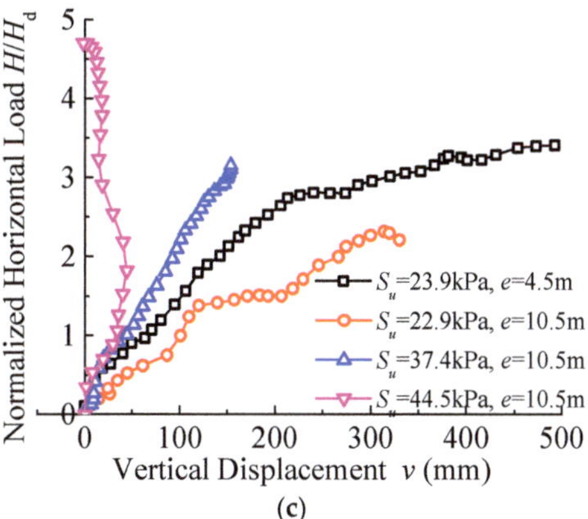

(c)

Figure 6. Load–displacement relations for breakwater under monotonic load; (**a**) Rotational angle; (**b**) Horizontal displacement; (**c**) Vertical displacement.

The relationship between the rotation center depth and the horizontal load of the composite bucket foundation is illustrated in Figure 7, which considers different load eccentricities and soil strengths. The test result shows that the rotation center is located above the seabed level when the horizontal load is small. As the momentum increases, the rotation center moves downward below the seabed level, and the rotation center stays stable around the bucket tip when the bearing capacity load is approached. Consequently, the depth of the rotation center can be considered to be located at a depth of L (L is the length of the bucket skirt) below the seabed level at failure [33,35].

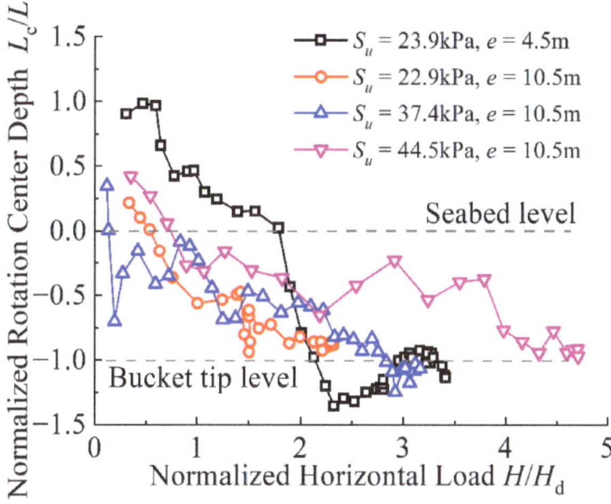

Figure 7. Rotation center depth with horizontal load for breakwater under monotonic load.

3.1.2. Excess Pore Pressure in the Foundation under Monotonic Load

The variation in excess pore pressure in the clay foundation at a depth of 6m outside of the sea-ward and harbor-ward buckets are illustrated in Figure 8, and the pore pressure transducers' position is demonstrated in Figure 4a. It is observed that positive excess pore pressure is generated outside of the harbor-ward bucket. Negative excess pore pressure was generated outside the sea-ward bucket, although minor positive excess pressure arose under low-load levels in stiff clay. This is attributed to the stress variation in clay foundations outside the foundation bucket under monotonic load. Vertical stress was kept constant, and radial stress increased outside the harbor-ward bucket, causing soil contraction behavior, which induced positive excess pore pressure. In contrast, radial stress decreased outside of the sea-ward bucket, caused dilatant behavior in the soil, and generated negative excess pore pressure. Additionally, the contraction of stiff clay under low stress levels could induce minor positive excess pore pressure [41,52].

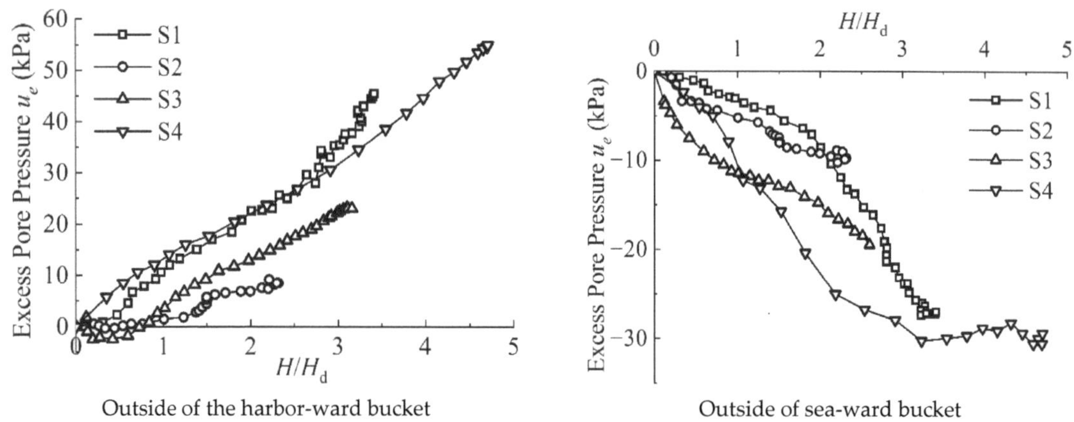

Outside of the harbor-ward bucket   Outside of sea-ward bucket

Figure 8. Excess pore pressure in foundation under monotonic load.

The variation in excess pore pressure with the loading increase is also plotted in Figure 8. The results show that excess pore pressure increases when soil strength increases. When soil strength rises from 23.9 kPa (test S2) to 44.5 kPa (test S4), positive excess pore pressure outside the harbor-ward bucket increases from 4.1 kPa to 38.6 kPa, and negative excess pore pressure outside the sea-ward bucket increases from 7.3 kPa to 30.0 kPa. Regarding the load eccentricity effect, the magnitude of excess pore pressure decreases when loading eccentricity rises from 4.5 m (test S1) to 10.5 m (test S2). In detail, positive excess pore pressure outside the harbor-ward bucket decreases from 29.6 kPa to 4.1 kPa, and negative excess pore pressure outside the sea-ward bucket decreases from 16.1 kPa to 7.3 kPa.

3.2. Behavior of Breakwater under Cyclic Load

3.2.1. Cyclic Displacement of Breakwater

Since the total number of cyclic loads applied on the breakwater reaches 46,800, the analysis of bucket foundation movement by processing the raw data of several cycles as proposed by Zhang would be more efficient and concise [7,27]. All the displacement and pore pressure values in the cyclic centrifuge test are the average result of every 13 cycles.

The relationship between three critical displacements (rotation, horizontal, and vertical displacement) and the cyclic loading process are depicted in Figure 9, and all displacement signs are illustrated in Figure 1b. Notably, rotation and vertical displacement roughly increased linearly with cycle number; the symmetric load test (test C1) has a larger rotation angle and vertical displacement than the non-symmetric load test (test C2). The development of rotation and vertical displacement are significantly affected by the loading magnitude. Horizontal displacement has different trends for two types of cyclic loading: the bucket foundations moves towards the sea-ward horizontally and increases with cycle number under symmetric loading condition. Regarding the non-symmetric loading tests, the horizontal displacement of the foundation increases towards the harbor-ward direction, containing two stages. In detail, the displacement increases when the cycle number is below 22,000 and decreases afterwards. It is evident that the displacement in the cyclic test is significantly smaller than in the monotonic test, indicating that the displacement of the breakwater might be overestimated when designed based on the monotonic test result.

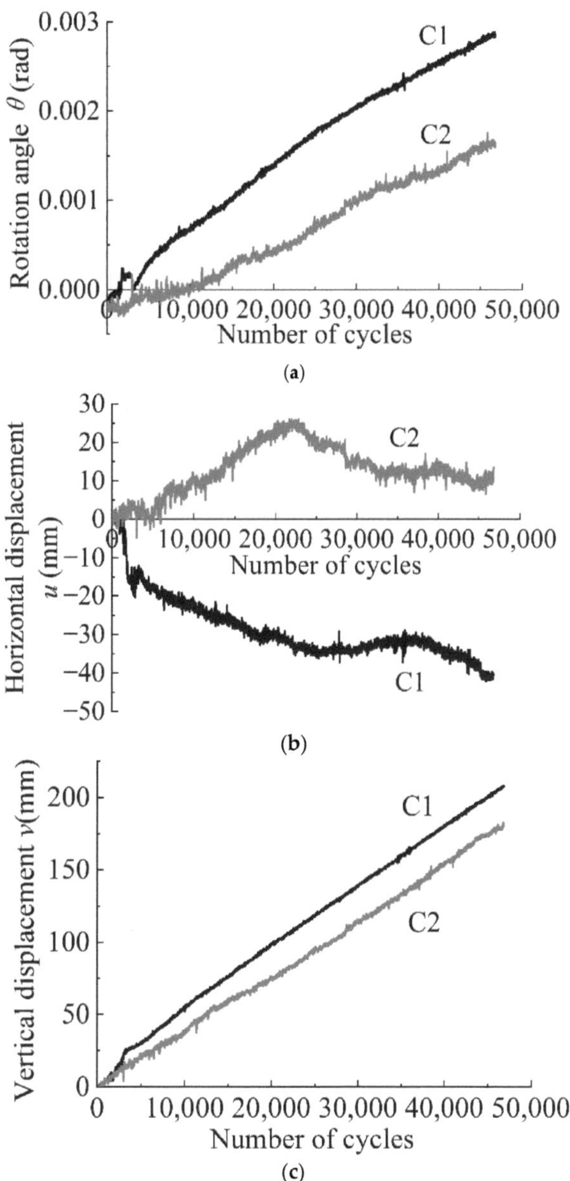

Figure 9. Average displacement of breakwater under cyclic load; (**a**) Rotational angle; (**b**) Horizontal displacement; (**c**) Vertical displacement.

The variations in rotation center depth under cyclic loads are shown in Figure 10. It is demonstrated that the rotation center mobilized above the seabed level constantly in the symmetric load test (C1); however, in the non-symmetric load test (C2), the rotation center moved downward below the seabed level after about 7000 cycles. After about 30,000 cycles, the rotation center depth came to a stable stage, approximately 0.5 L above seabed level in the symmetric test and 0.3 L below the seabed level in the non-symmetric test. The movement of the rotation center in cyclic tests is complicated, and has upward and downward tendencies. Specifically, the rotation center would rise above the seabed level in symmetric centrifuge tests, while it would move downward from the original seabed level in non-symmetric tests.

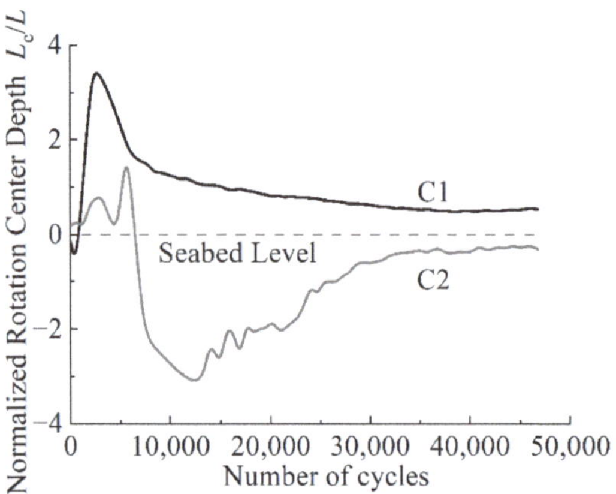

Figure 10. Rotation center depth under cyclic load.

3.2.2. Excess Pore Pressure under Cyclic Load

The excess pore pressure developments in clay foundations around buckets at a depth of 6m are presented for the symmetric load tests (test C1) and non-symmetric load tests (test C2), as shown in Figure 11. It was observed that positive excess pore pressure was generated outside both sea-ward and harbor-ward buckets under cyclic loading conditions, except that a small negative value was monitored outside the sea-ward bucket of the C1 test at the initial loading stage. The excess pore pressures for both tests increased significantly with the increasing load cycles, and the increase in excess pore pressure under non-symmetric loading was higher than under symmetric loading. In the non-symmetric test, the excess pore pressures outside sea-ward and harbor-ward sides attained 9.1 kPa and 8.5 kPa, respectively. When the bucket foundation was subjected to a symmetric load, the pressures were registered with minimum values of 0.7 kPa and 1.5 kPa, respectively. The variations in excess pore pressure demonstrate that non-symmetric loading would lead to considerable pore pressure, indicating that a more robust interaction was developed between the bucket foundation and surrounding soils. This is also consistent with previous observations of rotation centers in non-symmetric conditions. When the rotation center moves downward, the deep soils provide more resistance in sustaining the cyclic loading, and the horizontal displacement can be restricted considerably. In addition, it was found that the magnitude of excess pore pressure at the sea-ward (PPT1) side is close to that at the harbor-ward (PPT2) side for both symmetric and non-symmetric load tests.

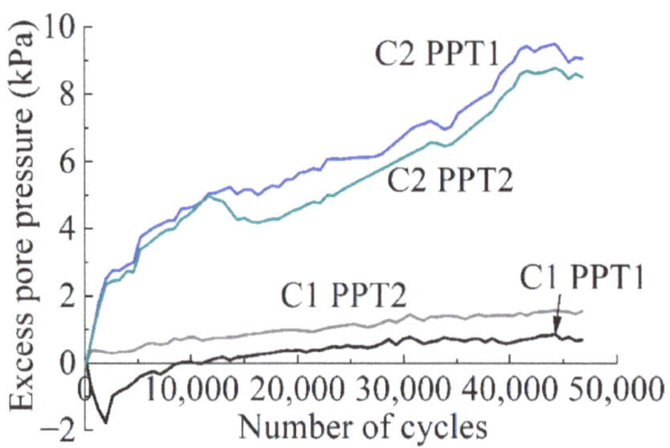

Figure 11. Excess pore pressure under cyclic load.

4. Discussions

The failure mode of bucket foundations under monotonic load is further discussed to interpret the response of the bucket foundation. A sketch of foundation displacement for test S1 is illustrated in Figure 12. Under horizontal load, the breakwater rotates and moves horizontally toward the harbor side, and the harbor-ward bucket settles vertically. While the sea-ward bucket heaves, it tends to uplift with soil, and the harbor-ward bucket compresses with surrounding soil. The displacement pattern for a bucket foundation is composed of three components: rotational movement, horizontal movement, and vertical movement. As listed in Table 4, the displacements, considering different soil strength and load eccentricity, are generalized. In test S1 (soil strength 23.9 kPa and load eccentricity 4.5 m), the rotational angle, normalized horizontal displacement, and normalized vertical displacement were 0.035 rad, 0, and 0.023, respectively, which indicates the failure mode of test S1 is a combination of rotational and vertical movement. When the load eccentricity increased from 4.5 m (test S1) to 10.5 m (test S2), the horizontal displacement increased significantly, expressing that the failure mode converted to a combination of rotational, horizontal, and vertical movement. When soil undrained shear strength increased from about 23.9 kPa (test S2) to 44.5 kPa (test S4), the rotation angle and horizontal displacement continuously increased, and vertical displacement decreased considerably. This indicates that the failure mode was converted to rotational and horizontal movement, as soil undrained shear strength increased two times from test S2 to test S4.

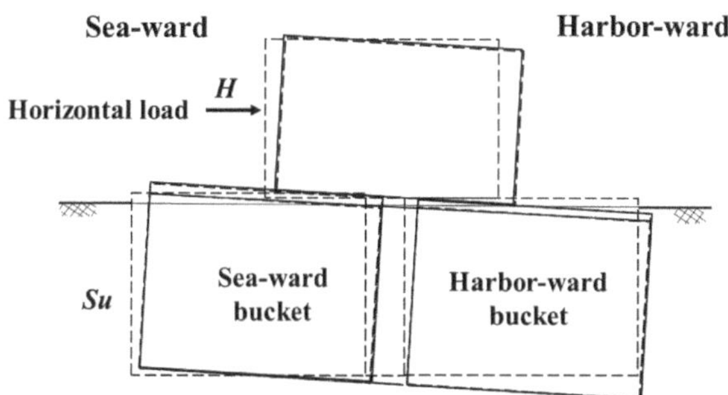

Figure 12. Schematic of movement of breakwater under bearing capacity load for test S1.

Table 4. Failure modes of monotonic tests.

Test No.	H_u	θ_f (rad)	u_f (mm)	v_f (mm)
S1	2.69 H_d	0.035	2	207
S2	1.31 H_d	0.028	110	162
S3	3.05 H_d	0.044	189	144
S4	4.29 H_d	0.049	153	18

5. Conclusions

This study investigated the monotonic and cyclic performance of composite foundation breakwater in clay through a series of centrifuge tests. Based on a monotonic loading device, a novel non-contact loading device and a real-time control system were developed to apply cyclic load. The effects of soil undrained shear strength and load eccentricity on monotonic failure mode were investigated. Two types of cyclic load (symmetric and non-symmetric) were applied to the bucket foundation to investigate cyclic deformation and excess pore pressure, and the results were compared with the monotonic test results.

When the soil strength increased in monotonic centrifuge tests, the foundation failure pattern varied from a complicated mode (including rotation, horizontal displacement, and vertical displacement) to a simplified mode (only including rotation, and horizontal displacement). As expected, the ultimate bearing capacity H_u of the bucket foundation experienced a significant growth (from 1.31 H_d to 4.29 H_d) as the soil strength increased. The foundation's rotation center was constantly moving downward during the loading process, indicating that the deeper soil would be activated to resist the horizontal loading. In discussing the influence of wave force height, centrifuge models with two loading eccentricities were tested. It is shown that the ultimate capacity (2.69 H_d) of the lower eccentricity case is about two times that of the higher case (1.31 H_d) because the higher wave load would induce a larger bending moment on the structure, as tested by S2. Under the horizontal loading in all tests, the bucket foundation moved towards the harbor-ward direction, and the surrounding soil was compressed, resulting in positive excess pore pressure. The negative pore pressure was generated on the outside of the sea-ward bucket foundation due to the unloading effect.

In the cyclic loading tests, the variation in rotation center depth for the symmetric loading test (test C1) and non-symmetric loading test (test C2) shows a noticeable difference. In detail, the rotation center rose above the seabed level in the symmetric centrifuge test, while the center moved downward in the non-symmetric test. The low-position rotation center observed in the non-symmetric test resulted from the deeper soil required to provide resistance to balance more severe load by non-symmetric loading. Consequently, all parameters involving rotation angle, horizontal displacement, and vertical displacement in the non-symmetric centrifuge test (C2) were smaller than in the symmetric test (C1), as the utilization of deep-soil resistance in the non-symmetric test was beneficial in controlling deformation. The non-symmetric loading test also produced considerable pore pressure, indicating that a stronger interaction between the bucket foundation and surrounding soils was developed.

From the results of centrifuge tests, the stability of coastal structures can be significantly improved by the replacement of foundation soils because the ultimate bearing capacity in this study shows a significant growth with soil strength. Wave force height is one crucial factor in influencing the failure pattern of composite bucket foundations, which needs to be considered in designing coastal structures. It should be noted that non-symmetric loading does not impact the bucket foundation as seriously as symmetric loading. Utilizing deep-soil resistance in non-symmetric tests is beneficial in controlling deformation; however, future work in comparing the symmetric and non-symmetric conditions is suggested to be undertaken by using numerical simulation and experiment studies.

Author Contributions: Conceptualization, M.J. and Z.C.; methodology, M.J., Z.L. and G.X.; validation, Z.C. and G.X.; formal analysis, M.J. and G.X.; investigation, M.J., Z.L., G.X. and Z.C.; resources, Z.C. and G.X.; data curation, M.J. and Z.L.; writing—original draft preparation, M.J. and Z.L.; writing—review and editing, Z.C. and G.X.; visualization, M.J. and Z.C.; supervision, Z.C. and G.X.; project administration, Z.C. and G.X.; funding acquisition, M.J., Z.C. and Z.L. All authors have read and agreed to the published version of the manuscript.

Funding: This research was financially supported by the National Natural Science Foundation of China (Grant No. 51179105, 51408197), Henan Key Laboratory of Grain and Oil Storage Facility & Safety, HAUT, Zhengzhou, 450001, China (grant number: 2021KF-B03), Key Project of Science and Technology Research of Henan Education Department (grant number: 23B560002), Project of Hetao Shenzhen-Hong Kong Science and Technology Innovation Cooperation Zone (HZQB-KCZYB-2020083), Shenzhen Science and Technology Program (No. KCXFZ202110020163816023). The authors also acknowledge the help of the geotechnical centrifuge laboratory in Nanjing Hydraulic Research Institute.

Institutional Review Board Statement: Not applicable.

Informed Consent Statement: Not applicable.

Data Availability Statement: Data are contained within the article.

Conflicts of Interest: The authors declare that they have no known competing financial interests or personal relationships that could have appeared to influence the work reported in this paper.

References

1. Andersen, K.H.; Dyvik, R.; Schrøder, K.; Hansteen, O.E.; Bysveen, S. Field tests of anchors in clay II: Predictions and interpretation. *J. Geotech. Eng.* **1993**, *119*, 1532–1549. [CrossRef]
2. Kim, S.-R.; Hung, L.C.; Oh, M. Group effect on bearing capacities of tripod bucket foundations in undrained clay. *Ocean Eng.* **2014**, *79*, 1–9. [CrossRef]
3. Liu, X.; Wang, Y.; Zhang, H.; Guo, X. Susceptibility of typical marine geological disasters: An overview. *Geoenvironmental Disasters* **2023**, *10*, 10. [CrossRef]
4. Liu, X.-L.; Jia, Y.-G.; Zheng, J.-W.; Hou, W.; Zhang, L.; Zhang, L.-P.; Shan, H.-X. Experimental evidence of wave-induced inhomogeneity in the strength of silty seabed sediments: Yellow River Delta, China. *Ocean Eng.* **2013**, *59*, 120–128. [CrossRef]
5. Wang, Y.Z.; Yan, Z.; Wang, Y.C. Numerical analyses of caisson breakwaters on soft foundations under wave cyclic loading. *China Ocean. Eng.* **2016**, *30*, 1–18. [CrossRef]
6. Xiao, Z.; Tian, Y.; Gourvenec, S. A practical method to evaluate failure envelopes of shallow foundations considering soil strain softening and rate effects. *Appl. Ocean Res.* **2016**, *59*, 395–407. [CrossRef]
7. Zhang, X.; Leung, C.; Lee, F. Centrifuge modelling of caisson breakwater subject to wave-breaking impacts. *Ocean Eng.* **2009**, *36*, 914–929. [CrossRef]
8. Zhang, X.; Lee, F.; Leung, C. Response of caisson breakwater subjected to repeated impulsive loading. *Géotechnique* **2009**, *59*, 3–16. [CrossRef]
9. Pan, Z.; Guan, Y.; Han, X. Reliability analysis of bucket foundation breakwater considering water-level variations under complex natural conditions. *Eng. Fail. Anal.* **2023**, *149*, 107250. [CrossRef]
10. Wang, Y.Z.; Xiao, Z.; Chi, L.H.; Xie, S.W.; Li, Y.Y. A simplified calculation method for stability of bucket foundation breakwater. *Rock Soil Mech.* **2009**, *30*, 1367–1372. [CrossRef]
11. Xiao, Z.; Wang, Y.Z.; Ji, C.N. Stability analysis of bucket foundation breakwaters based on limit equilibrium method. *Chin. J. Geotech. Eng.* **2013**, *35*, 828–833.
12. Zhu, X.; Chen, Z.; Guan, Y.-F.; Ni, P.; Fan, K.-F.; Jing, Y.-X.; Yang, C.-J. Field test on the mechanism of composite bucket foundation penetrating sandy silt overlying clay. *Ocean Eng.* **2023**, *288*, 116102. [CrossRef]
13. Guo, X.; Fan, N.; Zheng, D.; Fu, C.; Wu, H.; Zhang, Y.; Song, X.; Nian, T. Predicting impact forces on pipelines from deep-sea fluidized slides: A comprehensive review of key factors. *Int. J. Min. Sci. Technol.* **2024**, *34*, 187–201. [CrossRef]
14. Guo, X.; Liu, X.; Li, M.; Lu, Y. Lateral force on buried pipelines caused by seabed slides using a CFD method with a shear interface weakening model. *Ocean Eng.* **2023**, *280*, 114663. [CrossRef]
15. Guo, X.; Fan, N.; Liu, Y.; Liu, X.; Wang, Z.; Xie, X.; Jia, Y. Deep seabed mining: Frontiers in engineering geology and environment. *Int. J. Coal Sci. Technol.* **2023**, *10*, 23. [CrossRef]
16. Bransby, M.F.; Yun, G.-J. The undrained capacity of skirted strip foundations under combined loading. *Géotechnique* **2009**, *59*, 115–125. [CrossRef]
17. Achmus, M.; Akdag, C.; Thieken, K. Load-bearing behavior of suction bucket foundations in sand. *Appl. Ocean Res.* **2013**, *43*, 157–165. [CrossRef]
18. Thieken, K.; Achmus, M.; Schröder, C. On the behavior of suction buckets in sand under tensile loads. *Comput. Geotech.* **2014**, *60*, 88–100. [CrossRef]
19. Hong, Y.; Chen, X.Y.; Wang, L.Z.; Wang, L.L.; He, B. A bounding-surface based cyclic 'p-y+M-θ' model for unified description of laterally loaded piles with different failure modes in clay. *Can. Geotech. J.* **2024**. [CrossRef]
20. Hong, Y.; Yao, M.H.; Wang, L.Z. A multi-axial bounding surface p-y model with application in analyzing pile responses under multi-directional lateral cycling. *Comput. Geotech.* **2023**, *157*, 105301. [CrossRef]
21. Park, J.-S.; Park, D.; Yoo, J.-K. Vertical bearing capacity of bucket foundations in sand. *Ocean Eng.* **2016**, *121*, 453–461. [CrossRef]
22. Park, J.-S.; Park, D. Vertical bearing capacity of bucket foundation in sand overlying clay. *Ocean Eng.* **2017**, *134*, 62–76. [CrossRef]
23. Wang, X.; Yang, X.; Zeng, X. Centrifuge modeling of lateral bearing behavior of offshore wind turbine with suction bucket foundation in sand. *Ocean Eng.* **2017**, *139*, 140–151. [CrossRef]
24. Wang, X.; Yang, X.; Zeng, X. Lateral response of improved suction bucket foundation for offshore wind turbine in centrifuge modelling. *Ocean Eng.* **2017**, *141*, 295–307. [CrossRef]
25. Choo, Y.W.; Kim, D.-J.; Youn, J.-U.; Hossain, M.S.; Seo, J.; Kim, J.-H. Behavior of a Monopod Bucket Foundation Subjected to Combined Moment and Horizontal Loads in Silty Sand. *J. Geotech. Geoenvironmental Eng.* **2021**, *147*, 107250. [CrossRef]
26. Tasiopoulou, P.; Chaloulos, Y.; Gerolymos, N.; Giannakou, A.; Chacko, J. Cyclic lateral response of OWT bucket foundations in sand: 3D coupled effective stress analysis with Ta-Ger model. *Soils Found.* **2021**, *61*, 371–385. [CrossRef]
27. Zhang, J.; Zhang, L.; Lu, X. Centrifuge modeling of suction bucket foundations for platforms under ice-sheet-induced cyclic lateral loadings. *Ocean Eng.* **2007**, *34*, 1069–1079. [CrossRef]
28. Barari, A.; Ibsen, L. Undrained response of bucket foundations to moment loading. *Appl. Ocean Res.* **2012**, *36*, 12–21. [CrossRef]
29. Liu, M.; Yang, M.; Wang, H. Bearing behavior of wide-shallow bucket foundation for offshore wind turbines in drained silty sand. *Ocean Eng.* **2014**, *82*, 169–179. [CrossRef]
30. Byrne, B.W.; Houlsby, G.T. Experimental investigations of the response of suction caissons to transient combined loading. *J. Geotech. Geoenvironmental Eng.* **2004**, *130*, 240–253. [CrossRef]

31. Cox, J.A.; O'Loughlin, C.D.; Cassidy, M.; Bhattacharya, S.; Gaudin, C.; Bienen, B. Centrifuge study on the cyclic performance of caissons in sand. *Int. J. Phys. Model. Geotech.* **2014**, *14*, 99–115. [CrossRef]

32. Zhu, B.; Byrne, B.W.; Houlsby, G.T. Long-term lateral cyclic response of suction caisson foundations in sand. *J. Geotech. Geoenvironmental Eng.* **2013**, *139*, 73–83. [CrossRef]

33. Zhu, B.; Kong, D.-Q.; Chen, R.-P.; Kong, L.-G.; Chen, Y.-M. Installation and lateral loading tests of suction caissons in silt. *Can. Geotech. J.* **2011**, *48*, 1070–1084. [CrossRef]

34. Grecu, S.; Ibsen, L.B.; Barari, A. Winkler springs for axial response of suction bucket foundations in cohesionless soil. *Soils Found.* **2021**, *61*, 64–79. [CrossRef]

35. Ding, H.; Liu, Y.; Zhang, P.; Le, C. Model tests on the bearing capacity of wide-shallow composite bucket foundations for offshore wind turbines in clay. *Ocean Eng.* **2015**, *103*, 114–122. [CrossRef]

36. Kim, D.-J.; Choo, Y.W.; Kim, J.-H.; Kim, S. Investigation of monotonic and cyclic behavior of tripod suction bucket foundations for offshore wind towers using centrifuge modeling. *J. Geotech. Geoenvironmental Eng.* **2014**, *140*, 04014008. [CrossRef]

37. Kim, S.-W.; Suh, K.-D. Determining the stability of vertical breakwaters against sliding based on individual sliding distances during a storm. *Coast. Eng.* **2014**, *94*, 90–101. [CrossRef]

38. *DNV-OS-J101*; Design of Offshore Wind Turbine Structures. Det Norske Veritas (DNV): Oslo, Norway, 2007.

39. Wang, Y.Z.; Xiao, Z.; Li, Y.Y.; Xie, S.W. Finite element analysis for earth pressure on bucket foundation of breakwater. *Chin. J. Geotech. Eng.* **2009**, *31*, 622–627.

40. Xiao, Z.; Wang, Y.Z.; Ji, C.N.; Li, Y.Y.; Xie, S.W. Finite element analysis of the stability of bucket foundation breakwater. *China Civ. Eng. J.* **2009**, *42*, 119–125.

41. Cai, Y.; Hao, B.; Gu, C.; Wang, J.; Pan, L. Effect of anisotropic consolidation stress paths on the undrained shear behavior of reconstituted Wenzhou clay. *Eng. Geol.* **2018**, *242*, 23–33. [CrossRef]

42. Ha, J.G.; Lee, S.-H.; Kim, D.-S.; Choo, Y.W. Simulation of soil–foundation–structure interaction of Hualien large-scale seismic test using dynamic centrifuge test. *Soil Dyn. Earthq. Eng.* **2014**, *61–62*, 176–187. [CrossRef]

43. Loh, C.K.; Tan, T.S.; Lee, F.H. Three-dimensional excavation test. *Centrifuge* **1998**, *98*, 85–90.

44. Garnier, J.; Gaudin, C.; Springman, S.M.; Culligan, P.J.; Goodings, D.; Konig, D.; Kutter, B.; Phillips, R.; Randolph, M.R.; Thorel, L. Catalogue of scaling laws and similitude questions in geotechnical centrifuge modelling. *Int. J. Phys. Model. Geotech.* **2007**, *8*, 1–23. [CrossRef]

45. Kutter, B.L. Recent Advances in Centrifuge Modeling of Seismic Shaking. In Proceedings of the 3rd International Conference on Recent Advances in Geotechnical Earthquake Engineering and Soil Dynamics, St. Louis, MO, USA, 5 April 1995; Volume 2, pp. 927–942.

46. Ling, H.I.; Cheng, A.H.-D.; Mohri, Y.; Kawabata, T. Permanent displacement of composite breakwaters subject to wave impact. *J. Waterw. Port Coast. Ocean Eng.* **1999**, *125*, 1–8. [CrossRef]

47. Mustapa, M.; Yaakob, O.; Ahmed, Y.M.; Rheem, C.-K.; Koh, K.; Adnan, F.A. Wave energy device and breakwater integration: A review. *Renew. Sustain. Energy Rev.* **2017**, *77*, 43–58. [CrossRef]

48. Leblanc, C.; Houlsby, G.T.; Byrne, B.W. Response of stiff piles in sand to long-term cyclic lateral loading. *Géotechnique* **2010**, *60*, 79–90. [CrossRef]

49. Xiao, Z.; Ge, B.; Wang, Y. Capacities and failure modes of suction bucket foundation with internal bulkheads. *J. Ocean Univ. China* **2017**, *16*, 627–634. [CrossRef]

50. Hung, L.C.; Kim, S.-R. Evaluation of undrained bearing capacities of bucket foundations under combined loads. *Mar. Georesources Geotechnol.* **2014**, *32*, 76–92. [CrossRef]

51. Villalobos, F.A.; Byrne, B.W.; Houlsby, G.T. An experimental study of the drained capacity of suction caisson foundations under monotonic loading for offshore applications. *Soils Found.* **2009**, *49*, 477–488. [CrossRef]

52. Tanaka, H.; Tsutsumi, A.; Ohashi, T. Unloading behavior of clays measured by CRS test. *Soils Found.* **2014**, *54*, 81–93. [CrossRef]

Journal of
Marine Science and Engineering

Propagation Velocity of Excitation Waves Caused by Turbidity Currents

Guohui Xu [1,2], Shiqing Sun [1,2], Yupeng Ren [3,*], Meng Li [1,2] and Zhiyuan Chen [1,2]

[1] Shandong Provincial Key Laboratory of Marine Environment and Geological Engineering, Ocean University of China, Qingdao 266100, China
[2] Key Laboratory of Marine Environment and Ecology, Ministry of Education, Ocean University of China, Qingdao 266100, China
[3] Key Laboratory of Submarine Geosciences and Prospecting Techniques, Ministry of Education, Ocean University of China, Qingdao 266100, China
* Correspondence: renyupeng@ouc.edu.cn; Tel.: +86-156-6682-8355

Abstract: Turbidity currents are important carriers for transporting terrestrial sediment into the deep sea, facilitating the transfer of matter and energy between land and the deep sea. Previous studies have suggested that turbidity currents can exhibit high velocities during their movement in submarine canyons. However, the maximum vertical descent velocity of high-concentration turbid water simulating turbidity currents does not exceed 1 m/s, which does not support the understanding that turbidity currents can reach speeds of over twenty meters per second in submarine canyons. During their movement, turbidity currents can compress and push the water ahead, generating propagating waves. These waves, known as excitation waves, exert a force on the seafloor, resuspending bottom sediments and potentially leading to the generation of secondary turbidity currents downstream. Therefore, the propagation distance of excitation waves is not the same as the initial journey of the turbidity currents, and the velocity of excitation waves within this journey has been mistakenly regarded as the velocity of the turbidity currents. Research on the propagation velocity of excitation waves is of great significance for understanding the sediment supply patterns of turbidity currents and the transport patterns of deep-sea sediments. In this study, numerical simulations were conducted to investigate the velocity of excitation waves induced by turbidity currents and to explore the factors that can affect their propagation velocity and amplitude. The relationship between the velocity and amplitude of excitation waves and different influencing factors was determined. The results indicate that the propagation velocity of excitation waves induced by turbidity currents is primarily determined by the water depth, and an expression ($v^2 = 0.63gh$) for the propagation velocity of excitation waves is provided.

Keywords: turbidity current; excitation wave; propagation speed; flume test; FLOW-3D

Citation: Xu, G.; Sun, S.; Ren, Y.; Li, M.; Chen, Z. Propagation Velocity of Excitation Waves Caused by Turbidity Currents. *J. Mar. Sci. Eng.* **2024**, *12*, 132. https://doi.org/10.3390/jmse12010132

Academic Editor: Angelo Rubino

Received: 20 November 2023
Revised: 3 January 2024
Accepted: 5 January 2024
Published: 9 January 2024

1. Introduction

Submarine turbidity currents, often referred to as underwater rivers, are important carriers that transport terrestrial sediments to the deep sea [1–7]. These turbidity currents, carrying a large amount of silt and sand, not only have strong erosive capabilities on the seabed [8–10], but also pose a threat to underwater communication cables, resulting in significant economic losses [11–13]. For example, the 2006 Pingdong earthquake in Taiwan caused the rupture of 11 submarine cables within the Kaoping Canyon, resulting in a slowdown in network speed in Southeast Asia for 49 days and requiring the deployment of 11 cable ships for repairs [13–15]. Investigating the velocity and patterns of turbidity currents in submarine canyons is of great significance for the protection of infrastructure such as pipelines and cables in these canyons.

One of the main methods for quantitatively studying the velocity of turbidity currents in submarine canyons is to infer their speed through cable ruptures. The first confirmed occurrence of cable rupture caused by a turbidity current was in 1929, when the Grand Banks earthquake triggered the continuous rupture of 12 submarine cables. Inferred maximum turbidity current velocities reached 28 m/s [16–18]. Subsequently, multiple cable rupture incidents caused by turbidity currents have occurred worldwide. Table 1 summarizes the inferred maximum turbidity current velocities from these cable rupture incidents.

Table 1. Cable breakage events caused by turbidity currents worldwide.

Event	Maximum Turbidity Velocity	References
18 November 1929 Grand Banks earthquake	28 m/s	[16,19–21]
1953 Suva earthquake in the Fiji Islands	5.1 m/s	[22]
The Orleansville earthquake of 9 September 1954, Algeria	20.6 m/s	[23]
Earthquake, Solomon Islands, Western Pacific, 23 December 1966	10.3 m/s	[24]
Incident at Nice airport, France, 16 October 1979	7 m/s	[25]
Taitung earthquake, 22 August 2002	9.8 m/s	[26]
21 May 2003 earthquake in Algeria	15.8 m/s	[27]
The Taitung earthquake of 10 December 2003	16.5 m/s	[26]
The Taitung earthquake of 18 December 2003	18.6 m/s	[26]
Pingtung earthquake on 26 December 2006	20 m/s	[28]
Typhoon Morakot on 7–9 August 2009	16.6 m/s	[29]
The 15 January 2022 eruption of Hunga volcano	33.9 m/s	[30]

Previous studies have shown that the maximum vertical velocity of high-concentration turbidity currents in water does not exceed 1 m/s, and the maximum downward velocity of spherical particles in water does not exceed 10 m/s [31]. The maximum velocity of professional athlete Usain Bolt in the 100 m sprint on land is 9.58 m/s, while dolphins in the ocean can reach speeds of up to 20 m/s. Deep-sea turbidity currents, characterized by a small density difference compared to water, are primarily driven by the gravitational component along the direction of flow. However, factors such as bed friction also need to be considered. The driving force behind turbidity currents is primarily the density difference between the turbulent flow and the surrounding water, as well as the gravitational downslope component. Previous studies have detected a maximum sediment concentration of 12% in the basal layer of turbidity currents [32]. However, even high concentrations of suspended sediment, such as 1720 g/L, in seawater with a density of 1020 g/L, do not exceed a maximum vertical velocity of 1 m/s [33]. Similarly, spherical particles also have a maximum settling velocity in water of less than 10 m/s [33]. Turbidity currents, being density-driven flows, have relatively low density differences compared to water, and the gentle slope of submarine canyons also contributes to a smaller gravitational downslope force. Additionally, the influence of bed friction and other factors related to sediment deposition needs to be considered. It is incredible to think that turbidity currents can achieve flow velocities as high as 28 m/s [16,18,28,34,35].

When submarine landslides occur on continental slopes, the sliding mass entering the bottom of submarine canyons can cause the destruction of soft sediment beds. The mixing of sliding or flowing sediment with water forms turbidity currents. Turbidity currents exert pressure and propel the water ahead, forming an excitation wave. This aligns with Paull's hypothesis that in the course of turbidity currents, a high-pressure zone is formed ahead, capable of causing an increase in pore water pressure in the sediment ahead [36]. Similar to surging waves, the excitation waves generated can propagate downstream along the submarine canyon, with a propagation velocity much greater than the velocity of turbidity currents [31]. The rapid propagation of excitation waves can exert a force on the seafloor of the submarine canyon, causing the resuspension of sediment in front of the head of the turbidity currents, which may lead to the formation of secondary turbidity currents at some downstream locations. The distance between the secondary and initial turbidity currents is actually the propagation distance of the excitation waves, rather than the journey of the initial turbidity currents. Therefore, the speed of the excitation waves within this distance is mistakenly considered as the velocity of the turbidity currents (see Figure 1). This may explain why the velocity of the turbidity currents as deduced from cable breakages is so high.

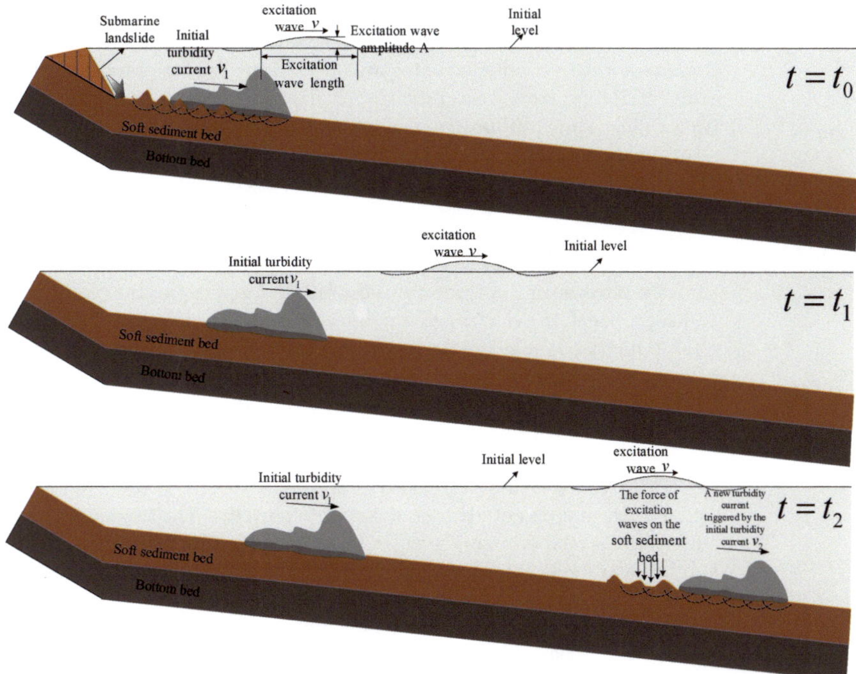

Figure 1. Diagram of excitation wave propagation due to turbidity current (v_1 is the velocity of turbidity current. This refers to the ratio of distance to time experienced by a turbidity current mass moving underwater. v_2 is the velocity of secondary turbidity current: the rapidly propagating excitation wave applies a force on the submarine canyon floor, leading to the destruction of the soft sediment floor and the secondary turbidity current. v is the propagation velocity of the excitation wave; this refers to the propagation velocity of the turbidity current excitation wave. This speed is not the velocity of the motion of the water mass. At time t_0, the initial turbidity current moves underwater, pushing the stationary water in front to generate an excitation wave. At time t_1, the excitation wave is propagating. At time t_2, the rapidly propagating excitation wave exerts pressure on the soft bottom bed, resulting in the destruction of the bottom bed and secondary turbidity current).

Turbidity currents are mass movements composed of sediment particles, with a high concentration of the dense basal layer near the seabed. Depending on their density and granulometric composition, turbidity currents can move along submarine canyons through mechanisms such as diffusion, collapse, and flow [37], which differ from the downward movement as a single entity of landslide bodies after slope failure (this distinguishes them from surges). Additionally, during the long-distance movement of turbidity currents in canyons, the completion of subsequent water replenishment may generate multiple excitation waves. Furthermore, secondary excitation waves may also occur during the movement of secondary turbidity currents triggered by the initial turbidity current, which differs significantly from the surges caused by submarine landslides. Furthermore, previous studies [38–41] on sediment supply during turbidity current movements have mostly focused on the scouring action on the seabed, whereas the resuspension of sedimentary deposits in front of the initial turbidity current caused by excitation waves may serve as an effective mode of sediment supply during the long-distance transport of turbidity currents.

In 2023, Ren et al. proposed that the cause of the long-distance high-speed motion of turbidity currents is due to the excitation waves caused by the primary turbidity currents. However, only preliminary research has been conducted on the comparison of excitation wave velocity and solitary wave velocity, and there has been no specific discussion on the reasons for the excitation wave velocity being much greater than that of the turbidity current. In an experiment conducted using an indoor flume, it was observed that the wavelength of the excitation waves was much larger than the water depth, similar to shallow water waves [33]. The amplitude of excitation waves in proportion to their wavelength was small, consistent with the theory of small-amplitude waves. Similar to the velocity model of shallow water waves, it is expected that the propagation speed of excitation waves is also influenced by the water depth. However, since excitation waves are triggered by sediment-laden turbidity currents, the velocity model may differ from that of surface waves induced by gravitational flows.

The purpose of this study is to simulate and investigate the effects of different factors on the propagation velocity and amplitude of excitation waves through a validated numerical model based on laboratory experiments. The study aims to determine the maximum propagation velocity of excitation waves at a field scale and whether there is attenuation in the long-distance propagation after their formation. In recent studies, seafloor sediment flows have been collectively referred to as turbidity currents [42]. Therefore, we simulated the movement of turbidity currents by sediment flow.

This study uses the CFD-based fluid computation software FLOW-3D to simulate the underwater movement process of turbidity currents. The numerical model is validated against indoor experimental results. During the simulation process, a velocity model for surging wave generation triggered by submarine landslides is used as a reference, and multiple factors that may affect the propagation velocity of the excitation wave are considered. By controlling a single variable, the main factors influencing the excitation wave propagation velocity are determined, and the corresponding expression for excitation wave propagation velocity is provided. The results indicate that the propagation velocity of the excitation wave induced by turbidity currents is primarily determined by the water depth. This research provides a new perspective for understanding the high-speed movement of turbidity currents in submarine canyons and enriches the understanding of the movement patterns of turbidity currents in submarine canyons. In addition, studying the propagation speed of excitation waves is highly significant for the resuspension of underwater sediments, as well as the re-circulation of carbon sequestration, nutrients, heavy metals, and microplastics.

2. Experimental Study on Excitation Waves Induced by Turbidity Currents

2.1. Experimental Design and Apparatus

The experimental apparatus used for the turbidity current-induced excitation wave tests is a straight water tank [33]. The water tank is 12.5 m long, 0.5 m wide, and 0.7 m high. A turbidity source area is located on the right side of the tank to generate turbidity currents. The tank is equipped with a terrain with a certain slope.

Turbidity currents are generated underwater using a weir. The mass ratio of silt and clay used in the experimental turbid water solution was 8:2, with a density of 1600 kg/m^3. Previous experiments have shown that this turbid mixture can reach a maximum flow velocity of 18.7 cm/s [31]. Three pressure sensors are placed along the straight section of the tank at intervals of 0.4 m. These sensors continuously monitor the bottom shear stress caused by the turbidity current-induced excitation wave, as well as the force exerted by the turbidity current itself on the bed. The monitoring frequency is set at 100 Hz.

2.2. Experimental Phenomenon and Results

In the laboratory water tank experiments, it was observed that as the turbidity current propagates, a wave is generated ahead of the turbidity front, moving in the same direction as the current and with a velocity greater than the turbidity current velocity [33]. By monitoring the pressure changes on the bed during the turbidity current motion [33], the propagation velocity of the excitation wave, the head movement velocity of the turbidity current, and the amplitude of the excitation wave (obtained from the measured surface elevation changes caused by the wave) can be estimated based on the distances between the sensors and the time when the pressure change peaks occur.

The results of indoor experiments on turbidity currents indicate that they can compress and propel the water ahead of them, generating excitation waves similar to pulses. The propagation speed of these excitation waves caused by turbidity currents is found to be much greater than the velocity of the turbidity current movement at its head, as determined by pressure sensors installed on the seabed.

3. Numerical Simulation of Excitation Waves Induced by Turbidity Currents

FLOW-3D is a powerful computational fluid dynamics (CFD) software that excels in making accurate calculations of free surface and six-degrees-of-freedom motions of objects. Similar to other CFD software, FLOW-3D consists of three modules: pre-processing, solver, and post-processing. In recent years, there have been many simulations of turbidity currents using FLOW-3D due to its superior capabilities. For example, Heimsund (2007) simulated turbidity currents in the Monterey Canyon system using FLOW-3D based on high-resolution bathymetry and flow data [43]. Zhou et al. (2017) used FLOW-3D software to simulate turbidity currents in a flume with obstacles, analyzing the impact of the proportion between obstacle height and flume height on the movement of turbidity currents, including their velocity, flow state, and morphological evolution [44]. In this study, using the CFD software FLOW-3D, the underwater motion process of turbidity currents is simulated. The model is validated by comparing it with experimental results, and the motion of the waves induced by turbidity currents is simulated based on this validation.

3.1. Control Equations

FLOW-3D, a mature three-dimensional fluid simulation software, is used in this study. It employs the RNG turbulence model, which is capable of handling high strain rate flows and is suitable for simulating excitation waves. The research focus of this paper is on sediment gravity flows (turbulent flows), and the control equations used in the calculations include the basic continuity equation, the momentum equation, the turbulent kinetic energy k equation, and the turbulent kinetic energy dissipation rate ε equation.

The continuity equation:

$$\frac{\partial(uA_x)}{\partial x} + \frac{\partial(vA_y)}{\partial y} + \frac{\partial(wA_z)}{\partial z} = 0 \tag{1}$$

The momentum equation:

$$\frac{\partial u}{\partial t} + \frac{1}{V_F}\left\{uA_x\frac{\partial u}{\partial x} + vA_y\frac{\partial u}{\partial y} + wA_z\frac{\partial u}{\partial z}\right\} = -\frac{1}{\rho}\frac{\partial p}{\partial x} + G_x + f_x \tag{2}$$

$$\frac{\partial u}{\partial t} + \frac{1}{V_F}\left\{uA_x\frac{\partial u}{\partial x} + vA_y\frac{\partial u}{\partial y} + wA_z\frac{\partial u}{\partial z}\right\} = -\frac{1}{\rho}\frac{\partial p}{\partial y} + G_y + f_y \tag{3}$$

$$\frac{\partial u}{\partial t} + \frac{1}{V_F}\left\{uA_x\frac{\partial u}{\partial x} + vA_y\frac{\partial u}{\partial y} + wA_z\frac{\partial u}{\partial z}\right\} = -\frac{1}{\rho}\frac{\partial p}{\partial z} + G_z + f_z \tag{4}$$

The turbulence model:

k equation:

$$\frac{\partial(\rho k)}{\partial t} + \frac{\partial(\rho k u_i)}{\partial x_i} = \frac{\partial}{\partial x_j}\left[\sigma_k(\mu + \mu_t)\frac{\partial k}{\partial x_j}\right] + G_k - \rho\varepsilon \tag{5}$$

ε equation:

$$\frac{\partial(\rho\varepsilon)}{\partial t} + \frac{\partial(\rho\varepsilon u_i)}{\partial x_i} = \frac{\partial}{\partial x_j}\left[\sigma_\varepsilon(\mu + \mu_t)\frac{\partial\varepsilon}{\partial x_j}\right] + C_{\varepsilon 1}^{*}\frac{\varepsilon}{k}G_k - C_{\varepsilon 1}\rho\frac{\varepsilon^2}{k} \tag{6}$$

where u, v and w is the flow velocity component in x, y and z directions; A_x, A_y and A_z represent the area fraction that can flow in x, y and z directions; G_x, G_y and G_z are the gravitational acceleration in x, y and z directions; f_x, f_y and f_z are the viscous forces in the three directions; V_F is the fraction of the volume that can flow; ρ is the fluid density; p is the pressure acting on the fluid element; k is the turbulence energy; ε is the turbulence kinetic energy dissipation rate; μ is turbulence viscosity coefficient $\mu_t = \rho C_\mu \frac{k^2}{\varepsilon}$, where $C_\mu = 0.0845$; G_k is the turbulent kinetic energy generation term, expressed as $G_k = \mu_t\left(\frac{\partial u_i}{\partial x_j} + \frac{\partial u_j}{\partial x_i}\right)\frac{\partial u_i}{\partial x_j}$; and σ_k and σ_ε are the Prandtl numbers corresponding to the turbulent kinetic energy and dissipation rate, respectively, both of which are 1.39. In addition, $C_{\varepsilon 1}^{*} = C_{\varepsilon 1} - \frac{\eta(1-\eta/\eta_0)}{1+\beta\eta^3}$ where $C_{\varepsilon 1}$ and $C_{\varepsilon 2}$ are the empirical constants, 1.42 and 1.68, respectively. Furthermore, $\eta = \left(2E_{ij}E_{ij}\right)^{1/2}\frac{k}{\varepsilon}$ where $E_{ij} = \frac{1}{2}\left(\frac{\partial u_i}{\partial x_j} + \frac{\partial u_j}{\partial x_i}\right)$, $\eta_0 = 4.377$, $\beta = 0.012$.

The general mass continuity equation is as follows:

$$V_F\frac{\partial\rho}{\partial t} + \frac{\partial}{\partial x}(\rho u A_x) + R\frac{\partial}{\partial y}(\rho v A_y) + \frac{\partial}{\partial z}(\rho w A_z) + \xi\frac{\rho u A_x}{x} = R_{DIF} + R_{SOR} \tag{7}$$

where V_F is the fractional volume open to flow, ρ is the fluid density, R_{DIF} is a turbulent diffusion term, and R_{SOR} is the mass source.

3.2. Model Validation

To determine the factors affecting the velocity of the turbidity-induced excitation wave and its velocity expression, first, the indoor flume test was taken as the prototype. Then, a 1:1 geometric solid model was established, and the simulation parameters were set to be consistent with the flume test parameters [33]. Finally, the simulation results were compared with the laboratory test results.

The computational domain employs the method of unstructured grid and is entirely divided into structured orthogonal grids. Nested grids are used for local refinement at the interfaces of straight sections, resulting in a total of 800,000 grid cells after refinement.

The simulation results were compared with the indoor experimental results, with the velocity of the excitation wave and the turbidity current head being represented by changes in surface elevation and water density. The experimental and simulation results are shown in Table 2, and the calculation formula for the error is $\frac{|\text{Calculated value} - \text{Test value}|}{\text{Test value}} \times 100\%$.

Table 2. The test results of the propagation velocity of the excitation wave, the turbidity current velocity, and the excitation wave amplitude are compared with the simulation results.

Result	Propagation Velocity of Excitation Wave (m/s)		Velocity of Turbidity Current (m/s)		Excitation Wave Amplitude (m)	
	Sensor 1 to 2	Sensor 2 to 3	Sensor 1 to 2	Sensor 2 to 3	Sensor 1 to 2	Sensor 2 to 3
Test results	1.54	1.48	0.24	0.23	0.029	0.03
Computed results	1.55	1.52	0.25	0.23	0.03	0.03
Error range	0.6%	2.7%	4.2%	0%	3.4%	0%

From the above comparison, it can be observed that the simulated velocities of the excitation wave and the head of the turbidity current align well with the experimental results, indicating the rationality of using the numerical model established in this study for simulating the propagation velocity of the excitation wave induced by turbidity currents.

3.3. Analysis of Factors Affecting the Propagation Velocity of Excitation Waves

An analysis of the factors influencing the propagation velocity of excitation waves was conducted using numerical simulation. The reference model for wave velocity was based on the surge velocity model. The main factors affecting the propagation velocity of excitation waves were summarized, including the turbidity current density ρ, the thickness of the turbidity current source area d, the length of the turbidity current source area L, the depth at the initial flow of turbidity currents h, the canyon width l, and the initial velocity of the turbidity current v_0 (as shown in Figure 2). The simulations were performed using a controlled variable approach for different parameters, and the velocity changes of the excitation wave were obtained, as shown in Table 3. The slope angle was fixed at $3°$, and sensors were placed at intervals of 100 m starting from a distance of 500 m from the turbidity current source area (named Sensors 1, 2, 3). These sensors were used to extract surface elevation, density, and other relevant parameters at their respective locations. We can obtain the propagating velocity of excitation waves by measuring the time difference in surface elevation changes at the monitoring points. Similarly, we can determine the propagation velocity of turbidity currents by measuring the time difference in density changes.

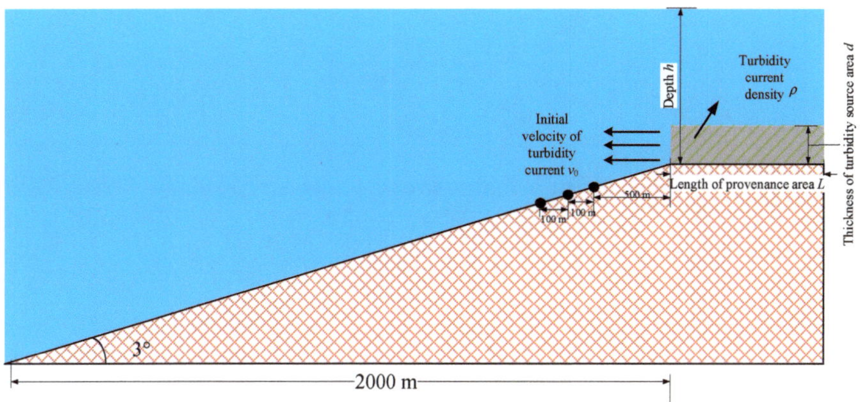

Figure 2. Excitation wave velocity simulation model and parameters.

Table 3. Simulation results under different variables conditions.

Group Order	Turbidity Current Density (kg/m³)	Length of Turbidity Source Area (m)	Canyon Width (m)	Thickness of Turbidity Source Area (m)	Depth (m)	Initial Velocity of Turbidity Current (m/s)	Propagation Velocity of Excitation Wave (m/s)	Excitation Wave Amplitude (m)	Velocity of Turbidity Current (m/s)
1	1600	1000	200	20	200	0	33.43	0.345	5.88
2	1500	1000	200	20	200	0	33.09	0.304	5.41
3	1400	1000	200	20	200	0	33.35	0.223	4.99
4	1300	1000	200	20	200	0	33.33	0.177	4.35
5	1200	1000	200	20	200	0	33.86	0.092	3.74
6	1600	1000	200	40	200	0	33.05	1.109	9.09
7	1600	1000	200	60	200	0	33.39	2.689	10.79
8	1600	1000	200	80	200	0	33.21	4.828	12.91
9	1600	1000	200	100	200	0	36.43	7.744	13.79
10	1600	200	200	20	200	0	32.93	0.181	5.58

Table 3. *Cont.*

Group Order	Turbidity Current Density (kg/m³)	Length of Turbidity Source Area (m)	Canyon Width (m)	Thickness of Turbidity Source Area (m)	Depth (m)	Initial Velocity of Turbidity Current (m/s)	Propagation Velocity of Excitation Wave (m/s)	Excitation Wave Amplitude (m)	Velocity of Turbidity Current (m/s)
11	1600	400	200	20	200	0	33.49	0.25	5.71
12	1600	600	200	20	200	0	33.06	0.278	5.79
13	1600	800	200	20	200	0	33.17	0.31	5.72
14	1600	1000	200	20	100	0	26.67	0.56	5.72
15	1600	1000	200	20	300	0	39.65	0.169	5.80
16	1600	1000	200	20	400	0	45.98	0.12	5.80
17	1600	1000	200	20	500	0	49.97	0.08	5.96
18	1600	1000	100	20	200	0	33.60	0.354	5.72
19	1600	1000	300	20	200	0	32.98	0.338	5.97
20	1600	1000	400	20	200	0	33.27	0.356	5.87
21	1600	1000	500	20	200	0	33.31	0.365	5.86
22	1600	1000	200	20	200	2	33.50	0.532	4.35
23	1600	1000	200	20	200	5	33.12	1.389	6.56
24	1600	1000	200	20	200	8	33.52	2.271	8.10
25	1600	1000	200	20	200	10	33.33	2.878	8.99

3.3.1. The Influence of Turbidity Current Density on the Propagation Velocity and Amplitude of Excitation Waves

The variations in surface elevation at three sensor locations in the simulated results of five different turbidity current density groups are presented in Figure 3.

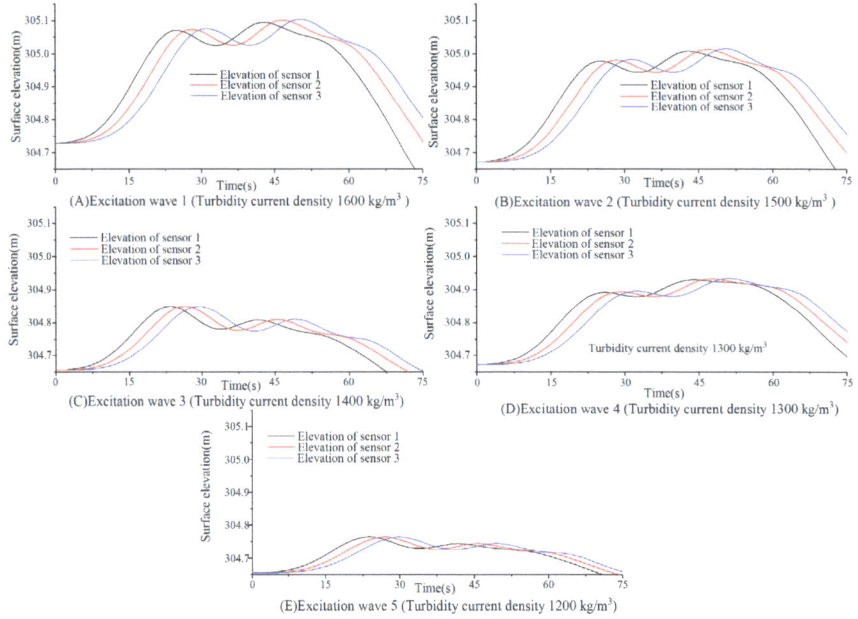

Figure 3. Simulation of propagating velocity of excitation wave under the sole variable condition of turbulent current density. (Length of turbidity source area: 1000 m; canyon width: 200 m; thickness of turbidity source area: 20 m; depth: 200 m; initial velocity of turbidity current: 0 m/s).

Based on the simulation results described above, while keeping all other conditions constant, the impact of a single variable, namely, the turbidity current density, on the propagation velocity and amplitude of the excitation wave was analyzed. By fitting the data, the relationship between turbidity current density and the propagation velocity of turbidity currents as well as the amplitude of the excitation wave was obtained, as shown in Figure 4.

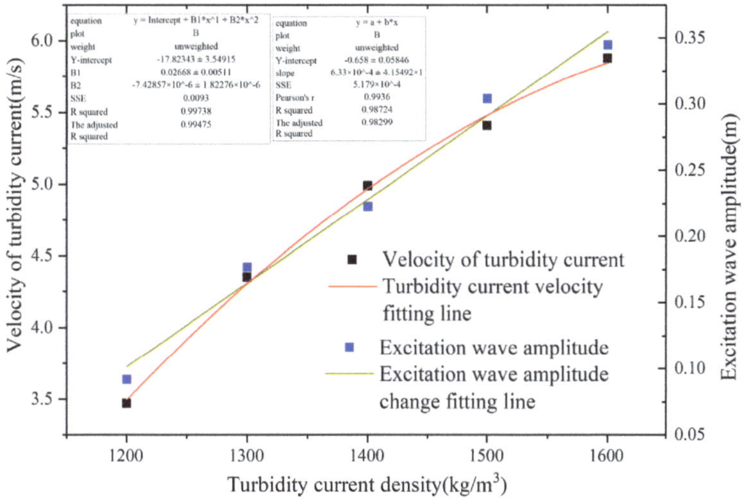

Figure 4. Relationship between turbidity current density and turbidity current velocity, as well as excitation wave amplitude.

The simulation results indicate that changes in turbidity current density, while keeping the other conditions constant, do not result in a change in the propagation velocity of the excitation waves. However, they do affect the amplitude of the excitation waves and the velocity of the turbidity current itself. The simulation reveals that within the selected density range, both the amplitude of the excitation waves and the velocity of the turbidity current increase with increasing turbidity current density. When the turbidity current density is equal to that of water ($\rho_{\text{Turbidity current}} = \rho_{\text{Water}}$), there is no turbidity current or excitation wave generation. Thus, the relationship between the turbidity current velocity (v) and density (ρ) is expressed as $v = -34.80643 + 0.05082 \bullet \rho - 1.59286 \times 10^{-5} \rho^2$ ($\rho > 1000$, $R^2 = 0.994$). Additionally, the relationship between the amplitude of the excitation waves (A) caused by turbidity currents and density (ρ) is expressed as $A = -0.6021 + 5.9729 \times 10^{-4} \rho$ ($\rho > 1000$, $R^2 = 0.991$).

3.3.2. The Influence of the Thickness of the Turbidity Source Area on the Propagation Velocity and Amplitude of Excitation Waves

The variations in surface elevation at three sensor locations in the simulated results of five different thickness of turbidity source area groups are presented in Figure 5.

Based on the simulation results described above, while keeping all other conditions constant, the impact of a single variable, namely, the thickness of the turbidity source area, on the propagation velocity and amplitude of the excitation wave was analyzed. By fitting the data, the relationship between the thickness of the turbidity source area and the propagation velocity of the turbidity current as well as the amplitude of the excitation wave was obtained, as shown in Figure 6.

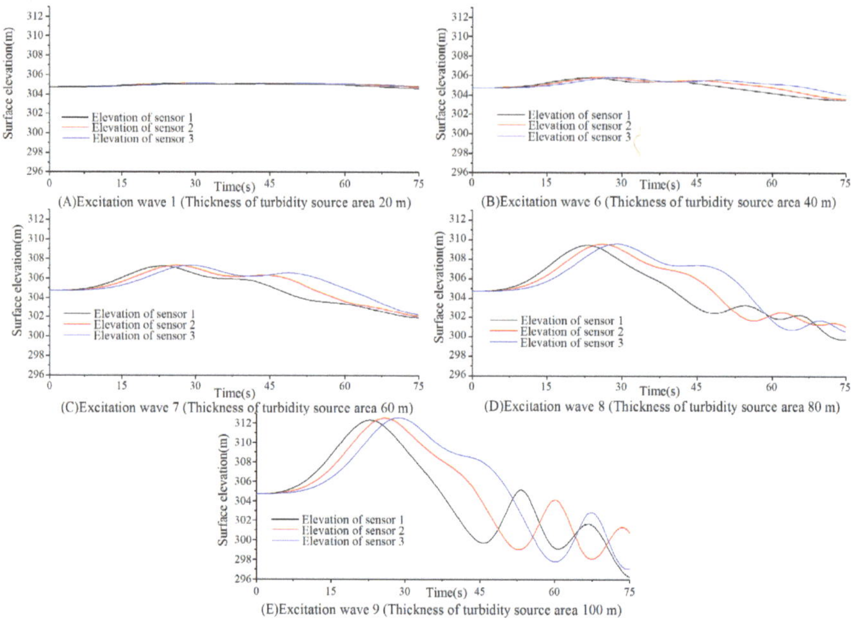

Figure 5. Simulation of propagating velocity of excitation wave under the sole variable condition of thickness of turbidity source area. (Turbidity current density: 1600 kg/m^3; length of turbidity source area: 1000 m; canyon width: 200 m; depth: 200 m; initial velocity of turbidity current: 0 m/s).

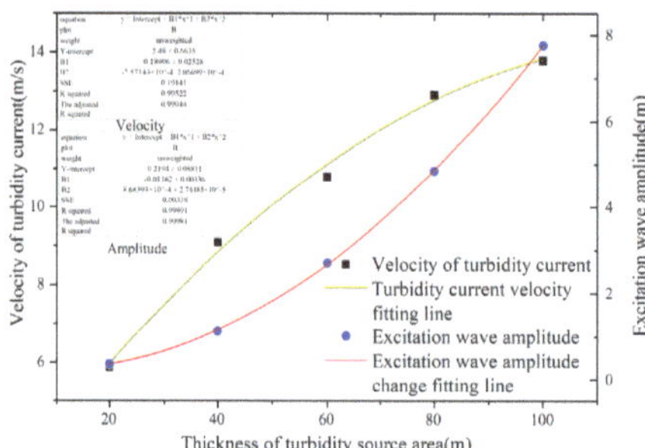

Figure 6. Relationship between thickness of turbidity source area and turbidity current velocity, as well as excitation wave amplitude.

Based on the simulated results mentioned above, it can be concluded that, while keeping the other conditions constant, changing only the thickness of the turbidity current source area does not affect the propagation velocity of the excitation waves. However, it does impact both the amplitude of the excitation waves and the velocity of the turbidity current itself. The simulation reveals that within the selected range of thickness values for the turbidity current source area, both the amplitude of the excitation waves and the velocity of the turbidity current increase with an increase in the thickness of the source area. Additionally, it is observed that when the length of the turbidity current source area is zero, neither the turbidity current nor the excitation waves are generated (i.e., no turbidity current is produced when $h_{\text{Turbidity current}} = 0$). Therefore, the relationship between the velocity (v) of the turbidity current and its thickness (h) is expressed as $v = 0.27983 \bullet h - 0.00146 \bullet h^2$ ($h \geq 0$, $R^2 = 0.999$). Similarly, the relationship between the amplitude (A) of the excitation waves caused by the turbidity current and its thickness (h) is $A = -0.00375 \bullet h - 0.0008 \bullet h^2$ ($h \geq 0$, $R^2 = 0.999$).

3.3.3. The Influence of the Length of the Turbidity Source Area on the Propagation Velocity and Amplitude of Excitation Waves

The variations in surface elevation at three sensor locations in the simulated results of five different length of turbidity source area groups are presented in Figure 7.

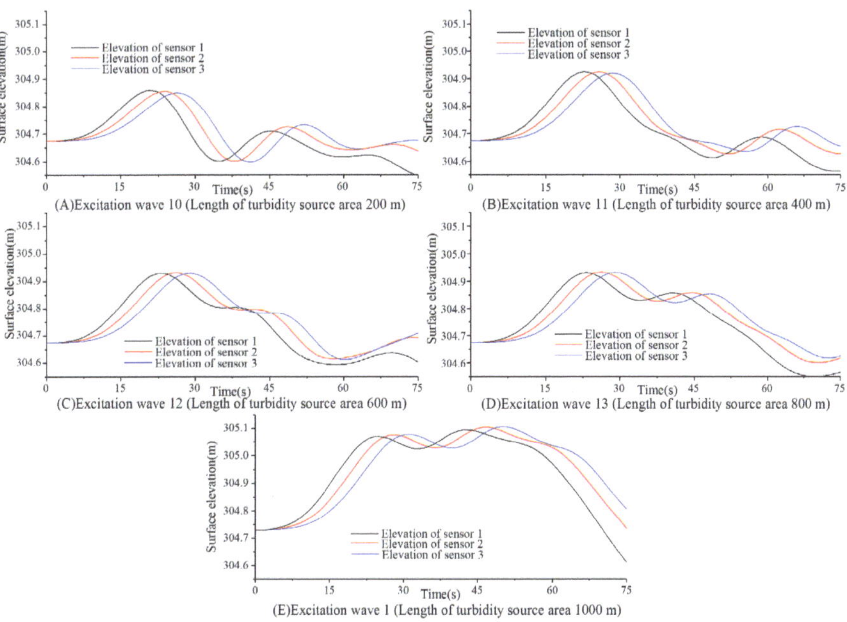

Figure 7. Simulation of propagating velocity of excitation wave under the sole variable condition of length of turbidity source area. (Turbidity current density: 1600 kg/m^3; canyon width: 200 m; thickness of turbidity source area: 20 m; depth: 200 m; initial velocity of turbidity current: 0 m/s).

Based on the simulation results described above, while keeping all other conditions constant, the impact of a single variable, namely, the length of the turbidity source area, on the propagation velocity and amplitude of the excitation wave was analyzed. By fitting the data, the relationship between the length of the turbidity source area and the amplitude of the excitation wave was obtained, as shown in Figure 8.

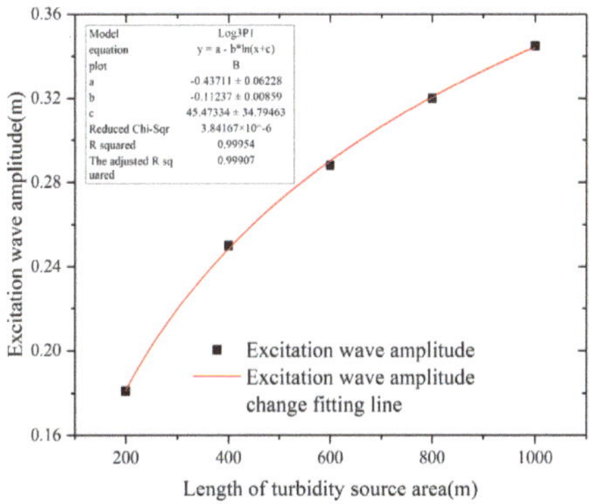

Figure 8. Relationship between length of turbidity source area and excitation wave amplitude.(Amplitude refers to the surface elevation change caused by the excitation wave).

Through simulations, it has been determined that within the chosen range of the length of the turbidity source area, the amplitude of the excitation waves increases with an increase in the length of the turbidity source area. When the length of the turbidity source area is zero, there is no turbidity current and no generation of excitation waves (i.e., when $L_{\text{Turbidity current}} = 0$). Additionally, for large lengths of the turbidity source area, under the condition of sufficient sediment supply, the variations in surface elevation caused by the waves generated by turbidity currents are negligible. Therefore, the relationship between the amplitude of the excitation waves (A) generated by turbidity currents and the length of the turbidity source area (L) is expressed as follows: $A = -0.3624 + 0.10305 \bullet \ln(L - 6.15619)$ ($L \geq 0$, $R^2 = 0.997$).

3.3.4. The Influence of Depth on the Propagation Velocity and Amplitude of Excitation Waves

The variations in surface elevation at three sensor locations in the simulated results of five different depth groups are presented in Figure 9.

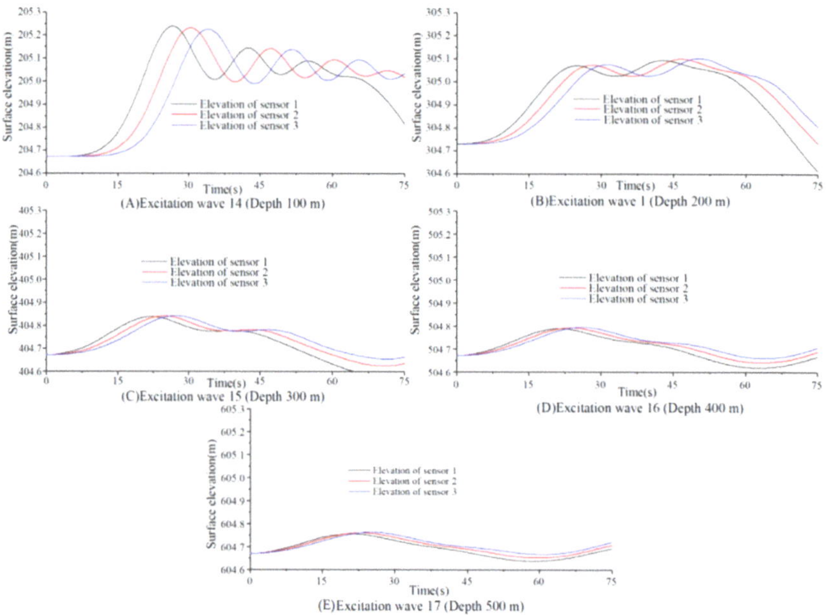

Figure 9. Simulation of propagation velocity of excitation wave under the sole variable condition of depth. (Turbidity current density: 1600 kg/m^3; length of turbidity source area: 1000 m; canyon width: 200 m; thickness of turbidity source area: 20 m; initial velocity of turbidity current: 0 m/s).

Based on the simulation results described above, while keeping all other conditions constant, the impact of a single variable, namely, depth, on the propagation velocity and amplitude of the excitation wave was analyzed. By fitting the data, the relationship between depth and the propagation velocity of the excitation wave as well as the amplitude of the excitation wave was obtained, as shown in Figure 10.

As the water depth approaches infinity, the excitation wave amplitude can only approach zero but cannot reach zero. Therefore, the characteristics of the excitation wave amplitude change with the water depth are similar to those of the velocity propagation of the excitation wave. The relationship between the velocity of the excitation wave induced by turbidity currents ($v_{\text{Excitation wave}}$) and the water depth ($H$) can be described as $v_{\text{Excitation wave}} = -287.05446 + 48.59211 \bullet \ln(H + 535.14863)$ ($R^2 = 0.998$). The relationship between the excitation wave amplitude (A) and the water depth (H) can be expressed as $A = 1.46573 - 0.22816 \bullet \ln(H - 47.67563)$ ($R^2 = 0.985$).

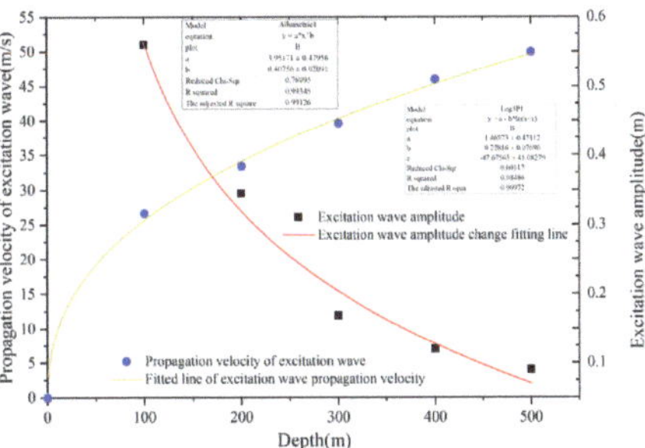

Figure 10. Relationship between depth and propagating velocity of excitation wave, as well as excitation wave amplitude.

3.3.5. The Influence of the Canyon Width on the Propagation Velocity and Amplitude of Excitation Waves

The variations in surface elevation at three sensor locations in the simulated results of five different canyon width groups are presented in Figure 11.

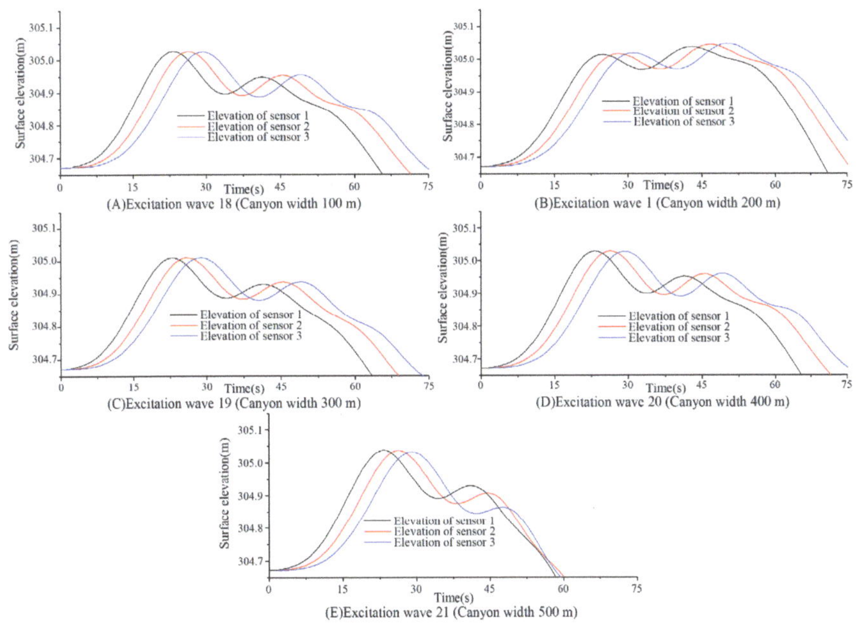

Figure 11. Simulation of propagating velocity of excitation wave under the sole variable condition of canyon width. (Turbidity current density: 1600 kg/m^3; length of turbidity source area: 1000 m; thickness of turbidity source area: 20 m; depth: 200 m; initial velocity of turbidity current: 0 m/s).

When the canyon width is taken as the single variable condition, changing the canyon width does not significantly affect the propagation velocity of excitation waves, the amplitude of excitation waves, and the velocity of turbidity currents. Therefore, it can be concluded that, without considering the impact of the differences in the terrain and sediment on the canyon width, the canyon width has no impact on the propagation of excitation waves and the movement of turbidity currents.

3.3.6. The Influence of the Initial Velocity of the Turbidity Current on the Propagation Velocity and Amplitude of Excitation Waves

The variations in surface elevation at three sensor locations in the simulated results of five different initial velocity of turbidity current groups are presented in Figure 12.

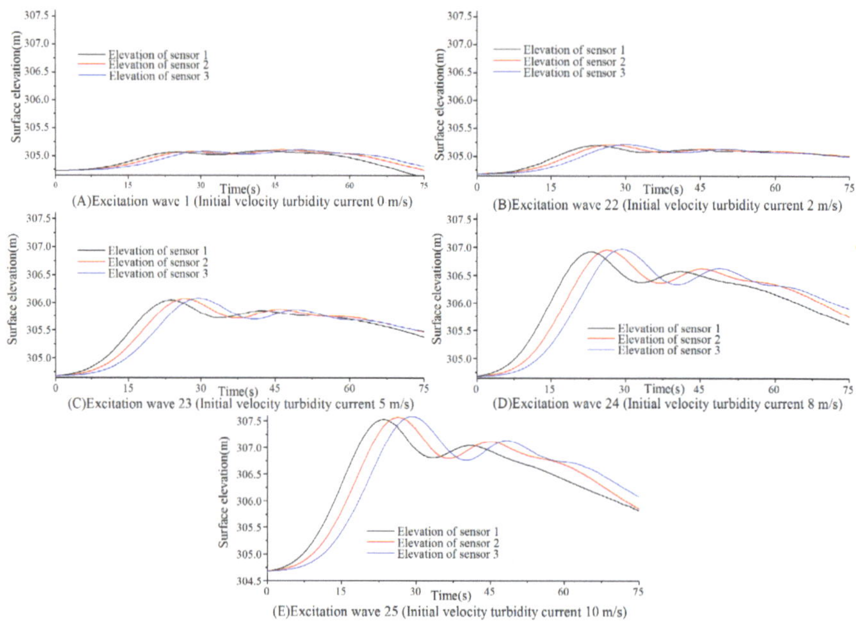

Figure 12. Simulation of propagating velocity of excitation wave under the sole variable condition of initial velocity of turbidity current. (Turbidity current density: 1600 kg/m^3; length of turbidity source area: 1000 m; canyon width: 200 m; thickness of turbidity source area: 20 m; depth: 200 m).

Based on the simulation results described above, while keeping all other conditions constant, the impact of a single variable, namely, the initial velocity of the turbidity current, on the propagation velocity and amplitude of the excitation wave was analyzed. By fitting the data, the relationship between the initial velocity of the turbidity current and the amplitude of the excitation wave was obtained, as shown in Figure 13.

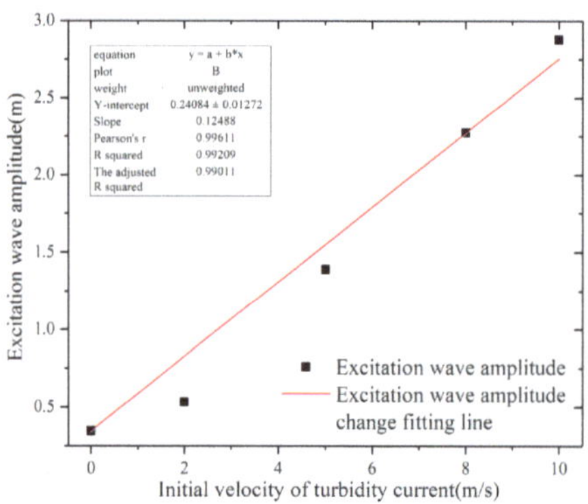

Figure 13. Relationship between initial velocity of turbidity current and excitation wave amplitude.

Based on the simulation, it is observed that within the selected range of the initial velocity of the turbidity current, the amplitude of the excitation wave increases linearly with the increase in the initial velocity of the turbidity current. Therefore, the relationship between the amplitude (A) of the excitation wave caused by the turbidity current and the initial velocity of the turbidity current (v_0) can be expressed as $A = 0.34 + 0.24084 \bullet v_0$ ($A \geq 0$, $R^2 = 0.992$).

Through controlling the simulation calculation of a single variable, it was found that there are several factors that can affect the amplitude of the excitation wave. These factors include the turbidity current density ρ, the thickness of the turbidity current source area d, the length of the turbidity current source area L, the water depth h, and the initial velocity of the turbidity current v_0. In contrast, there are relatively few factors that influence the propagation velocity of the excitation wave. Within the selected parameter range, only the water depth can affect the propagation velocity of the excitation wave. The physical parameters of the turbidity current, including the turbidity current density ρ, the thickness of the turbidity current source area d, the length of the turbidity current source area L, the canyon width l, and the initial velocity of the turbidity current v_0, have no direct influence on the propagation velocity of the excitation wave. Therefore, the turbidity current only serves as a triggering factor for the excitation wave and is not directly related to the propagation velocity of the excitation wave.

3.4. Analyze the Changes in Propagation Velocity of Excitation Waves along a Path

In order to further investigate the underlying truth behind the variation in the propagation velocity of the excitation wave, a discussion on whether there is velocity attenuation along the propagation path of the excitation wave is conducted. Since the seventh group of the excitation wave causes significant changes in surface elevation, the seventh group of the excitation wave is selected as the research object in order to study the variations in surface elevation along the propagation path of the excitation wave. The changes in surface elevation are extracted every 200 m along the sediment slope (with the first extraction point located 400 m away from the source area of the turbidity current). A total of six sets of surface elevation data are extracted (ranging from 400 m to 1400 m distance from the source area of the turbidity current), as shown in Figure 14.

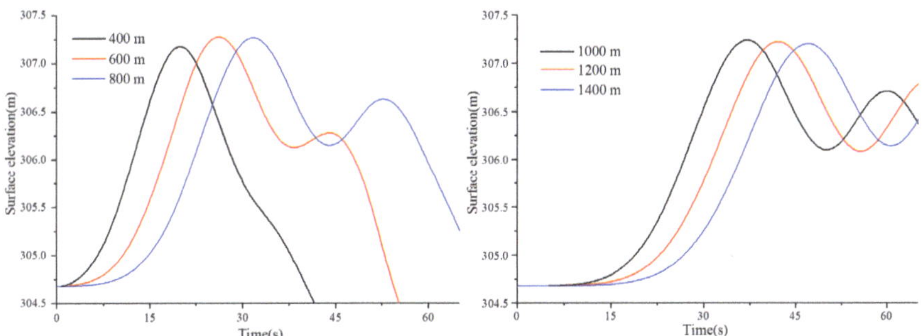

Figure 14. Surface elevation changes during excitation wave propagation along sediment slopes.

The amplitudes and propagation velocities of the excitation wave at each point are shown in Table 4.

From the table above, it can be observed that the amplitude of the excitation wave does not change while traveling along the slope. This indicates that the change in surface elevation caused by the propagation of the excitation wave does not attenuate. Furthermore, the propagation velocity of the excitation wave gradually increases, although the change is not very pronounced. This variation may be attributed to the change in the water depth caused by the sloping bed. To investigate this, a simulation was conducted in a straight channel with a length of 3000 m. Six sampling points were established from 400 m to 1400 m away from the turbidity current source area to extract the amplitude of the excitation wave. The results of the simulation are presented in Figure 15.

Table 4. Excitation wave velocity during the excitation wave propagation along the sediment slope.

Distance from Turbidity Current Source Area (m)	Propagation Velocity of Excitation Wave (m/s)	Excitation Wave Amplitude (m)
400	33.34	2.524
600	36.79	2.596
800	37.13	2.589
1000	39.99	2.566
1200	40.04	2.542
1400	40.13	2.523

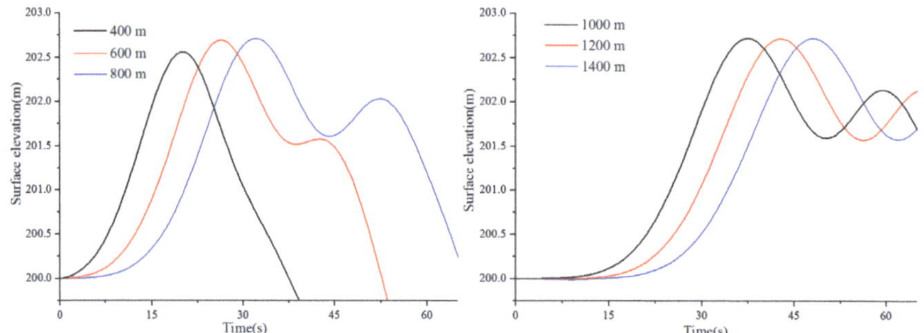

Figure 15. Surface elevation changes during wave propagation along a straight channel.

The amplitudes and propagation velocities of the excitation wave at each point are shown in Table 5.

Table 5. Excitation wave velocity during the propagation along the straight channel.

Distance from Turbidity Current Source Area (m)	Propagation Velocity of Excitation Wave (m/s)	Excitation Wave Amplitude (m)
400	33.89	2.559
600	37.66	2.692
800	37.12	2.712
1000	36.92	2.717
1200	37.09	2.715
1400	37.48	2.718

The data from the table above indicate that during the propagation of the excitation wave along a straight water channel, its velocity remains constant, except for a slight decrease at the initial point. This phenomenon may be attributed to the fact that in the starting phase, the excitation wave is not fully developed, and hence its velocity is relatively smaller. However, once it is fully developed, the propagation velocity of the excitation wave does not decrease in subsequent processes. Therefore, the propagation velocity of the excitation wave is only dependent on the real-time water depth of the wave. In future studies, we aim to explore the relationships between these influencing factors and other physical parameters, such as the speed of wave propagation, using the effective and accurate method of machine learning algorithms [45].

3.5. Expression of the Propagation Velocity of the Excitation Wave

The propagation of the excitation wave along a long distance does not experience an attenuation in velocity, as is the case with the propagation velocity of solitary waves.

Referring to the estimated wave propagation velocity (the square of the propagation velocity is directly proportional to the water depth amplitude) [46], the wavelengths under different water depth conditions were extracted, as shown in Table 6.

Table 6. Physical parameters of excitation wave under different water depth conditions.

Depth (m)	Propagation Velocity of Excitation Wave (m/s)	Excitation Wave Amplitude (m)	Excitation Wave Length (m)
100	26.67	0.56	2580
200	33.43	0.35	2850
300	39.65	0.17	3250
400	45.98	0.12	3600
500	49.97	0.08	4150
1000	66.67	0.04	6000
2000	90.91	0.02	9500
4000	165.84	0.3	17600

From the simulation results of a single variable, the water depth, it could be seen that the wavelengths of the excitation waves were much larger than the water depth. Therefore, further simulations were conducted under water depth conditions ranging from 1000 m to 4000 m. Due to the minimal change in wave amplitude when the water depth reached 4000 m, it was not possible to observe a distinct waveform. However, through simulations with the thickness of the turbidity current source area as the single variable, it was found that an increase in the thickness of the source region led to a larger amplitude of the excitation waves, but it did not affect the wavelength of the excitation waves. Therefore, in order to better extract the wavelength of the excitation waves, the thickness of the source region in the simulation with a water depth of 4000 m was set to 200 m.

Through simulations at water depths of 1000 m and 4000 m, it is observed that the wavelengths of the excitation waves are much larger than the water depth, indicating that these waves belong to the category of shallow water waves. The amplitude of the excitation waves is relatively small compared to their wavelength, aligning with the small amplitude wave theory [47]. According to this theory, the wave velocity of shallow water waves is only dependent on the water depth (h) and gravity acceleration (g), regardless of the wave period. In the case of excitation waves induced by turbidity currents in deep water, the amplitudes of these waves are relatively small compared to the water depth. Referring to the expression for shallow water waves (when the relative water depth, which is the ratio of water depth to wavelength, is much smaller than 1/2), the wave velocity is denoted as '$C_s = \sqrt{gh}$'. This implies that the propagation velocity of the excitation waves is also solely related to the water depth. Therefore, a fitting of the square of the propagation velocity of the excitation waves (v^2) and the water depth (h) was conducted (Figure 16).

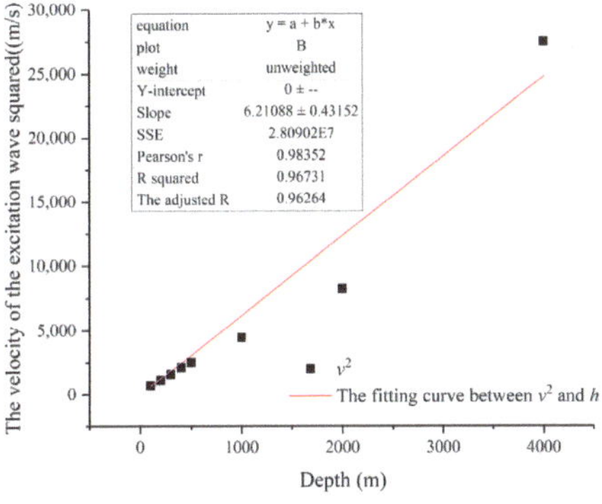

Figure 16. The relationship between the propagation velocity of excitation wave and the depth.

Through fitting, the following can be obtained:

$$v^2 = 0.63gh(R^2 = 0.967) \tag{8}$$

Through fitting, it can be discovered that the propagation model of the velocity of excitation waves is different from the shallow water wave theory. This is because turbidity currents, as granular materials, generate excitation waves by pushing the water in front of them with sediment particles underwater, which is different from the surges formed by solid blocks entering the ocean. Additionally, excitation waves formed by turbidity currents occur in an underwater environment, which may be the reason why the propagation velocity equation for the excitation waves behaves as if the velocity squared is equal to half the Earth's gravity. This equation reveals the variation in the propagation velocity of the excitation wave with depth, explaining why the average velocity between the monitoring points in the field is greater than the instantaneous velocity measured at these points [41]. Further theoretical research on the propagation velocity of excitation waves requires subsequent field monitoring and the deployment of monitoring systems to more thoroughly investigate the fundamental causes.

4. Conclusions

This study aimed to investigate the velocity of turbidity current-induced excitation waves through numerical simulation. By fixing a single variable, different factors that could affect the propagation velocity and amplitude of the excitation waves were analyzed and discussed, leading to the following three conclusions:

1. Within the selected parameter range, there are several factors that can influence the amplitude of the excitation waves, including the turbidity current density ρ, the thickness of the turbidity current source area d, the length of the turbidity current source area L, the water depth h, and the initial velocity of the turbidity current v_0. The amplitude of the excitation waves is positively correlated with the turbidity density, the thickness of the source area, the length of the source area, and the initial velocity, while it is negatively correlated with the water depth.
2. Within the selected parameter range, only the water depth can affect the propagation velocity of the excitation waves. As the water depth increases, the propagation velocity of the excitation waves also increases, and a relationship of $v^2 = 0.63gh$ ($R^2 = 0.967$) is established between the square of the propagation velocity v^2 and the water depth h.
3. During the propagation of the excitation waves, both the propagation velocity and the changes in surface elevation caused by the waves do not attenuate. Considering the relatively calm deep-sea environment, the high-speed propagation of the excitation waves and the resuspension of bottom sediments they cause not only complement the understanding of turbidity current motion patterns in canyons, but also provide new research directions for deep-sea sediment transport.

Author Contributions: Conceptualization and methodology, G.X., S.S. and Y.R.; formal analysis, Z.C. and Y.R.; data curation, S.S. and M.L.; writing—original draft preparation, S.S. and G.X.; writing—review and editing, S.S. and G.X.; visualization, Y.R. and M.L.; project administration, G.X.; funding acquisition, G.X. All authors have read and agreed to the published version of the manuscript.

Funding: The project is supported by the National Natural Science Foundation of China (Grant Nos. 42206055 and 41976049).

Institutional Review Board Statement: Not applicable.

Informed Consent Statement: Not applicable.

Data Availability Statement: The data presented in this study are available on request from the corresponding author.

Acknowledgments: The author would like to thank Yang Li for his help with the work.

Conflicts of Interest: The authors declare no conflicts of interest.

References

1. Azpiroz-Zabala, M.; Cartigny, M.J.B.; Talling, P.J.; Parsons, D.R.; Sumner, E.J.; Clare, M.A.; Simmons, S.M.; Cooper, C.; Pope, E.L. Newly recognized turbidity current structure can explain prolonged flushing of submarine canyons. *Sci. Adv.* **2017**, *3*, e1700200. [CrossRef] [PubMed]
2. Daly, R.A. Origin of submarine canyons. *Am. J. Sci.* **1936**, *5*, 401–420. [CrossRef]
3. Winterwerp, J. Stratification effects by fine suspended sediment at low, medium, and very high concentrations. *J. Geophys. Res. Oceans.* **2006**, *111*, C5. [CrossRef]
4. Nilsen, T.H.; Shew, R.D.; Steffens, G.S.; Studlick, J.R.J. *Atlas of Deep-Water Outcrops*; American Association of Petroleum Geologists: Tulsa, OK, USA, 2008. [CrossRef]
5. Xu, J. Turbidity Current Research in the Past Century: An Overview. *J. Ocean Univ. China* **2014**, *44*, 98–105. [CrossRef]
6. Talling, P.J.; Allin, J.; Armitage, D.A.; Arnott, R.W.C.; Cartigny, M.J.B.; Clare, M.A.; Felletti, F.; Covault, J.A.; Girardclos, S.; Hansen, E.; et al. Key future directions for research on turbidity currents and their deposits. *J. Sediment. Res.* **2015**, *85*, 153–169. [CrossRef]
7. Maier, K.L.; Gales, J.A.; Paull, C.K.; Rosenberger, K.; Talling, P.J.; Simmons, S.M.; Gwiazda, R.; McGann, M.; Cartigny, M.J.; Lundsten, E. Linking direct measurements of turbidity currents to submarine canyon-floor deposits. *Front. Earth Sci.* **2019**, *7*, 144. [CrossRef]
8. Hughes Clarke, J.E. First wide-angle view of channelized turbidity currents links migrating cyclic steps to flow characteristics. *Nat. Commun.* **2016**, *7*, 11896. [CrossRef]
9. Normandeau, A.; Bourgault, D.; Neumeier, U.; Lajeunesse, P.; St-Onge, G.; Gostiaux, L.; Chavanne, C. Storm-induced turbidity currents on a sediment-starved shelf: Insight from direct monitoring and repeat seabed mapping of upslope migrating bedforms. *Sedimentology* **2020**, *67*, 1045–1068. [CrossRef]
10. Hill, P.R.; Lintern, D.G. Turbidity currents on the open slope of the Fraser Delta. *Mar. Geol.* **2022**, *445*, 106738. [CrossRef]
11. Summers, M. Review of deep-water submarine cable design. In Proceedings of the SubOptic 2001, Kyoto, Japan, 20–24 May 2001; p. 4.
12. Carter, L.; Gavey, R.; Talling, P.J.; Liu, J.T. Insights into Submarine Geohazards from Breaks in Subsea Telecommunication Cables. *Oceanography* **2014**, *27*, 58–67. [CrossRef]
13. Gavey, R.; Carter, L.; Liu, J.T.; Talling, P.J.; Hsu, R.; Pope, E.; Evans, G. Frequent sediment density flows during 2006 to 2015, triggered by competing seismic and weather events: Observations from subsea cable breaks off southern Taiwan. *Mar. Geol.* **2017**, *384*, 147–158. [CrossRef]
14. Carter, L.; Burnett, D.; Drew, S.; Hagadorn, L.; Marle, G.; Bartlett-Mcneil, D.; Irvine, N. *Submarine Cables and the Oceans: Connecting the World*; UNEP-WCMC Biodiversity Series 31; UNEP World Conservation Monitoring Centre: Cambridge, UK, 2010.
15. Qiu, W. Submarine cables cut after Taiwan earthquake in Dec 2006. *Submar. Cable Netw.* **2011**, *19*.
16. Heezen, B.C.; Ewing, W.M. Turbidity currents and submarine slumps, and the 1929 Grand Banks [Newfoundland] earthquake. *Am. J. Sci.* **1952**, *250*, 849–873. [CrossRef]
17. Kuenen, P.H. Estimated size of the Grand Banks [Newfoundland] turbidity current. *Am. J. Sci.* **1952**, *250*, 874–884. [CrossRef]
18. Heezen, B.C.; Ericson, D.; Ewing, M. Further evidence for a turbidity current following the 1929 Grand Banks earthquake. *Deep-Sea Res.* **1954**, *1*, 193–202. [CrossRef]
19. Heezen, B.C. Whales entangled in deep sea cables. *Deep-Sea Res.* **1957**, *4*, 105–115. [CrossRef]
20. Piper, D.J.; Shor, A.N.; Farre, J.A.; O'Connell, S.; Jacobi, R. Sediment slides and turbidity currents on the Laurentian Fan: Sidescan sonar investigations near the epicenter of the 1929 Grand Banks earthquake. *Geology* **1985**, *13*, 538–541. [CrossRef]
21. Piper, D.J.; Cochonat, P.; Morrison, M.L. The sequence of events around the epicentre of the 1929 Grand Banks earthquake: Initiation of debris flows and turbidity current inferred from sidescan sonar. *Sedimentology* **1999**, *46*, 79–97. [CrossRef]
22. Houtz, R.; Wellman, H. Turbidity current at Kadavu Passage, Fiji. *Geol. Mag.* **1962**, *99*, 57–62. [CrossRef]
23. Heezen, B.C.; Ewing, M. Orleansville earthquake and turbidity currents. *AAPG Bull.* **1955**, *39*, 2505–2514. [CrossRef]
24. Krause, D.C.; White, W.C.; PIPER, D.J.W.; Heezen, B.C. Turbidity currents and cable breaks in the western New Britain Trench. *Geol. Soc. Am. Bull.* **1970**, *81*, 2153–2160. [CrossRef]
25. Piper, D.J.; Savoye, B. Processes of late Quaternary turbidity current flow and deposition on the Var deep-sea fan, north-west Mediterranean Sea. *Sedimentology* **1993**, *40*, 557–582. [CrossRef]
26. Soh, W.; Machiyama, H.; Shirasaki, Y.; Kasahara, J. Deep-sea floor instability as a cause of deepwater cable fault, off eastern part of Taiwan. *AGU Fall Meet. Abstr.* **2004**, *2*, 1–8.
27. Cattaneo, A.; Babonneau, N.; Ratzov, G.; Dan-Unterseh, G.; Yelles, K.; Bracène, R.; De Lepinay, B.M.; Boudiaf, A.; Déverchère, J. Searching for the seafloor signature of the 21 May 2003 Boumerdès earthquake offshore central Algeria. *Nat. Hazards Earth Syst. Sci.* **2012**, *12*, 2159–2172. [CrossRef]
28. Hsu, S.-K.; Kuo, J.; Chung-Liang, L.; Ching-Hui, T.; Doo, W.-B.; Ku, C.-Y.; Sibuet, J.-C. Turbidity currents, submarine landslides and the 2006 Pingtung earthquake off SW Taiwan. *Terr. Atmos. Ocean. Sci.* **2008**, *19*, 7. [CrossRef]
29. Carter, L.; Milliman, J.D.; Talling, P.J.; Gavey, R.; Wynn, R.B. Near-synchronous and delayed initiation of long run-out submarine sediment flows from a record-breaking river flood, offshore Taiwan. *Geophys. Res. Lett.* **2012**, *39*, L12603. [CrossRef]
30. Clare, M.A.; Yeo, I.A.; Watson, S.; Wysoczanski, R.; Seabrook, S.; Mackay, K.; Hunt, J.E.; Lane, E.; Talling, P.J.; Pope, E.; et al. Fast and destructive density currents created by ocean-entering volcanic eruptions. *Science* **2023**, *381*, 1085–1092.

31. Ren, Y.; Zhang, Y.; Xu, G.; Xu, X.; Wang, H.; Chen, Z. The failure propagation of weakly stable sediment: A reason for the formation of high-velocity turbidity currents in submarine canyons. *J. Ocean. Limnol.* **2023**, *41*, 100–117. [CrossRef]
32. Wang, Z.; Xu, J.; Talling, P.J.; Cartigny, M.J.B.; Simmons, S.M.; Gwiazda, R.; Paull, C.K.; Maier, K.L.; Parsons, D.R. Direct evidence of a high-concentration basal layer in a submarine turbidity current. *Deep-Sea Res. Part I Oceanogr. Res. Pap.* **2020**, *161*, 103300. [CrossRef]
33. Ren, Y.; Tian, H.; Chen, Z.; Xu, G.; Liu, L.; Li, Y. Two Kinds of Waves Causing the Resuspension of Deep-Sea Sediments: Excitation and Internal Solitary Waves. *J. Ocean Univ. China* **2023**, *22*, 429–440. [CrossRef]
34. Lambert, A.M.; Kelts, K.R.; Marshall, N.F. Measurements of density underflows from Walensee, Switzerland. *Sedimentology* **1976**, *23*, 87–105. [CrossRef]
35. Piper, D.J.W.; Shor, A.N.; Hughes Clarke, J.E. The 1929 "Grand Banks" earthquake, slump, and turbidity current. In *Sedimentologic Consequences of Convulsive Geologic Events*; Geological Society of America: Boulder, CO, USA, 1988; pp. 77–92. [CrossRef]
36. Paull, C.K.; Talling, P.J.; Maier, K.L.; Parsons, D.; Xu, J.; Caress, D.W.; Gwiazda, R.; Lundsten, E.M.; Anderson, K.; Barry, J.P. Powerful turbidity currents driven by dense basal layers. *Nat. Commun.* **2018**, *9*, 4114. [CrossRef] [PubMed]
37. Ren, Y.; Zhou, H.; Wang, H.; Wu, X.; Xu, G.; Meng, Q. Study on the critical sediment concentration determining the optimal transport capability of submarine sediment flows with different particle size composition. *Mar. Geol.* **2023**, *464*, 107142. [CrossRef]
38. Bagnold, R.A. Auto-suspension of transported sediment; turbidity currents. *Proc. R. Soc. Lond. Ser. A Math. Phys. Sci.* **1962**, *265*, 315–319.
39. Parker, G. Conditions for the ignition of catastrophically erosive turbidity currents. *Mar. Geol.* **2003**, *46*, 307–327. [CrossRef]
40. Pantin, H.M. Interaction between velocity and effective density in turbidity flow: Phase-plane analysis, with criteria for autosuspension. *Mar. Geol.* **1979**, *31*, 59–99. [CrossRef]
41. Heerema, C.J.; Talling, P.J.; Cartigny, M.J.; Paull, C.K.; Bailey, L.; Simmons, S.M.; Parsons, D.R.; Clare, M.A.; Gwiazda, R.; Lundsten, E.; et al. What determines the downstream evolution of turbidity currents? *Earth Planet. Sci. Lett.* **2020**, *532*, 116023. [CrossRef]
42. Talling, P.J.; Cartigny, M.J.B.; Pope, E.; Baker, M.; Clare, M.A.; Heijnen, M.; Hage, S.; Parsons, D.R.; Simmons, S.M.; Paull, C.K.; et al. Detailed monitoring reveals the nature of submarine turbidity currents. *Nat. Rev. Earth Environ.* **2023**, *4*, 642–658. [CrossRef]
43. Heimsund, S. *Numerical Simulation of Turbidity Currents: A New Perspective for Small-and Large-Scale Sedimentological Experiments*; Sedimentology/Petroleum Geology; University of Bergen: Bergen, Norway, 2007.
44. Zhou, J.; Cenedese, C.; Williams, T.; Ball, M.; Venayagamoorthy, S.K.; Nokes, R.I. On the Propagation of Gravity Currents Over and through a Submerged Array of Circular Cylinders. *J. Fluid Mech.* **2017**, *831*, 394–417. [CrossRef]
45. Saha, S.; De, S.; Changdar, S. An Application of Machine Learning Algorithms on the Prediction of the Damage Level of Rubble-Mound Breakwaters. *J. Offshore Mech. Arct. Eng.* **2024**, *146*, 011202. [CrossRef]
46. Russell, J.S. *Report on Waves: Made to the Meetings of the British Association*; Richard and John E Taylor: London, UK, 1845.
47. Airy, G.B. *Tides and Waves*; B. Fellowes: London, UK, 1845.

Journal of
Marine Science and Engineering

Article

Convenient Method for Large-Deformation Finite-Element Simulation of Submarine Landslides Considering Shear Softening and Rate Correlation Effects

Qiuhong Xie [1,2], Qiang Xu [2], Zongxiang Xiu [1,2,*], Lejun Liu [1], Xing Du [1], Jianghui Yang [3] and Hao Liu [1]

[1] First Institute of Oceanography, Ministry of Natural Resources, Qingdao 266061, China; xqh@fio.org.cn (Q.X.); liulj@fio.org.cn (L.L.); duxing@fio.org.cn (X.D.); liuh@fio.org.cn (H.L.)
[2] State Key Laboratory of Geohazard Prevention and Geoenvironment Protection, Chengdu University of Technology, Chengdu 610059, China; xq@cdut.edu.cn
[3] China Offshore Oil Engineering Corp CNOOC Ltd., Tianjin 300461, China; yangjh24@cooec.com.cn
* Correspondence: xzx@fio.org.cn

Abstract: Submarine landslides pose a serious threat to the safety of underwater engineering facilities. To evaluate the safety of undersea structures, it is important to estimate and analyze the sliding processes of potential submarine landslides. In this study, a convenient model for simulating submarine landslide processes is established by using Abaqus Eulerian large deformation technology with an explicit finite element framework. The VUSDFLD Fortran subroutine is used to consider the strain-softening and rate-dependency characteristics of soil shear strength. The proposed method is validated by comparing its results with experimental data and those of mainstream numerical methods. Then, the results of a dynamic analysis of typical potential submarine landslides in the Shenhu sea area are analyzed using the proposed method. Case studies are carried out under different soil shear strength distributions, and the influence of initial stress is also analyzed. The shear strain-softening and rate-dependency effects are highly involved in the runout process. The simulated landslide's failure mode is consistent with the geophysical interpretation of existing landslide characteristics.

Keywords: submarine landslide; sliding characteristics; Eulerian analysis; large deformation; explicit finite element analysis

Citation: Xie, Q.; Xu, Q.; Xiu, Z.; Liu, L.; Du, X.; Yang, J.; Liu, H. Convenient Method for Large-Deformation Finite-Element Simulation of Submarine Landslides Considering Shear Softening and Rate Correlation Effects. *J. Mar. Sci. Eng.* **2024**, *12*, 81. https://doi.org/10.3390/jmse12010081

Academic Editor: Assimina Antonarakou

Received: 10 November 2023
Revised: 20 December 2023
Accepted: 26 December 2023
Published: 29 December 2023

1. Introduction

Submarine landslides are common in continental margins and island slope areas, and the large number of landslide mass movements, involving the movement of sand, clay, and gravel, can cause great damage to submarine pipelines, cables, and other underwater facilities [1–5]. Large submarine landslides even pose a risk of triggering catastrophic tsunamis [6,7]. How to reasonably simulate and predict the movement processes of submarine landslides has thus become a key aspect of assessing the risk of the impacts of submarine landslides [8], as well as in the study of the transport and sedimentation system of continental slope areas.

The study of the sliding process of submarine landslides includes multiple disciplines, such as geotechnical mechanics and fluid mechanics (as shown in Figure 1), making it difficult to simulate the entire process using a unified model [9].

As matters stand, the main methods for simulating the mass movement of submarine landslides are as follows: analytical methods based on rigid bodies [10,11], single-phase plastic fluid methods [12,13], multiphase flow methods [14–17], large deformation finite element methods [18], material point methods [19,20], discrete element methods/smoothed particle hydrodynamics [21,22], and CFD-DEM coupling methods [23,24].

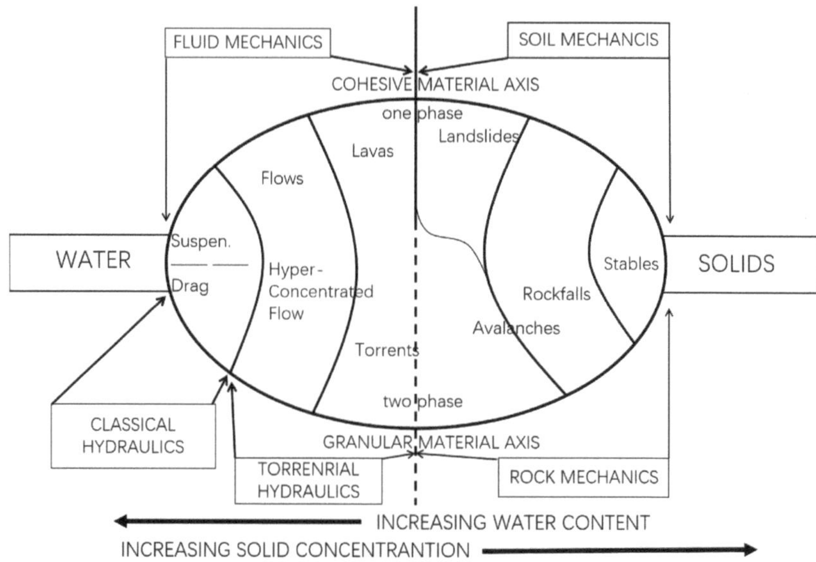

Figure 1. Classification of submarine mass movements (modified from Locat and Lee, 2002 [9]).

Analytical methods based on rigid body assumption can achieve high efficiencies at low computational costs since they do not require complicated theoretical derivation. However, the sliding soils of most landslides actually undergo large deformation. Therefore, using a rigid body assumption would cause a certain degree of difference in the results.

Imran et al. [12] developed a numerical program (BING) to simulate the runout of submarine landslides, which incorporated the Bingham, Herschel–Bulkley, and bilinear rheologies of viscoplastic fluids. Using the depth-averaged method, the two-dimensional runout is essentially simplified as a one-dimensional problem. On this basis, the Bing program has been further expanded to include the whole submarine landslide process, consisting of four phases: the initial flow stage, the water wedge development stage, the hydroplaning stage, and the post-hydroplaning stage [13]. However, several artificially assumed parameters, which need to be subjectively assigned, were introduced to distinguish different stages. It is not clear if the program can be used for complex morphologies, as the sliding material might not be divided into the shear and plug layers [25].

Multiphase flow methods can handle interactions between sliding soil and the environmental water to a certain extent [14–17]. However, they cannot reasonably reflect the complex stress–strain relationship of the landslide mass, especially in the early phase of sliding. In the early and middle stages, the landslide mass is more aligned with the scope of soil mechanics. Multiphase flow methods cannot accurately describe the shear softening process of sliding soil. Discrete element methods (DEMs) and smoothed particle hydrodynamics (SPHs) can deal with mesh distortion problems well since they are mesh-free techniques. However, difficulties still exist in the determination of boundary conditions when they are applied to the simulation of actual landslide problems, and the artificial stress technology can lead to high stress [21,22]. Furthermore, CFD-DEM coupling methods have also been used for the runout simulation of submarine landslides [23,24], showing good prospects. However, their computational cost is high when particle size increases, especially in the case of real-scale submarine landslide simulations.

The shear strength of sliding soil plays an important role in the runout of submarine landslides. Cohesive marine soils, especially those with medium to high sensitivity, exhibit not only shear strength softening characteristics under shear loads but also rate-dependency effects during the mass movement of the submarine landslide [26–28]. The aforementioned studies did not take these complex shear characteristics into account quantitatively. The large deformation finite element method (RITSS) [18,29] and material point method (MPM) [19,20] are two prominent methods used in large deformation analysis. The RITSS approach is based on the "remeshing and interpolation technique with small strain." It was initially proposed by Hu and Randolph [30] and was then extended to an Abaqus-based dynamic RITSS approach by Wang et al. [18]. In the dynamic RITSS approach, a small strain incremental analysis step is executed first; then, the configuration is updated by the new boundary node positions. After that, the new configuration is re-meshed, and all the variables, such as velocities, accelerations, stresses, and material properties, are mapped to the new mesh. The MPM originates from the particle-in-cell method in computational fluid dynamics [31]. The continua are discretized by Lagrangian particles moving over a fixed Eulerian background grid. The materials' mechanical and kinematic properties (mass, volume, density, velocities, momentum, deformation gradients, and stresses) are recorded and updated within the particles, and the Eulerian mesh is only used for calculation at each incremental step. Both of these methods have been adopted and validated in runout simulations of submarine landslides in which the shear strain-softening and rate-dependency effects were also incorporated [18–20,29]. However, in these two methods, some key but unpublished in-house programs regarding remeshing, interpolation, and numerical algorithms are needed. This gap poses high theoretical and technical requirements for the engineering community.

This article proposes a convenient simulation method for the runout simulation of submarine landslides based on the Eulerian analysis technology of Abaqus/Explicit. The used Eulerian analysis technology provides a large deformation calculation method in an explicit finite element scheme [32]. In each incremental step of the Eulerian analysis, a Lagrangian calculation is first performed, and then the new variables are mapped to the fixed Eulerian background mesh to achieve material flow [33–35]. This technology is effective for applications involving extreme deformation, up to and including fluid flow. The VUSDFLD Fortran subroutine provided in Abaqus can be applied to conveniently take into account the complex stress–strain relationships. So, the dynamic runout process of submarine landslides involving the shear strain-softening and rate-dependency characteristics of sliding soil can be implemented with VUSDFLD. The proposed convenient method can facilitate the common engineers to quickly analyze the runout process of submarine landslides.

2. Materials and Methods

2.1. Eulerian Analysis Technique

The Eulerian large deformation analysis technique implemented in Abaqus/Explicit uses an explicit time integration scheme [32]. In each time increment, the Eulerian analysis is internally divided into two steps, a Lagrangian analysis step and an Eulerian analysis step; the steps are shown in Figure 2.

At the beginning of each time increment, Δt, a traditional Lagrangian analysis step is first performed, in which the movement of the continuum is treated as a function of the material coordinates and time. The nodes of the Lagrangian mesh move together with the material, and elements deform as the material deforms (see Figure 2, left side). At the end of the Lagrangian step, a tolerance is adopted to determine which elements are significantly deformed. Elements with little or no deformation will remain inactive during the Eulerian step to improve computational performance. In the Eulerian analysis step, an Eulerian reference mesh (see Figure 2 right side), which remains undistorted is used to trace the motion of the material in the Eulerian domain. Deformation is suspended during the Eulerian analysis step. The elements with significant deformation are automatically remeshed, and the corresponding material flow between neighboring elements is then computed.

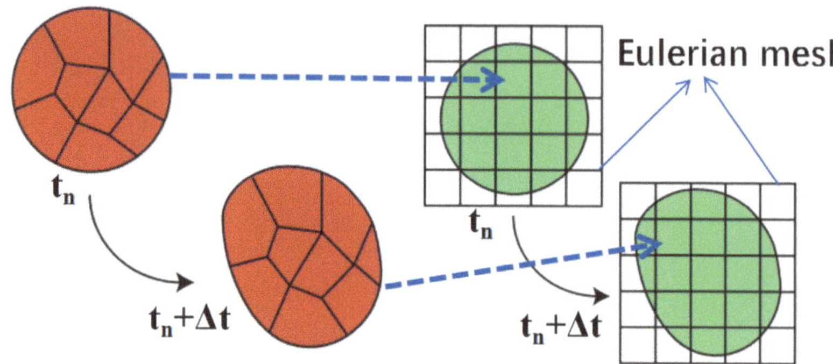

Figure 2. Deformation of a continuum in an Eulerian analysis.

The central difference rule is adopted for the solution of the non-linear differential equations. The unknown solution for the time $t + \Delta t$ can be calculated directly according to the solution at time t without iteration. Explicit calculations are not stringently stable, but their numerical stability can be guaranteed via the introduction of the critical time step size Δt_{crit}. For Eulerian analysis, the stable time increment size is adjusted automatically to prevent material from flowing across more than one element in each increment. Δt_{crit} is calculated in each time step and can be described using the characteristic element length L_e and the dilatory wave speed c_d:

$$\Delta t_{crit} = L_e / c_d \tag{1}$$

Compared with the traditional Lagrangian method, the Eulerian analysis technique is more suitable for large deformation calculations. For the runout simulation of a submarine landslide, cohesive soils with medium to high sensitivity exhibit significant strength-softening characteristics under shear loads [26–28]. The strain-softening and rate-related effects of landslide soil should be taken into account quantitatively in the soil's constitutive model. However, this feature is not directly provided in the Abaqus/Explicit module. Thus, this paper proposed a convenient method (more on that later) to achieve this function by using the VUSDFLD Fortran subroutine interface.

2.2. Detail Methodology

A long preset rectangular computational domain consisting of a soil domain and a void domain is first chosen to ensure that the submarine landslide mass cannot flow out of the boundary. The entire computational domain was set as the Eulerian part, and the first-order reduced integration hexahedral Eulerian element, EC3D8R, was used for calculation. The total stress method is adopted since EC3D8R does not involve pore pressure, and, therefore, the buoyant unit weight, γ', is used for the landslide soil. Velocity-constrained conditions are applied on the boundary of the Eulerian domain.

The Tresca soil model, which simulates linearly elasto-plastic behavior, was adopted to describe the mechanical properties of a submarine landslide mass. It can be defined by three parameters: undrained shear strength, S_u; elastic modulus, E; and Poisson's ratio, v. The undrained shear strength is governed by shear strain rate and soil remolding. The peak strength increases with the shearing rate but decreases with the absolute plastic shear strain. Then, the Tresca model can be considered through an enhanced Herschel–Bulkley (H-B) model [18–20,26–28]. During the runout process of a submarine landslide, the soil's strength is gradually reduced as the absolute plastic shear strain accumulates. This process continues until a fully remolded condition is achieved. At the same time, the current undrained strength changes with the dynamic shear rate. The current undrained strength,

S_u, at the Gauss points can be expressed in terms of shear strain rate, $\dot{\gamma}$, and accumulated absolute plastic shear strain, ξ:

$$s_u = s_{u0}\alpha\beta = s_{u0}\left[\delta_{rem} + (1 - \delta_{rem})e^{-3\xi/\xi_{95}}\right]\left\{1 + \eta\left(\frac{\dot{\gamma}}{\dot{\gamma}_{ref}}\right)^n\right\} \tag{2}$$

where δ_{rem} is the inverse of the soil sensitivity, S_t, denoting the ratio between the fully remolded and initial shear strengths; ξ_{95} is the accumulated shear strain when a 95% soil strength degradation occurs between the intact and fully remolded conditions; η is the viscosity coefficient and n is the shear-thinning index; $\dot{\gamma}_{ref}$ is the reference shear strain rate; and α, β are the coefficients of the strain softening and the strain rate dependency, respectively.

The shear strain-softening and rate-dependency model (Equation (2)) was incorporated into ABAQUS based on the Tresca soil constitutive model using the user subroutine VUSDFLD. Because the field variables of Eulerian element EC3D8R cannot be called directly in VUSDFLD, the temperature, as a dummy variable, is called in the VUSDFLD subroutine as an initial state variable to allow a variation in temperature that is identical to the initial undrained shear strength, S_{u0}. This approach has been used in the study of the penetration resistance of bucket foundations and has proven to be effective [35]. In the computational domain, an initial temperature field that is identical to S_{u0} is provided by an "*initial conditions" keyword, while in the void domain, the temperature is zero. The accumulated absolute plastic shear strain, ξ, in Equation (2) is given by the sum of the accumulated absolute plastic shear strain, $\Delta\varepsilon_{p1} - \Delta\varepsilon_{p3}$. $\Delta\varepsilon_{p1}$ and $\Delta\varepsilon_{p3}$ are the cumulative major and minor plastic shear strains, respectively, in each increment (dt). The shear strain rate, $\dot{\gamma}$, in Equation (1) can be computed using Equation (3):

$$\dot{\gamma} = \frac{(\Delta\varepsilon_1 - \Delta\varepsilon_3)}{\Delta t} \tag{3}$$

where $\Delta\varepsilon_1$ and $\Delta\varepsilon_3$ are the cumulative major and minor principal strains over the duration of Δt. All involved strain components and temperature dummy variables can be obtained using the utility routine VGETVRM. The variations in ξ and $\dot{\gamma}$ can be captured through simple algebraic operations with the help of state variables.

2.3. Approach Validation

A flume experiment (Case 1) and a numerical model (Case 2) are chosen for the validation of the proposed approach.

A laboratory study of slurry runout induced by a dam break was carried out by Wright and Krone [36] by instantaneously releasing a bentonite slurry from an upstream reservoir into a rectangular flume with a constant slope of 3.44°. The initial length of the slurry was 1.8 m; the initial height was 0.3 m; the yield strength s_{u0} was 42.5 Pa; Young's modulus was taken as $100s_{u0}$; the slurry density was 1073 kg/m³; and the reference strain rate was 193.2 s⁻¹, which can be expressed by the yield strength divided by the dynamic viscosity. The base of the flume was assumed to be a no-slip boundary. The model parameters for Equation (2) in Case 1 are shown in Table 1. The element sizes of 0.005 m, 0.004 m, and 0.0025 m show almost the same results, so the element size 0.005 m with the time step $t = 3.3 \times 10^{-4}$ s can satisfy the computing requirement.

Table 1. Model parameters of simulation cases.

Case	$\theta/(°)$	S_{u0}/kPa	η	n	$\dot{\gamma}_{ref}$	S_t	ξ_{95}	τ/kPa	$V_0/m.s^{-1}$	$\gamma'/kN.m^{-3}$
1	3.44	0.0425	1	1	193.2	1	-	0.0425	0	10.73
2	2	5	0	-	-	5	10	1.0	1.0	6.0

Figure 3 shows that the runout results predicted by the proposed approach agreed reasonably well with the experimental observation and the numerical results of MPM and

CFD [18]. The divergence is 3.3% compared with the experimental data and 8.6% compared with the CFD results.

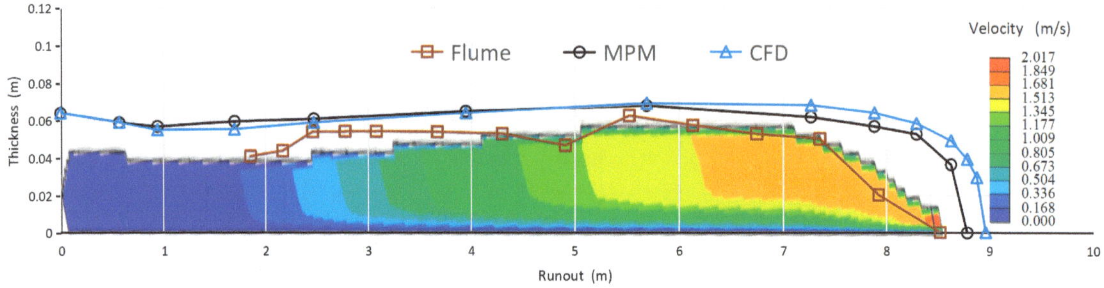

Figure 3. Runout of Case 1 at t = 4.1 s.

A simple designed submarine landslide case, which has also been used in other studies [18–20], is adopted here for model validation (Case 2). The simulation results are then compared with those of the RITSS method [18]. The geometry of the submarine landslide (showed in Figure 4) is trapezoidal and is preset on a slope with an inclination angle $\theta = 30°$. The height and base width of the trapezoidal landslide are 5 m and 48.66 m, respectively. The seabed is modeled as a rigid body. The contact between the seabed and sliding soil is implemented as "general contact," with a frictional strength, τ, on the seabed–sliding mass interface. Young's modulus was taken as $500s_{u0}$. The soil strength parameters of the landslides used in the validation model are shown in Table 1. The element sizes of 0.25 m, 0.2 m, and 0.15 m show almost the same results, so the 0.25 m element size with the time step $t = 3.86 \times 10^{-4}$ s can satisfy the computing requirements. The comparison of the results of the two case studies is shown in Figure 5. The simulated runout basically presents a similar trend to the RITSS method, with a maximum divergence of 10.7% at the final runout distance. Figure 6 shows the softened effect of soil due to the shear disturbance during the sliding process. The softening ratio distribution is also consistent with that of the RITSS method. The inclusion of soil softening causes wedges to form at a 45-degree inclination, with the shearing becoming concentrated within the weakened soil between the wedges. Both Case 1 and Case 2 indicate that the proposed method can provide reasonable simulation results.

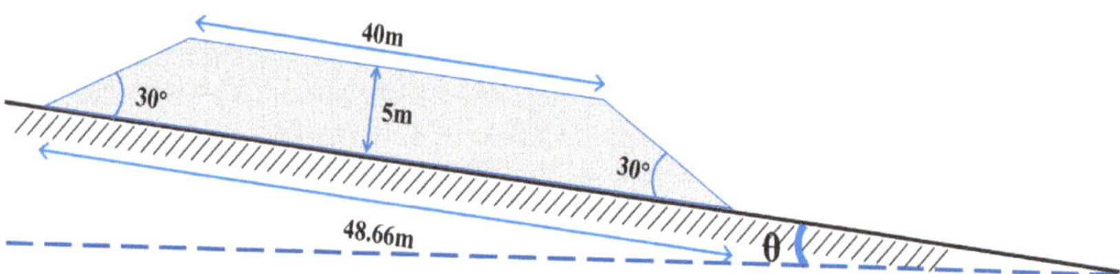

Figure 4. Schematic diagram of landslide geometric model.

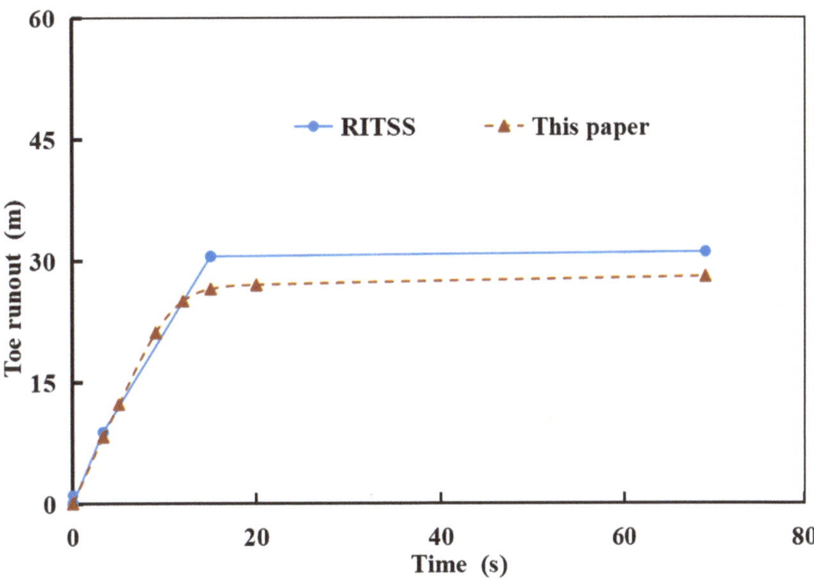

Figure 5. Runout of the front toe at different time points.

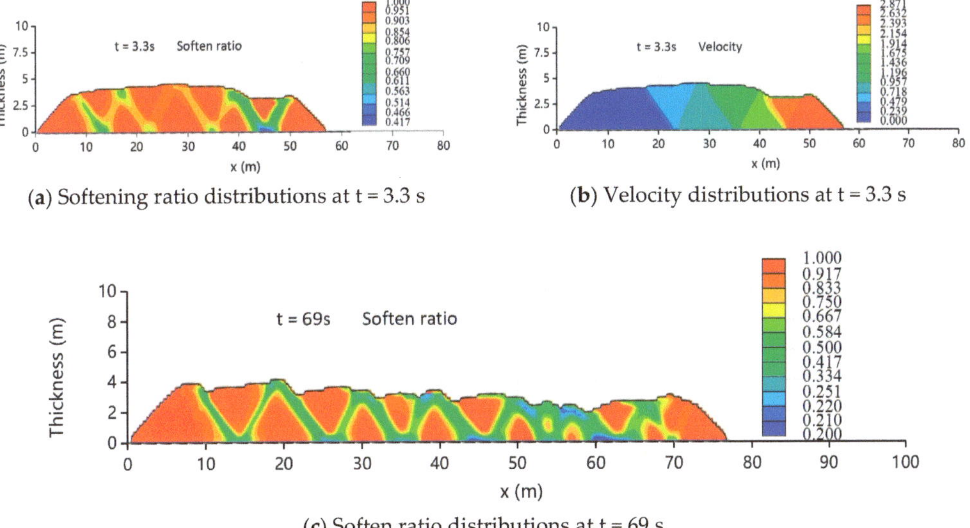

(**a**) Softening ratio distributions at t = 3.3 s   (**b**) Velocity distributions at t = 3.3 s

(**c**) Soften ratio distributions at t = 69 s

Figure 6. Typical variable distribution of Case 2 at different time points.

3. Numerical Modeling of Submarine Landslide in Shenhu Sea Area

3.1. Engineering Geological Conditions

The Shenhu sea area, which is on the northern slope of the South China Sea, is not only rich in oil and gas resources [37] but also prone to submarine landslides [38–40]. A great number of small submarine landslides have been identified in this area [41,42]. The great potential for submarine landslides would pose a great threat to the safety of undersea pipelines across this area, as shown in Figure 7. Multibeam topography has shown that the heights of the landslide scarps, which are mainly located at the sidewall and the interfluve of the canyon, are basically between 8 and 125 m, and the maximum local slope can reach

26.6° [43]. The soil in the Shenhu sea area has medium or high soil sensitivity based on the drilling test and the in situ CPT test [44,45], and the undrained shear strength exhibits linear and segmented linear growth with increasing burial depth. The intact undrained shear strength of the soil, S_{u0}, can be expressed as

$$S_{u0} = S_{um} + kz \tag{4}$$

where S_{um} is the shear strength at the mudline, k is the shear strength gradient, and z is the burial depth.

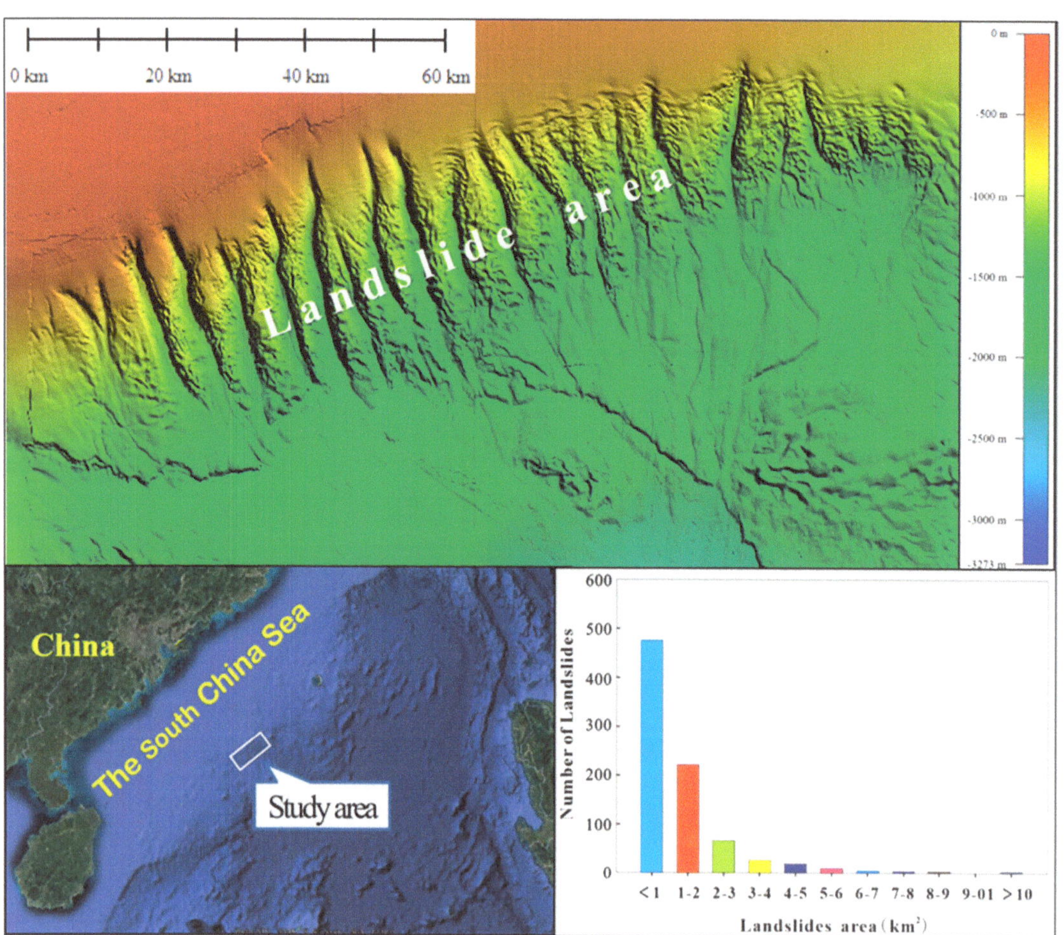

Figure 7. Multibeam topography and submarine landslides in the Shenhu sea area.

3.2. Numerical Modeling of Typical Submarine Landslides

According to the geological engineering background of the Shenhu sea area, a typical submarine landslide model is chosen for runout analysis. The terrain includes horizontal sections, local steep slope sections, and gentle slope sections. A 6.5° gentle slope is used to simulate the regional terrain slope. Local steep slopes are generally located at the landslide scarp. Here, a 30° slope to the horizontal (23.5° compared to the 6.5° seabed), which is close to the maximum local slope of 26.6°, is used to simulate a slope erosion trigger mechanism. The initial submarine slope adopted here is actually in an unstable state (i.e., it has a slope stability coefficient less than 1.0), since otherwise, a submarine landslide will not occur. Thus, the stability analysis of the submarine slope has not been performed here. A schematic diagram of the submarine landslide model is presented in Figure 8.

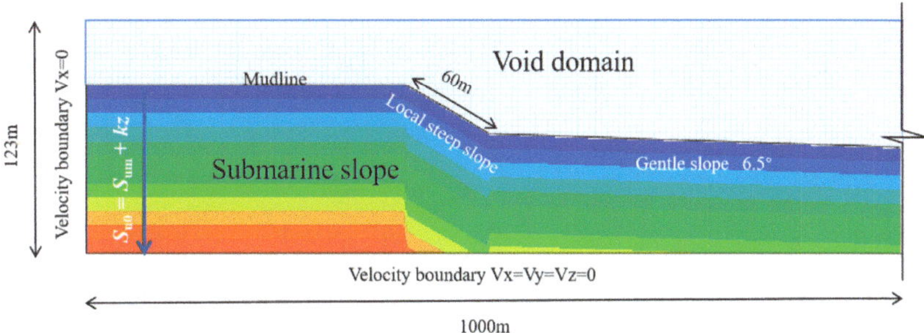

Figure 8. Typical submarine landslide model.

The Eulerian element size is 1 m $\times$ 1 m $\times$ 1 m, with the time step t = 4.0 $\times$ 10^{-4} s. The intact undrained strength at the mudline is S_{um} = 2~5 kPa; the shear strength gradient is k = 1.15~1.25 kPa/m; the soil sensitivity, S_t, is chosen as 2.0, 3.0, 4.0, and 5.0, corresponding to medium sensitivity to extra-high sensitivity, respectively; the elastic modulus is E = 500 S_u; and Poisson's ratio is υ = 0.49. The shear-softening and rate-dependency parameters are η = 0.3, n = 0.5, and ξ_{95} = 10.0. A soil density ρ_s = 600 kg/m^3 (ρ_s = Saturation density – water density) is used in the model, so the weight of the soil is actually the submerged weight. In this study, we conducted runout simulations of various cases based on the typical submarine landslide model. The following section presents a detailed analysis and discussion of the results obtained.

4. Results and Discussion

The initial in situ stress distribution in the seabed may have an impact on a landslide's mass movement process. In this paper, the in situ stress field under self-weight is taken as the initial stress. In order to illustrate this, two cases with and without in situ stress are first calculated. Figures 9 and 10 show the Tresca stress and velocity distributions at different time points, respectively, in the case of S_{um} = 5 kPa, K = 1.15 kPa/m, and S_t = 4.0. It can be seen that both cases show a shallow landslide failure mode, which is consistent with the stability analysis of the submarine slope and the geophysical interpretation of existing landslide characteristics [16,40–42,45]. In the case that the initial stress has not been taken into account, the self-weight load can lead to large fluctuations in the stress field. The additional stress fluctuation in turn has a negative effect on the accuracy of the velocity field. As shown in Figure 10, the velocity of the landslide in the early stage (for example, t = 1 s) is higher than when initial stress is considered. This is because the gravitational load poses an overall dynamic vertical additional velocity at early stages. However, the velocity decreases quickly over time due to the subsequent stress oscillations, eventually leading to a rather smaller landslide runout distance. Therefore, in the subsequent calculations, an initial stress field is adopted to balance the influence of the gravitational load.

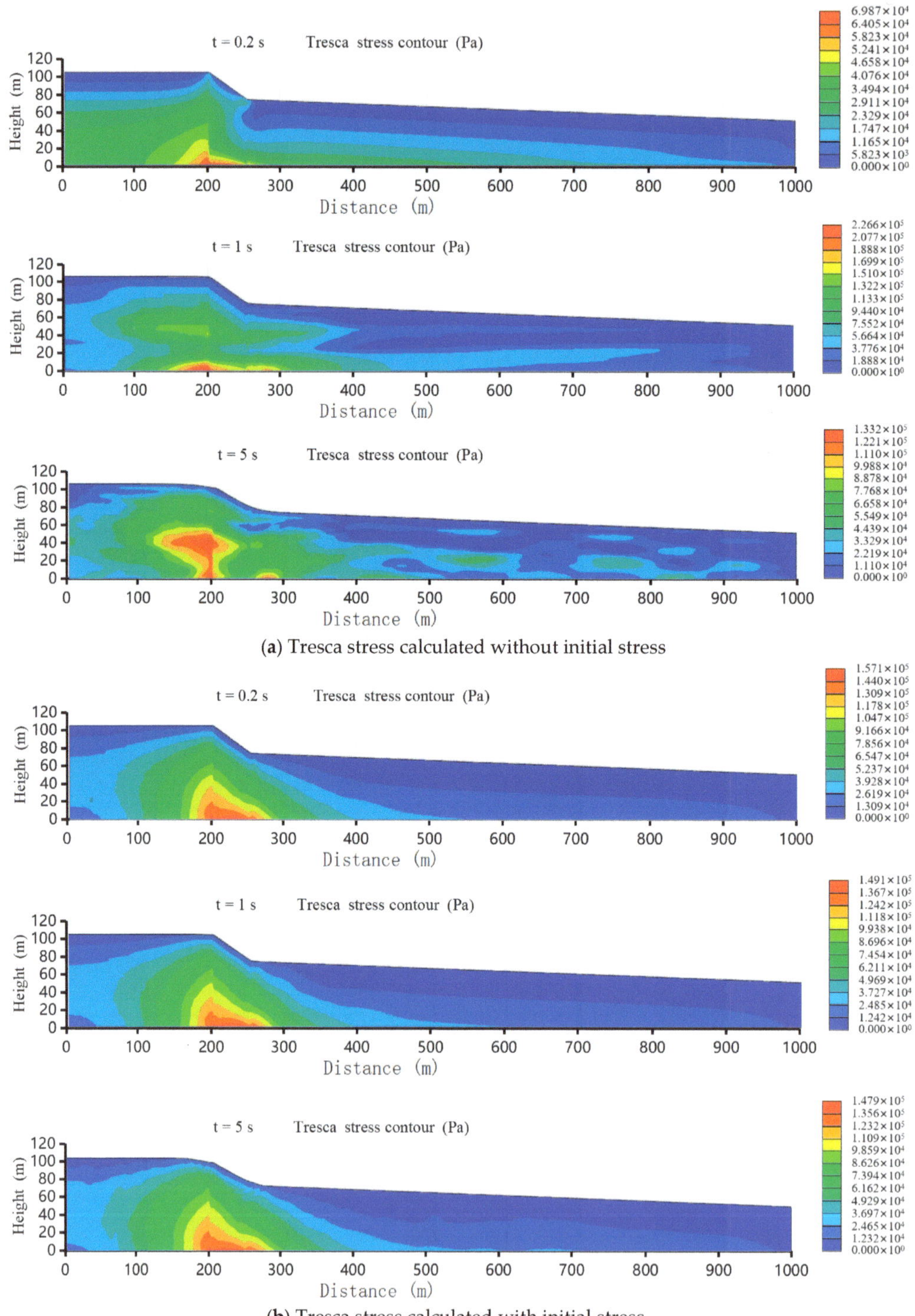

(**a**) Tresca stress calculated without initial stress

(**b**) Tresca stress calculated with initial stress

Figure 9. Tresca stress during the landslide process.

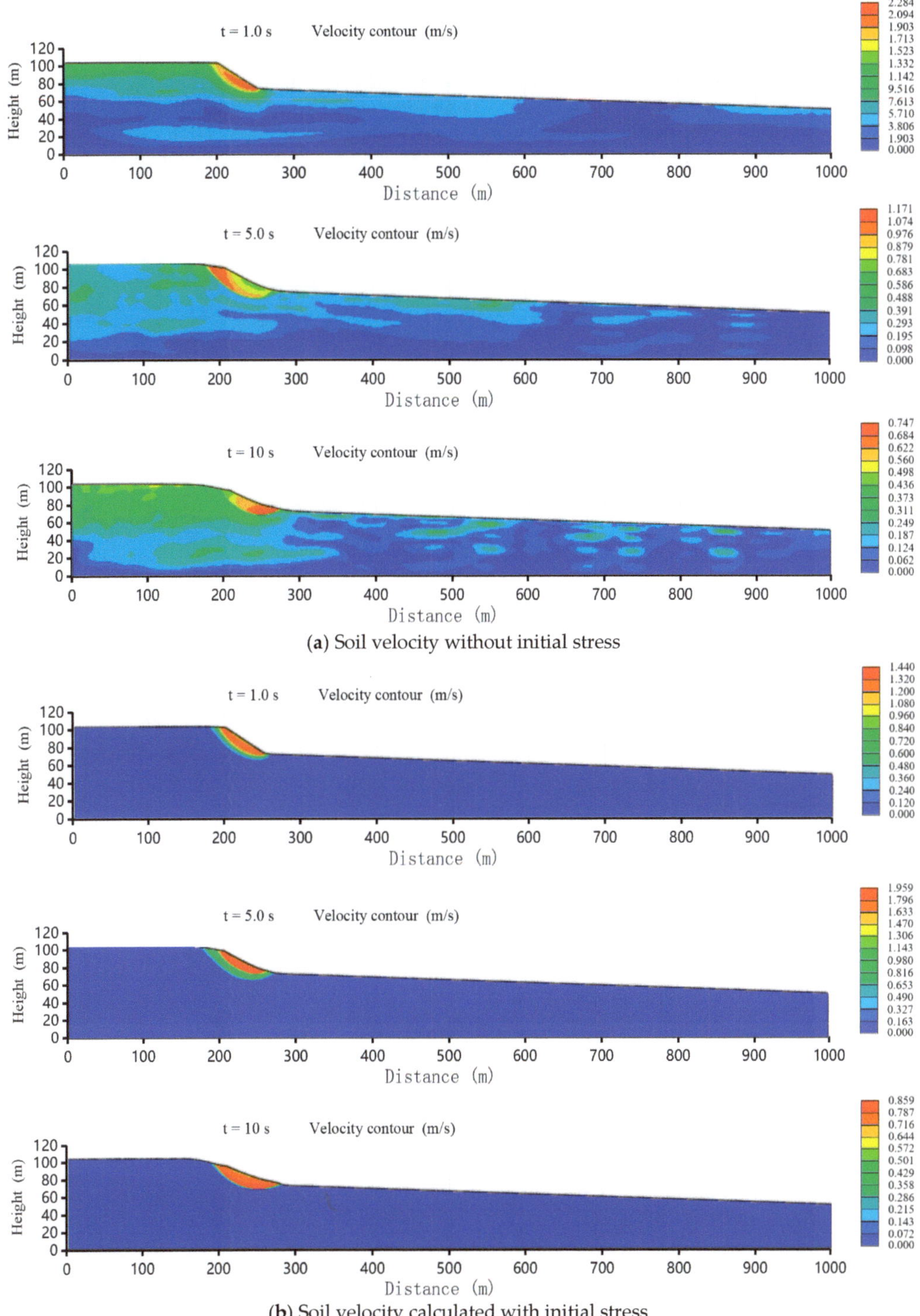

(**a**) Soil velocity without initial stress

(**b**) Soil velocity calculated with initial stress

Figure 10. Soil velocity during the landslide process.

Figure 11 shows the equivalent plastic strain during the landslide process for the cases both with and without the strain-softening and rate-dependency effects. It can be seen that when the strain-softening and rate-dependency effects are incorporated, a distinct difference can be seen in the submarine landslide runout process, although both cases have almost the same initial landslide sliding surface depth (about 25 m). The shear zone presents similar features in the early stage (for example, t = 1 s). As time goes on, the strain-softening and rate-dependency effects gradually come into play, corresponding to the accumulation of absolute plastic shear strain and the shear strain rate. The sliding soil without the strain-softening and rate-dependency effects decelerates quickly after sliding down the local steep slope, and a much thicker deposit can be found. In contrast, the sliding soil with the strain-softening and rate-dependency effects shows much better ductility, which means a much larger runout distance. The distance from the slope toe to the front of the final mass transport deposition is 48 m and 166 m, respectively, for the two cases.

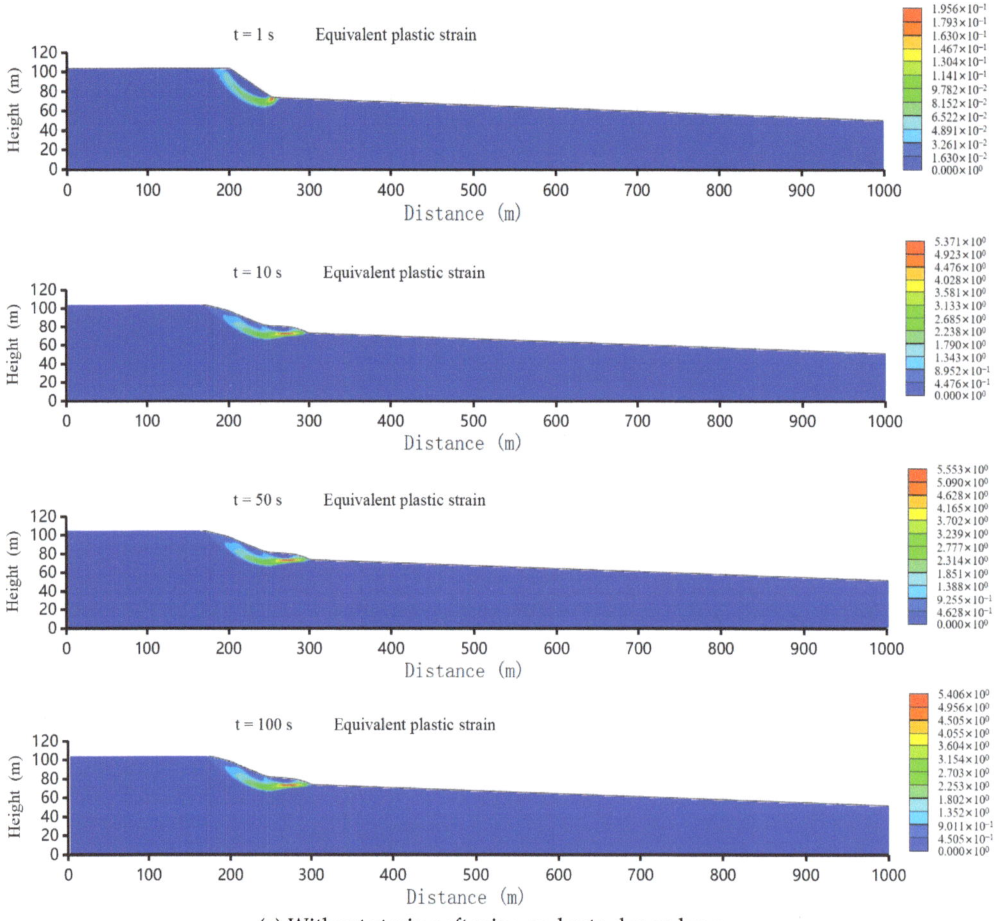

(a) Without strain-softening and rate-dependency

Figure 11. *Cont.*

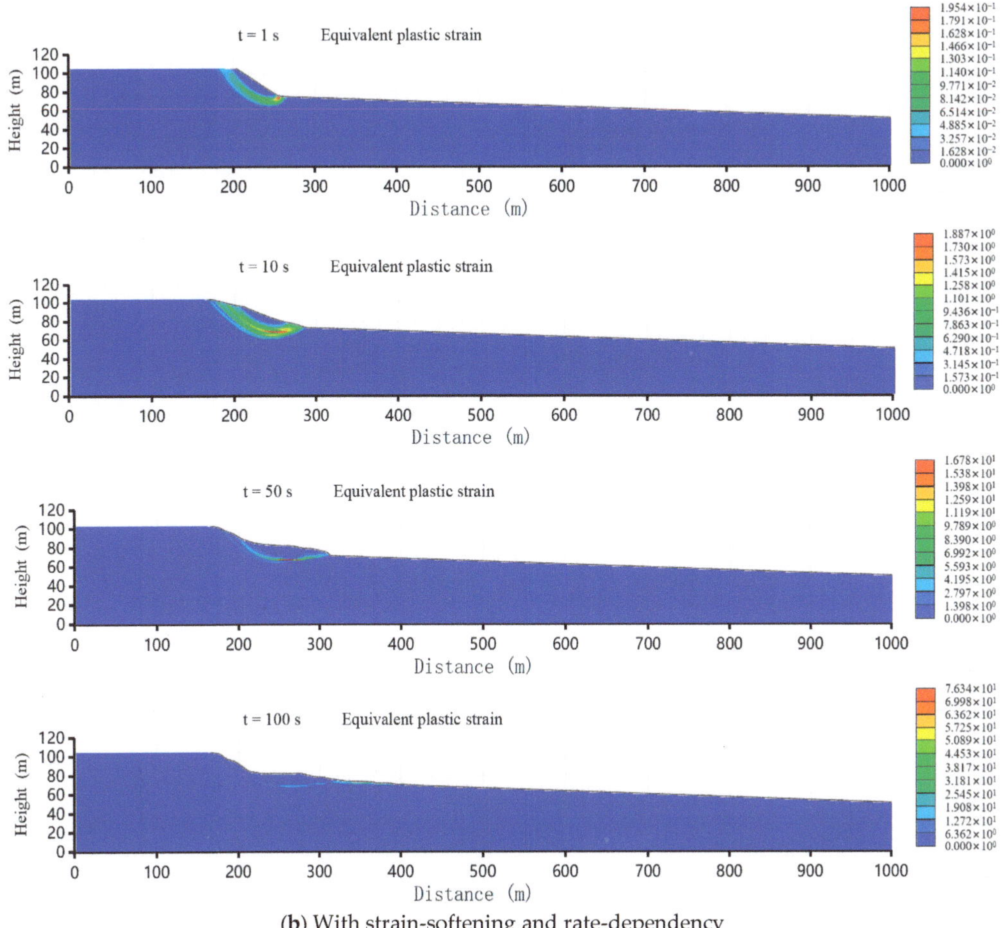

(b) With strain-softening and rate-dependency

Figure 11. Equivalent plastic strain during the landslide process.

The accumulated absolute plastic shear strain induces a continuous downward trend in the soil's shear strength (strain-softening), while the shear strain rate has a positive correlation with the shear strength. The shear strength of the sliding soil and the strain shear rate effect are shown in Figures 12 and 13, respectively. It can be seen that S_u in the shear zone changes gradually with the shear remolding and shear strain rate-dependency. At the early stage of the landslide process, the soil accelerates at a high velocity when sliding down the local steep slope. At the same time, the accumulated absolute plastic shear strain of the soil is not large enough, and, therefore, a local high shear strength occurs in the shear zone due to the combination of a higher rate-dependency effect and a lower strain-softening effect. This also leads to a relatively small sliding velocity when the shear layer has not been significantly remolded. At the middle to late stages of the sliding process, the strain-softening effect caused by the shear disturbance gradually increases.

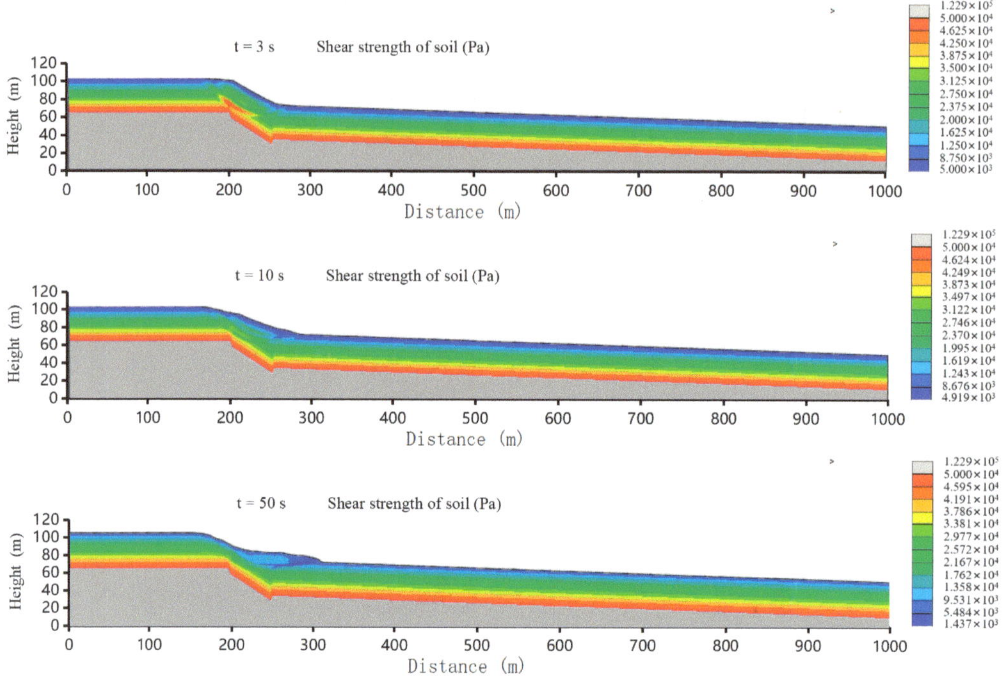

Figure 12. Shear strength distribution at different time points.

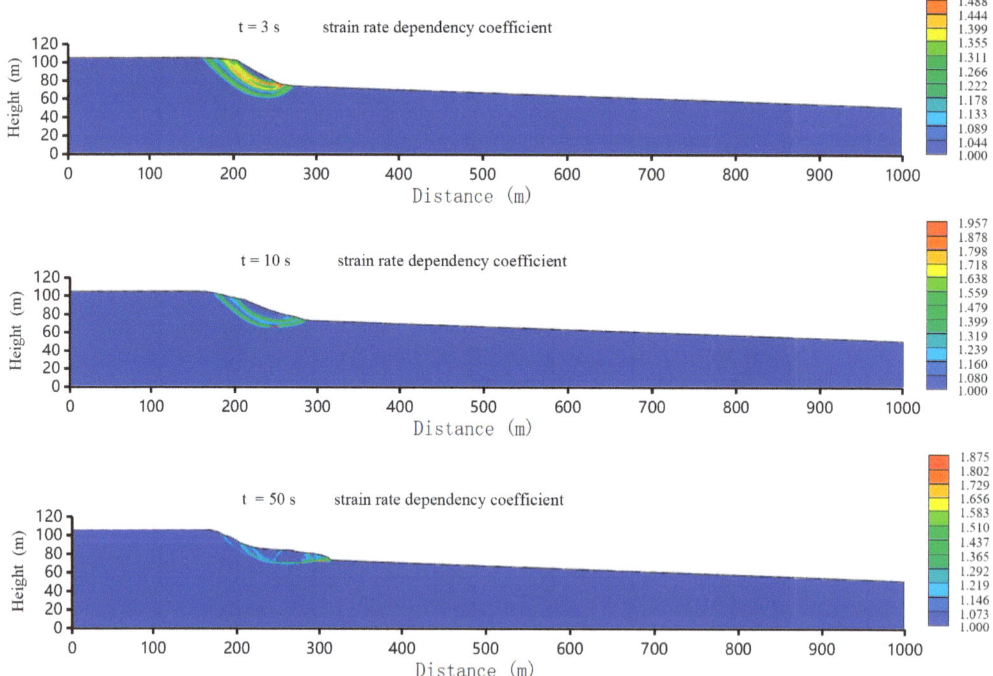

Figure 13. Strain shear rate at different time points.

The shear strength distribution of the soil and the slope angle are the main controlling factors of the mass movement of a submarine landslide. The effect of the shear strength distribution on the landslide process is analyzed by comparison of different combinations of S_{um} and k, which is described by Equation (3), when the initial geometric configuration is the same. This is an approximate equivalence of submarine areas that have similar slope angles but different sediments. Figures 14 and 15 show the shapes of the sliding soils at different time points. In Figure 14, the soil has the same mudline shear strength, $S_{um} = 2$ kPa, and three different shear strength gradients k are adopted. It can be seen that at a larger k, the landslide has a relatively shallow slide surface depth, as well as a relatively small runout distance at the time points of 15s and 30s. In Figure 15, the landslide has the same $k = 1.15$ kPa/m but a different S_{um}. The differences in the initial slide surfaces compared with those in Figure 14 are not obvious for the three cases of S_{um}. This is because the S_{um} in the cases in Figure 15 have a small range of variation, with the maximum difference being 3 kPa. Meanwhile, in Figure 14, the variation in k can result in a relatively larger range of variation in soil shear strength. For instance, k changing from 1.0 kPa/m to 1.25 kPa/m leads to a difference in shear strength from 2.5 kPa to 7.5 kPa linearly from a depth of 10 m to 30 m.

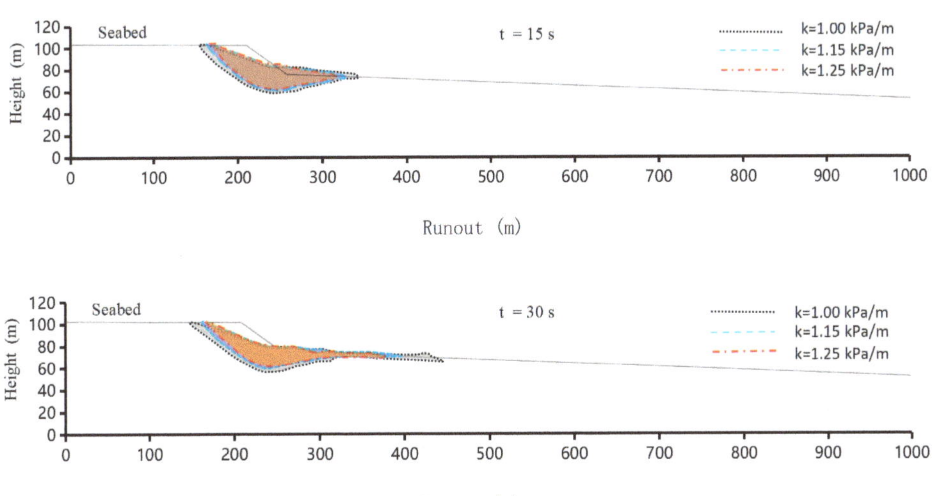

Figure 14. Runout morphologies with $S_{um} = 2$ kPa, St = 4.0.

According to the simulation results of all the above cases, the sliding soil first accelerates along the local steep slope before a large proportion of the sliding soil deposits rapidly on the seabed of the lower slope near the slope's toe due to the friction resistance of the seabed surface. The higher the soil's shear strength is, the greater the proportion of the sliding soil deposited near the slope's toe. When the shear strength of the upper sliding soil is low, it can still rapidly slide forward under the inertial effect. For example, in case $S_{um} = 2$ kpa and k = 1.15 kPa/m, after sliding down the steep slope, the upper soil with lower shear strength keeps sliding with a relatively high velocity. The maximum velocity occurs at the front part of the sliding soil, characterized by the elongation of the length of the slide and the subsequent reduction in thickness. During the elongation process, the shear-softening degree of the soil is continuously strengthened as the shear disturbance accumulates. This further increases the deformability of the soil, causing the upper soil to continue sliding on the low slope seabed. As can be seen from Figure 16, soils with a lower shear strength show a rapid sliding process with larger runout distance, while soils with a higher shear strength show a slow sliding process with a smaller runout distance.

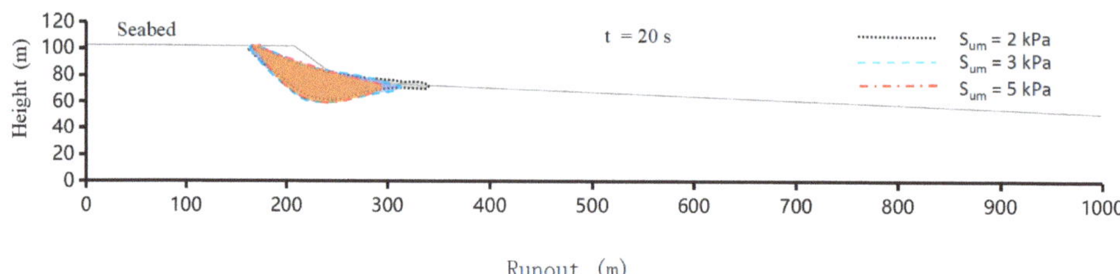

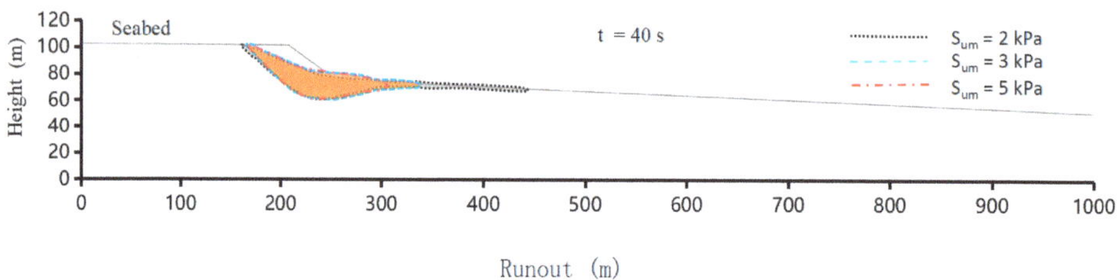

Figure 15. Runout morphologies with k = 1.15 kPa/m, S_t = 4.0.

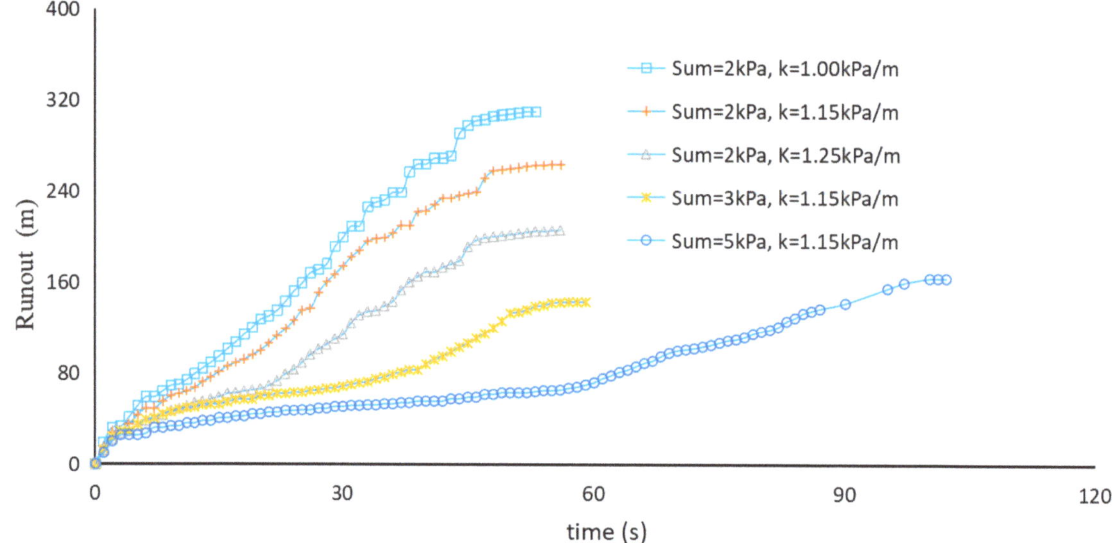

Figure 16. Runout of different Sum and k.

Several combinations of parameters pertaining to shear softening and rate dependency, such as S_t ranging from 2.0 to 5.0, η from 0.3 to 0.5, and n from 0.3 to 0.5, are also calculated for comparison. However, the differences in numerical values between the combinations used here are not significant in our landslide model. However, this does not mean that the effects of shear softening and low rate dependency are not important in the runout process. The model used here is just for the runout of small landslides, and the mass movement tends to be of the slump type [41,42], which is also a common type in canyon head areas and terraces on canyon flanks. The typical mass transport deposits are shown in Figure 17.

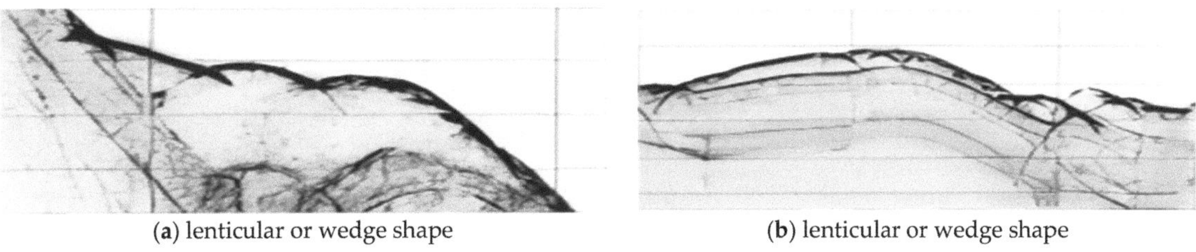

(**a**) lenticular or wedge shape   (**b**) lenticular or wedge shape

Figure 17. Slump-type submarine landslides derived from SBP data.

It should be pointed out that a natural Eulerian material interface is adopted here for the interaction between the sliding soil and seabed. This is different from the "general contact" method [18–20], in which a frictional strength, τ, is used between the rigid seabed and sliding soil. The "general contact" method can provide a clear measure of friction resistance using common friction models. However, it does not involve the deformation of the seabed induced by the sliding soil. Therefore, an Eulerian-to-Eulerian material interface is more suitable for this situation. The friction resistance is calculated based on both the shear strength of the sliding soil and the seabed topsoil, since Abaqus adopts a single strain field when soils with different shear strengths exist within the same element. The two methods each have their own strengths and weaknesses. A method reaching a compromise between the methods studied here can be considered to obtain a better simulation result, for example, by applying the two methods in different parts of the model. As this is not the focus of this study, we will not discuss this idea in detail here.

5. Conclusions

The mass movements of submarine landslides pose a great threat to undersea pipelines, cables, and other facilities. A convenient method for simulating submarine landslide processes is proposed in this work. The convenient model is established by using Eulerian large deformation analysis technology in the Abaqus/Explicit finite element framework. The strain-softening and rate-dependency characteristics of the soil shear strength are taken into account in the simple user subroutine VUSDFLD, employing the help of a dummy temperature field variable and several state variables.

A flume experiment and a simple submarine landslide case are used to validate the proposed model. The simulation results of this paper demonstrate good consistency with those of the flume experiment and other widely validated numerical methods. Focusing on the potential mass movements of submarine landslides in the Shenhu sea area on the northern slope of the South China Sea, the runout process is analyzed under different combinations of soil parameters cases by using the convenient method proposed. Typical examples indicated that the landslide failure model in the Shenhu sea area can mainly be manifested as shallow landslides, which is consistent with the stability analysis of the slope in the region and the geophysical interpretation and identification results of existing landslide characteristics.

The proposed convenient method poses relatively low technical requirements compared to the MPM and RITSS methods. It can help the engineering community to easily determine the mass movement of submarine landslides. The interaction between the sliding soil and the seabed can be simulated by using the "general contact" method or an Eulerian-to-Eulerian material interface according to the physical demands. Multiple materials are allowed in Eulerian elements, which gives the proposed method further potential to model the interaction between landslide and seawater. However, Abaqus adopts a single strain field when materials with different shear strengths exist within a single element. This means that no slip would occur on the two-phase interface, which could cause some deviation in the case of obvious interface sliding. In addition, Eulerian elements cannot provide material spatial displacement directly, and materials just flow through the elements with nodes fixed in space. Therefore, the runout distance is obtained using an additional calculation based on the original grid position and the updated spatial shape of the slidi"g so'l. This can lead to a certain degree of error. To improve the calculation accuracy, a small element size Is needed. However, the incremental time step would then reduce with the decrease in the element size. The element size should therefore be considered to balance the accuracy and computational cost of the solution.

Author Contributions: Conceptualization, Q.X. (Qiuhong Xie) and Z.X.; methodology, Z.X. and L.L.; software, Q.X. (Qiang Xu), Z.X., and J.Y.; validation, Z.X., L.L. and J.Y.; formal analysis, Q.X. (Qiuhong Xie) and H.L.; data curation, J.Y. and X.D.; writing—original draft preparation, Q.X. (Qiuhong Xie) and Z.X.; writing—review and editing, Q.X. (Qiuhong Xie) and Z.X.; visualization, H.L. and X.D.; supervision, Q.X. (Qiang Xu) and L.L.; funding acquisition, Z.X. and Q.X. (Qiang Xu) All authors have read and agreed to the published version of the manuscript.

Funding: This research was funded by the National Natural Science Foundation of China (grant number 41876066; 41606084) and the Opening Fund of State Key Laboratory of Geohazard Prevention and Geoenvironment Protection (Chengdu University of Technology) (grant number SKLGP2019K023).

Institutional Review Board Statement: No applicable.

Informed Consent Statement: No applicable.

Data Availability Statement: Data are contained within the article.

Conflicts of Interest: Author Zongxiang Xiu was employed by the company First Institute of Oceanography, Ministry of Natural Resources, China. The remaining authors declare that the research was conducted in the absence of any commercial or financial relationships that could be construed as a potential conflict of interest.

References

1. Hance, J.J. *Submarine Slope Stability*; The University of Texas: Austin, TX, USA, 2003.
2. Canals, M.; Lastras, R.G.; Urgeles Casamora, J.L.; Mienert, J.; Cattaneo, A.; De Batist, M.; Haflidason, H.; Imbo, Y.; Laberg, J.S.; Locat, J.; et al. Slope failure dynamics and impacts from seafloor and shallow sub-seafloor geophysical data: Case studies from the COSTA project. *Mar. Geol.* **2004**, *213*, 9–72. [CrossRef]
3. Liu, X.; Wang, Y.; Zhang, H.; Guo, X. Susceptibility of typical marine geological disasters: An overview. *Geoenviron. Disasters* **2023**, *10*, 1–31. [CrossRef]
4. Mosher, D.C.; Moscardelli, L.; Shipp, R.C.; Chaytor, J.D.; Baxter, C.D.P.; Lee, H.J.; Urgeles, R. Submarine mass movements and their consequences. In *Submarine Mass Movements and Their Consequences*; Mosher, D.C., Shipp, R.C., Moscardelli, L., Chaytor, J.D., Baxter, C.D.P., Lee, H.J., Urglese, R., Eds.; Springer: Berlin, Germany, 2010; pp. 1–8.
5. Guo, X.; Liu, X.; Li, M.; Lu, Y. Lateral force on buried pipeline scaused by seabed slides using a CFD method with a shear interface weakening model. *Ocean Eng.* **2023**, *280*, 114663. [CrossRef]
6. Sun, Y.; Huang, B. A Potential Tsunami impact assessment of submarine landslide at Baiyun Depression in Northern South China Sea. *Geoenviron. Disasters* **2014**, *1*, 1–7.
7. Sun, Q.L.; Wang, Q.; Shi, F.Y.; Alves, T.; Gao, S.; Xie, X.N.; Wu, S.G.; Li, J.B. Runup of landslide-generated tsunamis controlled by paleogeography and sea-level change. *Commun. Earth Environ.* **2022**, *3*, 244. [CrossRef]
8. Vanneste, M.; Sultan, N.; Garziglia, S.; Forsberg, C.F.; L'Heureux, J.-S. Seafloor instabilities and sediment deformation processes: The need for integrated, multi-disciplinary investigations. *Mar. Geol.* **2014**, *352*, 183–214. [CrossRef]
9. Locat, J.; Lee, H.J. Submarine landslides: Advances and challenges. *Can. Geotech. J.* **2002**, *39*, 193–212. [CrossRef]
10. Harbitz, C.B.; Parker, G.; Elverhøi, A.; Marr, J.G.; Mohrig, D.; Harff, P.A. Hydroplaning of subaqueous debris flows and glide blocks: Analytical solutions and discussion. *J. Geophys. Res.* **2003**, *108*, 2349. [CrossRef]
11. Bradshaw, A.S.; Tappin, D.R.; Rugg, D. The kinematics of a debris avalanche on the sumatra margin. In *Submarine Mass Movements and Their Consequences*; Mosher, D.C., Shipp, R.C., Moscardelli, L., Chaytor, J.D., Baxter, C.D.P., Lee, H.J., Urglese, R., Eds.; Springer: Berlin, Germany, 2010; pp. 117–125.
12. Imran, J.; Harff, P.; Parker, G. A numerical model of submarine debris flow with graphical user interface. *Comput. Geosci.* **2001**, *27*, 717–729. [CrossRef]
13. De Blasio, F.V.; Engvik, L.; Harbitz, C.B.; Elverhøi, A. Hydroplaning and submarine debris flows. *J. Geophys. Res.* **2004**, *109*, C01002.1–C01002.15. [CrossRef]
14. Gauer, P.; Elverhøi, A.; Issler, D.; De Blasio, F.V. On numerical simulations of subaqueous slides: Back-calculations of laboratory experiments. *Nor. J. Geol.* **2006**, *86*, 295–300.

15. Zakeri, A.; Høeg, K.; Nadim, F. Submarine debris flow impact on pipelines—Part II: Numerical analysis. *Coast. Eng.* **2009**, *56*, 1–10. [CrossRef]

16. Xiu, Z.X.; Liu, L.J.; Xie, Q.H.; Li, J.G.; Hu, G.H.; Yang, J.H. Runout prediction and dynamic characteristic analysis of potential submarine landslide in Liwan 3-1 gas field. *Acta Oceanol. Sin.* **2015**, *34*, 116–122. [CrossRef]

17. Xie, Q.H.; Xiu, Z.X.; Liu, L.J.; Li, X.S.; Li, J.G.; Hu, G.H.; Zhao, Y. Back analysis of large-scale submarine landslides in the northwest waters of Sumatra Island. *Eng. Mech.* **2006**, *33*, 241–247.

18. Wang, D.; Randolph, M.F.; White, D.J. A dynamic large deformation finite elementmethod based on mesh regeneration. *Comput. Geotech.* **2013**, *54*, 192–201. [CrossRef]

19. Dong, Y.K.; Wang, D.; Randolph, M.F. Runout of submarine landslide simulated with material point method. *J. Hydrodyn.* **2017**, *29*, 438–444. [CrossRef]

20. Dong, Y.; Wang, D.; Cui, L. Assessment of depth-averaged method in analysing runout of submarine landslide. *Landslides* **2020**, *17*, 543–555. [CrossRef]

21. Onyelowe, K.C.; Sujatha, E.R.; Aneke, F.I.; Ebid, A.M. Solving geophysical flow problems in Luxembourg: SPH constitutive review. *Cogent Eng.* **2022**, *9*, 2122158.1–2122158.15. [CrossRef]

22. Dai, Z.L.; Li, X.F.; Lan, B.S. Three-Dimensional Modeling of Tsunami Waves Triggered by Submarine Landslides Based on the Smoothed Particle Hydrodynamics Method. *J. Mar. Sci. Eng.* **2023**, *11*, 2015. [CrossRef]

23. Jiang, M.J.; Sun, C.; Crosta, G.B.; Zhang, W.C. A study of submarine steep slope failures triggered by thermal dissociation of methane hydrates using a coupled CFD-DEM approach. *Eng. Geol.* **2015**, *190*, 1–16. [CrossRef]

24. Nian, T.K.; Zhang, F.; Zheng, D.F.; Li, D.Y.; Shen, Y.Q.; Lei, D.Y. Numerical simulation on the movement behavior of viscous submarine landslide based on coupled CFD-DEM method. *Rock Soil Mech.* **2022**, *43*, 3174–3184.

25. Pastor, M.; Haddad, B.; Sorbino, G.; Cuomo, S.; Drempetic, V. A depth-integrated, coupled SPH model for flow-like landslides and related phenomena. *Int. J. Numer. Anal Methods Geomech.* **2009**, *33*, 143–172. [CrossRef]

26. Einav, I.; Randolph, M.F. Combining upper bound and strain path methods for evaluating penetration resistance. *Int. J. Numer. Methods Eng.* **2005**, *63*, 1991–2016. [CrossRef]

27. Zhou, H.; Randolph, M.F. Computational techniques and shear band development for cylindrical and spherical penetrometers in strain-softening clay. *Int. J. Geomech.* **2007**, *7*, 287–295. [CrossRef]

28. Boukpeti, N.; White, D.; Randolph, M.; Low, H. Strength of fine-grained soils at the solid–fluid transition. *Géotechnique* **2012**, *62*, 213–226. [CrossRef]

29. Shan, Z.G.; Zhang, W.C.; Wang, D.; Wang, L.Z. Numerical investigations of retrogressive failure in sensitive clays: Revisiting 1994 Sainte-Monique slide, Quebec. *Landslides* **2021**, *18*, 1327–1336. [CrossRef]

30. Hu, Y.X.; Randolph, M.F. A practical numerical approach for large deformation problems in soil. *Int. J. Numer. Anal. Methods Géoméch.* **1998**, *22*, 327–350. [CrossRef]

31. Harlow, F.H. The particle-in-cell computing method for fluid dynamics. *Methods Comput. Phys.* **1964**, *3*, 319–343.

32. Dassault Systèmes. *Abaqus Analysis Users' Manual*; Simula Corp: Providence, RI, USA, 2010.

33. Qiu, G.; Henke, S.; Grabe, J. Application of a Coupled Eulerian–Lagrangian approach on geomechanical problems involving large deformations. *Comput. Geotech.* **2011**, *38*, 30–39. [CrossRef]

34. Tho, K.K.; Leung, C.F.; Chow, Y.K.; Swaddiwudhipong, S. Eulerian finite element simulation of spudcan -pile interaction. *Can. Geotech. J.* **2013**, *50*, 595–608. [CrossRef]

35. Xiao, Z.; Fu, D.F.; Zhou, Z.F.; Lu, Y.M.; Yan, Y. Effects of strain softening on the penetration resistance of offshore bucket foundation in nonhomogeneous clay. *Ocean. Eng.* **2019**, *193*, 106594.1–106594.16. [CrossRef]

36. Wright, V.G.; Krone, R.B. Laboratory study of mud flows. In Proceedings of the National Conference on Hydraulic Engineering, New York, NY, USA, 3–7 August 1987; pp. 237–242.

37. Wu, N.Y.; Zhang, G.X.; Liang, J.Q.; Su, Z.; Wu, D.D.; Lu, H.L.; Lu, J.A.; Sha, Z.B.; Fu, S.Y. Progress of gas hydrate research in Northern South China Sea. *Adv. New Renew. Energy* **2013**, *1*, 80–94.

38. Feng, W.K.; Shi, Y.H.; Chen, L.H. Research for seafoor landslide stability on the outer continental shelf and the upper continental slope in the northern South China Sea. *Mar. Geol. Quat. Geol.* **1994**, *14*, 81–94. (In Chinese)

39. Chen, J.R.; Yang, M.Z. Research on the potential factors for geologic hazards in the South China Sea. *J. Eng. Geol.* **1996**, *4*, 34–39. (In Chinese)

40. Yang, J.H.; Wu, Q.Y.; Zhou, Y. Engineering geological zoning and evaluation along the deep water segment of the pipeline route in LW3-1 gasfled. *China Offshore Oil Gas* **2014**, *26*, 82–87. (In Chinese)

41. Li, X.S.; Liu, L.J.; Li, J.G.; Gao, S.; Zhou, Q.J.; Su, T.Y. Mass movements in small canyons in the northeast of Baiyun deepwater area, north of the South China Sea. *Acta Oceanol. Sin.* **2015**, *34*, 35–42. [CrossRef]

42. Zhou, Q.J.; Li, X.S.; Zhou, H.; Liu, L.J.; Xu, Y.Q.; Gao, S.; Ma, L. Characteristics and genetic analysis of submarine landslides in the northern slope of the South China Sea. *Mar. Geophys. Res.* **2018**, *40*, 303–314. [CrossRef]

43. He, Y.; Zhong, G.; Wang, L.L.; Kuang, Z.G. Characteristics and occurrence of submarine canyon-associated landslides in the middle of the northern continental slope, South China Sea. *Mar. Pet. Geol.* **2014**, *57*, 546–560. [CrossRef]

44. Wang, S.L.; Zhou, S.W.; Yao, S.L.; Shen, Z.M.; Wang, H.G.; Hu, X.M.; Dai, S.J.; Jiang, Z.B.; Zheng, X.Y.; Jiang, B.F.; et al. *Geological Survey of Deep Water Oil and Gas Fields on the Northern Slope of the South China Sea (Liwan 3-1 and Surrounding Areas)—Engineering Geological Survey Results Report Tianjin*; China Oilfield Services Co., Ltd.: Beijing, China, 2015; pp. 17–50.
45. Xiu, Z.X.; Liu, L.J.; Li, X.S.; Xie, Q.H.; Li, J.G.; Hu, G.H. Stability Analysis of Submarine Canyon Slope of Liwan 3-1 Gas Field Pipeline Route. *J. Eng. Geol.* **2016**, *24*, 535–541.

Journal of
Marine Science and Engineering

Article

Prediction Model of the Sound Speed of Seafloor Sediments on the Continental Shelf of the East China Sea Based on Empirical Equations

Guangming Kan [1,2,3], Junjie Lu [1,3], Xiangmei Meng [1,2,3,*], Jingqiang Wang [1,2,3], Linqing Zhang [1,3], Guanbao Li [1,2,3], Baohua Liu [2,3], Qingfeng Hua [1,2,3] and Mujun Chen [4]

[1] Key Laboratory of Marine Geology and Metallogeny, First Institute of Oceanography, Ministry of Natural Resources, Qingdao 266061, China; kgming135@fio.org.cn (G.K.); lujunjie@fio.org.cn (J.L.); wangjqfio@fio.org.cn (J.W.); zlq@fio.org.cn (L.Z.); gbli@fio.org.cn (G.L.); hqf@fio.org.cn (Q.H.)
[2] Laboratory for Marine Geology, Laoshan Laboratory, Qingdao 266237, China; bhliu@ndsc.org.cn
[3] Key Laboratory of Submarine Acoustic Investigation and Application of Qingdao (Preparatory), Qingdao 266061, China
[4] College of Marine Geoscience, Ocean University of China, Qingdao 266100, China; chenmujun@fio.org.cn
* Correspondence: mxmeng@fio.org.cn

Abstract: Based on the data of the acoustic and physical properties of seafloor sediments obtained on the continental shelf area of the East China Sea (ECS), prediction equations of the sediment sound speed based on single and dual parameters were established, and the correlation of the sound speed with physical parameters was discussed. The results show that the sediment sound speed (c) is strongly correlated with water content (w), density (ρ), void ratio (e), mean grain size (M_z) and median grain size (M_d), and the coefficient of determination R^2 of the empirical regression equation is generally greater than 0.80, while the empirical regression equation coefficient of determination R^2 of the compression coefficient (a) and compression modulus (E) are slightly lower, with 0.79 and 0.73, respectively. The coefficients of determination of the dual parameter regression equations between sediment sound speed with physical property parameters are generally higher than those of the single parameter equations, which are all higher than 0.90 indicating better prediction performance. The sensitivity of the physical parameters to the sound speed was analyzed, and the result shows that the sequence of sensitivity from high to low influence on sediment sound speed is void ratio, density, compressibility coefficient, median grain size and mean grain size. The prediction equations established in this study are a good extension and supplement to marine geoacoustic models and is of great significance for obtaining the acoustic properties of the seafloor sediments on the shelf of the East China Sea.

Keywords: seafloor sediments; sound speed; regression equations; single and dual parameter equations; East China Sea

Citation: Kan, G.; Lu, J.; Meng, X.; Wang, J.; Zhang, L.; Li, G.; Liu, B.; Hua, Q.; Chen, M. Prediction Model of the Sound Speed of Seafloor Sediments on the Continental Shelf of the East China Sea Based on Empirical Equations. *J. Mar. Sci. Eng.* **2024**, *12*, 27. https://doi.org/10.3390/jmse12010027

Academic Editors: Timothy S. Collett and George Kontakiotis

Received: 17 October 2023
Revised: 23 November 2023
Accepted: 14 December 2023
Published: 21 December 2023

1. Introduction

The seafloor, as an important boundary of the underwater sound field, is a common concern of various disciplines, including marine acoustics [1,2], oceanography [3], marine geology and marine geophysics [4,5]. The acoustic properties of seafloor sediments have significant practical value in many fields such as marine environmental security guarantees, underwater target detection, submarine engineering survey and seafloor resource exploration [6–11]. Furthermore, the research of seafloor sediment acoustic properties represents the forefront and focal points of marine acoustics, marine geology and marine geophysics. In general, technologies for obtaining the acoustic properties of seafloor sediments mainly include the in situ measurement technique, laboratorial measurement technique, acoustic inversion technique and model-based prediction technique [12–15]. The in situ measurement technique, laboratorial measurement technique and acoustic inversion technique all usually use instruments and equipment to obtain the seafloor sediment acoustic properties on site at sea or in the laboratory. On the other hand, the model-based prediction method usually utilizes previously obtained physical and mechanical properties of seafloor sediments to predict the acoustic properties of the sediment, which is an effective method to obtain the acoustic properties of seafloor sediments with the merit of allowing for rapid and large-scale acquisition of seafloor sediment acoustic properties.

Analyzing the correlation between the acoustic properties and physical characteristics of seafloor sediments and constructing a prediction model is a key step to predicting the acoustic properties of sediments. Some research works have shown the relation between the acoustic properties of sediments and physical parameters and established prediction models for sediment acoustic properties in relevant sea areas [14,16–21]. The correlation between acoustic properties and physical parameters and models for the relationship between sediment acoustic properties and physical characteristics were constructed in many sea areas around the world [14,16,18,22–30]. Bachman [26,31] and other researchers [14,32] analyzed the correlation of the acoustic properties and physical parameters of continental shelves, deep-sea plains, and deep-sea ridges based on sediment environment and sediment type, and established the corresponding prediction equations. Lu Bo et al. [18,27,33], Tang Yonglu [28], and Zou Dapeng et al. [29] studied the acoustic properties of shallow-sea sediments in the Yellow Sea and the northern South China Sea. They established single parameter prediction equations for sediment sound speed based on one certain physical parameters of seafloor sediments in the corresponding research areas and analyzed the relevant prediction curves. Pan Guofu [34] conducted correlation analyses between shear wave speed and sediment physical parameters in the East China Sea continental shelf and Hangzhou Bay, but the sound speeds of longitudinal waves were not involved. Wang Jingqiang et al. [30] constructed single parameter prediction equations for sound speed and sound attenuation coefficients based on in situ measured data, but dual parameter prediction equations were not studied. At present, the prediction models for the acoustic properties of seafloor sediments in the East China Sea are mainly single parameter prediction equations. This paper aims to establish single parameter and dual parameter prediction equations for sediment sound speed to expand the marine geoacoustic model database and present models for seafloor sediment sound speed prediction on the shelf of the East China Sea. This paper also analyzes the sensitivity of the different physical parameters to sediment sound speed and compares the differences between the prediction equations of this research and those established by other researchers.

2. Study Area and Methodology

2.1. Location of Study Area

The study area is located on the shelf of the East China Sea at a water depth of 30 m to 107 m. As one of the widest continental shelves worldwide, it has considerable terrigenous input and is an important area for studying land–ocean interactions and source–sink processes [35]. The deposition of terrigenous sediments in this area is primarily controlled by the coastal upwelling and down welling of the southern continental shelf of the East China Sea [36]. Owing to the inflow of small coastal rivers such as the Yangtze River and the influence of the Yellow River, the continental shelf of the East China Sea has received a high input of terrigenous materials. The sediment types in the study area can be classified into three main regions: the clayey sand subregion in the north, the fine sand subregion in the central area, and clayey silt subregion in the south. The specific sediment types include fine sand, silty sand, clayey sand, coarse silt, sandy clay, sandy silt and clayey silt.

2.2. Data Source

The sediment samples were collected at 44 stations in the study area using box corers or gravity cores. The sound speed and physical mechanical properties of the collected sediment samples were measured at the Seafloor Acoustic Laboratory in First Institute of Oceanography, Ministry of Natural Resources with a measurement frequency of 100 kHz, using the longitudinal transmission method. The measurement equipment includes a seafloor sediment cylindrical sample sound speed measurement platform, an integrated sediment acoustic measurement system and acoustic transducers [15]. The formula for calculating seafloor sediment sound speed is as follows:

$$c = L/(t - t_0).\tag{1}$$

In Equation (1), c is the sound speed of sediment samples in meters per second (m/s), L is the sample length in meters (m), t is the traveling time for the sound wave to penetrate through the sample in seconds (s) and t_0 is the zero-time correction value in seconds (s). During the sound speed measurement, the sediment sample was first cut and segmented according to actual requirements. Typically, it was cut into 30 cm segments and placed on the cylindrical sample measurement platform for sound speed measurement. After completing the sound speed measurement for the 30 cm section, a 10 cm section was then cut from the 30 cm section and the remaining 20 cm section of sediment continued to be measured for sound speed. After sound speed measurement, the physical property parameters of both segments were tested, including sediment water content, density, void ratio, compressibility coefficient, compressibility modulus, mean grain size, median grain size. The average values of the 10 cm and 20 cm segments are taken to obtain the physical properties of the 30 cm sediment sample. The measurement accuracy of the sediment core length of the measurement platform was 0.1 mm, and the integrated sediment acoustic measurement system had a sampling rate of 10 MHz (that is, the accuracy of travel time is ±0.1 μs). So, the accuracy of the sound speed measurement equipment is estimated to be better than ±0.1% for a typical sample with a length of 30 cm and velocity of 1500 m/s. In order to eliminate the influence of temperature, the sound speed is corrected to the in situ temperature of about 18 °C according to the sea water temperature near the seafloor measured using CTD.

A total of 284 datasets pertaining to sediment sound speed and physical properties were meticulously acquired, with a comprehensive synthesis presented in Table 1. The analytical examination underscores the predominant classification of sediment within the study area into seven distinct types: fine sand, silty sand, clayey sand, coarse silt, sandy clay, sandy silt and clayey silt. Notably, clayey silty sand and medium silty sand manifest as comparatively prevalent, while fine sand sediment assumes a less prominent occurrence. The sound speed of sediment spans a range from 1709.78 m/s to 1492.86 m/s, with an average sound speed of 1642.96 m/s. Remarkably, the extremes in sound speed correspond to silty sand and clayey silty sand, respectively. Silty sand sediment attains maximal values for density (2.00 g/cm^3), compressibility modulus (11.37 MPa), mean grain size (3.50 φ) and median grain size (2.20 φ), concurrently registering minimal values for water content (21.00%), void ratio (0.62), and compressibility coefficient (0.86 MPa^{-1}). Conversely, clayey silt exhibits the highest compressibility coefficient (2.06 MPa^{-1}) and concurrently, the lowest values for density and compressibility modulus (1.56 MPa). Coarse silty sand manifests as the extreme in water content (81.10%) and void ratio (2.14). Notably, the minimum values for mean grain size (8.23 φ) and median grain size (8.22 φ) correspond to sandy clay and silty clay, respectively.

Table 1. Statistical summary of the acoustic and physical-mechanical properties of sediment on the shelf of the East China Sea.

Sediment Type		c (m/s)	w (%)	ρ (g/cm^3)	e	a (MPa^{-1})	E (MPa)	M_z (ϕ)	M_d (ϕ)
Fine Sand	Maximum	1692.31	27.65	1.99	0.79	0.22	9.88	3.78	2.91
	Minimum	1597.16	26.45	1.92	0.71	0.18	8.28	4.29	3.00
	Average	1649.53	26.98	1.96	0.74	0.20	8.86	3.98	2.95
Silty Sand	Maximum	1636.47	39.50	1.92	1.08	0.86	5.46	4.77	4.12
	Minimum	1583.60	28.30	1.81	0.80	0.34	2.45	5.41	4.99
	Average	1610.51	32.11	1.89	0.89	0.49	4.36	5.14	4.59
Clayey Sand	Maximum	1673.75	32.40	2.00	0.89	0.61	6.40	4.17	2.73
	Minimum	1596.91	25.70	1.90	0.70	0.27	3.19	4.80	2.98
	Average	1650.60	28.48	1.97	0.77	0.36	5.42	4.34	2.86
Coarse Silt	Maximum	1598.47	81.10	1.88	2.14	1.62	6.53	5.21	4.88
	Minimum	1509.69	28.75	1.59	0.86	0.28	1.58	7.53	7.53
	Average	1537.90	53.23	1.73	1.42	1.16	2.28	6.32	6.08
Sand Clay	Maximum	1533.33	65.60	1.73	1.86	1.69	1.89	7.03	7.09
	Minimum	1503.61	54.80	1.61	1.44	1.39	1.69	8.29	8.23
	Average	1513.73	60.97	1.65	1.67	1.54	1.77	7.67	7.74
Sandy Silt	Maximum	1709.78	37.90	2.04	1.09	0.86	11.37	3.49	2.24
	Minimum	1587.96	21.00	1.78	0.62	0.15	2.51	5.60	5.01
	Average	1642.96	29.35	1.93	0.82	0.40	5.40	4.49	3.36
Clayey Silt	Maximum	1638.27	76.85	1.90	2.03	2.06	4.56	5.70	5.06
	Minimum	1492.86	32.05	1.56	0.90	0.42	1.25	7.74	7.97
	Average	1524.68	56.29	1.69	1.53	1.31	2.02	6.94	6.89

Note: c: sound speed; w: water content; ρ: density; e: void ratio; a: compressibility coefficient; E: compressibility modulus; M_z: mean grain size; M_d: median grain size.

2.3. Methods

Using the least squares method, regression equations for sediment sound speed were derived as a function of water content, density, void ratio, compressibility coefficient, compressibility modulus, mean grain size and median grain size, and the single parameter prediction equations for sediment sound speed were established. Among the seven physical and mechanical parameters mentioned above, the void ratio is derived from the water content, and they are closely related. The compressibility coefficient and compressibility modulus are reciprocal, reflecting the compressibility of the sediment. Considering the correlation between sediment sound speed and physical parameter as well as the correlation among different physical properties, five physical parameters including density, void ratio, compressibility coefficient, mean grain size and median grain size were selected from the seven physical parameters mentioned above to conduct the dual parameter prediction equations by using the least squares method.

By employing the Sobol method, the sensitivity of each physical-mechanical parameter to sound speed in the dual parameter equations were analyzed to assess the magnitude of the sensitivity for each parameter on sound speed, consequently to determine the extent of the influence and the key factors among the various physical parameters [37,38]. The main effect index S in the Sobol method characterizes the impact of each physical parameter on the sound speed. The larger the value of S, the greater the impact of this parameter on the sound speed, indicating that it is a sensitive parameter. The total effect index ST in the Sobol method characterizes the main effect of an individual physical and its interaction effect between itself and other parameters. A higher value of ST implies a stronger influence of the physical parameter on the sound speed, and the higher inter-action effect with another input physical parameter [37,38].

3. Result

3.1. Single Parameter Prediction Equations for Sediment Sound Speed

The single parameter prediction equations and the correlation diagrams between sediment physical parameters and sound speed are shown Table 1 and Figure 1. According to Figure 1 and Table 2, it can be seen that sediment sound speed exhibits a good correlation with water content, density, void ratio, mean grain size, median grain size, compressibility coefficient and compressibility modulus, with the determination coefficients (R^2) greater than 0.70 for all these parameters. Especially, the determination coefficients (R^2) of the prediction equations for sound speed with respect to water content, density, void ratio, mean grain size and median grain size are greater than 0.8, indicating stronger correlations. As shown in Figure 1, sediment sound speed is positively correlated with density, median grain size and mean grain size, meaning that as these parameters increase, sediment sound speed also increases. It's important to note that in Figure 1, median grain size and mean grain size are represented in φ values, where an increase in φ values corresponds to a decrease in median and mean grain size. Conversely, sound speed is negatively correlated with water content and void ratio. As water content and void ratio increase, sediment sound speed decreases. This is because sediment sound speed is closely related to its compactness. The increase of water content and void ratio results in more pores between sediment particles, reducing the density of the sediment, which, in turn, increases the compactness of the sediment and leads to a decrease in sediment sound speed. Saturated seafloor sediment is composed of particles and pore water and exhibits the characteristics of a two-phase medium. Wet bulk density and porosity indicate the ratio of particle specific gravity and pore water and the degree of compaction [39]. The calculation results based on the theory of elastic wave propagation in two-phase media indicate that the fast longitudinal wave velocity (i.e., sediment sound speed) of sediment decreases with the increase of porosity, which is consistent with the results of this study [40].

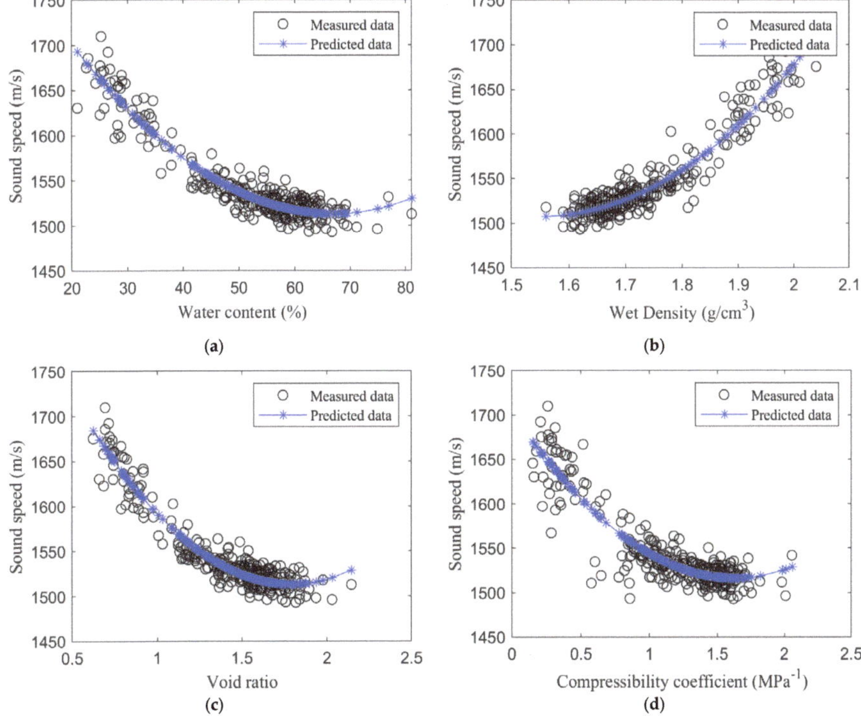

Figure 1. *Cont.*

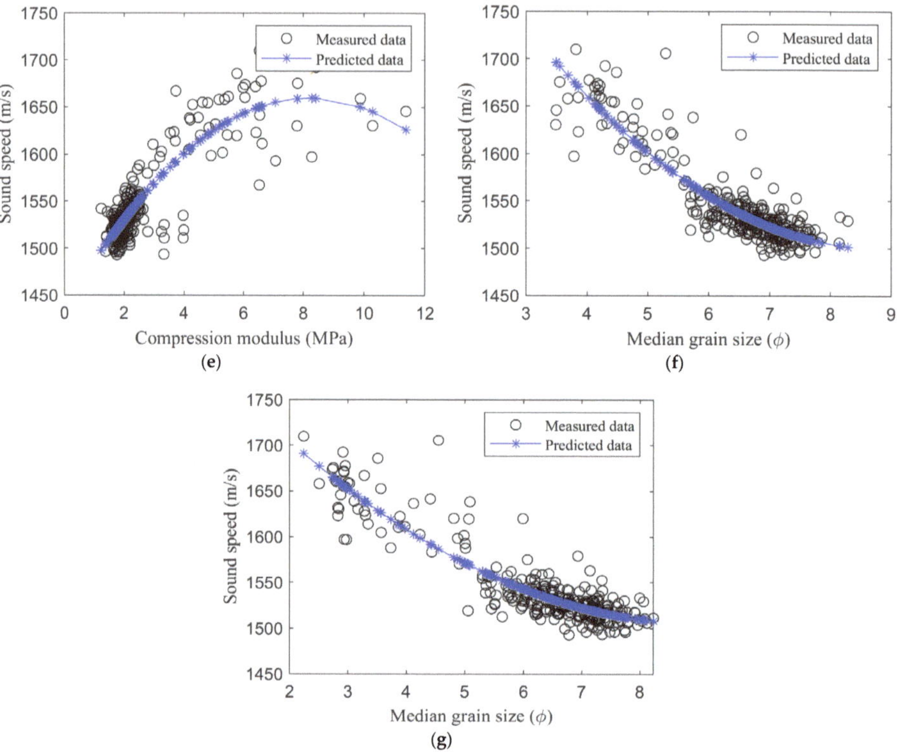

Figure 1. Two–dimensional correlation diagrams between sediment sound speed and single physical parameters. (**a**) Sound speed versus water content; (**b**) sound speed versus density; (**c**) sound speed versus void ratio; (**d**) sound speed versus compressibility coefficient; (**e**) sound speed versus compressibility modulus; (**f**) sound speed versus mean grain size; (**g**) sound speed versus median grain size.

Table 2. Single parameter prediction equations of sound speed on the shelf of the East China Sea.

Parameters	Equations	R^2
Water content (w)	$c = 0.085w^2 - 11.419w + 1895.432$	0.886
Density (ρ)	$c = 274.743\rho^2 - 443.644\rho + 1682.547$	0.868
Void ratio (e)	$c = 125.867e^2 - 450.420e + 1916.038$	0.894
Compressibility coefficient (a)	$c = 70.492a^2 - 229.339a + 1702.653$	0.792
Compressibility modulus (E)	$c = -3.344E^2 + 54.941E + 1433.321$	0.730
Mean grain size (M_z)	$c = 6.797M_z^2 - 120.551M_z + 2033.322$	0.808
Median grain size (M_d)	$c = 3.944M_d^2 - 71.822M_d + 1831.579$	0.827

3.2. Dual Parameter Prediction Equations for Sediment Sound Speed

Based on the measured data of 284 sets of sound speed and physical property data, dual parameter prediction equations for sediment sound speed were established, including the sound speed-density-void ratio, sound speed-density-compressibility coefficient, sound speed-density-mean grain size, sound speed-density-median grain size, sound speed-void ratio-compressibility coefficient, sound speed-void ratio-mean grain size, sound speed-void ratio-median grain size, sound speed-compressibility coefficient-mean grain size, and sound speed-compressibility coefficient-median grain size (Table 3). The determination coefficients (R^2) for the dual parameter regression equations all exceed 0.85, which are higher than the single parameter prediction equations. The three-dimensional correlation diagrams between sediment sound speed and two parameters were drawn (Figure 2).

Table 3. Dual parameter prediction equations of sound speed on the shelf of the East China Sea.

Parameters	Equations	R^2
Density-void ratio (ρ, e)	$c = 145.321\rho^2 - 414.143\rho + 92.615e^2 - 253.597e - 56.950\rho e + 2125.148$	0.895
Density-compressibility coefficient (ρ, a)	$c = 251.544\rho^2 - 404.277\rho - 3.098a^2 + 321.840a - 189.758\rho a + 1497.356$	0.873
Density-mean grain size (ρ, M_z)	$c = 467.958\rho^2 - 1212.456\rho - 0.846M_z^2 + 47.442M_z - 28.129\rho M_z + 2277.411$	0.885
Density-median grain size (ρ, M_d)	$c = 269.877\rho^2 - 540.511\rho + 0.152M_d^2 + 38.781M_d - 28.918\rho M_d + 1728.658$	0.891
Void ratio-compressibility coefficient (e, a)	$c = 148.337e^2 - 506.679e + 2.712a^2 + 31.706a - 18.500ea + 1939.992$	0.896
Void ratio-mean grain size (e, M_z)	$c = 86.574e^2 - 420.296e - 2.511M_z^2 + 3.539M_z + 15.014eM_z + 1899.443$	0.900
Void ratio-median grain size (e, M_d)	$c = 72.368e^2 - 330.327e + 0.011M_d^2 - 18.337M_d + 8.364eM_d + 1895.848$	0.904
Compressibility coefficient-mean grain size (a, M_z)	$c = 9.141a^2 - 207.244a - 2.269M_z^2 - 15.281M_z + 22.939aM_z + 1786.873$	0.855
Compressibility coefficient-median grain size (a, M_d)	$c = 5.865a^2 - 152.869a - 0.514M_d^2 - 25.968M_d + 16.696aM_d + 1766.918$	0.869

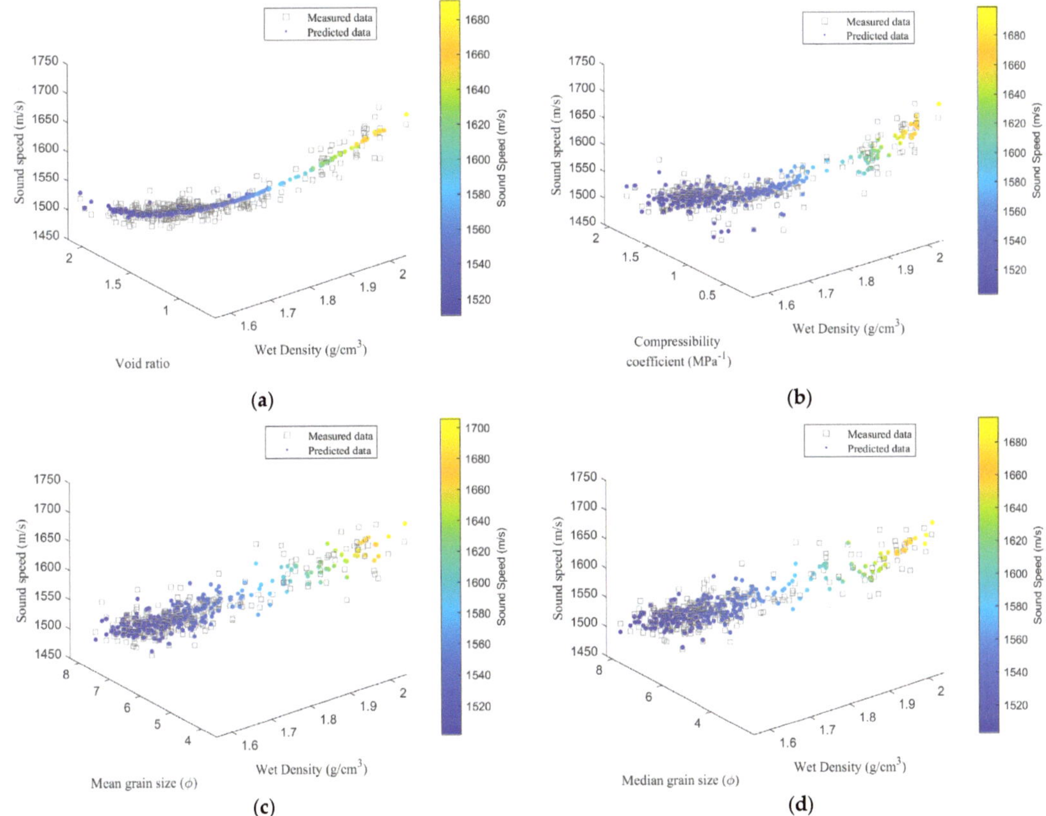

Figure 2. *Cont.*

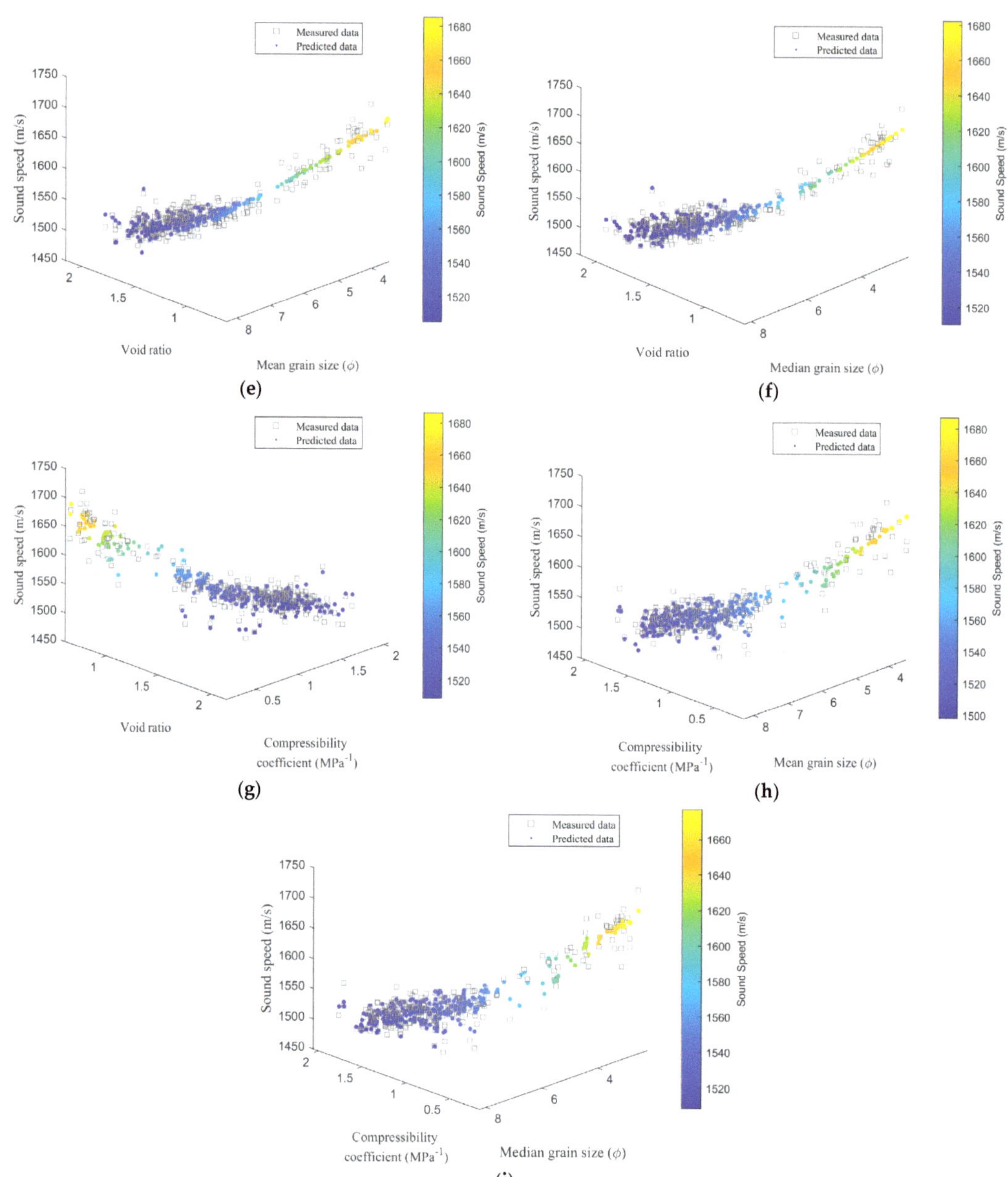

Figure 2. Three−dimensional correlations between sediment sound speed and dual-parameter pairs: (**a**) sound speed versus the dual-parameter pair of density and void ratio; (**b**) sound speed versus the dual-parameter pair of density and compressibility coefficient; (**c**) sound speed versus the dual-parameter pair of density and mean grain size; (**d**) sound speed versus the dual-parameter pair of density and median grain size; (**e**) sound speed versus the dual-parameter pair of void ratio and mean grain size; (**f**) sound speed versus the dual-parameter pair of void ratio and median grain size; (**g**) sound speed versus the dual-parameter pair of void ratio and compressibility coefficient; (**h**) sound speed versus the dual-parameter pair of compressibility coefficient and mean grain size; (**i**) sound speed versus the dual-parameter pair of compressibility coefficient and median grain size.

The density is a physical parameter that measures the compactness of the sediment, while the void ratio represents the ratio of pore volume to the volume of solid particles within the sediment. For saturated seafloor sediment, the calculation formula for sediment density is:

$$\rho = n\rho_w + (1-n)\rho_s. \tag{2}$$

In Equation (2), n, ρ_w and ρ_s represent sediment porosity, pore water density and sediment particle density, respectively. From Equation (2), it can be seen that there is a certain correlation between density and porosity (or void ratio). Generally, as porosity (void ratio) increases, density decreases, indicating a negative correlation between the two parameters. However, the density and void ratio have a certain degree of independence. Due to differences in material source, mineral content, and other factors, there are significant differences in particle density among different types of sediment even when the void ratio is the same. Therefore, establishing a dual parameter sound speed prediction equation based on density and void ratio provides a more comprehensive reflection of the influence of different sediment parameters on sound speed. The compressibility coefficient is a parameter that describes the compressibility of sediment, and the higher the compressibility coefficient, the slower the speed at which sound waves propagate. According to the difference between sediment properties characterized by the two parameters, the compressibility coefficient is an indicator of sediment mechanical properties, while density represents the physical property indicator of sediment compactness. Although they exhibit a certain degree of correlation, they are relatively independent mechanical and physical indicators. Hence, a dual parameter sound speed prediction equation based on the density and compressibility coefficient was established to predict the sediment sound speed more accurately. Hamilton provided a regression relationship between density and mean grain size [24,32]:

$$\rho = 2.374 - 0.175M_z + 0.008M_z^2. \tag{3}$$

The relationship shown in Formula (3) indicates that there is a certain correlation between density and particle size. However, Hamilton [22] also pointed out that sediments with the same mean grain size may have significant variations in density due to differences in sediment sources, mineral content, and sedimentary environments. For example, sediments with the same mean grain size increase density through compaction, while the grain size remains unchanged [41]. Therefore, density and mean grain size characterize different aspects of sediment physical properties. Based on this, this paper establishes a dual parameter prediction equation for sound speed based on density and mean grain size. The median grain size is also a parameter that characterizes sediment granularity, usually referring to the size of the particle located in the middle position when particles are sorted by size, which differs from the mean grain size [41]. Therefore, a dual parameter prediction equation based on the density and median grain size is necessary.

As shown in Figure 2, the variation trend of sound speed and dual parameters can be divided into two regions: a slow change region with small variation and a rapid change region with high variation. It can be seen that sound speed increases with an increase of density and decreases as the void ratio increases as a whole. When the void ratio is about 1.6–2.2 and sediment density is about 1.56–1.8 g/cm^3, the sediment sound speed exhibits relatively small variations, ranging from about 1490–1530 m/s. Subsequently, as the void ratio further decreases and sediment density further increases, the sediment sound speed enters a high-value and high-variation region with a variation range of 1520–1700 m/s (Figure 2a). When the density is in the range of 1.56–1.8 g/cm^3 and the sediment compressibility coefficient is about 1.1–2.0 MPa^{-1}, the sediment sound speed exhibits relatively small variations, with a range of about 1490–1530 m/s. As the compressibility coefficient further decreases and the sediment density further increases, the sediment sound speed starts to enter a rapid change region with a significant range of variation between 1520–1700 m/s (Figure 2b). As shown in Figure 2c, when density falls within the range of 1.56–1.8 g/cm^3 and sediment mean grain size ranges from 5.7–8.3 ϕ, the sediment sound speed exhibits relatively small variation, with a range of about 1490–1530 m/s. As the mean grain size and the density further begin to increase, the sediment sound speed enters a high-value region with a significant range of variation between 1520–1700 m/s. Similarly, when the density is in the range of 1.56–1.8 g/cm^3 and the median grain size of sediment is between 5.1–8.2 φ, the variation in sediment sound speed is relatively small, with a range of about 1490–1530 m/s. When the median grain size and density of sediment continue to increase further, the sediment sound speed is in the high-value zone, with a larger range of variation, about 1520–1700 m/s.

The void ratio, a critical parameter, characterizes the proportion of pores within sediment, whereas the mean grain size delineates the size of sediment particles. These parameters represent distinct sediment properties, albeit exhibiting correlation to some extent. As mentioned earlier, the compressibility coefficient, a mechanical parameter, characterizes sediment compressibility, differing from both the void ratio and particle size parameters. A void ratio within the range of 1.6–2.2 and a mean grain size within the range of 5.7–8.3 φ corresponds to a relatively minor change in sediment sound speed, ranging from approximately 1490–1530 m/s (Figure 2e). A decrease in the void ratio and a simultaneous increase in the mean grain size of sediment lead to higher values in sediment sound speed, exhibiting a variation range of approximately 1520–1700 m/s, signifying a more substantial variation (Figure 2e). The sediment sound speed resides in the slow change region when the median grain size falls within the range of 5.1–8.2 φ, and the void ratio is approximately 1.6–2.2. It transitions into the rapid change region, featuring a high variation of 1520–1700 m/s, as the median grain size increases and the sediment void ratio decreases (Figure 2f). Figure 2g illustrates the range of the compressibility coefficient (1.1–2.1 MPa^{-1}) and void ratio (1.6–2.2) associated with the slow change region of sound speed.

Although there is some correlation between the compressibility coefficient and grain size, they are relatively independent determinants of sediment physical properties. The compressibility coefficient, ranging from 1.1 to 2 MPa^{-1}, and the sediment mean grain size, approximately 5.7–8.3 φ, exhibit a relatively small variation in sediment sound speed, with the range stabilizing around 1490–1530 m/s (Figure 2h). However, when the compressibility coefficient starts to decrease from 1.1 to 0.2 MPa^{-1} and the sediment mean grain size increases from 3.5 to 5.7 φ, the sediment sound speed enters a higher value range with a variation range of about 1520–1700 m/s, indicating a larger change in sound speed (Figure 2h). The slow change region of sound speed is situated within the compressibility coefficient range of 1.1–2.1 MPa^{-1} and the median grain size range of 5.1–8.2 φ (Figure 2i). Conversely, the rapid change region is delineated by the compressibility coefficient ranging from 1.1 to 0.2 MPa^{-1} and the median grain size ranging from 5.1 to 2.2 φ (Figure 2i).

4. Discussion

4.1. Sensitivity of Physical Parameters to Sediment Sound Speed

Normalization was performed on these parameters to eliminate the influence of dimensionality and the dual parameter prediction equations for sediment sound speed were established after parameter normalization, as shown in Table 4. For the sake of comparing the sensitivity of mean grain size and median grain size, an additional normalized dual-parameter prediction equation for sediment sound speed with mean grain size and median grain size was established. The main effect index (S) and total effect index (ST) for each parameter in the ten normalized dual parameter prediction equations are shown in Figure 3.

According to the value of the evaluation indicator among ten equations, it is evident that the total effect index of the void ratio remains relatively stable, basically remaining in the range of 0.8 to 1. This indicates that the sediment void ratio not only has high influence on sediment sound speed but also has favorable interaction effects with other parameters. The total effect index of sediment density exhibits stability among ten equations, and the equation involving density and the compressibility coefficient has the highest total effect index. It signifies that the density has a high impact on sound speed and demonstrates a higher degree of interaction with the compressibility coefficient than other physical parameters. The compressibility coefficient, relative to the median grain size and the mean grain size, exhibits higher impact on speed and displays a greater level of interaction with median grain size (Figure 3h,i). The total effect index of the median grain size is slightly higher than that of the mean grain size (Figure 3j). By synthesizing the main and total effect indices of the physical parameters in the ten equations listed in Table 4, the sensitivity of physical parameters to sediment sound speed can be ranked in descending order as follows: void ratio, density, compressibility coefficient, median grain size, and mean grain size.

Table 4. Normalized dual parameter prediction equations of the sound speed on the shelf of the shelf of the East China Sea.

Parameters	Equations	R^2
Density-void ratio (ρ, e)	$c = 33.482\rho^2 + 1.813\rho + 214.260e^2 - 345.330e - 41.578\rho e + 1655.344$	0.895
Density—compressibility coefficient (ρ, a)	$c = 57.956\rho^2 + 167.573\rho - 11.256a^2 + 47.438a - 173.606\rho a + 1478.017$	0.873
Density—mean grain size (ρ, M_z)	$c = 107.817\rho^2 + 71.741\rho - 19.490M_z{}^2 - 11.240M_z - 64.805\rho M_z + 1526.925$	0.885
Density—median grain size (ρ, M_d)	$c = 62.180\rho^2 + 113.604\rho + 5.438M_d{}^2 - 33.846M_d - 83.136\rho M_d + 1528.803$	0.891
Void ratio-compressibility coefficient (e, a)	$c = 1343.170e^2 - 493.755e + 9.851a^2 + 40.015a - 53.632ea + 1684.994$	0.896
Void ratio—mean grain size (e, M_z)	$c = 200.283e^2 - 395.549e - 57.854M_z{}^2 - 22.201M_z + 109.605eM_z + 1685.618$	0.900
Void ratio—median grain size (e, M_d)	$c = 167.418e^2 - 336.757e + 0.403M_d{}^2 - 78.317M_d + 76.196eM_d + 1688.771$	0.904
Compressibility coefficient—mean grain size (a, M_z)	$c = 33.208a^2 - 237.282a - 52.265M_z{}^2 - 132.789M_z + 209.853aM_z + 1687.095$	0.855
Compressibility coefficient—median grain size (a, M_d)	$c = 21.306a^2 - 216.671a - 18.445M_d{}^2 - 154.340M_d + 190.601aM_d + 1688.933$	0.869
Mean grain size—median grain size (M_z, M_d)	$c = -407.073M_z{}^2 - 29.894M_z - 179.921M_d{}^2 - 307.723M_d + 731.483M_zM_d + 1695.085$	0.829

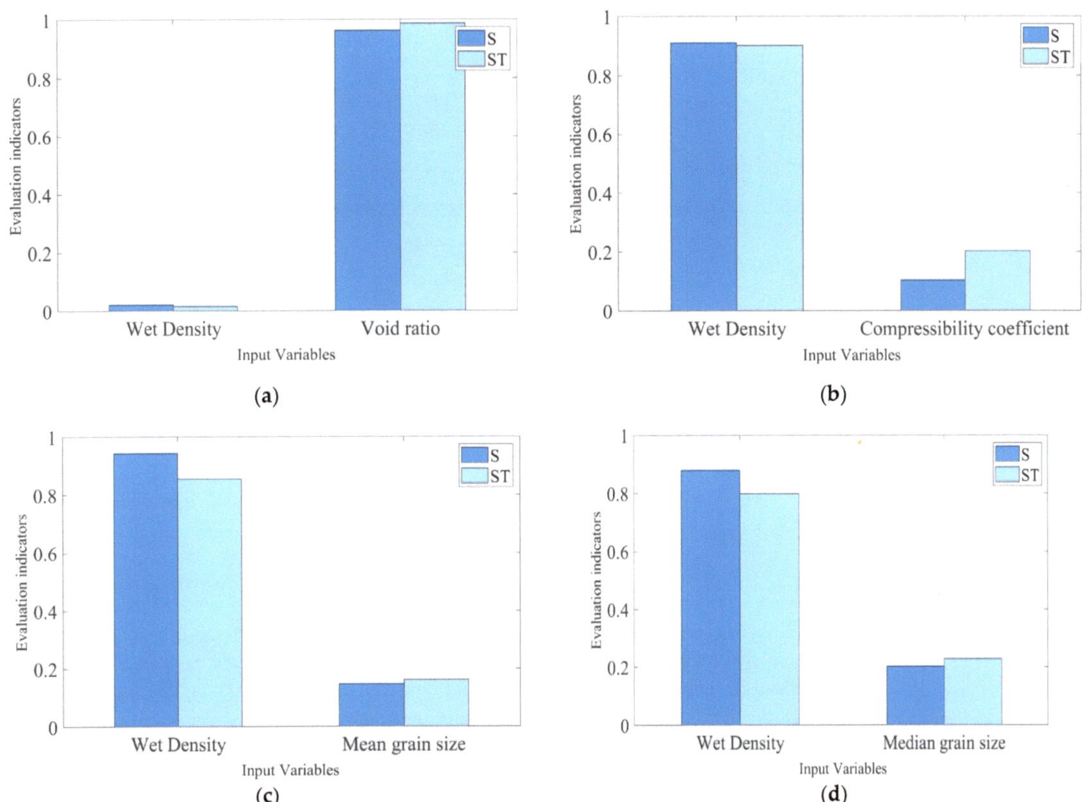

Figure 3. *Cont.*

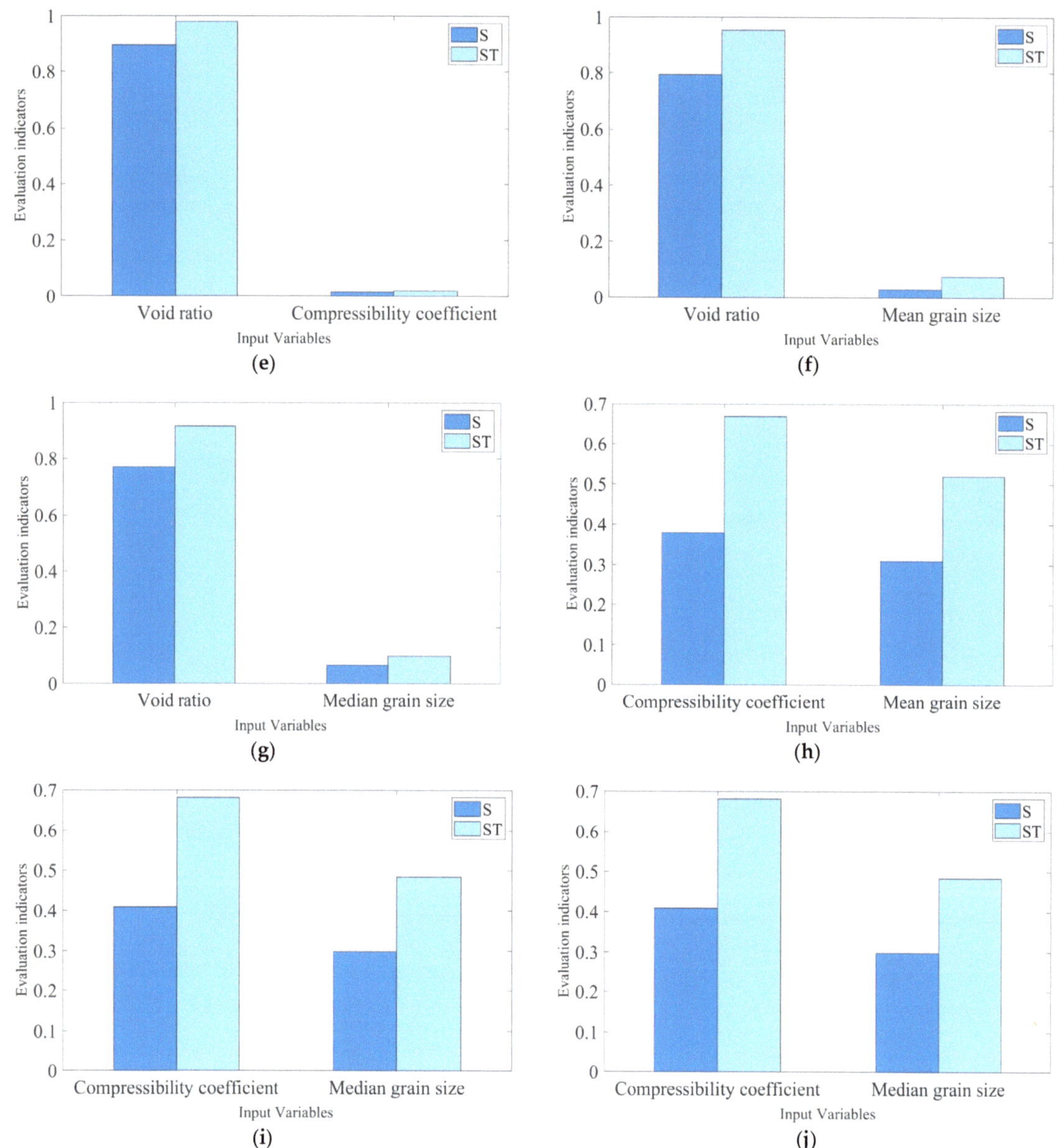

Figure 3. Evaluation indicators of sensitivity (S and ST) of the physical parameters on sound speed for the normalized dual-parameter prediction equation. (**a**) density and void ratio; (**b**) density and compressibility coefficient; (**c**) density and mean grain size; (**d**) density and median grain size; (**e**) void ratio and compressibility coefficient; (**f**) void ratio and mean grain size; (**g**) void ratio and median grain size; (**h**) compressibility coefficient and mean grain size; (**i**) compressibility coefficient and median grain size; (**j**) mean grain size and median grain size.

4.2. Comparison between Different Prediction Equations

Due to the close relationship between the acoustic properties of seafloor sediments and the physical-mechanical properties, researchers have conducted extensive research on the relationship between them and established a lot of prediction equations. Based on the sensitivity analysis of physical parameters discussed above, the void ratio, which can be used to calculate the porosity, exhibits the highest sensitivity to sediment sound speed. Therefore, several single parameter sound speed prediction equations based on porosity established by Hamilton [14], Anderson [19], Lubo [27], Thomas [20] and Sun [42] are selected to compare and analyze the differences between the prediction equations established in this study and those established by them (Table 5).

Table 5. Single parameter prediction equations of sound speed in the shelf of the ECS.

Author	Single Parameter Prediction Equation	Background of Equation Establishment
Hamilton [14]	$c = 2502.0 - 24.45n + 0.14n^2$	Based on data collected in continental shelf
Anderson [19]	$c = 2506 - 27.58n + 0.1868n^2$	Based on data of sediments from the less than 1500 m deep seafloor
Thomas [20]	$c = 2527.4 - 27.132n + 0.1782n^2$	Based on sediment core samples collected from the offshore margin of the Brazilian continental shelf to the Pernambuco deep-sea plain, with an average water depth of 5047 m.
Bo Lu [27]	$c = 2470.7 - 32.2n + 0.25n^2$	Based on the data collected in the northern continental shelf of South China Sea
Sun [42]	$c = 3239.1 - 4358.7n + 2718.7n^2$	Based on sediment core Samples collected from the seabed at depths ranging from 3164 to 5592 m in the Philippine deep sea
This study	$c = 1823.5 - 2604.2n + 1823.5n^2$	Based on the data collected in the shelf of East China Sea

As shown in Figure 4, the prediction result of this study is closest to that of Anderson's equation with the maximum bias of 21.81 m/s at the porosity of 0.38, and their bias is very small in the porosity range of 0.53–0.62. The maximum bias of prediction result between this study and Hamilton's equation is 69.00 m/s, which is larger than that of Anderson's equation. The curves of the results predicted by Hamilton and Anderson are both above the prediction curve of this study when the porosity is lower than 0.63, while the prediction curve is below that of this study in the entire porosity range. Moreover, there is a significant difference between the prediction curve in this study and that of Lu's equation, with a maximum difference of 120 m/s. Lu's prediction equation was based on data from multiple sea areas, including the East China Sea continental shelf, Taiwan Strait, and the northern South China Sea, and the scatteration of data used in Lu's study is relatively high, which may explain the significant differences compared to this study's results. Another possible reason for the differences in the prediction results could be the difference in measurement frequencies, for example, measurement frequency of the sediment sound speed of Hamilton's equation is at 200 kHz, while that of this study is at 100 kHz. In addition, different marine environmental conditions such as water depth and in situ temperature of seafloor caused by different sea areas are also reasons for the above differences.

In order to compare the prediction results of different dual parameter prediction equations, we chose dual parameter prediction equations based on the porosity and density established by Lu and Hou to compare and analyze the differences between the prediction equations established in this study and those established by them, which are listed in Table 6 [18,21]. The prediction results of the prediction equation established in this study is relatively close to those of Hou's equation with a maximum, minimum and average bias of 30.63 m/s, 0.0442 m/s, and 4.68 m/s, respectively, but differ significantly from those of Lu's equation, with a maximum, minimum and average bias of 173.67 m/s, 28.80 m/s, and 80.27 m/s (Figure 5).

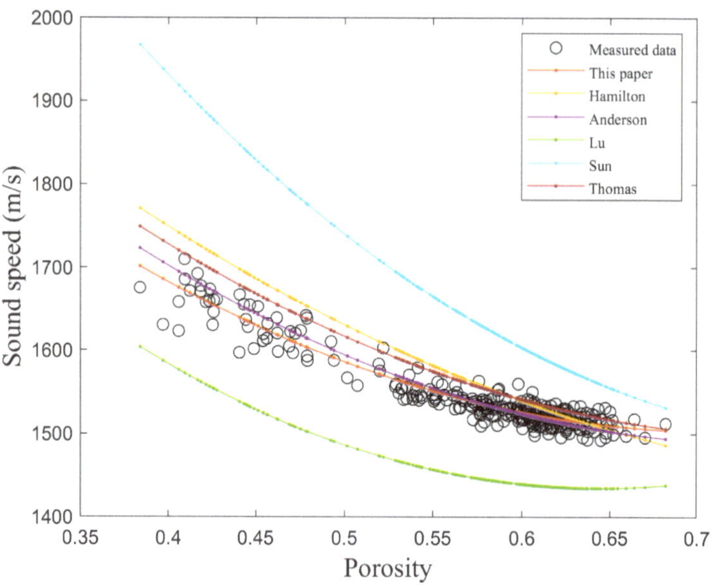

Figure 4. Comparison of prediction results of different single parameter sound speed prediction equations.

Table 6. Dual parameter prediction equations of sound speed in the study area.

Author	Single Parameter Prediction Equation	Background of Equation Establishment
Lu [18]	$c = 1895.78 - 2525.29\rho + 1014.88\rho^2$ $+51.05n - 0.31n^2 - 9.42\rho n$	Based on the data collected in continental shelf of South China Sea
Hou [21]	$c = 3342 - 1177\rho - 29.23n + 269.5\rho^2$ $+7.018\rho n + 0.1236n^2$	Based on the data of data in the southern South China Sea
This study	$c = 265.50 + 1516.39\rho + 255.03n - 262.37\rho^2$ $-1016.00\rho n + 899.87n^2$	Based on the data collected in the shelf of the East China Sea

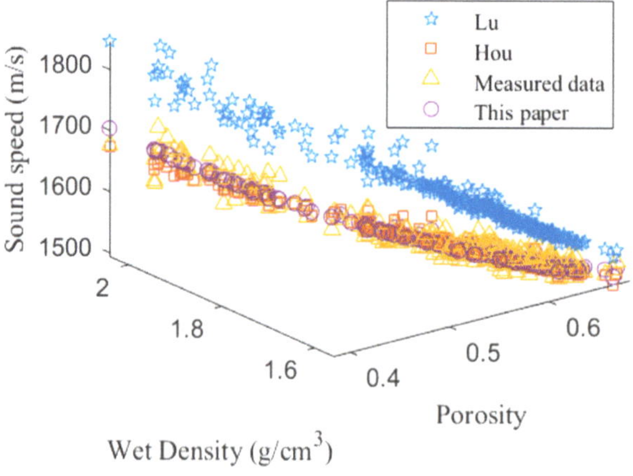

Figure 5. Comparison of predictions from different dual parameter sediment sound speed prediction equations.

5. Conclusions

1. Establishment of Single and Dual Parameter Prediction Equations: Through regression analysis, single parameter and dual parameter prediction equations for sediment sound speed in the East China Sea shelf area have been established. Judging from the coefficient of determination (R^2) of the single equations, sediment sound speed is strongly correlated with water content, density, void ratio, mean grain size and median grain size with the R^2 greater than 0.80. The dual parameter prediction equations all have R^2 values greater than 0.85, which is higher than the R^2 values of the single parameter prediction equations. This indicates that the dual parameter prediction equations have a better performance for the sediment sound speed prediction. The prediction equations established in this study serve as a valuable supplement to the acoustic property prediction equations for coastal seafloor sediments.

2. Order of Sensitivity of Physical-Mechanical Parameters to Sound Speed: In terms of sensitivity to sediment sound speed, the sensitivity of physical parameters ranks from the highest to the lowest as follows: void ratio, density, compressibility coefficient, median grain size and mean grain size. This is helpful for the selection of sediment physical parameters when establishing the sound velocity prediction equation for seafloor sediments.

3. Comparative Analysis of Prediction Equations: The comparison of the prediction results of sediment sound speed among different prediction equations indicates that there may be significant differences between different prediction equations. The maximum difference of the single parameter prediction equations based on porosity is 120 m/s. For the dual parameter prediction equations based on density and porosity, it reaches 173.67 m/s. These differences remind us that it is necessary to select appropriate prediction equations which are suitable to the study sea area when using prediction equations for sound velocity prediction.

Author Contributions: Conceptualization, G.K., X.M. and J.L.; methodology, G.K., X.M. and J.L.; software, Q.H.; validation, M.C., J.L. and G.K.; formal analysis, G.K., J.L. and M.C.; investigation, J.W. and L.Z.; resources and data curation, J.W. and L.Z.; writing—original draft preparation, G.K. and J.L.; writing—review and editing, G.L.; visualization, J.L.; supervision, B.L. All authors have read and agreed to the published version of the manuscript.

Funding: This study was supported by the National Natural Science Foundation of China under Grant No. 42376076, 42049902, U2006202, 42176191; and by the Central Public-interest Scientific Institution Basal Research Fund under Grant No. GY0220Q09.

Institutional Review Board Statement: Not applicable.

Informed Consent Statement: Not applicable.

Data Availability Statement: Data are contained within the article.

Acknowledgments: Data acquisition and sample collections were supported by the National Natural Science Foundation of China Open Research Cruise (Cruise No. NORC2021-02+NORC2021-301), funded by the Ship Time Sharing Project of the National Natural Science Foundation of China. This cruise was conducted onboard the R/V "XiangYang Hong 18" by The First Institute of Oceanography, Ministry of Natural Resources, China.

Conflicts of Interest: The authors declare no conflict of interest.

References

1. Beebe, J.; McDaniel, S.; Rubano, L. Shallow-water transmission loss prediction using the Biot sediment model. *J. Acoust. Soc. Am.* **1982**, *71*, 1417–1426. [CrossRef]
2. Collins, M.D. A higher-order parabolic equation for wave propagation in an ocean overlying an elastic bottom. *J. Acoust. Soc. Am.* **1989**, *86*, 1459–1464. [CrossRef]
3. Chen, S.; Wu, L.; Zhang, R.; Li, Z.; James, F.L.; Timothy, F.D.; Arthur, E.N. Internal wave characteristics in the South China Sea and the associated fluctuations in the acoustic field. *Prog. Nat. Sci.* **2004**, *14*, 1163–1170.

4. Kim, D.C.; Kim, G.Y.; Yi, H.I.; Seo, Y.K.; Lee, G.S.; Jung, J.H.; Kim, J.C. Geoacoustic provinces of the South Sea shelf off Korea. *Quat. Int.* **2012**, *263*, 139–147. [CrossRef]
5. Schock, S.G. Remote estimates of physical and acoustic sediment properties in the South China Sea using chirp sonar data and the Biot model. *IEEE J. Ocean. Eng.* **2004**, *29*, 1218–1230. [CrossRef]
6. Yang, K.; Lei, B.; Ma, Y. Subcritical scattering field and signal-to-reverberation ratio from buried objects. *Tech. Acoust.* **2007**, *26*, 1081–1088.
7. Hodges, R.P. *Underwater Acoustics: Analysis, Design and Performance of Sonar*; John Wiley & Sons: Hoboken, NJ, USA, 2011.
8. Zhang, S. Geoacoustics: An important subject in the study of the seabed. *Physics* **1997**, *26*, 280–285.
9. Panda, S.; LeBlanc, L.R.; Schock, S.G. Sediment classification based on impedance and attenuation estimation. *J. Acoust. Soc. Am.* **1994**, *96*, 3022–3035. [CrossRef]
10. Kim, G.Y.; Richardson, M.D.; Bibee, D.L.; Kim, D.C.; Wilkens, R.H.; Shin, S.R.; Song, S.T. Sediment types determination using acoustic techniques in the Northeastern Gulf of Mexico. *Geosci. J.* **2004**, *8*, 95–103. [CrossRef]
11. Pszonka, J.; Schulz, B. SEM Automated Mineralogy applied for the quantification of mineral and textural sorting in submarine sediment gravity flows. *Gospod. Surowcami Miner.-Miner. Resour. Manag.* **2022**, *38*, 105–131.
12. Stoll, R.D. Acoustic waves in ocean sediments. *Geophysics* **1977**, *42*, 715–725. [CrossRef]
13. Buckingham, M.J. Compressional and shear wave properties of marine sediments: Comparisons between theory and data. *J. Acoust. Soc. Am.* **2005**, *117*, 137–152. [CrossRef] [PubMed]
14. Hamilton, E.L.; Bachman, R.T. Sound velocity and related properties of marine sediments. *J. Acoust. Soc. Am.* **1982**, *72*, 1891–1904. [CrossRef]
15. Meng, X.; Liu, B.; Kan, G.; Li, G. An experimental study on acoustic properties and their influencing factors of marine sediment in the southern Huanghai Sea. *Acta Oceanol. Sin.* **2012**, *34*, 74–83.
16. Kan, G.; Liu, B.; Han, G.; Li, G.; Zhao, Y. Application of in-situ measurement technology to the survey of seafloor sediment acoustic properties in the Huanghai Sea. *Acta Oceanol. Sin.* **2010**, *32*, 88–94.
17. Wang, J.; Li, G.; Liu, B.; Kan, G.; Sun, Z.; Meng, X.; Hua, Q. Experimental study of the ballast in situ sediment acoustic measurement system in South China Sea. *Mar. Georesources Geotechnol.* **2018**, *36*, 515–521. [CrossRef]
18. Lu, B.; Ganxian, L.; Shaojian, H. Correlation between sound speed and physical parameters of shallow sediments in Nansha sea area. In *Proceedings on Sound and Light Fields in Nansha Sea Area*; China Ocean Press: Beijing, China, 1996; pp. 9–22.
19. Anderson, R.S. Statistical correlation of physical properties and sound velocity in sediments. In *Physics of Sound in Marine Sediments*; Springer: Berlin/Heidelberg, Germany, 1974; pp. 481–518.
20. Orsi, T.H.; Dunn, D.A. Sound velocity and related physical properties of fine-grained abyssal sediments from the Brazil Basin (South Atlantic Ocean). *J. Acoust. Soc. Am.* **1990**, *88*, 1536–1542. [CrossRef]
21. Hou, Z. *The Correlation of Seafloor Sediment Acoustic Properties and Physical Parameters in the Southern South China Sea*; Insitute of Oceannology, Chinese Academy of Sciences: Qingdao, China, 2016.
22. Hamilton, E.L. Sound velocity as a function of depth in marine sediments. *J. Acoust. Soc. Am.* **1985**, *78*, 1348–1355. [CrossRef]
23. Hamilton, E.L. Geoacoustic modeling of the sea floor. *J. Acoust. Soc. Am.* **1980**, *68*, 1313–1340. [CrossRef]
24. Hamilton, E.L. Elastic properties of marine sediments. *J. Geophys. Res.* **1971**, *76*, 579–604. [CrossRef]
25. Hamilton, E.L. Low sound velocities in high-porosity sediments. *J. Acoust. Soc. Am.* **1956**, *28*, 16–19. [CrossRef]
26. Bachman, R.T. Acoustic and physical property relationships in marine sediment. *J. Acoust. Soc. Am.* **1985**, *78*, 616–621. [CrossRef]
27. Lu, B.; Li, G.; Liu, Q.; Huang, S.; Zhang, F. Sea floor sediment and its acouso-physical properties in the southeast open sea area of Hainan Island in China. *Mar. Georesources Geotechnol.* **2008**, *26*, 129–144. [CrossRef]
28. Tang, Y. The Relationship between porosity of sea bed sediment and sound velocity. *Acta Oceanol. Sin.* **1998**, *20*, 39–43.
29. Zou, D.; Wu, B.; Lu, B. Analysis and study on the sound velocity empirical equations of seafloor sediments. *Acta Oceanol. Sin.* **2007**, *29*, 43–50.
30. Wang, J.; Kan, G.; Li, G.; Meng, X.; Zhang, L.; Chen, M.; Liu, C.; Liu, B. Physical properties and in situ geoacoustic properties of seafloor surface sediments in the East China Sea. *Front. Mar. Sci.* **2023**, *10*, 1195651. [CrossRef]
31. Bachman, R.T. Estimating velocity ratio in marine sediment. *J. Acoust. Soc. Am.* **1989**, *86*, 2029–2032. [CrossRef]
32. Hamilton, E.L. Prediction of in-situ acoustic and elastic properties of marine sediments. *Geophysics* **1971**, *36*, 266–284. [CrossRef]
33. Lu, B.; Liang, Y. Statistical correlation between physical parameters of marine sediments and sound velocity in the southeast coast of China. *Sci. Sin.* **1994**, *24*, 556–560.
34. Pan, G.; Ye, Y.; Lai, X.; Chen, X.; Lü, X. Shear wave velocity of seabed sediment from laboratory measurements and its relationship with physical properties of sediment. *Acta Oceanol. Sin.* **2006**, *28*, 64–68.
35. Liu, S.; Shi, X.; Fang, X.; Dou, Y.; Liu, Y.; Wang, X. Spatial and temporal distributions of clay minerals in mud deposits on the inner shelf of the East China Sea: Implications for paleoenvironmental changes in the Holocene. *Quat. Int.* **2014**, *349*, 270–279. [CrossRef]
36. Zhang, K.; Li, A.; Huang, P.; Lu, J.; Liu, X.; Zhang, J. Sedimentary responses to the cross-shelf transport of terrigenous material on the East China Sea continental shelf. *Sediment. Geol.* **2019**, *384*, 50–59. [CrossRef]
37. Saltelli, A.; Andres, T.; Homma, T. Sensitivity analysis of model output: An investigation of new techniques. *Comput. Stat. Data Anal.* **1993**, *15*, 211–238. [CrossRef]

38. Saltelli, A. Making best use of model evaluations to compute sensitivity indices. *Comput. Phys. Commun.* **2002**, *145*, 280–297. [CrossRef]
39. Jackson, D.; Richardson, M. *High-Frequency Seafloor Acoustics*; Springer Science & Business Media: Berlin/Heidelberg, Germany, 2007.
40. Williams, K.L.; Jackson, D.R.; Thorsos, E.I.; Tang, D.; Schock, S.G. Comparison of sound speed and attenuation measured in a sandy sediment to predictions based on the Biot theory of porous media. *IEEE J. Ocean. Eng.* **2002**, *27*, 413–428. [CrossRef]
41. Pszonka, J.; Schulz, B.; Sala, D. Application of mineral liberation analysis (MLA) for investigations of grain size distribution in submarine density flow deposits. *Mar. Pet. Geol.* **2021**, *129*, 105109. [CrossRef]
42. Sun, Z.; Sun, L.; Li, G.; Kan, G.; Guo, C.; Wang, J.; Meng, X. The relationship between the acoustic characteristics and physical properties of deep-sea sediments in the Philippine Sea. *Mar. Sci.* **2018**, *5*, 12–22.

Journal of
Marine Science and Engineering

MDPI

Article

Determination of Formulae for the Hydrodynamic Performance of a Fixed Box-Type Free Surface Breakwater in the Intermediate Water

Guoxu Niu [1], Yaoyong Chen [1], Jiao Lv [2], Jing Zhang [1] and Ning Fan [2,3,4,*]

[1] Wenzhou Academy of Agricultural Sciences, Wenzhou 325006, China
[2] College of Civil Engineering and Architecture, Wenzhou University, Wenzhou 325035, China
[3] Key Laboratory of Engineering and Technology for Soft Soil Foundation and Tideland Reclamation of Zhejiang Province, Wenzhou 325035, China
[4] Zhejiang International Science and Technology Cooperation Base of Ultra-Soft Soil Engineering and Smart Monitoring, Wenzhou 325035, China
* Correspondence: fanning@wzu.edu.cn

Abstract: A two-dimensional viscous numerical wave tank coded mass source function in a computational fluid dynamics (CFD) software Flow-3D 11.2 is built and validated. The effect of the core influencing factors (draft, breakwater width, wave period, and wave height) on the hydrodynamic performance of a fixed box-type free surface breakwater (abbreviated to F-BW in the following texts) are highlighted in the intermediate waters. The results show that four influence factors, except wave period, impede wave transmission; the draft and breakwater width boost wave reflection, and the wave period and wave height are opposite; the draft impedes wave energy dissipation, and the wave height is opposite; the draft and wave height boost the horizontal extreme wave force; four influence factors, except the draft, boost the vertical extreme wave force. Finally, new formulas are provided to determine the transmission, reflection, and dissipation coefficients and extreme wave forces of the F-BW by applying multiple linear regression. The new formulas are verified by comparing with existing literature observation datasets. The results show that it is in good agreement with previous datasets.

Keywords: breakwater; regular wave; numerical wave tank; wave force; prediction formulae

Citation: Niu, G.; Chen, Y.; Lv, J.; Zhang, J.; Fan, N. Determination of Formulae for the Hydrodynamic Performance of a Fixed Box-Type Free Surface Breakwater in the Intermediate Water. *J. Mar. Sci. Eng.* **2023**, *11*, 1812. https://doi.org/10.3390/jmse11091812

Academic Editors: Achilleas Samaras and Giuseppe Roberto Tomasicchio

Received: 25 July 2023
Revised: 30 August 2023
Accepted: 13 September 2023
Published: 17 September 2023

1. Introduction

A breakwater dissipates wave energy and reflects waves from the open sea, representing a crucial protective structure for the exploitation and utilization of marine resources. It is also an essential auxiliary marine structure that improves offshore engineering construction conditions and shortens ship berthing times [1–3]. With the development and utilization of ocean space and resources, the demand for breakwaters has also varied. The construction of breakwaters has shifted from onshore to offshore. Because most wave energy is concentrated near the water surface, a fixed box-type free surface breakwater (F-BW, Figure 1) was created [4,5]. The F-BW is a type of reflective breakwater with simple structure and high efficiency, which reduces the transmitted wave height by reflecting the incident wave energy [6,7]. Compared with the traditional bottom-founded breakwater, F-BW does not influence water exchange inside and outside the breakwater while maintaining a high wave attenuation efficiency, and has a high application prospect.

The hydrodynamic performance of the breakwater is important for the research and development of the F-BW, which mainly comprises two aspects. One aspect is the wave attenuation performance, including wave transmission coefficient C_t, reflection coefficient C_r, and dissipation coefficient C_d (hereinafter referred to as RTD coefficients). The other is the wave force, which concerns the safety and stability of the breakwater, including the horizontal wave force and vertical wave force.

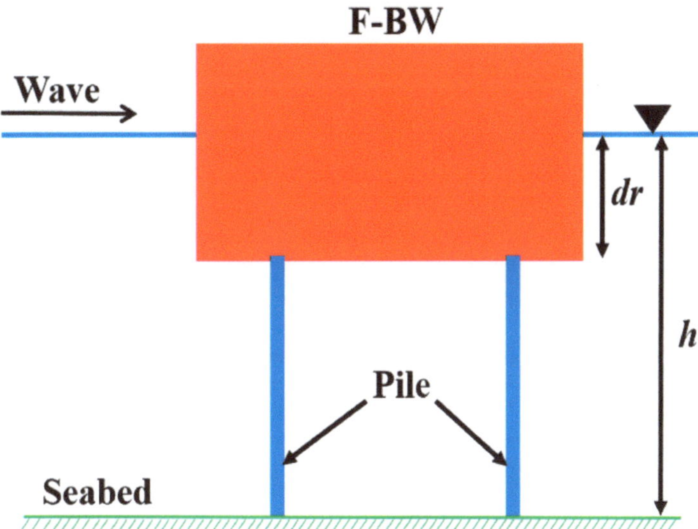

Figure 1. Two-dimensional schematic sketch of the F-BW models.

In terms of research on the RTD coefficients of an F-BW, some scholars have studied the influence of the breakwater width and draft on the reflection and transmission coefficients when energy dissipation was ignored. [8–11]. In order to provide some judgement for the needy practitioners, a closed-form formula has been created to predict the transmission coefficient in deep water [8–10]. A study by Kolahdoozan et al. [12] showed the poor prediction performance of the formula proposed by Macagno [8] for intermediate water. Therefore, it is necessary to explore a proposed formula for the transmission coefficient under intermediate-water conditions. Different from the analytical solution of potential flow theory, other scholars studied the influence of the draft, breakwater width, and wave height on the performance of wave reflection, transmission, and dissipation of the F-BW via experimental tests conducted in intermediate waters [13–16]. The computational fluid dynamics (CFD) technique provides us an alternative way to interpret the interaction between wave and F-BW. Koftis and Prinos [16] applied the two-dimensional unsteady Reynolds Averaged Navier–Stokes model to study the influence of the dimensionless draft on the transmission and reflection coefficients of an F-BW. Elsharnouby et al. [17] studied the influence of the draft on the wave transmission of the F-BW by using Flow-3D 11.2 software. Their results showed that the increasing draft impedes wave transmission.

Some researchers carried out earlier work on wave force of F-BW due to concerns regarding safety and stability of the F-BW. Guo et al. [11] confirmed that draft, breakwater width and wave period also influenced the horizontal and vertical wave forces by adopting mathematical analysis based on linear potential flow theory. Chen et al. [18] investigated the effects of wave height and wave period on the horizontal and vertical wave forces of F-BW through a series of experiments. The results showed that the horizontal and vertical wave forces increase with increasing wave height. Limited by the fact that the mathematical analysis tends to ignore flow viscosity [19–21] and the physical model test is complicated and costly, considerable effort has been devoted to studying the hydrodynamic performance of an F-BW through numerical simulation in recent years. Zheng et al. [22] and Ren et al. [23] used the smoothed particle hydrodynamics (SPH) method to numerically simulate the horizontal and vertical wave forces of F-BWs under regular waves. Unlike previous studies which overlooked the nonlinearity of wave forces, the positive and negative maximum wave force could be observed in the studies of Zheng et al. [22] and Ren et al. [23].

Human activities are less involved in deep water, and the cost-effectiveness of F-BW construction is poorer in deep water than intermediate water. Reflection coefficient C_t, and dissipation coefficient C_d are also an indispensable part of the wave attenuation performance of F-BW. The horizontal and vertical wave forces are related to the security of the F-BW. However, the prediction formulas based on tests or numerical simulations for horizontal and vertical wave forces of the F-BW in the above studies were rare. Therefore, an attempt is necessary to present a proposed formula for the prediction of RTD coefficients and wave forces, which will provide design judgments for the relevant practitioners in intermediate waters.

The objective of this paper is to provide the prediction formulas for RTD coefficients and wave forces in the intermediate waters under the condition that waves do not overtop the breakwater. With the rapid development of the CFD technique, Kurdistani et al. [24] proposed a formula for submerged homogeneous rubble mound breakwaters based on a large dataset from the CFD model, and the proposed formula was verified by using the literature observation datasets. Inspired by their research method, a numerical wave flume is built through a grid convergence test and validated with the existing experimental results. The prediction equations of RTD coefficients and wave forces are provided by applying multiple linear regression and verified by comparing with existing literature observation datasets. The major conclusions are finally summarized, and some prospects are proposed.

2. Theoretical Introduction

2.1. Governing Equations

Flow-3D 11.2 is widely used in coastal engineering as a powerful CFD software program [25]. The interaction of waves and breakwaters is simulated in a numerical wave tank by using Flow-3D 11.2 software in this paper. The numerical wave tank adopts an incompressible viscous fluid in the wave and F-BW interaction. The Reynolds averaged Navier–Stokes (RANS) equation was applied as the governing equation for turbulent flow. Assuming that the Cartesian coordinate system o-xyz originates from the still water surface, the continuity equation is shown in Equation (1), and the momentum equation is expressed in Equation (2).

$$\frac{\partial (u_i A_i)}{\partial x_i} = 0 \tag{1}$$

$$\frac{\partial \rho u_i}{\partial t} + \frac{\partial \rho u_i u_j}{\partial x_j} = -\frac{\partial p}{\partial x_i} + \frac{\partial}{\partial x_j}\left[v\left(\frac{\partial u_i}{\partial x_j} + \frac{\partial u_j}{\partial x_i} \right) - \rho u_i' u_j' \right] + \rho g_i \tag{2}$$

where i,j = 1,2 for two-dimensional flows, x_i represents the Cartesian coordinate, and u_i is the fluid velocity along the x- and z-axes. A_x and A_z are the area fractions open to flow in the x and z directions, respectively, ρ is the fluid density, p is the pressure, v is the dynamic viscosity, and g is the gravity force. The Reynolds stresses term, $\rho u_i' u_j'$, is modeled by the renormalized-group (RNG) turbulence model.

2.2. RNG Turbulence Model

The interaction of waves and an F-BW induces turbulence. The RNG turbulence model is adopted to close the governing equations [26], and the discrete governing equations are solved by the finite difference method. The transport equations of turbulent kinetic energy k_T and dissipation rate ε_T in this model are as follows:

$$\frac{\partial k_T}{\partial t} + \frac{\partial (k_T u_i)}{\partial x_i} = \frac{1}{\rho}\frac{\partial}{\partial x_j}\left[\left(\mu + \frac{\mu_t}{\sigma_k} \right) \frac{\partial k_T}{\partial x_j} \right] + \frac{P_k}{\rho} + (-\varepsilon_T) \tag{3a}$$

$$\frac{\partial \varepsilon_T}{\partial t} + \frac{\partial (\varepsilon_T u_i)}{\partial x_i} = \frac{1}{\rho}\frac{\partial}{\partial x_j}\left[\left(\mu + \frac{\mu_t}{\varepsilon_T}\right)\frac{\partial \varepsilon_T}{\partial x_j}\right] + \frac{1}{\rho}C_{\varepsilon 1}\frac{\varepsilon}{k_T}P_k - C_{\varepsilon 2}\frac{\varepsilon^2}{k_T}\frac{1}{\rho}R_\varepsilon \tag{3b}$$

where $\mu_t = C_\mu \rho \frac{k_T^2}{\varepsilon}$ is the turbulent viscosity, $P_k = \frac{1}{2}\mu_t\left(\frac{\partial u_i}{\partial x_j} + \frac{\partial u_j}{\partial x_i}\right)\left(\frac{\partial u_i}{\partial x_j} + \frac{\partial u_j}{\partial x_i}\right)$, and the constant values are $\sigma_k = \sigma_\varepsilon = 1.39, C_{\varepsilon 1} = 1.42, C_{\varepsilon 2} = 1.92,$ and $C_\mu = 0.085$.

The volume of fluid (VOF) method was developed to track the evolution of the free surface [27]. The governing equation is shown as follows:

$$\frac{\partial F}{\partial t} + \frac{\partial (Fu_i)}{\partial x_i} = 0 \tag{4}$$

where F represents the fractional volume of water fluid, $F = 1$ indicates that the numerical cell is full of water, and $F = 0$ corresponds to the cell fully occupied by air. Numerical cells with a value of $0 < F < 1$ represent a water surface.

Furthermore, the generalized minimal residual (GMRES) method was used to solve the velocity-pressure term [28], and the first-order upwind scheme and Split Lagrangian method were used to solve the volume of fluid advection. The structure of the F-BW is directly imported into Flow-3D 11.2 by the software built-in drawing function. The appearance of an F-BW depicted by the mesh could be viewed with the fractional area volume obstacle representation (FAVOR) method. All numerical simulations were run in parallel using an Intel Core (TM) i5-4460 processor (3.20 GHz). Furthermore, to ensure the accuracy of the numerical solution, the maximum iteration time step was set to 0.001 s, and the results were output at 0.01-s intervals.

2.3. Principle of Mass Source Wavemaker

The present study emerged from the interest shown in the use of F-BW in a specific zone at an actual project in East China Sea. The detailed structural design dimensions of F-BW and wave characteristics are shown in Table 1. All the incident waves are considered to be regular waves. The regular waves used in the study contain a large range of wave periods and wave heights, which represent the majority of wave parameters in real-world problems, making this study of great practical importance. The interaction between the second-order Stokes wave and the current is not considered in the twelve major wave parameters, due to the differing time and spatial scales between the wave and the current [29]. The twelve waves in this research are all in the range of either linear or nonlinear second-order Stokes waves. Figure 2 shows the suitability range of different wave theories. According to Figure 2, the F-BW at this project is located in intermediate waters. Equation (5) presents the wave elevation equation η of the second-order Stokes wave and the wave elevation equation of the linear wave is the first term on the right side of this equation.

$$\eta(t) = \frac{H_i}{2}cos(kx - \sigma t) + \frac{H_i^2 k}{16}\frac{cosh(kh)}{sinh^3(kh)}[2 + cosh(2kh)]cos(kx - \sigma t) \tag{5}$$

where H_i is the incident wave height, k is the wavenumber, σ is the wave frequency, and h is the still water depth.

The boundary wavemaker method produces re-reflection waves. Lin and Liu [30] proposed a popular mass source wave generation method [31–36]. In the present method, numerical wave generation is achieved by importing a given volume flow rate V_{fr} into the mass source model. The expression of the volume flow rate V_{fr} is as follows:

$$V_{fr} = 2C\eta(t)W \tag{6}$$

where C is the phase velocity, W is the tank width, $\eta(t)$ is the wave surface elevation by solving Equation (5).

To effectively reduce the calculation divergence caused by excessive waves in the NWT at the initial stage, the volume flow rate V_{fr} is multiplied by an increasing envelope function to make the wave increase gradually in the first three wave periods. The equation of the increasing function is as follows:

$$R = \begin{cases} 1 - exp\left(-\frac{2t}{T}\right), & \frac{t}{T} \leq 3 \\ 1, & \frac{t}{T} > 3 \end{cases} \tag{7}$$

where t is time and T is the wave period.

Table 1. Summary of the simulated scenarios.

	Variable 1	Variable 2	Variable 3	Variable 4
	dr	B	T	H_i
	[m]	[m]	[s]	[m]
Case 1	0.07			0.05
Case 2				0.07
Case 3	0.14			0.05
Case 4				0.07
Case 5	0.21	0.05	1.2	0.05
Case 6				0.07
Case 7	0.28			0.05
Case 8				0.07
Case 9	0.35			0.05
Case 10				0.07
Case 11		0.2		0.05
Case 12				0.07
Case 13		0.3		0.05
Case 14	0.14		1.2	0.07
Case 15		0.4		0.05
Case 16				0.07
Case 17		0.6		0.05
Case 18				0.07
Case 19			1	0.05
Case 20				0.07
Case 21			1.4	0.05
Case 22	0.14	0.5		0.07
Case 23			1.6	0.05
Case 24				0.07
Case 25			1.8	0.05
Case 26				0.07
Case 27	0.14			0.03
Case 28	0.28	0.5	1.2	
Case 29	0.14			0.09
Case 30	0.28			

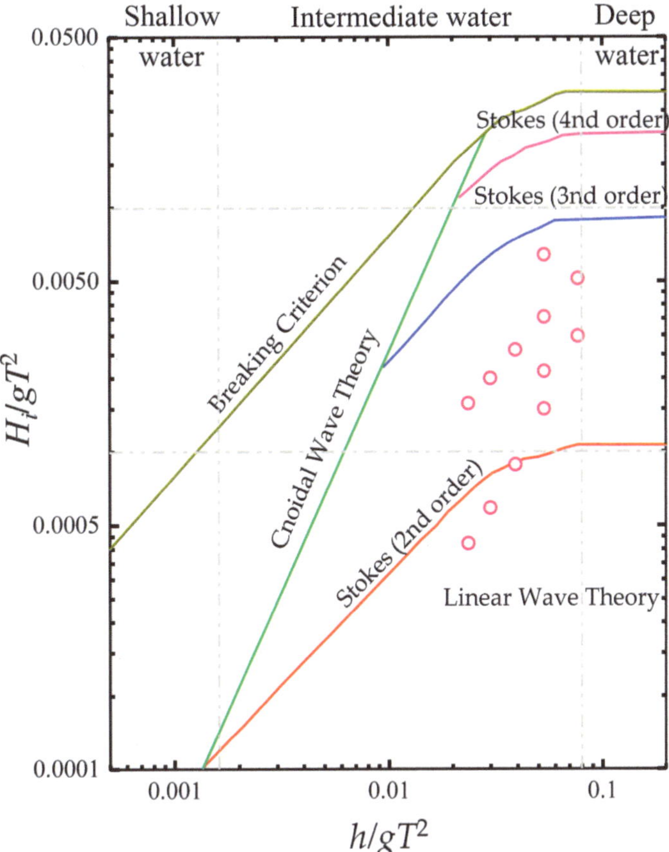

Figure 2. Wave parameter conditions analyzed in this study and their relations in the Le Méhauté diagram.

2.4. Principle of Numerical Solution

In this paper, the time series of wave elevations were recorded at five different locations (i.e., WG1–WG5) on the onshore and offshore sides of the F-BW (Figure 3a). Furthermore, the current WG spacings are selected according to the water depth and wave period. The distances between the wave source and WG1, WG1 and WG2, WG2 and F-BW, and F-BW and WG5 are set at 1.5 m, 0.2 m, 1.8 m, and 1.435 m, respectively. Note that the distance between wave gauges WG1 and WG2 is more than 0.05 L and less than 0.45 L, and the distances between wave gauges WG2 and F-BW and between WG5 and F-BW are less than 0.25 L and more than 0.2 L (wavelength), as recommended by the two-point method [37]. Two wave gauges (WG1 and WG2) are mounted in a line on the offshore side of the F-BW to separate the incident wave heights H_i and the reflected wave heights H_r by using this method. To prove that the horizontal wave force of the F-BW is related to the free surface onshore and offshore of the breakwater, probe WG3 is placed 0.02 m in front of the F-BW, while probe WG4 is placed 0.02 m behind the F-BW to measure the wave profile at the front ($\eta3$) and back ($\eta4$) of the F-BW. The wave gauge (WG5) is mounted on the onshore side of the F-BW to obtain the surface elevation of the transmitted wave heights H_t. The wave transmission, reflection, and wave energy dissipation coefficients are defined by solving Equation (8a)–(8c).

$$C_t = H_t/H_i \tag{8a}$$

$$C_r = H_r / H_i \tag{8b}$$

$$C_t^2 + C_r^2 + C_d^2 = 1 \tag{8c}$$

where C_t is the transmission coefficient; C_r is the reflection coefficient; and C_d is the wave energy dissipation coefficient.

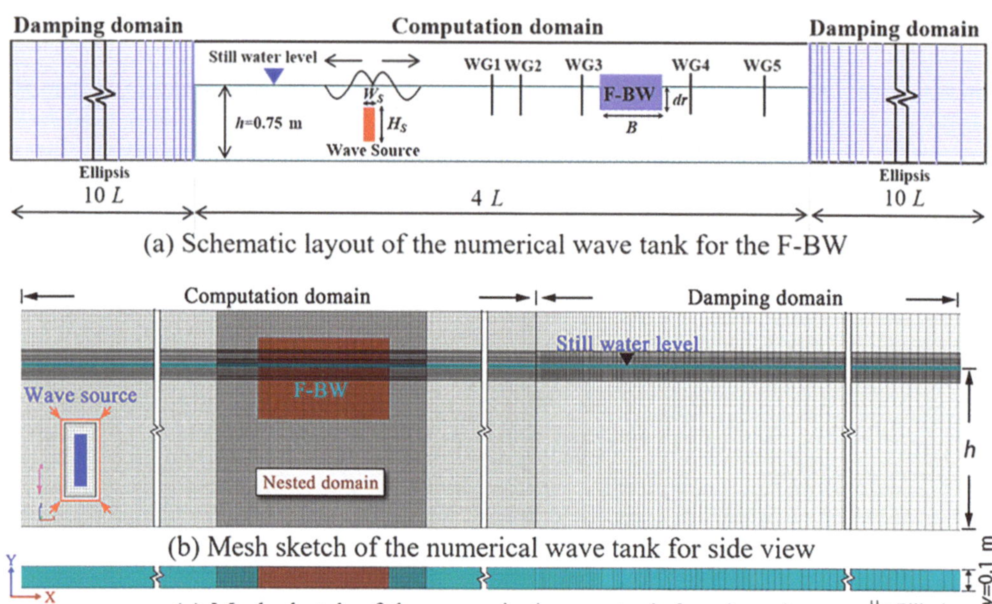

Figure 3. Schematic layout and mesh sketch of the numerical wave tank for the F-BW.

Furthermore, the horizontal and vertical wave forces are simulated by the integration of the water pressure p at the wet surface of the F-BW. The two kinds of wave forces include the hydrostatic force and hydrodynamic force according to the FLOW-3D theory manual [25]. Because the F-BW is always fixed at the free surface, the vertical wave force needs to remove part of the hydrostatic force (the value up to $\rho V g$, where ρ is the density of water and V is the volume of the F-BW). The shear stress is small enough to be ignored in this paper relative to the wave force. The horizontal wave force and the vertical wave force are denoted by F_x and F_z, respectively. The horizontal wave force is consistent with the direction of wave propagation, and the vertical wave force is vertically upward. To facilitate the research, obtaining the extreme value of the steady part of the wave force time series, we define the average value of the horizontal wave force positive and negative peak as the horizontal positive maximum wave force F_x^{+max} and horizontal negative maximum wave force F_x^{-max}, the vertical wave force positive and negative peak as the vertical positive maximum wave force F_z^{+max} and vertical negative maximum wave force F_z^{-max}. The representative time series of the dimensionless wave elevation, horizontal, and vertical wave forces are shown in Figure 4. The numerical results of H_i, H_r, H_t, $F_x^{\pm max}$, $F_z^{\pm max}$ were acquired based on the stable elevations in this figure. To facilitate discussion, we define $F_x^{\pm max}/0.005\ \rho g h^2$ and $F_z^{\pm max}/0.005\ \rho g h^2$ as the dimensionless horizontal and vertical maximum wave forces on the F-BW, respectively. The crest and trough values of the time series of the wave forces are studied because the extreme values of the horizontal and vertical wave forces on the F-BW under the Stokes second-order wave have a slightly sharper crest and flatter trough.

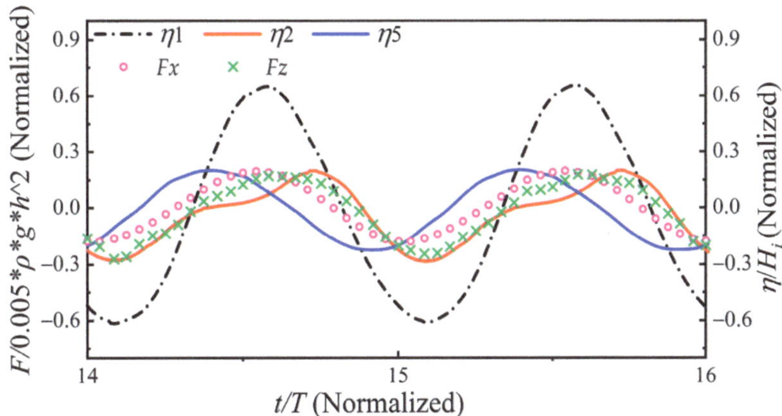

Figure 4. Time histories of wave elevation η measured by WG1, WG2, and WG5 and horizontal and vertical wave forces of F-BW at H_i = 0.07 m, T = 1.4 s, B = 0.5 m, dr = 0.14 m, and h = 0.75 m.

The integral formula of the horizontal and vertical wave force is shown in Equation (9).

$$F = \int p\vec{n}\,ds \tag{9}$$

where $\vec{n}$ is the unit normal vector of the object surface s and the water pressure p is determined by the Bernoulli equation.

3. Model Setup and Validation

3.1. Numerical Wave Tank Setup

The detailed numerical wave tank (NWT) setup is shown in Figure 3b,c. The total length of the NWT was twenty-four wavelengths L long in the x-axis direction, 0.1 m wide in the y-axis direction, and 1 m deep in the z-axis direction. A scale ratio of 1:40 and a constant water depth h of 0.75 m are adopted based on the Froude similarity law. The mesh consisted of two distinct regions. The first region was the computation domain, four wavelengths length, with a width of 0.1 m and a depth of 1 m. The unit grid size of the total NWT was $L/100 \sim L/200$ in the x and z directions, and ten grids were partitioned in the y directions in this domain. The second region was two identical damping domains with ten wavelength lengths. The Sommerfeld radiation method was employed to bate the secondary reflection of waves at both ends of the NWT. The grid size along the x-axis direction was gradually extended with an identical ratio of 1.01, and one grid was set for the lateral width of the NWT [38].

To describe the F-BW more accurately, nested grids were applied in the domain around the F-BW. The uniform nested grid was equal to half of the compute domain grid in the x, y and z-axis directions. Furthermore, the finer mesh resolution of 0.0035 cm in z direction was nested near the still water level (SWL), The region extends ±0.07 m from the SWL to ensure that the expected wave heights (0.03 m, 0.05 m, 0.07 m, 0.09 m) are encompassed within the region.

The boundary conditions of this NWT were set as follows: both ends of the NWT were defined as outflow boundaries, two sides of the domain were defined as symmetry boundaries, the atmospheric pressure was utilized at the upper boundary, and the lower surface of the computational domain was a no-slip wall boundary without surface roughness.

A mass source model with wide W_S and high H_S was added to the numerical flume. The symmetry boundaries were used overspreading with the mass source form, and the y-direction width of the mass source block was consistent with the width of the NWT. Pledging each edge of the mass block to coincide with the grid line of the NWT is shown in Figure 3b,c.

3.2. Numerical Model Validation

The present research is mainly implemented under the framework of CFD technology. To demonstrate the accuracy of the simulation results, it is essential to compare them with the extant results. The model is verified by the following three aspects in this section.

3.2.1. Grid Independent Verification

The mesh partition is a crucial procedure in CFD numerical simulation and needs much attention. The number and size of grids are essential criteria for evaluating the convergence of numerical results. Poor grid quality will directly affect the accuracy of numerical results and computation time. Consider that the proposed calculation cases $H_i = 0.06$ m, $T = 1.2$ s, and $h = 1.2$ m by Ren et al. are close to the target cases in this paper [23]. This wave condition is applied to complete the grid independence verification. Different grid arrangements can be seen in Table 2, and the time series of the wave profiles under the three grid sizes are compared with the theoretical results by solving Equation (5), as shown in Figure 5. The error of the numerical simulation results was calculated according to Equation (10). The wave profile deviations among the coarse mesh, medium mesh, and fine mesh are compared. The wave profiles under the medium mesh and the fine mesh are closer, and the deviation from the theoretical value is less than 5%, which meets the requirements of Det Norske Veritas (DNV) [39]. It can be judged that only medium meshes and refined meshes meet the requirements of numerical simulation. Considering the balance between calculation accuracy and calculation efficiency, the following numerical simulations always chose a medium mesh.

$$error = \frac{H_{theoretical} - H_{numerical}}{H_{theoretical}} \tag{10}$$

where $H_{theoretical}$ is the wave height of the theoretical result and $H_{numerical}$ is the wave height of the numerical result.

Table 2. Mesh independence check results.

Mesh Type	Computation Domain Grid Size (cm)	Nested Domain Grid Size (cm)	Cell Number	Elapsed Time ($\times 10^4$ s)	Wave Height (cm)	Error %
Coarse	2	1	701460	0.6496	5.642	5.96
Middle	1	0.5	3411180	7.6832	5.768	3.87
Fine	0.5	0.25	13350960	48.1437	5.769	3.85
Theoretical	-	-	-	-	6.000	-

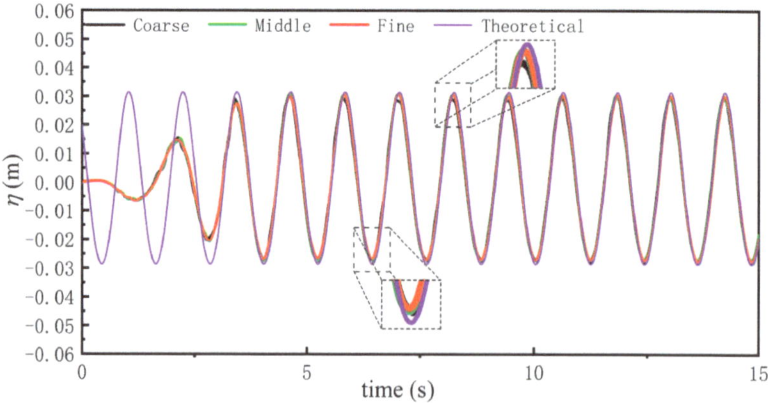

Figure 5. Grid independent verification: influence of mesh size on wave profile.

3.2.2. Validation of Wave Forces

In this section, to further inspect the accuracy of the numerical results of wave forces in this paper, according to the wave conditions of $H_i = 0.06$ m, $T = 1.2$ s, $h = 1.2$ m, and draft $dr = 0.2$ m, a rectangular box of width $B = 0.8$ m and wave height $H_i = 0.4$ m is fixed and semi-immersed, as proposed by Ren et al. [23]. The horizontal and vertical wave forces of the F-BW were verified by comparison with the theoretical results of Mei and Black [40] and the numerical simulation results of Ren et al. [23]. The time series of the wave forces are compared in Figure 6. The total simulation time of this case is 16 wave cycles. Since it takes some time for the progressive wave to arrive at the F-BW from the source, the horizontal and vertical wave forces begin to reach the stable state at $t = 7\ T$ seconds in Figure 6. By comparison, the simulated time series of horizontal and vertical wave forces are almost consistent with those presented by Ren et al. [23] and Mei and Black [40]. This result indicated that the present NWT could meet the calculation accuracy.

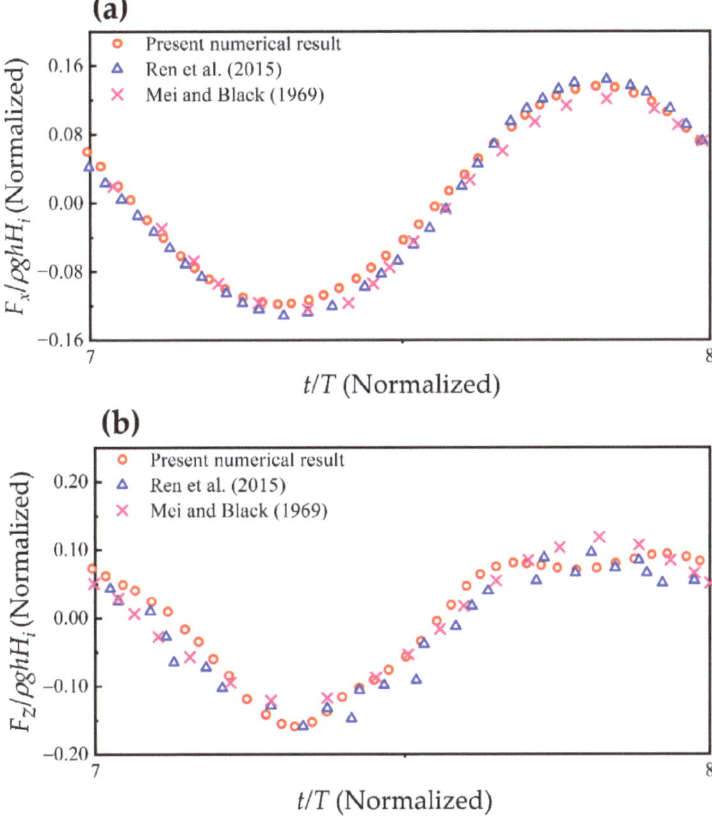

Figure 6. Comparison of the normalized wave force on an F-BW with previous studies (Mei and Black [40]; Ren et al. [23]). (**a**) Normalized horizontal wave force; (**b**) Normalized vertical wave force.

4. Results and Discussion

4.1. Influence Analysis of Four Factors on the Hydrodynamic Performance of F-BW

Among all the influencing factors (refer to Appendix A), the hydrodynamic performance of the F-BW is significantly affected by the following four factors: draft (dr/h), breakwater width (B/h), wave period (T^*sqrt(g/h)), and wave height (H_i/h). For the mechanism analysis of the interaction between waves and breakwater, the mechanism study of the horizontal wave force is rather complicated. Since the breakwater is in a semisubmerged state, the Morison formula is no longer applicable to the guidance of the calculation of the horizontal wave force. The horizontal wave force is studied separately from the water particle velocity; see the free surface difference ($\eta3$–$\eta4$) in the front and back sides of the F-BW and the water particle streamline in Figures 7 and 8 for details. Among them, five representative cases are selected from all cases in this article for comparative analysis corresponding to Figure 7a–e. Note that case (a) T1.2dr0.14B0.5Hi0.07 represents a wave period of 1.2 s, draft of 0.14 m, breakwater width of 0.5 m and incident wave height of 0.07 m. Due to the effect of water blockage, flow separation is generated at the bottom corner of the offshore side of the breakwater, and the generated clockwise vortex destroys the original motion path of the wave water particles without structure in Figure 8a and allows the free surface difference in the front and back of the F-BW to gradually reach a maximum. At time instant t_0 in Figure 7, the horizontal wave force also reaches a maximum. It can be seen in Figure 8b that the vertical wave force is easier to analyze. When the vertical wave force is at its maximum, the streamline realizes complete diffraction, and no vortex is generated. Furthermore, to understand the mechanism and contribution of each influencing factor on the hydrodynamic performance of the F-BW in detail, the statistical results are shown in Figures 9–12.

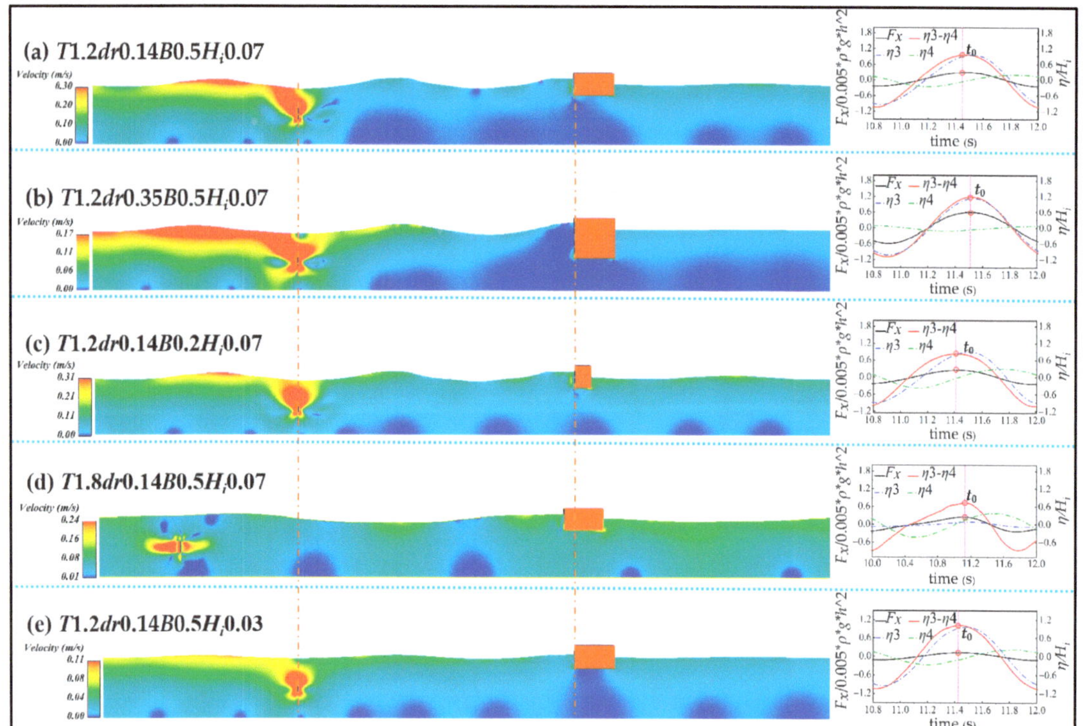

Figure 7. Comparative analysis of five different cases under the interaction between waves and breakwater: First column: numerically obtained snapshots of free surface profile and velocity field; Second column: time history of free surface and horizontal wave force.

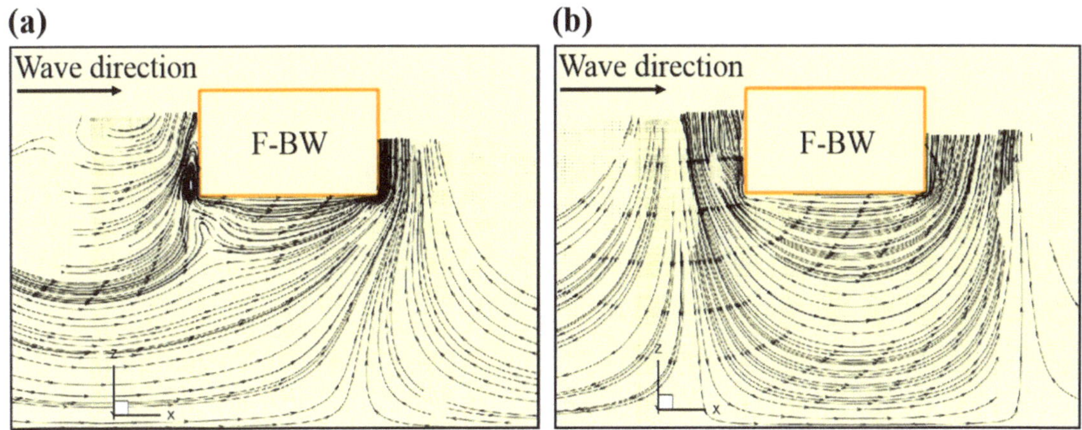

Figure 8. Snapshots of the velocity streamline field: (**a**) Time instant of the horizontal positive maximum wave force; (**b**) Time instant of the vertical positive maximum wave force.

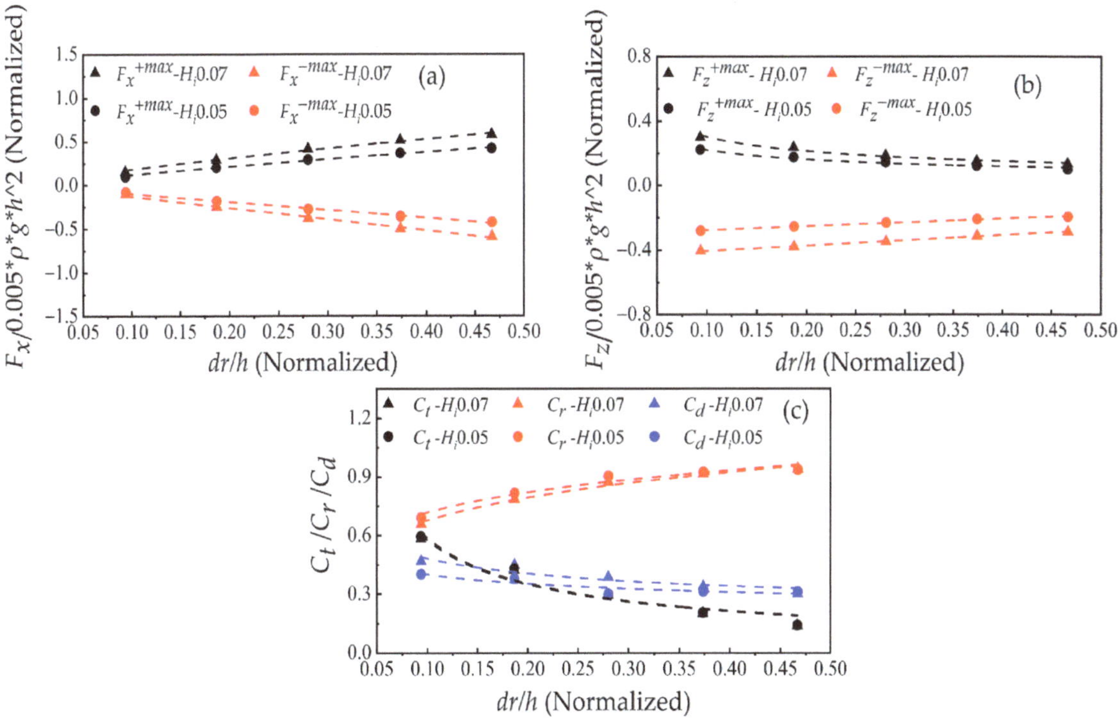

Figure 9. Effect of the draft *dr* on the hydrodynamic performance of the F-BW at wave heights $H_i = 0.05$ m and $H_i = 0.07$ m. (**a**) Horizontal positive and negative maximum wave forces F_x^{+max} and F_x^{-max}; (**b**) Vertical positive and negative maximum wave forces F_z^{+max} and F_z^{-max}; (**c**) Transmission coefficient C_t, reflection coefficient C_r, and dissipation coefficient C_d.

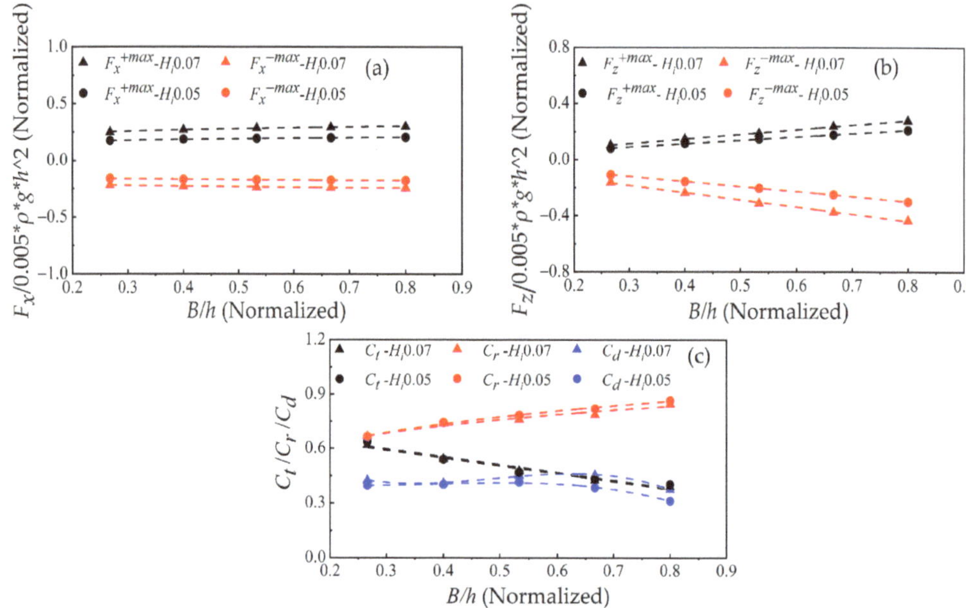

Figure 10. Influence of the breakwater width B on the hydrodynamic performance of the F-BW at wave heights $H_i = 0.05$ m and $H_i = 0.07$ m. (**a**) Horizontal positive and negative maximum wave forces F_x^{+max} and F_x^{-max}; (**b**) Vertical positive and negative maximum wave forces F_z^{+max} and F_z^{-max}; (**c**) Transmission coefficient C_t, reflection coefficient C_r, and dissipation coefficient C_d.

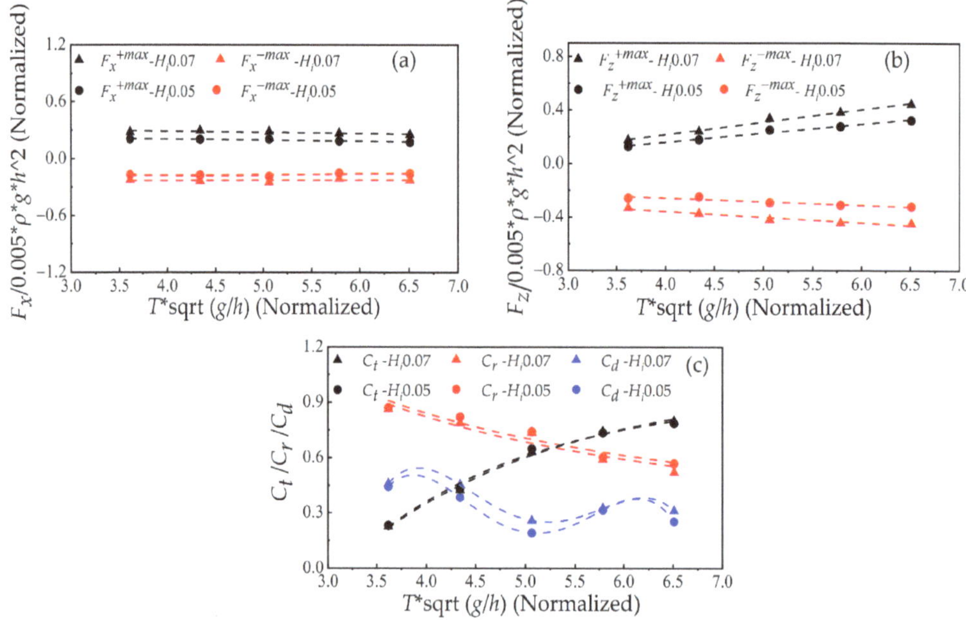

Figure 11. Influence of the wave period T on the hydrodynamic performance of the F-BW at wave heights $H_i = 0.05$ m and $H_i = 0.07$ m. (**a**) Horizontal positive and negative maximum wave forces F_x^{+max} and F_x^{-max}; (**b**) Vertical positive and negative maximum wave forces F_z^{+max} and F_z^{-max}; (**c**) Transmission coefficient C_t, reflection coefficient C_r, and dissipation coefficient C_d.

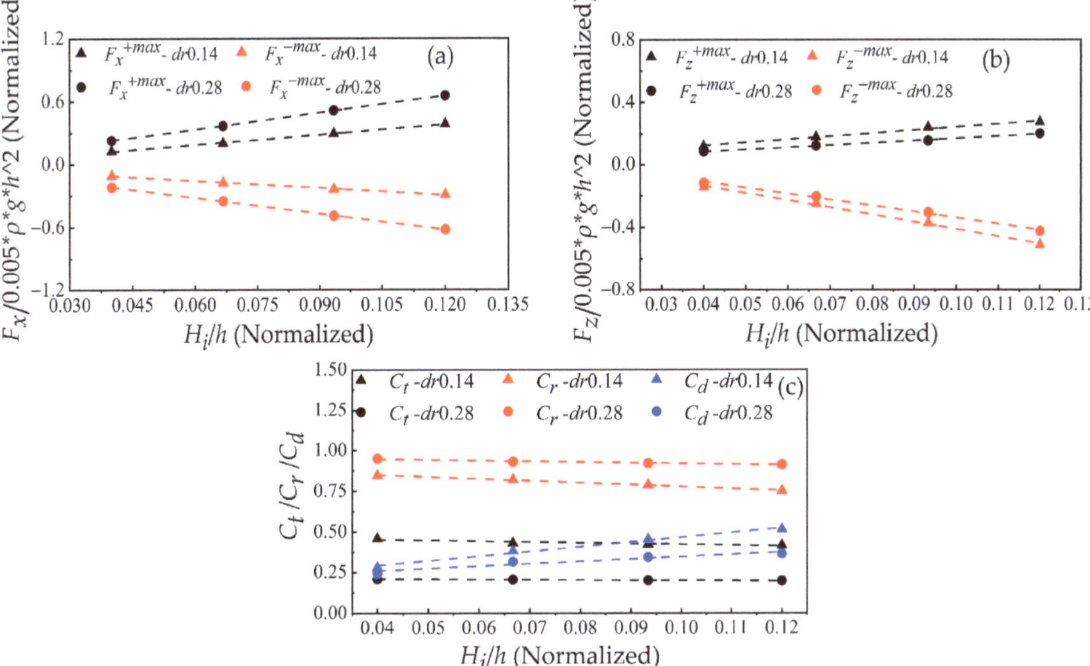

Figure 12. Influence of the wave height H_i on the hydrodynamic performance of the F-BW at draft $dr = 0.14$ m and $dr = 0.28$ m. (**a**) Horizontal positive and negative maximum wave forces F_x^{+max} and F_x^{-max}; (**b**) Vertical positive and negative maximum wave forces F_z^{+max} and F_z^{-max}; (**c**) Transmission coefficient C_t, reflection coefficient C_r, and dissipation coefficient C_d.

4.1.1. Effect of Draft

Figure 7 lists the distribution diagram of the free surface difference and water particle velocity under cases (a) and (b) at the time instant of the horizontal wave force maximum. Except for the draft being different, the two cases are consistent. Among them, case (a) has a wave period of 1.2 s, draft of 0.14 m, wave height of 0.07 m and breakwater width of 0.5 m. Case (b) has a period of 1.2 s, draft of 0.35 m, wave height of 0.07 m and breakwater width of 0.5 m.

In the second column of Figure 7a, when time $t_0 = 11.48$ s, the maximum free surface difference is 0.068 m, and the maximum horizontal wave force is 7.98 N. In the second column of Figure 7b, when time $t_0 = 11.52$ s, the maximum free surface difference is 0.083 m, and the maximum horizontal wave force is 15.91 N. Obviously, the increase in the draft enhances the water blockage action in front of the F-BW, weakens the diffraction effect of the wave, and delays the time for the horizontal wave force to reach its maximum. Figure 9a shows that F_x^{+max} increases with increasing draft under wave heights of $H_i = 0.05$ m and $H_i = 0.07$ m. Similarly, the absolute values of F_x^{-max} exhibit a similar law. The absolute values of F_z^{-max} and F_z^{+max} decrease with increasing draft under wave heights of $H_i = 0.05$ m and $H_i = 0.07$ m in Figure 9b, which is related to the exponential decay of the wave kinetic energy along the water depth. It is not difficult to see in Figure 7a,b that the wave hydrodynamic pressure on the lower surface of the F-BW decreases with decreasing wave kinetic energy as the water depth increases. The effective action area increases as the draft reduces the penetration of waves. Figure 9c shows that the transmission coefficient decreases with increasing draft under wave heights of $H_i = 0.05$ m and $H_i = 0.07$ m. Due to the increase in the interaction area between waves and F-BW, the reflected wave energy increases in Figure 7, and Figure 7b is more obvious than Figure 7a. The wave energy dissipation coefficient increases with decreasing draft in Figure 9c. Since the wave energy is mainly concentrated on the still water level, the fluid particle velocity maximum at the lower corner of F-BW is 0.30 m/s in Figure 7a is more than the 0.17 m/s in Figure 7b, more wave energy is dissipated when the fluid particle with higher velocity collides with F-BW due to decreasing draft.

Overall, the increasing draft impedes incident waves cross F-BW and promotes the increase in horizontal wave force and wave reflection, which threatens the stability of the structure.

4.1.2. Effect of Breakwater Width

To clarify the mechanism of the breakwater width effect on the hydrodynamic performance of the F-BW, except that the breakwater width is different, cases (a) and (c) in Figure 7 are consistent. In case (c), the period is 1.2 s, the draft is 014 m, the wave height is 0.07 m, and the breakwater width is 0.2 m.

The free surface difference and vortex in Figure 7a,c are similar. Figure 10a shows that the breakwater width effect on F_x^{-max} and F_x^{+max} is not obvious. When the vertical wave force is at its maximum, the streamline realizes complete diffraction, and no vortex is generated in Figure 8b. Therefore, the vertical wave force is related to the acting area of the F-BW lower surface. Figure 10b shows that the absolute values of F_z^{-max} and F_z^{+max} increase with increasing breakwater width. In the second column of Figure 7c, when time $t_0 = 11.42$ s, the free surface difference and the horizontal wave force reach a maximum faster than in case (a). Obviously, the increase in the breakwater width increases the wave diffraction difficulty. Figure 10c shows that the transmission coefficient decreases with increasing breakwater width, and the reflection coefficient increases with increasing breakwater width. Due to fluid particle velocity maximum is similar between Figure 7a,c. The increase in breakwater width has little influence on wave energy dissipation.

In short, the increasing breakwater width is not conducive to incident wave cross F-BW, and promotes the increase of wave reflection and vertical wave force. Obviously, more weights need to be added to ensure the safety of the breakwater when breakwater width increases.

4.1.3. Effect of Wave Period

To clarify the mechanism of the wave period effects on the hydrodynamic performance of the breakwater, except that the wave period is different, cases (a) and (d) are consistent. Figure 7d shows that the wave period is 1.8 s, the draft is 0.14 m, the wave height is 0.07 m and the breakwater width is 0.5 m.

In the second column of Figure 7d, when time $t_0 = 11.13$ s, the maximum free surface difference is 0.051 m, and the maximum horizontal wave force is 6.90 N. According to Equation (9), because the wave energy is more abundant on the two sides of the breakwater in case (4), the horizontal wave force is comparable even if the free surface difference is smaller than that in case (1). Figure 11a shows that F_x^{-max} and F_x^{+max} are weakly related to the wave period under wave heights of $H_i = 0.05$ m and $H_i = 0.07$ m. Because the long-period waves possess a large wave energy in Figure 7d, they increase the wave pressure on the lower surface of the F-BW. Therefore, the absolute values of F_z^{-max} and F_z^{+max} increase linearly with the wave period in Figure 11b. Figure 11c shows that the transmission coefficient increases with increasing wave period under wave heights of $H_i = 0.05$ m and $H_i = 0.07$ m. Long-period waves have a better diffraction ability at the same depth, and more wave energy passes through the F-BW. The decreasing ratio of the breakwater width to wavelength weakens the ability to block progressive waves, and the reflection coefficient decreases accordingly. The wave energy dissipation coefficient shows an alphabetic symbol "M" distribution with the wave period. This indicates that the wave energy dissipation is more complex and requires further study. When the dimensionless wave period is 5.06, both the transmission and reflection coefficients are close to 0.71, the dissipation coefficient is at the minimum by applying Equation (8c).

In brief, the increasing wave period plays a significant role in increasing the wave transmission and the reducing wave reflection. Although it has little effect on the horizontal wave force, it promotes an increase in the vertical wave force, which is unfavorable to the security of the breakwater.

4.1.4. Effect of Wave Height

To clarify the mechanism of the wave height effects on the hydrodynamic performance of the breakwater, except that the wave height is different, cases (a) and (e) are consistent. Figure 7e shows that the wave period is 1.2 s, the draft is 014 m, the wave height is 0.03 m and the breakwater width is 0.5 m.

In the second column of Figure 7e, when time $t_0 = 11.44$ s, the maximum free surface difference is 0.031 m, and the maximum horizontal wave force is 3.43 N. Obviously, the increase in wave height increases the diffraction difficulty of the wave and delays the time when the horizontal wave force reaches its maximum. The higher the wave height, the more abundant the wave energy in Figure 7a,e. The water particle velocity maximum is 0.11 m/s in Figure 7e, which is much less than the water particle velocity maximum in Figure 7a. The larger wave height causes a larger wave elevation difference, and the larger horizontal wave force under other variable conditions is consistent by comparing Figure 7a,e. Therefore, F_x^{-max} and F_x^{+max} increase linearly with increasing wave height under drafts $dr = 0.14$ m and $dr = 0.28$ m in Figure 12a. The increase in wave height leads to increasing dynamic wave pressure, which in turn leads to increasing wave pressure on the F-BW lower surface and an increase in vertical wave force. Therefore, F_z^{-max} and F_z^{+max} increase linearly with increasing wave height under drafts $dr = 0.14$ m and $dr = 0.28$ m in Figure 12b. Figure 12c shows that the increasing wave height results in more wave reflection and less transmission due to the increasing blockage effect. The reflection ability weakens with decreasing interaction area (the ratio of the wetted surface height of the front wall of the F-BW to the wave height). The water particle velocity maximum of 0.11 m/s in Figure 7e is less than the water particle velocity maximum of 0.3 m/s in Figure 7a. The increasing water particle velocity with increasing wave height results in better vortex dissipation near the F-BW. Hence, the wave energy dissipation coefficient increases.

In conclusion, the increasing wave height reduces the wave reflection but increases horizontal and vertical wave forces, which is disadvantageous to the security of the breakwater.

4.2. Prediction Equations of F-BW Hydrodynamic Performance Parameters

To understand the contribution of each influencing factor to the hydrodynamic performance of the F-BW in detail, the factors affecting the RTD coefficients and wave force mainly include the wave period T, wave height H_i, draft d_r, breakwater width B, and still water depth h. In Equation (11), the RTD coefficients $C_{t,r,d}$ and wave force extremum $F_{x,z}^{\pm max}$ are expressed as follows:

$$\left(F_{x,z}^{\pm max}, C_{t,r,d} \right) = f(h, T, \rho, H_i, dr, B, g) \tag{11}$$

Using the dimensionless analysis method and the numerical simulation results of 30 groups of simulated conditions in Table 1 based on the Origin 2019b software platform, multiple linear regression was performed by the least squares method, and the prediction equations of the RTD coefficients and wave force are given in Equation (12a–g). The detailed formulas are shown in Table 3.

Note that $0.0933 \leq dr/h \leq 0.4667$, $0.26667 \leq B/h \leq 0.8$, $3.6166 \leq T*sqrt(g/h) \leq 6.5099$, and $0.04 \leq H_i/h \leq 0.12$.

Table 3. Statistics of prediction equation.

Equation Number	Equations	R^2
(12a)	$C_t = 0.003\left(\frac{dr}{h}\right)^{-0.935}\left(\frac{B}{h}\right)^{-0.519}\left(\frac{gT^2}{h}\right)^{1.039}\left(\frac{H_i}{h}\right)^{-0.064}$	0.948
(12b)	$C_r = 3.650\left(\frac{dr}{h}\right)^{0.213}\left(\frac{B}{h}\right)^{0.187}\left(\frac{gT^2}{h}\right)^{-0.436}\left(\frac{H_i}{h}\right)^{-0.074}$	0.958
(12c)	$C_d = 3.100\left(\frac{dr}{h}\right)^{-0.235}\left(\frac{B}{h}\right)^{-0.113}\left(\frac{gT^2}{h}\right)^{-0.493}\left(\frac{H_i}{h}\right)^{0.426}$	0.695
(12d)	$\frac{F_x^{+max}}{0.005\rho gh^2} = 21.398\left(\frac{dr}{h}\right)^{0.866}\left(\frac{B}{h}\right)^{0.139}\left(\frac{gT^2}{h}\right)^{-0.127}\left(\frac{H_i}{h}\right)^{1.027}$	0.992
(12e)	$\frac{F_x^{-max}}{0.005\rho gh^2} = -14.199\left(\frac{dr}{h}\right)^{1.062}\left(\frac{B}{h}\right)^{0.057}\left(\frac{gT^2}{h}\right)^{-0.042}\left(\frac{H_i}{h}\right)^{0.912}$	0.988
(12f)	$\frac{F_z^{+max}}{0.005\rho gh^2} = 0.079\left(\frac{dr}{h}\right)^{-0.519}\left(\frac{B}{h}\right)^{0.852}\left(\frac{gT^2}{h}\right)^{0.829}\left(\frac{H_i}{h}\right)^{0.798}$	0.988
(12g)	$\frac{F_z^{-max}}{0.005\rho gh^2} = -2.314\left(\frac{dr}{h}\right)^{-0.254}\left(\frac{B}{h}\right)^{0.890}\left(\frac{gT^2}{h}\right)^{0.282}\left(\frac{H_i}{h}\right)^{1.145}$	0.989

4.3. Deviation Analysis of the Prediction Equations

Inspired by Kurdistani et al.'s [24] research method, the current study uses their method to assess the reliability of each predictive formula. The literature observation datasets include the measured RTD coefficients from Koutandos [13] (three cases (R1, R2 and R3) in Figure 16 of his literature) and Liang et al. [14] (six cases in Figures 14a, 19a and 22a of their literature), the wave forces from Mei and Black [40] and Ren et al. [23] (a case in Figure 10 of their literature). The numerical results obtained by Flow-3D are plotted on the *x*-axis in Figure 13, and predicted values of the predictive equations are plotted on the *y*-axis in Figure 13. Figure 13a shows a 20% error for the application of Liang et al. [14] transmission coefficient datasets that are mostly lower-estimated values of transmission coefficient with respect to Equation (12a), an almost 10% error for the application of Liang et al.'s [14] reflection coefficient and dissipation coefficient datasets, and Koutandos's [13] RTD coefficients datasets. Figure 13b shows an almost 10% error for the application of Mei and Black [40] and Ren et al. [23] maximum wave force, which indicates that the present prediction equations are valid.

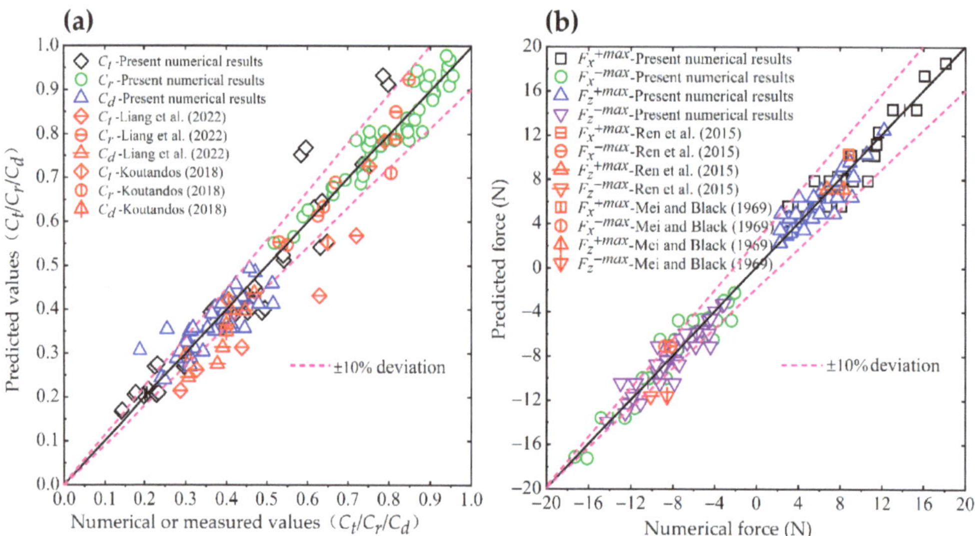

Figure 13. Comparison of the results between previous studies and the numerical results of this study. (**a**) Transmission coefficient C_t, reflection coefficient C_r, and dissipation coefficient C_d (Koutandos [13], Liang et al. [14]) and (**b**) Maximum wave force (Mei and Black [40]; Ren et al. [23]).

It is clearly found that the distribution points of the reflection coefficient and wave energy dissipation coefficient of the F-BW are relatively concentrated in a particular region in Figure 13a, indicating that the F-BW is dominant in reflecting waves and has stable wave dissipation ability. In addition, the horizontal negative maximum wave force of the F-BW is similar to the vertical negative maximum wave force, and the horizontal positive maximum wave force of the F-BW is slightly larger than the vertical positive maximum wave force in Figure 13b.

5. Conclusions

The present study investigated a high-accuracy numerical wave tank (NWT) based on the Flow-3D platform. A series of numerical simulations in the intermediate waters were carried out at a constant water depth (h) of 0.75 m under regular wave conditions, with a wave height range between 0.03–0.09 m, a wave period range between 1–1.8 s, and a breakwater width range between 0.2–0.6 m. The effects of four influencing factors (dr, B, T, H_i) on the hydrodynamic performance (RTD coefficients and wave forces) are highlighted. The vital conclusions are as follows:

(1) The performance of two-dimensional viscous numerical wave tanks (NWTs) with a mass source wave maker and small length scale (1:40) are analyzed. By comparison, the wave model employed in this paper is competent for the numerical simulation of the F-BW.

(2) The results show that the increase in the four influence factors, except the wave period, benefits the decrease in the wave transmission. The increase in draft and breakwater width is beneficial to the increase in the wave reflection, and the wave period and wave height are opposite. The increase in draft benefits the decrease in wave energy dissipation, and the wave height is opposite.

(3) The increase in the draft and wave height benefits the increase in the horizontal positive and negative maximum wave forces. In addition to the draft, the increase in the other three influence factors benefits the increase in the vertical positive and negative maximum wave forces.

(4) Applying multiple linear regression presents the prediction equations of RTD coefficients and the extreme wave force. The prediction equations are verified by comparing them with literature observation datasets.

This study provides insight into the relation of RTD coefficients and wave forces with parameters such as draft, breakwater width, wave period and wave height. The simulated results of the given predicted equations can be generalized to the prototype scale by using Froude's scaling law and can be used to guide the design of F-BWs in intermediate waters.

Author Contributions: Conceptualization and methodology, G.N., N.F. and Y.C.; software, J.L. and G.N.; visualization, J.Z. and N.F.; writing—original draft, G.N. and N.F.; writing—review and editing, G.N., N.F. and J.L.; supervision, N.F. and Y.C. All authors have read and agreed to the published version of the manuscript.

Funding: The work presented here was supported by the Science and Technology Innovation Serve Project of Wenzhou Association for Science and Technology (No. kjfw65), the General Scientific Research Project of Education Department of Zhejiang Province (Y202250703), the National Natural Science Foundation of China (52108337), the Basic Scientific Research Project of Wenzhou City (No. S20220033), the Wenzhou Science and Technology Commissioners Project (X20210087), and the Wenzhou Basic Public Scientific Research Project (2023S0109).

Institutional Review Board Statement: Not applicable.

Informed Consent Statement: Not applicable.

Data Availability Statement: The original contributions presented in the study are included in the article. Further inquiries can be directed to the corresponding author.

Acknowledgments: The assistance of Yongchun Yang and Baochang Zhang, of the Ocean University of China's College of Engineering, received during the course of the numerical experiments, is greatly appreciated.

Conflicts of Interest: The authors declare that the research was conducted in the absence of any commercial or financial relationships that could be construed as a potential conflict of interest.

Appendix A

The wave period T, wave height H_i, draft d_r, breakwater width B, and water depth h are the main factors that affect the wave dissipation performance and wave force of an F-BW in the intermediate waters. Therefore, the wave force of an F-BW can be expressed as a function of the above factors as follows:

$$F = f(h, T, \rho, H_i, dr, B, g) \tag{A1}$$

Taking water depth h, gravity acceleration g, and water density ρ as the repetitive parameters, the three dimensionless parameters are expressed as follows:

$[h] = [M^0 L^1 T^0]$, $[g] = [M^0 L^1 T^{-2}]$, $[\rho] = [M^1 L^{-3} T^0]$, where wave force per breakwater length in the vertical wave direction F, expressed as $[F = \rho g h^2]$, Equation (A1) can be written as follows:

$$\frac{F}{0.005 \rho g h^2} = f\left(\frac{dr}{h}, \frac{B}{h}, T\sqrt{\frac{g}{h}}, \frac{H_i}{h}\right) \tag{A2}$$

According to wave theory, there is a nonlinear relationship between the wave force and the four dimensionless parameters in Equation (A2). The relationship between the dependent variable and independent variable in nature is exponential. It can be expressed as follows:

$$\frac{F}{0.005 \rho g h^2} = \alpha \left(\frac{dr}{h}\right)^{x_1} \left(\frac{B}{h}\right)^{x_2} \left(T\sqrt{\frac{g}{h}}\right)^{x_3} \left(\frac{H_i}{h}\right)^{x_4} \tag{A3}$$

where α, x_1, x_2, x_3, and x_4 are the unknown coefficients.

Taking the natural logarithm of both sides of Equation (A3) to obtain the double logarithm function model, the equation can be written in linear form as follows:

$$ln\frac{F}{0.005 \rho g h^2} = ln\alpha + x_1 ln\left(\frac{dr}{h}\right) + x_2 ln\left(\frac{B}{h}\right) + x_3 ln\left(T\sqrt{\frac{g}{h}}\right) + x_4 ln\left(\frac{H_i}{h}\right) \tag{A4}$$

Using multiple function linear regression analysis, each unknown coefficient in the equations can be obtained and then substituted into Equation (A3) to obtain the wave force equations. Similarly,

$$C_t, C_r, C_d = f\left(\frac{dr}{h}, \frac{B}{h}, T\sqrt{\frac{g}{h}}, \frac{H_i}{h}\right) \tag{A5}$$

References

1. He, F.; Huang, Z.H.; Law, A.W.K. Hydrodynamic performance of a rectangular floating breakwater with and without pneumatic chambers: An experimental study. *Ocean Eng.* **2012**, *51*, 16–27. [CrossRef]
2. Zhan, J.; Chen, X.; Gong, Y.; Hu, W. Numerical investigation of the interaction between an inverse T-type fixed/floating breakwater and regular/irregular waves. *Ocean Eng.* **2017**, *137*, 110–119. [CrossRef]
3. Fu, D.; Zhao, X.Z.; Wang, S.; Yan, D.M. Numerical study on the wave dissipating performance of a submerged heaving plate breakwater. *Ocean Eng.* **2021**, *219*, 108310. [CrossRef]
4. Hales, L.Z. *Floating Breakwaters: State-of-the-Art Literature Review*; Coastal Engineering Research Center: London, UK, 1981.
5. Teh, H.M. Hydraulic performance of free surface breakwaters: A review. *Sains Malays.* **2013**, *42*, 1301–1310.
6. Liang, J.M.; Chen, Y.K.; Liu, Y.; Li, A.J. Hydrodynamic performance of a new box-type breakwater with superstructure: Experimental study and SPH simulation. *Ocean Eng.* **2022**, *266*, 112819. [CrossRef]
7. Zhao, X.L.; Ning, D.Z. Experimental investigation of breakwater-type WEC composed of both stationary and floating pontoons. *Energy* **2018**, *155*, 226–233. [CrossRef]

8. Macagno, A. Wave action in a flume containing a submerged culvert. In *La Houille Blanche*; Taylor and Francis: London, UK, 1954.
9. Wiegel, R.L. Transmission of waves past a rigid vertical thin barrier. *J. Waterw. Harb. Div.* **1960**, *86*, 1–12. [CrossRef]
10. Ursell, F. The effect of a fixed vertical barrier on surface waves in deep water. *Math. Proc. Camb. Philos. Soc.* **1947**, *43*, 374–382. [CrossRef]
11. Guo, Y.; Mohapatra, S.C.; Guedes Soares, C. Wave interaction with a rectangular long floating structure over flat bottom. In *Progress in Maritime Technology and Engineering*; CRC Press: Boca Raton, FL, USA, 2018; pp. 647–654.
12. Kolahdoozan, M.; Bali, M.; Rezaee, M.; Moeini, M.H. Wave-transmission prediction of π-type floating breakwaters in intermediate waters. *J. Coast. Res.* **2017**, *33*, 1460–1466. [CrossRef]
13. Koutandos, E. Regular-irregular wave pressures on a semi-immersed breakwater. *J. Mar. Environ. Eng.* **2018**, *10*, 109–145.
14. Liang, J.M.; Liu, Y.; Chen, Y.K.; Li, A.J. Experimental study on hydrodynamic characteristics of the box-type floating breakwater with different mooring configurations. *Ocean Eng.* **2022**, *254*, 111296. [CrossRef]
15. Fugazza, M.; Natale, L. Energy losses and floating breakwater response. *J. Waterw. Port Coast. Ocean Eng.* **1988**, *114*, 191–205. [CrossRef]
16. Koftis, T.; Prinos, P. 2 DV Hydrodynamics of a Catamaran-Shaped Floating Structure. *Iasme Trans.* **2005**, *2*, 1180–1189.
17. Elsharnouby, B.; Soliman, A.; Elnaggar, M.; Elshahat, M. Study of environment friendly porous suspended breakwater for the Egyptian Northwestern Coast. *Ocean Eng.* **2012**, *48*, 47–58. [CrossRef]
18. Chen, Y.; Niu, G.; Ma, Y. Study on hydrodynamics of a new comb-type floating breakwater fixed on the water surface. In Proceedings of the E3S Web of Conferences, Wuhan, China, 14–16 December 2018; EDP Sciences: Les Ulis, France, 2018; Volume 79, p. 02003.
19. Fan, N.; Nian, T.K.; Jiao, H.B.; Guo, X.S.; Zheng, D.F. Evaluation of the mass transfer flux at interfaces between submarine sliding soils and ambient water. *Ocean Eng.* **2020**, *216*, 108069. [CrossRef]
20. Fan, N.; Jiang, J.X.; Dong, Y.K.; Guo, L.; Song, L.F. Approach for evaluating instantaneous impact forces during submarine slide-pipeline interaction considering the inertial action. *Ocean Eng.* **2022**, *245*, 110466. [CrossRef]
21. Ghafari, A.; Tavakoli, M.R.; Nili-Ahmadabadi, M.; Teimouri, K.; Kim, K.C. Investigation of interaction between solitary wave and two submerged rectangular obstacles. *Ocean Eng.* **2021**, *237*, 109659. [CrossRef]
22. Zheng, X.; Lv, X.P.; Ma, Q.W.; Duan, W.Y.; Khayyer, A.; Shao, S.D. An improved solid boundary treatment for wave–float interactions using ISPH method. *Int. J. Nav. Archit. Ocean Eng.* **2018**, *10*, 329–347. [CrossRef]
23. Ren, B.; He, M.; Dong, P.; Wen, H.J. Nonlinear simulations of wave-induced motions of a freely floating body using WCSPH method. *Appl. Ocean Res.* **2015**, *50*, 1–12. [CrossRef]
24. Kurdistani, S.M.; Tomasicchio, G.R.; D'Alessandro, F.; Francone, A. Formula for Wave Transmission at Submerged Homogeneous Porous Breakwaters. *Ocean Eng.* **2022**, *266*, 113053. [CrossRef]
25. Hirt, C.W.; Nichols, B.D. *Flow-3D User's Manuals*; Flow science Inc.: Santa Fe, NM, USA, 2012.
26. Yakhot, V.; Orszag, S.A. Renormalization group analysis of turbulence. I. Basic theory. *J. Sci. Comput.* **1986**, *1*, 3–51. [CrossRef]
27. Hirt, C.W.; Nichols, B.D. Volume of fluid (VOF) method for the dynamics of free boundaries. *J. Comput. Phys.* **1981**, *39*, 201–225. [CrossRef]
28. Saad, Y.; Schultz, M.H. GMRES: A generalized minimal residual algorithm for solving nonsymmetric linear systems. *SIAM J. Sci. Stat. Comput.* **1986**, *7*, 856–869. [CrossRef]
29. Faraci, C.; Musumeci, R.E.; Marino, M.; Ruggeri, A.; Carlo, L.; Jensen, B.; Foti, E.; Barbaro, G.; Elsaßer, B. Wave-and current-dominated combined orthogonal flows over fixed rough beds. *Cont. Shelf Res.* **2021**, *220*, 104403. [CrossRef]
30. Lin, P.Z.; Liu, P.L.-F. Internal wave-maker for Navier-Stokes equations models. *J. Waterw. Port Coast. Ocean Eng.* **1999**, *125*, 207–215. [CrossRef]
31. Lara, J.L.; Garcia, N.; Losada, I.J. RANS modelling applied to random wave interaction with submerged permeable structures. *Coast. Eng.* **2006**, *53*, 395–417. [CrossRef]
32. Ha, T.; Lin, P.Z.; Cho, Y.S. Generation of 3D regular and irregular waves using Navier-Stokes equations model with an internal wave maker. *Coast. Eng.* **2013**, *76*, 55–67. [CrossRef]
33. Chen, Y.L.; Hsiao, S.C. Generation of 3D water waves using mass source wavemaker applied to Navier-Stokes model. *Coast. Eng.* **2016**, *109*, 76–95. [CrossRef]
34. Windt, C.; Davidson, J.; Schmitt, P.; Ringwood, J.V. On the assessment of numericalwave makers in CFD simulations. *J. Mar. Sci. Eng.* **2019**, *7*, 47. [CrossRef]
35. Wang, D.X.; Dong, S. Generating shallow-and intermediate-water waves using a line-shaped mass source wavemaker. *Ocean Eng.* **2021**, *220*, 108493. [CrossRef]
36. Wang, D.X.; Sun, D.; Dong, S. Numerical investigation into effect of the rubble mound inside perforated caisson breakwaters under random sea states. *Proc. Inst. Mech. Eng. Part M J. Eng. Marit. Environ.* **2022**, *236*, 48–61. [CrossRef]
37. Goda, Y.; Suzuki, Y. Estimation of incident and reflected waves in random wave experiments. In Proceedings of the 15th International Conference on Coastal Engineering, Honolulu, HI, USA, 11–17 July 1976; pp. 828–845.

38. Orlanski, I. A simple boundary condition for unbounded hyperbolic flows. *J. Comput. Phys.* **1976**, *21*, 251–269. [CrossRef]
39. Veritas, D.N. *Environmental Conditions and Environmental Loads: Recommended Practice*; DNV-RP-C205; Det Norske Veritas (DNV): Oslo, Norway, 2010.
40. Mei, C.C.; Black, J.L. Scattering of surface waves by rectangular obstacles in waters of finite depth. *J. Fluid Mech.* **1969**, *38*, 499–511. [CrossRef]

Journal of
Marine Science and Engineering

MDPI

Article

Study on the Relationship between Resistivity and the Physical Properties of Seafloor Sediments Based on the Deep Neural Learning Algorithm

Zhiwen Sun [1,2,3,4], Zhihan Fan [2,*], Chaoqi Zhu [2], Kai Li [2], Zhongqiang Sun [2], Xiaoshuai Song [2], Liang Xue [2], Hanlu Liu [2] and Yonggang Jia [2,*]

[1] Key Laboratory of Gas Hydrate, Ministry of Natural Resources, Qingdao Institute of Marine Geology, Qingdao 266237, China; zhiwensun91@163.com
[2] Shandong Key Laboratory of Marine Environmental Geology Engineering, Ocean University of China, Qingdao 266100, China
[3] Laboratory for Marine Mineral Resources, Pilot National Laboratory for Marine Science and Technology, Qingdao 266237, China
[4] Technology Innovation Center for Marine Methane Monitoring, Ministry of Natural Resources, Qingdao 266237, China
* Correspondence: fanzhihan@ouc.edu.cn (Z.F.); yonggang@ouc.edu.cn (Y.J.)

Citation: Sun, Z.; Fan, Z.; Zhu, C.; Li, K.; Sun, Z.; Song, X.; Xue, L.; Liu, H.; Jia, Y. Study on the Relationship between Resistivity and the Physical Properties of Seafloor Sediments Based on the Deep Neural Learning Algorithm. *J. Mar. Sci. Eng.* **2023**, *11*, 937. https://doi.org/10.3390/jmse11050937

Academic Editor: George Kontakiotis

Received: 1 March 2023
Revised: 9 April 2023
Accepted: 12 April 2023
Published: 27 April 2023

Abstract: The occurrence of deep-sea geohazards is accompanied by dynamic changes in the physical properties of seafloor sediments. Therefore, studying the physical properties is helpful for monitoring and early warnings of deep-sea geohazards. Existing physical property inversion methods have problems regarding the poor inversion accuracy and limited application scope. To address these issues, we establish a deep learning model between the resistivity of seafloor sediment and its density, water content, and porosity. Compared with empirical formulas, the deep learning model has the advantages of a more concentrated prediction range and a higher prediction accuracy. This algorithm was applied to invert the spatial distribution characteristics and temporal variation of the seafloor sediment density, water content, and porosity in the South China Sea hydrate test area for 12 days. The study reveals that the dynamic changes in the physical properties of seafloor sediments in the South China Sea hydrate zone exhibit obvious stratification characteristics. The dynamic changes in the physical properties of seafloor sediments are mainly observed at depths of 0–0.9 m below the seafloor, and the sediment properties remain stable at depths of 0.9–1.8 m below the seafloor. This study achieves the monitoring and early warning of dynamic changes in the physical properties of seafloor sediments and provides a guarantee for the safe construction of marine engineering.

Keywords: sediment resistivity; physical property; deep learning; the South China Sea

1. Introduction

Seabed geohazards in the South China Sea [1], such as landslides [2,3], turbidity currents [4,5], and liquefaction [6], pose a significant risk to marine engineering projects [7], such as gas hydrate extraction [8,9], offshore oil and gas platform construction [10], and submarine fiber optic cables [11,12]. This is an urgent issue for national deep-sea development [13]. The process of incubation and the occurrence of deep seabed geohazards is characterized by dynamic changes in the engineering geological properties of sediments [14]. Current studies on engineering geological properties of submarine sediments and other related topics are mostly based on empirical formulas [15] with low prediction accuracy.

The method of neural networks was proposed in 1950 [16], but it was not until 2010 and GPU studies were improved that it became one of the most important methods of deep learning [17]. At present, artificial neural networks are widely used in studies of seabed sediments [18]. In 2006, Singh et al. [19] used artificial neural networks to predict aquifers in sediments by inverting vertical resistivity bathymetry data. Similarly, in 2009, Luo et al. [20] used the backpropagation (BP) artificial neural network in Matlab to study the physical properties of submarine sediments. The backpropagation artificial neural network was established with porosity, density, and water content as the input parameters and sound velocity as the output parameter. In 2016, Mukherjee et al. [21] developed an artificial neural network to predict the porosity and saturation of gas hydrates in reservoirs. The input parameter was density logs and the output parameters were porosity and saturation. In 2018, Zhou et al. [22] reviewed the application of deep learning algorithms based on the Keras database in Python language in the field of geology. In 2021, Singh et al. [23] employed neural networks to estimate gas hydrate saturation in sediments using porosity, bulk density, and P wave velocity as the input parameters. They compared 12 different machine learning algorithms and obtained a prediction accuracy of approximately 84%, which was higher than the accuracies of seismic and resistivity methods, which do not exceed 75%. In 2021, Chen employed an artificial neural network algorithm based on machine learning to invert acoustic impedance and CPT data [24]. In 2022, Singh et al. [25] used machine learning methods and nuclear magnetic resonance (NMR) data from downhole gas hydrate reservoirs to classify lithology. AI automation also has applications in mineralogy, and in 2022, Pszonka et al. applied SEM automation to mineral and texture sorting in gravity flows from seafloor sediments [26].

Most of the studies in this field have focused on using artificial intelligence methods to establish the relationship between the physical properties of seafloor sediments and geophysical parameters. However, the predicted values are mostly single discrete labels, and there is a lack of research on the dynamic changes in the physical properties of seafloor sediments. This study aims to construct a single-input, multiple-output deep neural learning algorithm for predicting the physical properties of seafloor sediments in the hydrate test area in the South China Sea.

2. Geological Background

2.1. Overview of the South China Sea Hydrate Test Area

The focus of this study is a hydrate test area located in the Shenhu Sea region of the South China Sea (Figure 1). The study area is situated within the Baiyun Depression, which forms part of the Pearl River Estuary Basin and is positioned between the Xisha Trough and Dongsha Islands [27] at a water depth of 1342 m. The geological setting of the study area is characterized by the interaction of several tectonic plates, including the Eurasian, Pacific, and Sino-Indian plates. The region exhibits both passive and active continental features, and displays a complex seafloor topography with varying elevations, characterized by the presence of erosion channels, sea valleys, seamounts, steep slopes, plateaus, alluvial fans, landslides, and other geological hazards. The dominant substrate in the study area is argillaceous silt.

2.2. Study Material of Core Sediments from the South China Sea

In 2020, we boarded the "Ocean Geology No. 6" research vessel of the Guangzhou Marine Geological Survey Bureau and obtained an 8 m long sediment core sample from the natural gas hydrate test area on the northern continental slope of the South China Sea. The engineering geological properties of the sediment, including density, water content, liquid limit, particle size composition, consolidation, cohesion, and internal friction angle, were obtained through laboratory geotechnical tests (Figure 2). The core sediment test results showed that the density of the seabed sediment in the South China Sea hydrate test area ranged from 1.32 to 1.50 g/cm^3, the water content ranged from 119% to 148%, the liquid limit ranged from 83.35 to 101.54, and the plastic limit ranged from 29.60 to 43.05. The range of the liquidity index was 0.90 to 1.95 and the range of plasticity index was 40.83 to 65.46. The sediment in the study area has the physical characteristics of a low density, a high water content, and a high compressibility [28].

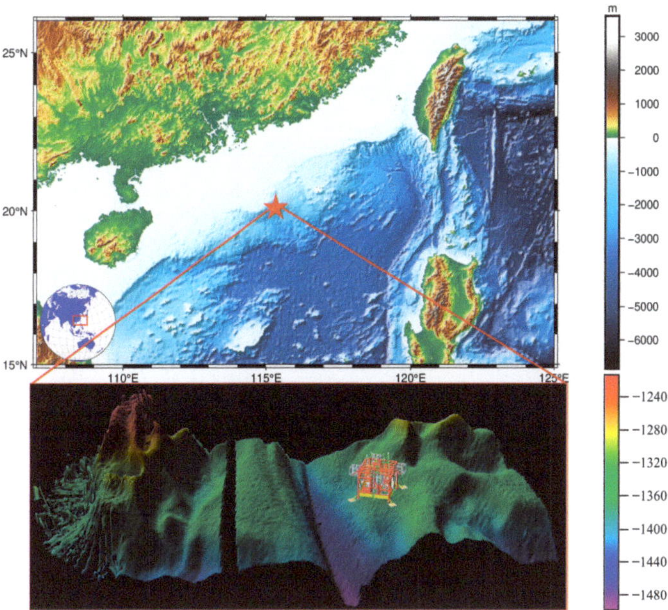

Figure 1. The study area for physical property inversion of seafloor sediments in the South China Sea hydrate test area.

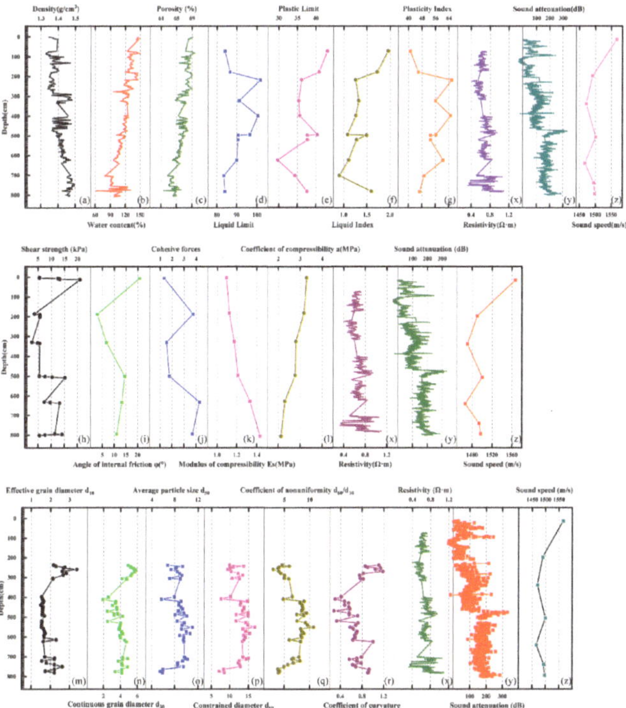

Figure 2. Physical and mechanical properties of sediments in the South China Sea (a, b, c, d, e, f, g, h, i, j, k, l, m, n, o, p, q, r, x, y, z are the sediment properties with depth, such as density, water content,

porosity, liquid limit, plastic limit, liquid index, plasticity index, shear strength, angle of internal friction, cohesive forces, modulus of compressibility, coefficient of compressibility, effective grain diameter, continuous grain diameter, average particle size, constrained diameter, coefficient of nonuniformity, coefficient of curvature, resistivity, sound attenuation, sound speed).

3. Deep Learning Inversion Method

3.1. Construction of Datasets

The present study utilizes a dataset consisting of multiple labels, including resistivity, density, water content, and porosity of sediments, for training and modeling. The training dataset was constructed using both actual measured data of seafloor sediments from the hydrate test area in the South China Sea, as well as the literature data collected from various locations such as the Yellow River estuary [29], offshore of Ningbo [30], and the northern land slope of the South China Sea [31]. The input parameter in the dataset is resistivity, while the output parameters are density, water content, and porosity.

3.1.1. Data Scatter Matrix

The scatter matrix visualization method is a highly effective tool for exploratory data analysis, typically using one feature as the x-axis and another feature as the y-axis. It facilitates the identification of relationships between features and enables the selection of optimal input and output parameters and the detection of outliers and other anomalies in the data [32]. The scatterplot matrix comprises two basic types of plots, namely a scatterplot and a histogram. The scatterplot above and below the diagonal presents the relationship between variables, while the histogram on the diagonal represents the variable distribution [33]. The scatterplot matrix of the actual measured data is presented in Figure 3.

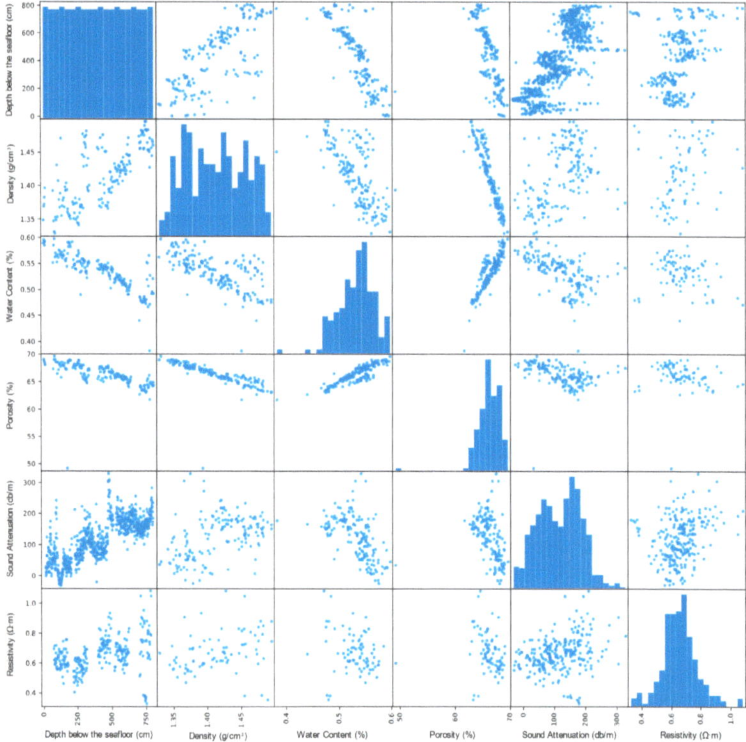

Figure 3. Data scatter matrix plot of burial depth below the seafloor, density, water content, porosity, sound attenuation, and resistivity of seafloor sediments in the South China Sea hydrate test area.

3.1.2. Data Preprocessing

Data with widely different value ranges can pose challenges for neural network learning, as the network may struggle to adapt to such variations. To address this issue, it is necessary to preprocess the input data by scaling each feature individually [34]. The most effective method is standardizing each feature, which entails subtracting the feature mean and dividing the result by the standard deviation for each feature of the input data (Equation (1)). This can result in data compression and distinctive data characteristics, namely a mean value of 0 and a standard deviation of 1 for each feature of the dataset.

$$z = \frac{x - \mu}{\sigma} \tag{1}$$

where x is the raw data, μ is the mean of the raw dataset, and σ is the standard deviation of the raw dataset.

3.1.3. Train–Test Dataset Split

The dataset in this study was partitioned into a training set and a test set, wherein the former comprises 75% of the data and the latter comprises 25%. To minimize the risk of errors due to arbitrary partitioning, the train_test_split() function was utilized for partitioning the dataset, with the random seed specified as RandomState = 0 to ensure the reproducibility of the results. Subsequently, the input and output parameters were separated for both the training and test sets.

3.2. Construction of the Deep Learning Model

This study presents a physical property prediction model for seafloor sediments in the hydrate test area of the South China Sea. The proposed model employs a nonlinear network topology and was developed using the functional APIs provided in the TensorFlow and Keras modules of Python. The functional API can directly manipulate the tensor, utilizing layers as functions that receive and return the tensor. This differs from the sequential model implementation of neural networks, which is limited to single-input and single-output models [35]. The functional API model is more general and flexible, allowing for the implementation of multiple input models, multiple output models, and graph-like models, and is therefore a suitable approach for addressing the problem at hand.

The neural network is comprised of layers, which function as data processing modules that convert input tensors into output tensors. The selection of layers depends on the tensor format and data processing requirements. In this study, a deep network learning model with seven layers was constructed, consisting of one input layer, one output layer, and five implicit neural network layers. The five implicit neural network layers contain four dense layers and one residual layer (Figure 4). The dense layer was used to train the neural network, while the last layer of the neural network was set as a one-unit linear layer without an activation function. The inclusion of an activation function would restrict the input range, limiting the prediction of density, water content, and porosity to within certain ranges. Additionally, a residual connection was incorporated into the model to prevent the loss of input resistivity information during data processing. The residual connection allows for the overlaying of the previous output tensor with the subsequent output tensor, ensuring the reinjection of previous resistivity information into the downstream data stream.

3.2.1. K-Fold Verification

When adjusting the parameters of a neural network, it is crucial to evaluate its performance. However, when the available data are scarce, a small validation set may result in significant fluctuations in the validation scores. Moreover, different ways of partitioning the validation set can further increase the variance of the validation scores, which hinders reliable model evaluation [36]. To address this issue, the K-fold cross-validation approach was employed in this study. The data in this paper were divided into five training and validation sets using the K-fold cross-validation method. Additionally, five identical models were built. Each model was trained on four partitions and evaluated on the remaining one. The validation score of the model was computed as the average of the five validation scores.

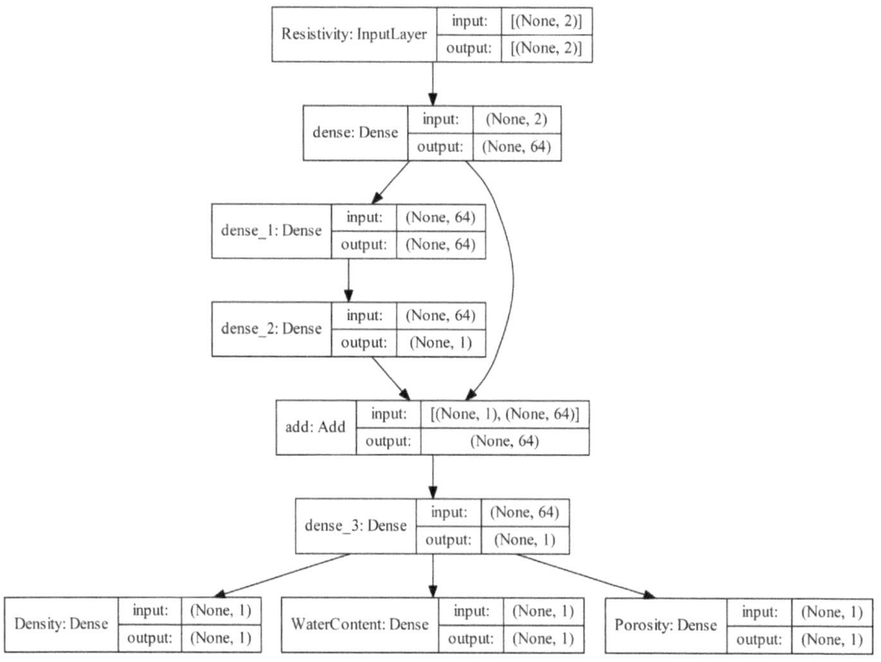

Figure 4. The framework of the deep learning neural network model based on resistivity inversion of seafloor sediment density, water content, and porosity.

3.2.2. Performance Evaluation Criteria

The loss function is one of the most important performance evaluation criteria and is used for the feedback signal of learning. The mean square error (MSE) loss function, which quantifies the squared difference between predicted and target values, was utilized in this prediction model (Equations (2) and (3)). In order to monitor the progress of the learning process, a new metric, the mean absolute error (MAE), was employed [37]. This metric measures the absolute value of the difference between the predicted and target values. By monitoring the MAE during the training process, the performance of the learning process can be evaluated and adjusted accordingly.

$$\text{MSE} = \frac{1}{N} \sum_{t=1}^{N} (y_t - f_t)^2 \tag{2}$$

$$\text{MSE} = \frac{1}{N} \sum_{t=1}^{N} \left| \frac{y_t - f_t}{y_t} \right| \tag{3}$$

3.3. Resistivity Observations of Seafloor Sediments in the South China Sea Hydrate Test Area

We have developed SEEGeo (In Situ Surveying Equipment of Engineering Geology in Complex Deep Sea) in order to realize the in situ long term observation of the engineering geological properties of seabed sediments. The SEEGeo (Figure 5) can simultaneously measure the resistivity, sound velocity, and excess pore pressure water pressure of the sediment [38]. Therefore, long-term observations of the resistivity of seafloor sediments were conducted in the hydrate test area of the South China Sea. The equipment was operated once a day, and a total of 12 days of data were collected from 15 to 27 September.

Figure 5. In Situ Surveying Equipment of Engineering Geology in Complex Deep Sea, SEEGeo.

4. Results and Discussion

4.1. Relationship between Sediment Resistivity and Physical Properties

4.1.1. Relationship between Sediment Resistivity and Density

Sediment density, which is the mass per unit volume of soil, displays considerable variability across diverse submarine sediments, and is closely linked to soil mineral composition, pore volume, and water content [39]. This study compares the relationship between resistivity and density derived from empirical equations, and the relationship derived from the deep learning model. We also explore the conditions and ranges of applicability of these distinct models.

The correlation based on empirical equations between resistivity and density is as follows: (1) the density increases monotonically with the increase in resistivity and (2) the seafloor sediment resistivity varies between 0.32 and 1.43 $\Omega\cdot$m, with a corresponding density range of 1.32~2.13 g/cm^3. By fitting the resistivity and density datasets of the seafloor sediment with the empirical equation, a quadratic polynomial relationship between these two variables can be derived (Equation (4)).

$$y = -0.31x^2 + 1.96x - 1.50 \qquad (4)$$

The result indicates that the correlation between seafloor sediment resistivity and density based on the empirical equation is poor, with an R^2 value of 0.33. The traditional empirical formula is inadequate as a fitting model for the resistivity–density correlation.

We explored the correlation between sediment resistivity and density based on deep learning (Figure 6). The results indicate that the relationship between these two properties follows a three-step trend: decreasing, increasing, and sharply increasing. More specifically, when the resistivity ranges from 0.32 to 0.46 $\Omega\cdot$m, the sediment density decreases as the resistivity increases, with a slope of -0.79 from 1.48 g/cm^3 to 1.37 g/cm^3. When the resistivity ranges from 0.46 to 1.15 $\Omega\cdot$m, the sediment density increases as the resistivity increases, with a slope of 0.25 from 1.37 g/cm^3 to 1.54 g/cm^3. Finally, when the resistivity is between 1.15 and 1.43 $\Omega \cdot$m, the sediment density increases sharply as the resistivity increases, with a slope of 1.04 from 1.54 g/cm^3 to 1.83 g/cm^3. These findings suggest that the deep learning model can reveal complex relationships between sediment properties, which may not be captured by traditional empirical formulas.

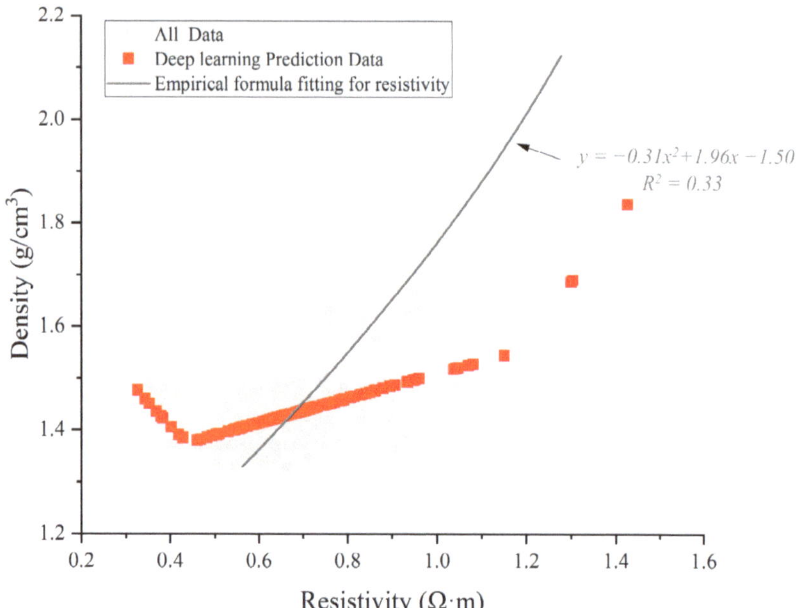

Figure 6. Correlation between sediment resistivity and density based on empirical equation and deep learning model (data source: measured data from the South China Sea and reference data of the Yellow River estuary sediment [29], Ningbo offshore sediment [30], the South China Sea [31]).

The present study investigated the relationship between sediment resistivity and density using both empirical and deep learning models. The results indicate that: (1) The deep learning model predicts that the sediment resistivity is not monotonically related to density, indicating that the same sediment density may have different resistivity values. (2) The resistivity cannot be solely determined based on density in the range of 1.38 to 1.47 g/cm^3 and the range of 0.33 to 0.83 $\Omega \cdot$m, as there are multiple solutions, but can be determined based on density when the density exceeds 1.47 g/cm^3. (3) The sediment density can be uniquely determined based on resistivity. (4) The sediment density is more sensitive to changes in resistivity when the resistivity is >1.15 $\Omega \cdot$m, where changes in density per unit of resistivity correspond to larger values. (5) The correlation coefficient between resistivity and density based on the empirical equation is very low, with a value of 0.33. The R^2 correlation coefficient is typically used in linear models and may not fully capture the predictive ability of the model in nonlinear models. Moreover, the correlation coefficient is closely related to the sample size of the dataset, and the model results can vary widely across different datasets.

4.1.2. Relationship between Sediment Resistivity and Water Content

Seafloor sediments are typically saturated or nearly saturated, with water content playing a crucial role in the design of seafloor engineering structures [40]. An increase in water content results in an improved fluid connectivity of sediment pores, leading to an enhanced electrical conductivity and, correspondingly, a lower resistivity. In contrast, sediments with a low water content exhibit a higher resistivity, and water with various ions functions as a good electrical conductor [30]. Despite a decline in resistivity with the decreasing water content in the saturated state, the magnitude of change diminishes [31]. The primary factors influencing sediment resistivity in descending order are water content, pore water conductivity, soil saturation, and soil type [41].

The present study investigated the correlation between sediment resistivity and the water content of seafloor sediment based on an empirical formula. The resistivity of the seafloor sediment varied between 0.32 and 2.17 $\Omega \cdot$m, while the water content ranged from 24% to 154%. By fitting the relationship between the resistivity and the water content of the seafloor sediment with the empirical equation, a power function relationship (Equation (5)) between these two variables can be derived.

$$y = 52.83x^{-1.745} \tag{5}$$

The results indicated a power function relationship between the resistivity and the water content of the seafloor sediment. In particular, the resistivity decreased sharply at first with the increase in water content, followed by a slow decrease. These findings suggest that water content plays an important role in determining the resistivity of the seafloor sediment and should be taken into consideration when modeling the resistivity. However, we need to develop more accurate models for predicting the sediment resistivity based on water content.

The deep learning model reveals a correlation between sediment resistivity and water content characterized by a three-step trend (Figure 7). Specifically, the water content exhibits a slow and linear decrease with increasing resistivity, followed by a rapid decrease expressed by a power function, and a subsequent slow and linear decrease. This trend is observed in three distinct resistivity ranges. For resistivities between 0.33 and 0.72 $\Omega \cdot$m, the water content decreases slowly and linearly with increasing resistivity, with a slope of -12.82, resulting in a decrease from 120% to 115%. For resistivities ranging from 0.72 to 1.16 $\Omega \cdot$m, the water content rapidly decreases with increasing resistivity in a power function with a negative exponent, with a slope of -12.82, resulting in a decrease from 115% to 115%. Finally, for resistivities between 1.16 and 2.17 $\Omega \cdot$m, the water content decreases slowly and linearly with increasing resistivity, with a slope of -24.75, resulting in a decrease from 41% to 16%.

The present study investigated the relationship between water content and resistivity using both empirical and deep learning models. The results indicate that: (1) The deep learning model shows a monotonically decreasing relationship between water content and resistivity, which is one-to-one and without multi-solution. (2) The correlation between water content and resistivity is characterized by both a linear decreasing function and a negative exponential power function in the deep learning model. For sediment resistivity values of <0.71 $\Omega \cdot$m (i.e., water content > 115%), a linear decreasing function of water content to resistivity is observed. The prediction result of the empirical formula model is an exponential function and differs significantly from the deep learning prediction model. For sediment resistivity values between 0.71 and 1.33 $\Omega \cdot$m (i.e., a water content between 34 and 115%), the prediction results of the empirical formula model are in good agreement with those of the deep-learning-based prediction model. For sediment resistivity values of >0.71 $\Omega \cdot$m (i.e., a water content < 115%), the deep learning prediction model fits better than the empirical formula. (3) The sensitivity of water content to resistivity is higher for resistivity values of >0.71 $\Omega \cdot$m (i.e., a water content < 115%), implying that a unit change in resistivity corresponds to a larger change in water content. Therefore, the inversion of water content in this interval is more accurate.

4.1.3. Relationship between Sediment Resistivity and Porosity

The sediment porosity is a critical physical parameter that characterizes the total volume of pores within the sediment. The influence of sediment porosity on the resistivity magnitude is significant. Sediments with a higher porosity possess a larger pore space that facilitates better connectivity within the pore network. As a result, the fluid within the pore has a higher mobility, leading to improved electrical conductivity and lower resistivity [42].

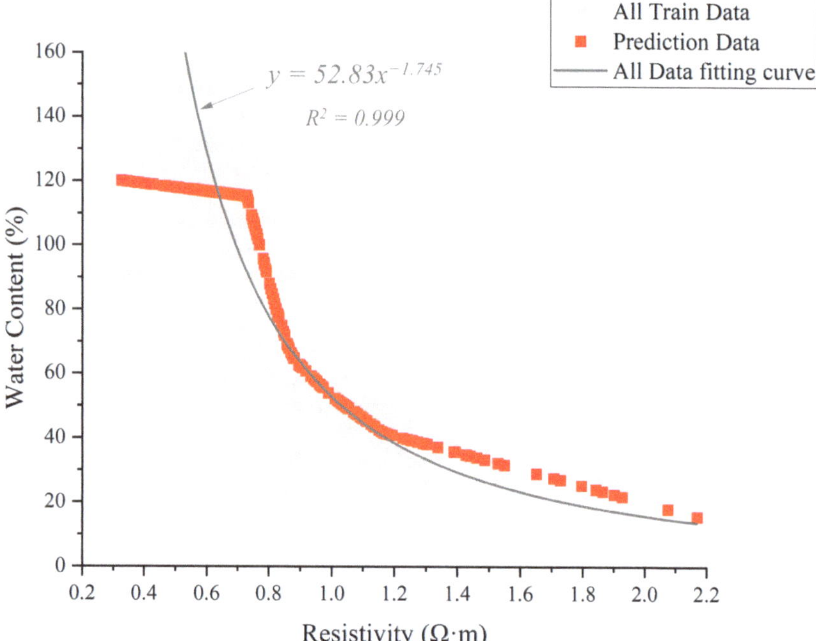

Figure 7. Correlation between sediment resistivity and water content based on an empirical equation and the deep learning model (data source: measured data from the South China Sea and reference data of the Yellow River estuary sediment [29], Ningbo offshore sediment [30], the South China Sea [31]).

The empirical equations reveal a correlation between sediment resistivity and porosity, where the seafloor sediment resistivity ranges between 0.32 and 1.93 Ω·m, while the porosity of the sediment ranges from 36.81 to 79.52%. The seafloor surface sediment exhibits a strong linear relationship between sediment resistivity and porosity (Equation (6)).

$$y = -26.58x + 84.08 \tag{6}$$

The present study demonstrates a favorable correlation between the resistivity and the porosity of seafloor sediments based on empirical formulas with an R^2 value of 0.97, indicating a high degree of correlation. The traditional empirical formulas exhibit good fits to the data, although discrepancies are observed in the predicted values at both low and high resistivity ranges.

This study employed deep learning to investigate the correlation between sediment resistivity and porosity (Figure 8). The results reveal that the porosity of seafloor sediments follows a three-step trend with increasing resistivity, characterized by extremely slow, slow, and rapid decreases. Specifically, for resistivities of <0.50 Ω·m (i.e., 0.33–0.50 Ω·m), the sediment porosity decreases extremely slowly from 69.37% to 69.13%, with a slope of −1.41. In the range of 0.50–0.70 Ω·m, the sediment porosity decreases slowly from 69.13% to 66.74%, with a slope of −11.95. Finally, when the resistivity is greater than 0.70 Ω·m (i.e., 0.70 to 1.93 Ω·m), the sediment porosity decreases rapidly and linearly from 66.74% to 29.19% with a slope of −30.53. These findings provide a valuable contribution to the understanding of the intricate relationship between sediment resistivity and porosity, and can facilitate the enhancement in the accuracy of seafloor sediment characterization.

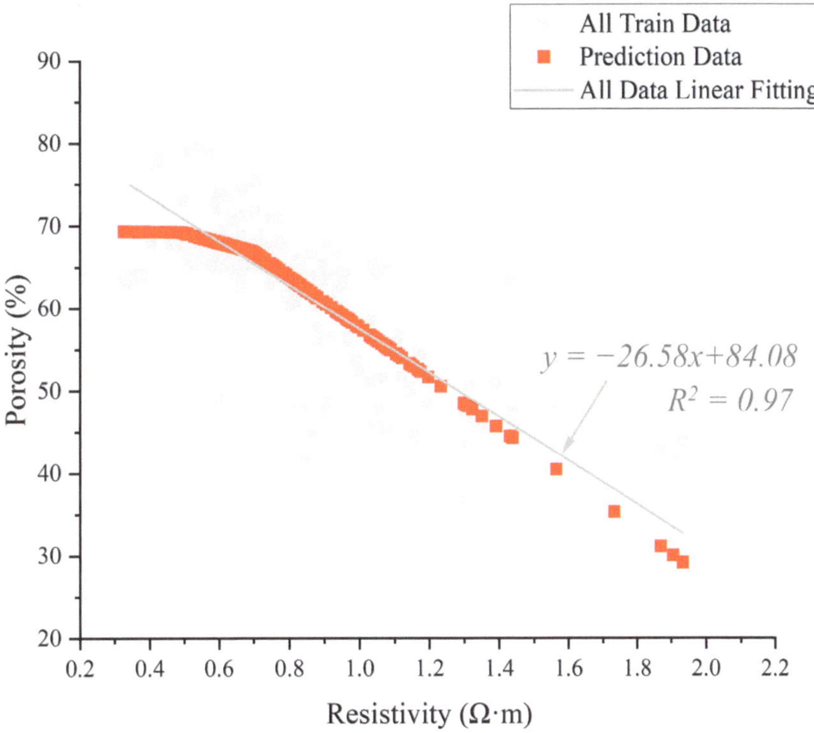

Figure 8. Correlation between sediment resistivity and porosity based on the empirical equation and the deep learning model (data source: measured data from the South China Sea and reference data of the Yellow River estuary sediment [29], Ningbo offshore sediment [30], the South China Sea [31]).

The present study investigated the relationship between sediment resistivity and porosity using both empirical and deep learning models. The results reveal that the sediment porosity and resistivity exhibit a monotonically decreasing function, indicating a one-to-one correspondence with no multi-solution. The deep learning model can be divided into three linearly decreasing functions, with the corresponding cutoff points of 0.5 Ω·m and 69.13% water content and 0.7 Ω·m and 66.74% water content. When the sediment resistivity is greater than 0.70 Ω·m, the sediment porosity is more sensitive to resistivity, and it is more accurate to invert the sediment density change by resistivity. Furthermore, when the sediment resistivity is less than 0.70 Ω·m and the water content is greater than 68%, the prediction values of the empirical formula model and the deep learning prediction model differ significantly, with the latter tending towards a constant value that is closer to the actual value. As the sediment porosity increases, the sediment resistivity approaches the seawater resistivity and exhibits a constant value. The correlation coefficient increases significantly with an increase in data samples, highlighting the superior performance of the deep learning model in describing the correlation between sediment resistivity and porosity compared to traditional models.

As a result, the deep neural network learning model exhibits superior performance and provides a more accurate prediction range compared to the empirical formula. The empirical formula inversion model is effective in analyzing linear relationships with simple correlations and high coefficients of determination. In contrast, the deep learning inversion model is suitable for examining nonlinear relationships with complex correlations and poor coefficients of determination. When training data are abundant, the deep learning inversion model outperforms the empirical formula model in terms of prediction accuracy. Specifically, the deep learning inversion model improves the accuracy of the prediction area by 50%, enhances the concentration of the prediction area, and increases the goodness-of-fit by 33%, thereby achieving a higher prediction accuracy.

4.2. In Situ Resistivity Observation Results in the South China Sea Hydrate Test Area

4.2.1. Spatial Structure of Seafloor Sediment Resistivity

The present study investigated the spatial structure of sediment resistivity within 1.80 m of the seafloor in the South China Sea hydrate test area (Figure 9). The resistivity of the seawater on 15 September was found to be 0.06 $\Omega \cdot$m, while the resistivity of the seafloor sediment ranged from 0.01 to 6.86 $\Omega \cdot$m, with a mean resistivity of 0.84 $\Omega \cdot$m and a median resistivity of 0.23 $\Omega \cdot$m. Most of the resistivity values were distributed between 0.01 and 2 $\Omega \cdot$m, with two extreme points at 0.9–1.1 m, 5.82, and 6.86 $\Omega \cdot$m, respectively. When focusing on the range of 0–2 $\Omega \cdot$m, most of the resistivity values were distributed between 0.01 and 1 $\Omega \cdot$m.

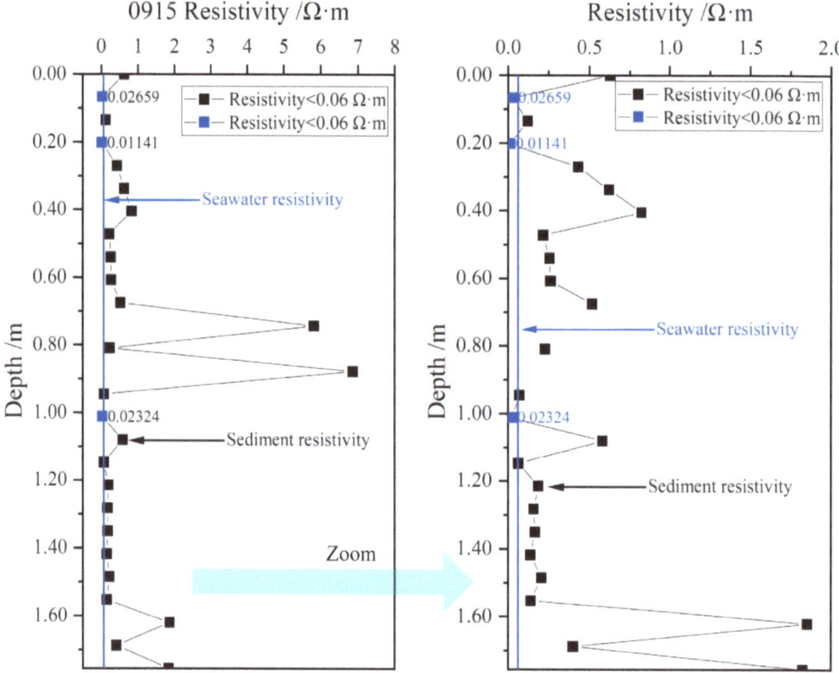

Figure 9. Spatial structure of sediment resistivity in the South China Sea (using 15 September as an example).

In this study, the first layer of sediment in the South China Sea hydrate test area was found to exhibit a low resistivity when buried at a depth of 0–0.07 m. The resistivity ranged from 0.01 to 0.83 $\Omega \cdot$m, with a variation rate of 0–71%. A total of 11 measurements were recorded in this layer, two of which exhibited resistivity values lower than that of seawater. Specifically, resistivity values of 0.03 $\Omega \cdot$m at a depth of 0.07 m and 0.01 $\Omega \cdot$m at a depth of 0.20 m were recorded.

The second layer of sediment was identified as a very high resistivity layer buried at depths ranging from 0.68 to 0.95 m. The resistivity of this layer varies from 0.07 to 6.86 $\Omega \cdot$m, exhibiting a variation rate of 5–600%. Within this layer, there were five measurement points, two of which had resistivity values significantly higher than the resistivity of the seafloor sediment (0.06–2 $\Omega \cdot$m). Specifically, the resistivity at a depth of 0.74 m was 5.82 $\Omega \cdot$m, which is 509% higher than the lowest value, and the resistivity at a depth of 0.88 m was 6.86 $\Omega \cdot$m, which is 600% higher than the lowest value.

The third layer of sediment was identified as a low resistivity layer, with a sediment depth ranging from 0.945 m to 1.5525 m. The resistivity values in this layer exhibit variations ranging from 0.02 $\Omega \cdot$m to 0.58 $\Omega \cdot$m, representing a variation rate of 1% to 59%. A total of 10 measurements were recorded within this layer, and it is noteworthy that one of these points situated at a depth of 1.01 m had a resistivity value of 0.02 $\Omega \cdot$m.

The fourth layer comprises high resistivity sediment buried at depths ranging from 1.55 to 1.76 m. The resistivity of the sediment varies from 0.14 to 1.85 $\Omega \cdot$m, with a variation rate of 11 to 161%. Four measurements were taken in this layer, out of which two had significantly higher resistivity values. At a depth of 1.62 m, the resistivity value was 1.85 $\Omega \cdot$m, indicating a variation rate of 161%. Similarly, at a depth of 1.76 m, the resistivity value was 1.82 $\Omega \cdot$m, indicating a variation rate of 159%.

4.2.2. The Variation in Seafloor Sediment Resistivity with Time

The present study investigated the dynamic changes i the resistivity of the seafloor sediment in the hydrate test area of the South China Sea over 12 days, from 15 to 27 September. The resistivity values of the seafloor sediment in the South China Sea hydrate test area ranged from 0.002 to 6.86 $\Omega \cdot$m, and were classified into three layers (Figure 10). The first layer, with resistivity values lower than the resistivity of seawater (<0.06 $\Omega \cdot$m), is shown in white. The second layer corresponds to the resistivity values of normal sediment (0.06–2 $\Omega \cdot$m) and is filled in blue. The third layer has resistivity values greater than the resistivity of normal sediment (>2 $\Omega \cdot$m) and is filled in green and red.

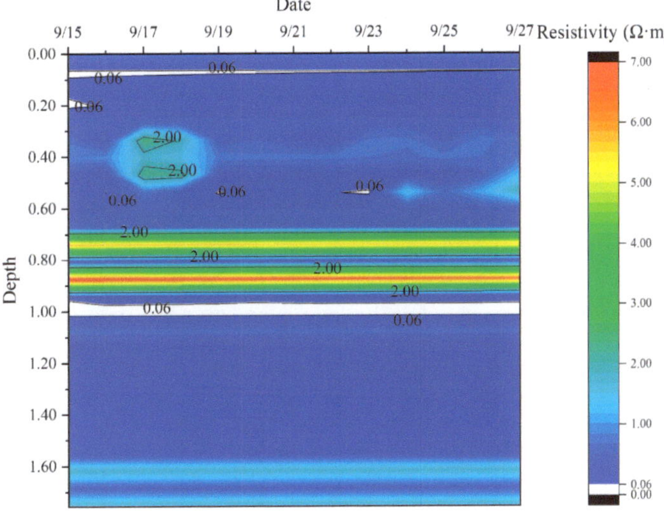

Figure 10. Time-varying curve of seafloor sediment resistivity in the South China Sea hydrate test area.

In the range of 0–1.755 m, the resistivity is divided into seven layers in the longitudinal structure, which are classified as a low resistivity layer, an ultra-low resistivity layer, a high resistivity layer, an ultra-high resistivity layer, an ultra-low resistivity layer, a low resistivity layer, and a high resistivity layer.

In the first layer, sediments are buried at a depth of 0–0.6 m in the normal low resistivity zone, and the resistivity variation ranges from 0.43 to 3.55 $\Omega \cdot$m. The resistivity does not change significantly over time.

The second layer, located at a sediment burial depth of 0.06~0.20 m, is characterized by ultra-low resistivity. Within this layer, there exist two distinct low resistivity zones, located at 0.06 and 0.20 m, respectively, with resistivity varying between 0.01 and 0.17 $\Omega \cdot$m. At a depth of 0.06 m, the resistivity gradually increases over time, ranging from 0.02 $\Omega \cdot$m to 0.06 $\Omega \cdot$m, ultimately approaching the resistivity value of seawater. Similarly, at a depth of 0.20 m, the resistivity also gradually increases over time, ranging from 0.01 $\Omega \cdot$m to 0.17 $\Omega \cdot$m, and reaches 0.08 $\Omega \cdot$m on the following day, 16 September, which is also close to the resistivity of seawater.

The third layer of sediment, with a burial depth of 0.33–0.47 m, is characterized by high resistivity. The resistivity in this layer ranges from 0.09 to 2.44 $\Omega \cdot$m and exhibits an increasing-then-decreasing trend. At a depth of 0.33 m, the resistivity increases from 0.62 $\Omega \cdot$m to 2.44 $\Omega \cdot$m and then decreases to 0.52 $\Omega \cdot$m. Similarly, at 0.40 m and 0.47 m, the resistivity increases and then decreases, with values changing from 0.82 $\Omega \cdot$m to 1.73 $\Omega \cdot$m to 0.64 $\Omega \cdot$m and from 0.21 $\Omega \cdot$m to 2.32 $\Omega \cdot$m to 1.22 $\Omega \cdot$m, respectively.

The fourth layer of sediment located at a depth of 0.67 to 0.94 m is characterized by ultra-high resistivity, which exceeds the normal sediment resistivity (>2 $\Omega \cdot$m). The resistivity ranges from 0.18 to 6.85 $\Omega \cdot$m, demonstrating a three-layer structure of high resistivity–low resistivity–high resistivity. The ultra-high resistivity layer at 0.74 m exhibits a gradual decrease in resistivity over time, from 5.82 $\Omega \cdot$m to 5.65 $\Omega \cdot$m. Likewise, the low resistivity layer at 0.81 m shows a decrease in resistivity with time, changing from 0.22 $\Omega \cdot$m to 0.18 $\Omega \cdot$m. In the ultra-high resistivity zone at 0.87 m, the resistivity gradually decreases with time from 6.85 $\Omega \cdot$m to 6.35 $\Omega \cdot$m.

The fifth layer of the sediment, a burial depth of 1.01 m, is an ultra-low resistivity zone with a resistivity even lower than that of seawater. The resistivity varies in a narrow range of 0.02~0.03 $\Omega \cdot$m and shows a trend of increasing initially and then decreasing, reaching a peak value of 0.03 $\Omega \cdot$m from 0.02 $\Omega \cdot$m before decreasing back to 0.02 $\Omega \cdot$m.

The fifth layer of the sediment, located at a burial depth of 1.08 to 1.55 m, exhibits a low resistivity. The resistivity values range between 0.06 and 0.58 $\Omega \cdot$m, and display little temporal variation, remaining in a stable state.

The seventh layer of sediment, located at a burial depth of 1.62–1.80 m, exhibits a high resistivity with a resistivity range of 0.38–1.88 $\Omega \cdot$m. At a depth of 1.62 m, the resistivity value is classified as a high resistivity and displays a gradually decreasing trend, declining from 1.85 $\Omega \cdot$m to 1.79 $\Omega \cdot$m. At 1.68 m, the resistivity is categorized as a low resistivity and exhibits little variation over time, with resistivity values oscillating between 0.38 and 0.40 $\Omega \cdot$m. The resistivity at a depth of 1.80 m is classified as a high resistivity and demonstrates a gradual increase from 1.82 $\Omega \cdot$m to 1.88 $\Omega \cdot$m.

The in situ observation structure of natural gas hydrates deposited on the seafloor exhibits a bimodal structure in resistivity. Jana et al. [43] pointed out that in the hydrate occurrence zone, the bimodal feature of the resistivity of the sediment layer indicates the presence of the hydrate layer. This paper presents resistivity observations of seafloor sediment in the South China Sea hydrate test area, which indicate that the resistivity at 0.7~0.9 m below the seafloor surface has a bimodal structure with a resistivity value of 7.0, indicating the presence of a thin layer of natural gas hydrate. The resistivity of the ultra-high resistivity layer at 0.74 m decreases gradually with time from 5.82 $\Omega \cdot$m to 5.65 $\Omega \cdot$m, while the resistivity of the low resistivity layer at 0.81 m decreases gradually from 0.23 $\Omega \cdot$m to 0.19 $\Omega \cdot$m over time. Similarly, the resistivity of the ultra-high resistivity zone at 0.88 m gradually decreases over time from 6.86 $\Omega \cdot$m to 6.35 $\Omega \cdot$m. The bimodal resistivity of the seafloor sediment is a typical characteristic of hydrate storage. Previous studies by Wu and Guo have also reported the bimodal structure of hydrate resistivity in seafloor sediments [44], with a resistivity value of approximately 7 $\Omega \cdot$m, consistent with the findings of this study.

4.3. Dynamic Changes in Physical Properties of Seafloor Sediments Based on Deep Learning Inversion
4.3.1. Temporal Variation in Seafloor Sediment Density in the South China Sea Hydrate Zone

The findings of the sediment density inversion of the seafloor in the South China Sea hydrate test area reveal distinct variations in sediment density at varying depths. The sediment within 0.7 m of the seafloor surface exhibited marked dynamic changes in density, whereas the density of sediment situated below 0.7 m remained essentially constant (Figure 11).

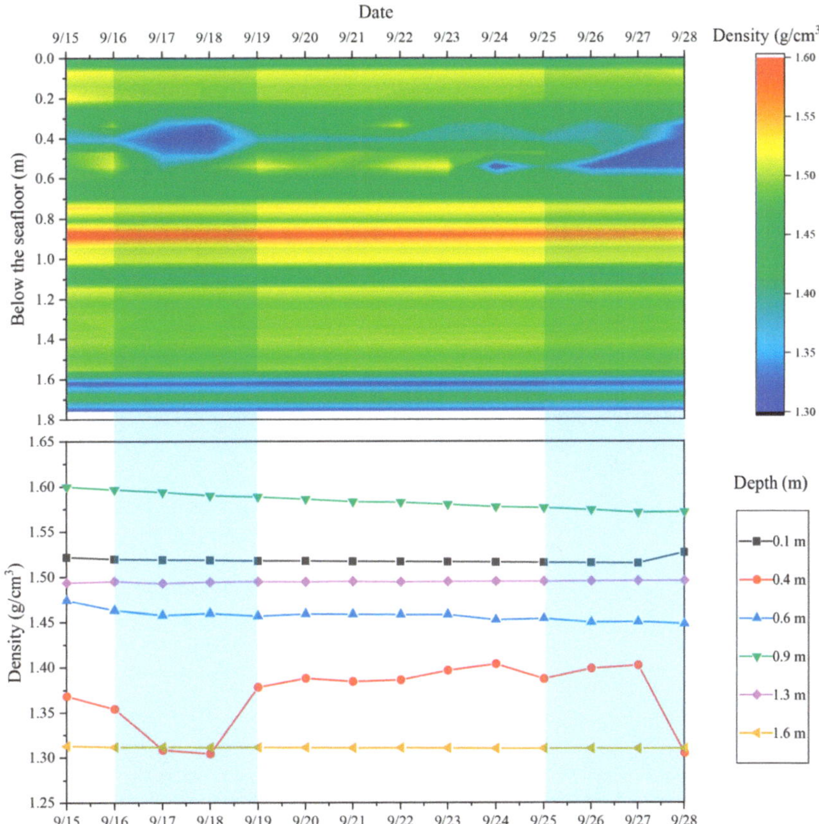

Figure 11. Temporal variation in density of the seafloor sediment in the South China Sea hydrate test area.

The present study depicts the temporal variation of sediment density, obtained through resistivity inversion of seafloor sediment in the South China Sea hydrate test mining area at a depth of 1.755 m below the seafloor surface. The sediment density exhibits a variation range of 1.30–1.60 g/cm^3, with distinct variation patterns observed among different sediment layers (Figure 11). Specifically, sediment samples collected from depths of 0.1, 0.4, 0.6, 0.9, 1.3, and 1.6 m below the seafloor surface were selected for analysis, and their density variation patterns were evaluated over time.

The results revealed varying patterns in sediment density with time at different depths. At a depth of 0.1 m, the sediment density fluctuated between 1.51 and 1.53 g/cm^3. Specifically, the density decreased gradually from 15 to 27 September, but then suddenly increased from 17 to 28 September. Within the depth range of 0.4 m, the sediment density varied between 1.30 and 1.40 g/cm^3. From 15 to 17 September, the density exhibited a rapid decrease, followed by a period of stability and no change until 27 September, when it suddenly decreased again. Within the depth range of 0.6 m, sediment density varied from 1.44 to 1.47 g/cm^3 and exhibited a gradual decrease from 15 to 28 September. At a depth of 0.9 m, the sediment density varied from 1.57 to 1.60 g/cm^3, representing a 2% decrease in density over time from 15 to 28 September. At the depth range of 1.3 m, the sediment density ranged from 1.49 to 1.50 g/cm^3, with a slow increase in density observed over the same time. At a depth of 1.6 m, the sediment density varied between 1.30 and 1.31 g/cm^3 and remained stable from 15 to 28 September.

At the depths of 0.1, 1.3, and 1.6 m below the seafloor surface, the sediment densities remained stable and unchanged, indicating that the sediment properties of these layers did not change during the measurement period. At the depth of 0.4 m below the seafloor surface, the sediment density exhibited a trend of "decrease–increase–decrease" with time. With increasing time, the sediment density of the sediment layer at a 0.4 m burial depth decreased, but the density of adjacent layers did not significantly change. It is assumed that there was no vertical transport of the sediment and that horizontal transport along the sediment layer may have occurred. A potential mechanism may be that when the horizontal fluid is transported, the fine sediment particles are transported and carried away first [45], leading to a reduction in sediment density. These observations suggest the possibility of fluid exchange between the soil and seawater at the site. The lack of changes in overlying sediment characteristics suggests that fluid transport occurs horizontally [46,47].

4.3.2. Temporal Dynamics of the Water Content in Seafloor Sediments in the South China Sea Hydrate Zone

The findings of the sediment water content inversion of the seafloor in the South China Sea hydrate test area reveal distinct variations in the sediment water content at varying depths. Specifically, the water content within 0.7 m below the seafloor showed pronounced dynamic changes, while the water content below 0.7 m remained relatively constant.

The present study investigated the temporal variations in water content in the seafloor sediment of the South China Sea hydrate test area at a depth of 1,755 m below the seafloor (Figure 12). The water content exhibited variations within the range of 42–125%, and different sediment layers displayed diverse patterns of variation. Specifically, the water content at depths of 0.1, 0.4, 0.6, 0.9, 1.3, and 1.6 m below the seafloor were selected for analysis to determine the time-dependent variation patterns.

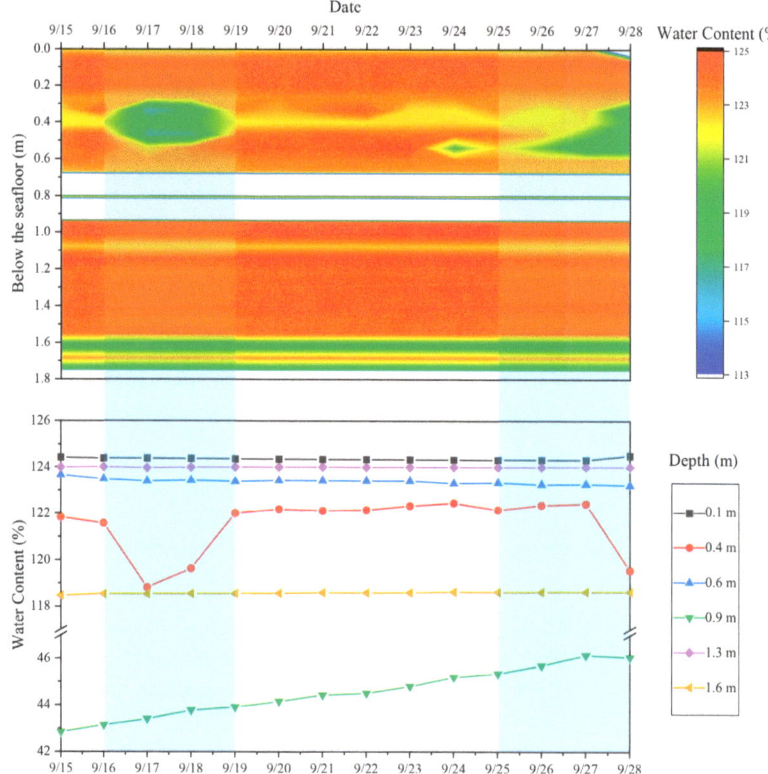

Figure 12. Temporal variation in water content of the seafloor sediment in the South China Sea hydrate test area.

In the South China Sea hydrate test area, the water content within specific depth intervals exhibited temporal variability. Within the depth interval of 0.1 m below the seafloor, the water content ranged between 124.3% and 124.5%. The water content gradually decreased from 15 to 27 September, with a sudden increase occurring between 17 and 28 September. Near the depth interval of 0.4 m below the seafloor surface, the water content ranged from 119.5% to 122.4%. Between 15 and 17 September, the water content decreased rapidly, followed by a stable and unchanged state between 17 and 27 September, and then a sudden decrease on 28 September. At the depth interval of 0.6 m below the seafloor surface, the water content ranged from 123.2% to 123.6%, gradually decreasing from 15 to 28 September. For the depth interval of 0.9 m below the seafloor surface, the water content ranged from 42.8% to 46.1%, exhibiting a 7% increase. Between 15 and 27 September, the water content gradually increased, followed by a slight decrease between 27 and 28 September. At the depth interval of 1.3 m below the seafloor surface, the water content ranged from 123.9% to 124.0%, remaining relatively stable from 15 to 28 September. Finally, for the depth interval of 1.6 m below the seafloor surface, the water content ranged from 118.4% to 118.6%, exhibiting little temporal variability over the same period.

In the water content study, stable sediment properties were found at the depths of 0.1, 1.3, and 1.6 m below the seafloor, as evidenced by the unchanging sediment porosity. However, at a depth of 0.4 m below the seafloor surface, the water content exhibited a temporal trend of "decrease–increase–decrease."

4.3.3. Temporal Dynamics of the Seafloor Sediment Porosity in the South China Sea Hydrate Test Area

The study of sediment porosity in the seafloor of the South China Sea hydrate test area using deep learning inversion techniques revealed varying patterns of variation at different depths (Figure 13). Specifically, the porosity of sediment below 0.7 m remained stable, whereas dynamic changes in the sediment porosity within 0.7 m of the seafloor surface are evident. These findings highlight the non-uniform distribution of sediment properties in the study area.

The temporal variation in the sediment porosity in the South China Sea hydrate test area was within the depth of 0~1.755 m below the seafloor. The sediment porosity varied within the range of 52~76%, and distinct layers of sediment porosity exhibited diverse variation patterns. The depths of 0.1, 0.4, 0.6, 0.9, 1.3, and 1.6 m below the seafloor were scrutinized to explore the sediment porosity variation patterns with time.

The variability results of sediment porosity indicate that the sediment porosity at a depth of 0.1 m ranged between 75.49% and 75.69%. Between 15 and 27 September, the porosity gradually decreased before abruptly increasing from 17 to 28 September. At a depth of 0.4 m, the porosity ranged from 69.63% to 73.49%. From 15 to 17 September, the porosity decreased rapidly before stabilizing until 27 September, and then abruptly decreased from 27 to 28 September. The sediment porosity at a depth of 0.6 m ranged between 74.31% and 74.78%, showing a gradual decrease from 15 to 28 September. At a depth of 0.9 m, the sediment porosity varied from 51.79% to 53.54%, and exhibited an overall increase of 3% from 15 to 27 September, followed by a gradual decrease from 27 to 28 September. The sediment porosity at a depth of 1.3 m remained essentially unchanged, ranging from 75.11% to 75.16% between 15 and 28 September. Finally, the sediment porosity at a depth of 1.6 m ranged between 69.24% and 69.45% and remained stable throughout the study period. These findings suggest that sediment porosity can vary significantly with depth and time and may provide valuable insights into sediment behavior and depositional processes in marine environments.

In this study, the sediment properties at various depths below the seafloor were investigated in order to better understand the behavior of sediment under pressure and over time. Specifically, the sediment porosity was monitored at depths of 0.1, 0.4, 0.6, 0.9, 1.3, and 1.6 m below the seafloor. The results showed that at depths of 0.1, 1.3, and 1.6 m below the seafloor, the sediment porosity remained stable and unchanged, indicating that the properties of the sediment in these layers did not change. However, at a depth of 0.4 m below the seafloor, the sediment porosity exhibited a trend of "decrease–increase–decrease" over time. At a depth of 0.6 m below the seafloor, the sediment porosity gradually increased over time, while at a depth of 0.9 m below the seafloor, the sediment porosity exhibited a clear increasing trend over time.

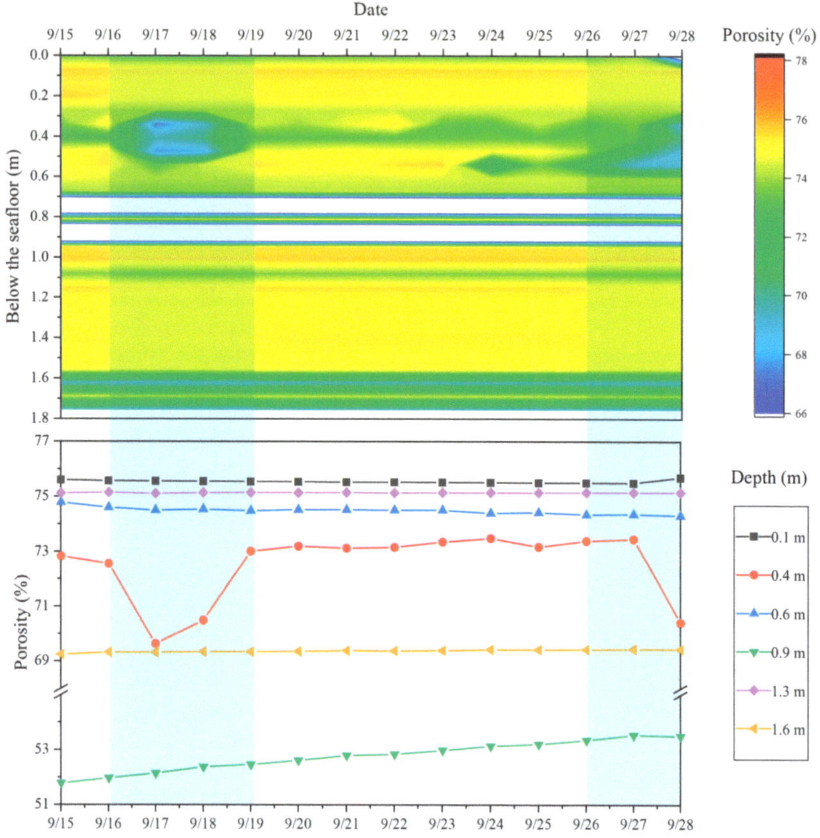

Figure 13. Temporal variation in porosity of the seafloor sediment in the South China Sea hydrate test area.

5. Conclusions

This study addresses the challenge of accurately inverting seafloor sediment physical properties and presents a model for the inversion of sediment physical properties based on a deep learning algorithm for seafloor sediment resistivity in offshore China. The deep learning algorithm establishes the relationship between seafloor sediment resistivity and key physical properties such as density, water content, and porosity in the hydrate test area of the South China Sea. This study clarifies the effect of the main controlling factors of density, water content, and porosity on resistivity and compares the differences between the deep learning inversion model and the empirical formula inversion model. The deep learning inversion algorithm is then used to invert the resistivity of seafloor sediments in the hydrate test area of the South China Sea, and the spatial distribution characteristics and dynamic changes in physical properties such as the density, water content, and porosity of the seafloor sediments are obtained. The results reveal the dynamic changes in sediment properties within 1.76 m below the seabed, providing valuable insights for future seafloor exploration and resource exploitation.

1. In this study, a novel model for the inversion of physical properties of seafloor sediments is presented. The model is based on a deep learning algorithm, which utilizes the TensorFlow and Keras deep learning databases to establish a six-layer neural network with a functional API. The mean squared error (MSE) is used as the loss function and the mean absolute error (MAE) is the supervisory function. The relationships between the resistivity of seafloor sediment and the density, water content, and porosity were established.

2. The use of seafloor sediment resistivity for the inversion of physical properties such as density, porosity, and water content has been explored. Empirical and deep learning relationships have been established, indicating that resistivity is strongly correlated with both porosity and water content in a power function, exhibiting a bipartite trend of "sharp decrease–slow decrease" with increasing water content. A good correlation between resistivity and density has also been identified, with density increasing in a "decrease–increase–sharp increase" three-stage trend as the resistivity increases.

3. The deep learning model exhibits a better fit and more accurate prediction capabilities than the empirical formula model. The empirical formula model is suitable for analyzing linear relationships with simple correlations and high correlation coefficients. In contrast, the deep learning inversion model is suitable for analyzing non-linear relationships with complex correlations and low correlation coefficients. When the model is trained with a large amount of data, the prediction results obtained using the deep learning inversion model are significantly better than those obtained using the empirical formula model. The deep learning inversion model improves the accuracy of the prediction range, resulting in a more concentrated prediction range and a higher prediction accuracy.

4. Based on in situ observations in the hydrate zone of the South China Sea, the physical properties of seafloor sediments in the hydrate zone of the South China Sea exhibit stratified variations. Specifically, in the range of 0–0.7 m below the seafloor, the density varies between 1.37 and 1.53 g/cm^3, the porosity varies between 73 and76%, and the water content varies between 122 and 124%. Within the 0.7–0.9 m depth range, the density varies between 1.43 and 1.60 g/cm^3, the porosity varies between 51 and 75%, and the water content varies between 43 and 124%. Lastly, in the range of 0.9–1.8 m, the density varies between 1.31 and 1.52 g/cm^3, the porosity varies between 69 and 76%, and the water content between 118 and 124%. These observations suggest that the physical properties of seafloor sediments in the hydrate zone exhibit distinct variations with depth, indicating the potential for stratified modeling of sediment properties in this region.

5. Based on long-term in situ observations in the hydrate zone of the South China Sea, the engineering properties of seafloor sediments in the region exhibit a four-layer feature that comprises low-resistivity, very high-resistivity, low-resistivity, and high-resistivity layers. Specifically, the resistivity of sediments at a depth of 0.3 m below the seafloor surface remains stable with time, while in the range of 0.3–0.7 m below the seafloor surface, the resistivity first increases and then decreases over time. In the range of 0.7–1 m below the seafloor surface, the resistivity decreases over time. At depths of 1–1.6 m below the seafloor surface, the resistivity remains stable over time, while at a depth of 1.6 m below the seafloor surface, the resistivity decreases over time. In addition, a layer containing hydrate is identified, displaying a resistivity bimodal structure with a value of 6.86 $\Omega \cdot$m.

6. By utilizing in situ resistivity observation data and a deep learning inversion sediment physical property model, the present study was able to identify a significant impact of hydrate decomposition on the sediment's physical properties. The observation period revealed that the resistivity increased in the range of 0–0.7 m below the seafloor, which coincided with a decrease in sediment density, water content, and porosity. Conversely, the resistivity decreased in the range of 0.7–0.9 m below the seafloor, which aligned with a decrease in sediment density and an increase in porosity and water content. In comparison to previous studies, this dynamic alteration was determined to be attributed to hydrate decomposition and the resultant gas production. It is hypothesized that the sediment dynamics at a depth of 0.7–0.9 m below the seafloor are instigated by the decomposition of

surface hydrate, whereas the sediment dynamics at a depth of 0–0.7 m below the seafloor are caused by the gas generated from hydrate decomposition.

Author Contributions: Conceptualization, Z.S. (Zhiwen Sun) and Y.J.; methodology, Z.F. and C.Z.; investigation, K.L.; resources, Z.S. (Zhongqiang Sun); writing—original draft preparation, Z.S. (Zhiwen Sun); writing—review and editing, X.S.; visualization, L.X. and H.L.; funding acquisition, Y.J. All authors have read and agreed to the published version of the manuscript.

Funding: This research was funded by the National Natural Science Foundation of China (nos. 41831280); the National Key R&D Program 2022YFC2803800; the National Natural Science Foundation of China (42207173, and 41427803, 42206078) , the Key R&D plan of Shandong Province 2022CXPT054 and Postdoctoral Applied Research Project of Qingdao (Assessment and Evaluation of the Carbon Sequestration Capacity of Jiaozhou Bay Salt Marshes).

Institutional Review Board Statement: Not applicable.

Informed Consent Statement: Not applicable.

Data Availability Statement: Available on request.

Acknowledgments: We acknowledge all experimenters, including Xiaomeng Li, Yue Qiao, Xianming Zhu, Na Zhu, Hong Zhang, Xiangqian Li, and Guan Yang, and all personnel of the Dongfanghong III and Marine Geology 6.

Conflicts of Interest: The authors declare no conflict of interest.

References

1. Gamboa, D.; Omira, R.; Terrinha, P. A Database of Submarine Landslides Offshore West and Southwest Iberia. *Sci. Data* **2021**, *8*, 185. [CrossRef]
2. Sun, Q.; Wang, Q.; Shi, F.; Alves, T.; Gao, S.; Xie, X.; Wu, S.; Li, J. Runup of Landslide-Generated Tsunamis Controlled by Paleogeography and Sea-Level Change. *Commun. Earth Environ.* **2022**, *3*, 244. [CrossRef]
3. Guo, X.; Stoesser, T.; Zheng, D.; Luo, Q.; Liu, X.; Nian, T. A Methodology to Predict the Run-out Distance of Submarine Landslides. *Comput. Geotech.* **2023**, *153*, 105073. [CrossRef]
4. Yin, S.; Pope, E.L.; Lin, L.; Ding, W.; Gao, J.; Wu, Z.; Yang, C.; Chen, J.; Li, J. Re-Channelization of Turbidity Currents in South China Sea Abyssal Plain Due to Seamounts and Ridges. *Mar. Geol.* **2021**, *440*, 106601. [CrossRef]
5. Zhu, C.; Li, S.; Chen, J.; Wang, D.; Song, X.; Li, Z.; Chen, B.; Shan, H.; Jia, Y. Nepheloid Layer Generation by Gas Eruption: Unexpected Experimental Results. *J. Ocean Limnol.* **2023**, *1*, 1–9. [CrossRef]
6. Jia, Y.; Zhang, L.; Zheng, J.; Liu, X.; Jeng, D.-S.; Shan, H. Effects of Wave-Induced Seabed Liquefaction on Sediment Re-Suspension in the Yellow River Delta. *Ocean Eng.* **2014**, *89*, 146–156. [CrossRef]
7. Guo, X.; Liu, Z.; Zheng, J.; Luo, Q.; Liu, X. Bearing Capacity Factors of T-Bar from Surficial to Stable Penetration into Deep-Sea Sediments. *Soil Dyn. Earthq. Eng.* **2023**, *165*, 107671. [CrossRef]
8. Bai, C.; Su, P.; Su, X.; Cui, H.; Shang, W.; Han, S.; Zhang, G. Characterization of the Sediments in a Gas Hydrate Reservoir in the Northern South China Sea: Implications for Gas Hydrate Accumulation. *Mar. Geol.* **2022**, *453*, 106912. [CrossRef]
9. Zhu, C.; Jiao, X.; Cheng, S.; Li, Q.; Liu, K.; Shan, H.; Li, C.; Jia, Y. Visualising Fluid Migration Due to Hydrate Dissociation: Implications for Submarine Slides. *Environ. Geotech.* **2020**, *40*, 1–9. [CrossRef]
10. Hui, Z. Study on the Reservoir Configuration of High Curvature Type Gravity Flow Channel in the Second Section of Yinggehai Formation, Dongfang Gas Field. Master Thesis, Northwestern University, Xi'an, China, 2021.
11. Vanneste, M.; Sultan, N.; Garziglia, S.; Forsberg, C.F.; L'Heureux, J.-S. Seafloor Instabilities and Sediment Deformation Processes: The Need for Integrated, Multi-Disciplinary Investigations. *Mar. Geol.* **2014**, *352*, 183–214. [CrossRef]
12. Guo, X.; Nian, T.; Fu, C.; Zheng, D. Numerical Investigation of the Landslide Cover Thickness Effect on the Drag Forces Acting on Submarine Pipelines. *J. Waterw. Port Coast. Ocean. Eng.* **2023**, *149*, 04022032. [CrossRef]
13. Zhou, S.; Li, Q. Developing ocean energy and building a strong ocean state. *Sci. Technol. Her.* **2020**, *38*, 17–26.
14. Sun, Z.; Jia, Y.; Shan, H.; Fan, Z.; Song, X.; Xue, L.; Li, K. Monitoring and Early Warning Technology of Hydrate-Induced Submarine Disasters. *IOP Conf. Ser. Earth Environ. Sci.* **2020**, *570*, 062030. [CrossRef]
15. Rezaei, S.; Shooshpasha, I.; Rezaei, H. Empirical Correlation between Geotechnical and Geophysical Parameters in a Landslide Zone (Case Study: Nargeschal Landslide). *Earth Sci. Res. J.* **2018**, *22*, 195–204. [CrossRef]
16. Zweiri, Y.H.; Whidborne, J.F.; Seneviratne, L.D. A Three-Term Backpropagation Algorithm. *Neurocomputing* **2003**, *50*, 305–318. [CrossRef]
17. Ciresan, D.C.; Meier, U.; Masci, J.; Gambardella, L.M.; Schmidhuber, J. Flexible, High Performance Convolutional Neural Networks for Image Classification. In Proceedings of the Twenty-Second International Joint Conference on Artificial Intelligence, Barcelona, Spain, 16–22 July 2011.

18. Ji, X.; Yang, B.; Tang, Q. Seabed Sediment Classification Using Multibeam Backscatter Data Based on the Selecting Optimal Random Forest Model. *Appl. Acoust.* **2020**, *167*, 107387. [CrossRef]
19. Singh, U.K.; Tiwari, R.K.; Singh, S.B.; Rajan, S. Prediction of Electrical Resistivity Structures Using Artificial Neural Networks. *J. Geol. Soc. India* **2006**, *67*, 234–242.
20. Luo, Z.F.; Lu, B.; Yang, Y. Application of artificial neural network technology to submarine sediment sound velocity prediction. *Mar. Technol.* **2009**, *28*, 40–43.
21. Mukherjee, B.; Sain, K. Prediction of Reservoir Parameters in Gas Hydrate Sediments Using Artificial Intelligence (AI): A Case Study in Krishna-Godavari Basin (NGHP Exp-02). *J. Earth Syst. Sci.* **2019**, *128*, 199. [CrossRef]
22. Zhou, Y.; Wang, J.; Zuo, R.; Xiao, F.; Shen, W.; Wang, S. Machine learning, deep learning and Python language in field of geology. *ACTA Petrol. Sin.* **2018**, *34*, 3173–3178.
23. Singh, H.; Seol, Y.; Myshakin, E.M. Prediction of Gas Hydrate Saturation Using Machine Learning and Optimal Set of Well-Logs. *Comput. Geosci.* **2021**, *25*, 267–283. [CrossRef]
24. Chen, J.; Vissinga, M.; Shen, Y.; Hu, S.; Beal, E.; Newlin, J. Machine Learning-Based Digital Integration of Geotechnical and Ultrahigh-Frequency Geophysical Data for Offshore Site Characterizations. *J. Geotech. Geoenvironmental Eng.* **2021**, *147*, 04021160. [CrossRef]
25. Singh, A.; Ojha, M. Machine Learning in the Classification of Lithology Using Downhole NMR Data of the NGHP-02 Expedition in the Krishna-Godavari Offshore Basin, India. *Mar. Pet. Geol.* **2022**, *135*, 105443. [CrossRef]
26. Pszonka, J.; Schulz, B. SEM Automated Mineralogy Applied for the Quantification of Mineral and Textural Sorting in Submarine Sediment Gravity Flows. *Gospod. Surowcami Miner. Miner. Resour. Manag.* **2022**, *38*, 105–131.
27. Xie, Y.; Lu, J.; Cai, H.; Deng, W.; Kuang, Z.; Wang, T.; Kang, D.; Zhu, C. The In-Situ NMR Evidence of Gas Hydrate Forming in Micro-Pores in the Shenhu Area, South China Sea. *Energy Rep.* **2022**, *8*, 2936–2946. [CrossRef]
28. Yang, Y.; Kou, H.; Li, Z.; Jia, Y.; Zhu, C. Normalized Stress–Strain Behavior of Deep-Sea Soft Soils in the Northern South China Sea. *J. Mar. Sci. Eng.* **2022**, *10*, 1142. [CrossRef]
29. Guo, X.-J.; Zhang, Z.-K.; Jia, Y.-G.; Huang, X.-Y. Electrical resistivity feature of saturated silty soil in Yellow River estuarine area and its engineering geology application. *J. Rock Soil Mech.* **2007**, *28*, 593–598.
30. Li, D.; Ye, Y.; Chen, P.; Chen, X.; Lv, X. Shelf Sediment Resistivity Property off Ningbo Coast. *Geotech. Investig. Surv.* **2010**, *38*, 19–22.
31. Yu, L. Relationship between Resistivity and Mechanical Properties of Shallow Soil in Gas Hydrate Mining Area of South China Sea. Master Thesis, Ocean University of China, Qingdao, China, 2019.
32. Nguyen, Q.V.; Miller, N.; Arness, D.; Huang, W.; Huang, M.L.; Simoff, S. Evaluation on Interactive Visualization Data with Scatterplots. *Vis. Inform.* **2020**, *4*, 1–10. [CrossRef]
33. Viau, C.; McGuffin, M.J.; Chiricota, Y.; Jurisica, I. The FlowVizMenu and Parallel Scatterplot Matrix: Hybrid Multidimensional Visualizations for Network Exploration. *IEEE Trans. Vis. Comput. Graph.* **2010**, *16*, 1100–1108. [CrossRef]
34. Wu, C.L.; Chau, K.W.; Fan, C. Prediction of Rainfall Time Series Using Modular Artificial Neural Networks Coupled with Data-Preprocessing Techniques. *J. Hydrol.* **2010**, *389*, 146–167. [CrossRef]
35. Grattarola, D.; Alippi, C. Graph Neural Networks in TensorFlow and Keras with Spektral [Application Notes]. *IEEE Comput. Intell. Mag.* **2021**, *16*, 99–106. [CrossRef]
36. Khan, M.A.; Zafar, A.; Farooq, F.; Javed, M.F.; Alyousef, R.; Alabduljabbar, H.; Khan, M.I. Geopolymer Concrete Compressive Strength via Artificial Neural Network, Adaptive Neuro Fuzzy Interface System, and Gene Expression Programming With K-Fold Cross Validation. *Front. Mater.* **2021**, *8*, 621163. [CrossRef]
37. Kim, T.; Oh, J.; Kim, N.; Cho, S.; Yun, S.-Y. Comparing Kullback-Leibler Divergence and Mean Squared Error Loss in Knowledge Distillation. *arXiv* **2015**, arXiv:2105.08919.
38. Sun, Z.; Jia, Y.; Quan, Y.; Guo, X.; Liu, T.; Meng, Q.; Sun, Z.; Li, K.; Fan, Z.; Chen, T.; et al. Development and Application of Long-Term in Situ Monitoring System for Complex Deep-Sea Engineering Geology. *Earth Sci. Front.* **2022**, *29*, 216–228. [CrossRef]
39. Hamilton, E.L.; Bachman, R.T. Sound Velocity and Related Properties of Marine Sediments. *J. Acoust. Soc. Am.* **1982**, *72*, 14. [CrossRef]
40. Sun, Y.F.; Sun, H.F.; Dong, L.F. Resistivity characteristics of submarine soils and their corrosion evaluation. *Coast. Eng.* **2005**, *14*, 48–53.
41. Liu, G.H.; Wang, Z.Y.; Huang, J.P. Resistivity properties of soils and their engineering applications. *J. Geotech. Eng.* **2004**, *1*, 83–87.
42. Archie, G.E. The Electrical Resistivity Log as an Aid in Determining Some Reservoir Characteristics. *Trans. AIME* **1942**, *146*, 54–62. [CrossRef]
43. Jana, S.; Ojha, M.; Sain, K.; Srivastava, S. An Approach to Estimate Gas Hydrate Saturation from 3-D Heterogeneous Resistivity Model: A Study from Krishna-Godavari Basin, Eastern Indian Offshore. *Mar. Petrol. Geol.* **2017**, *79*, 99–107. [CrossRef]
44. Wu, J.; Guo, X.; Jia, Y.G.; Sun, X.; Li, N. Simulation and analysis of the monitoring effect of sea-bed based high-density resistivity method for methane gas leakage during natural gas hydrate extraction. *J. Jilin Univ. (Earth Sci. Ed.)* **2018**, *48*, 1854–1864.
45. Cong, J.; Zhang, Y.; Hu, G.; Mi, B.; Kong, X.; Xue, B.; Ning, Z.; Yuan, Z. Textures, Provenances, and Transport Patterns of Sediment on the Inner Shelf of the East China Sea. *Cont. Shelf Res.* **2022**, *232*, 104624. [CrossRef]
46. Feng, M. Numerical Simulation of Hydrate Particle Migration and Deposition in Horizontal Pipe Flow. *Value Eng.* **2023**, *42*, 148–150.
47. Du, X.; Sun, Y.; Song, Y.; Zhu, C. In-situ observation of wave-induced pore water pressure in seabed silt in the yellow river estuary of China. *J. Mar. Environ. Eng.* **2021**, *10*, 305–317.

Journal of
Marine Science and Engineering

MDPI

Article

Experimental Study on the Bonding Performance of HIRA-Type Material Anchor Solids Considering Time Variation

Kun Wang [1], Qingsheng Meng [1], Yan Zhang [2,*], Huadong Peng [3] and Tao Liu [1,4,*]

[1] Shandong Provincial Key Laboratory of Marine Environment and Geological Engineering, Ocean University of China, Qingdao 266100, China
[2] College of Civil Engineering, Anhui Jianzhu University, Hefei 230009, China
[3] Shanghai Survey, Design and Research Institute (Group) Co., Ltd., Qingdao 266100, China
[4] Qingdao National Laboratory of Marine Science and Technology, Qingdao 266100, China
* Correspondence: avayan8006@163.com (Y.Z.); ltmilan@ouc.edu.cn (T.L.)

Abstract: In order to reveal the evolution of bonding performance of HIRA (High Intensity and Rapid Agent) anchor solids with maintenance time, the evolution characteristics of bond strength and stress distribution at the interface between HIRA-based anchor solids and a geotechnical body under different maintenance times and the fine damage pattern of anchor solids were studied by an indoor pull-out test of anchor solids. The comparative analysis was performed with 42.5 grade ordinary Portland cement (hereinafter referred to as P.O 42.5). The results show that the early strength and rapid setting characteristics of HIRA type material are obvious, and the difference between its average peak bond strength and that of cement is 10.45 times. The shear stress distribution has obvious stress concentration characteristics, and the peak value will appear and shift with the increase in load, and the peak shift of the HIRA anchor solid occurs earlier than that of cement. Due to different stress levels, the damage of the HIRA anchor solid after being pulled out increases with the increase in maintenance time, while that of cement gradually becomes more severe. The overall damage of the HIRA material is generally lower than that of cement in the same period.

Keywords: grouting material; anchor solid; maintenance time; bonding performance; destruction mode; pull-out test

Citation: Wang, K.; Meng, Q.; Zhang, Y.; Peng, H.; Liu, T. Experimental Study on the Bonding Performance of HIRA-Type Material Anchor Solids Considering Time Variation. *J. Mar. Sci. Eng.* **2023**, *11*, 798. https://doi.org/10.3390/jmse11040798

Academic Editor: José António Correia

Received: 10 March 2023
Revised: 27 March 2023
Accepted: 3 April 2023
Published: 7 April 2023

1. Introduction

With the implementation of China's strategy of strengthening transportation, more and more cities have accelerated the construction of rail transit [1], and the submarine tunnel has become one of the solutions to the subway planning problems in coastal cities along the river. Among the tunnel support methods, anchorage support makes up a high proportion, and anchor solids play a major role in the whole anchorage support system [2,3]. The material properties of anchor solids directly affect their mechanical properties [4–6], which, in turn, are related to the safe and efficient construction of marine projects [7]. The grouting material commonly used in the current anchorage support method of submarine tunnels is cement mortar, and there is less application and research on new, fast-setting, high-strength grouting materials [8,9].

Anchor cables are widely used in pit engineering and tunneling [10], and are divided into two main categories according to force characteristics: tension anchor cables and pressure anchor cables, both of which have different anchoring mechanisms, resulting in different force characteristics of anchor solids [11]. Pressure anchor cables have many advantages over the traditional tension anchor cables [12]. The theoretical analysis of the force in the anchorage section of pressure-type anchor cables is based on nonlinear slip relations [13–15], which can derive the theoretical solutions for the distribution of shear stress and axial force in the anchorage section. For the experimental studies, there are mainly scaled-down model tests and full-scale field tests [16–18]. The pressure-type anchor has good displacement ductility, and the shear stress increases with the increase in the load at locations farther away from the bearing plate. The compressive strength of the anchor solid increases significantly under the joint action of lateral pressure and surface shear stress [19]. A summary of typical previous studies is presented in Table 1. The anchor solid material in existing experimental studies is commonly cement mortar with a fixed maintenance time, but, for the construction of submarine tunnels, the special stratigraphic conditions [20,21] and complex physical and chemical fields [22,23] are a great challenge for conventional grouting materials.

Table 1. Summary of typical previous studies.

Analytical Methods	Material	Anchor Type	Authors
Nonlinear slip relations	/	Pressure type	[13–15]
Elastoplastic relationship	/	Pressure type	[11]
Kelvin solution	/	Pressure type	[12]
Model test	Cement mortar	Pressure type	[16]
Field test	Cement mortar	Pressure type	[17]
Field test	Cement mortar	Pressure type	[18]
Indoor test	Cement mortar	Pressure type	[19]
Field test	High polymer	Tensile type	[8]
Indoor test	High concrete	Tensile type	[9]

Currently, the existing research on pressure type anchor solids tends to be theoretical analysis, and the existing experimental research focuses more on the macroscopic failure mechanism during the tensioning process of the anchor system. The mesoscopic failure mode of the anchor solids needs to be further studied. In this study, for a new type of HIRA material, the influence law of different maintenance times on the evolution of tension load, bond strength, and stress distribution of anchor solids was studied by anchor solid pull-out tests, under the premise of comparative analysis, using P.O 42.5 material. The fine-scale damage mode of anchor solids was analyzed, aiming to provide a reference for anchorage engineering in sub-sea tunnels.

2. Materials and Methods

2.1. Test Materials

The test substrate is granite, and the basic physical property index is shown in Table 2. The specimen size is 150 mm × 150 mm × 180 mm with drilling in the center of the upper surface. The drilling diameter is 30 mm and the depth is 130 mm. The test tendon body, with the upper and lower extrusion sets of steel strand structure, is similar to the pressure anchor cable. The strand specifications for 7 × ∅5 mm are: nominal tensile strength of 1860 MPa, and the maximum force of the whole strand is greater than or equal to 260 kN.

Table 2. Basic physical property indexes of granite used in the test.

Volume Density (g/cm^3)	Water Absorption (%)	Bending Strength (MPa)	Compression Strength (MPa)	Antifreeze Coefficient (%)
2.87	0.853	28.63	83.27	83.27

The grouting material used is HIRA-type material, and P.O 42.5 material is used as a comparison. HIRA material is a new type of fast-setting, early-strength, mineral-based grouting material, which is a light grayish-yellow powder at room temperature. The particle size distribution curves of both are shown in Figure 1, from which it can be seen that the average particle size of HIRA is about three times that of P.O 42.5.

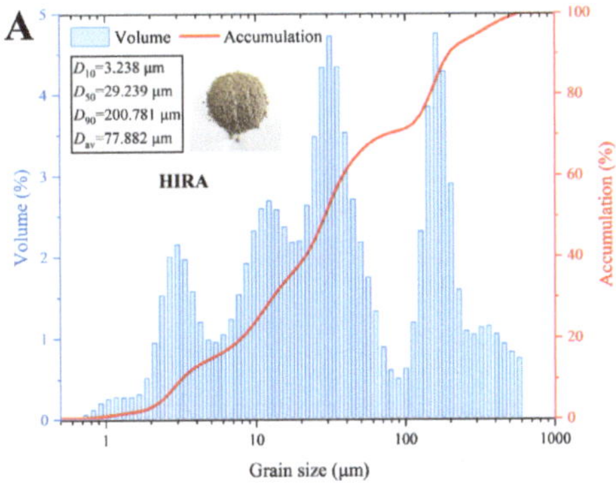
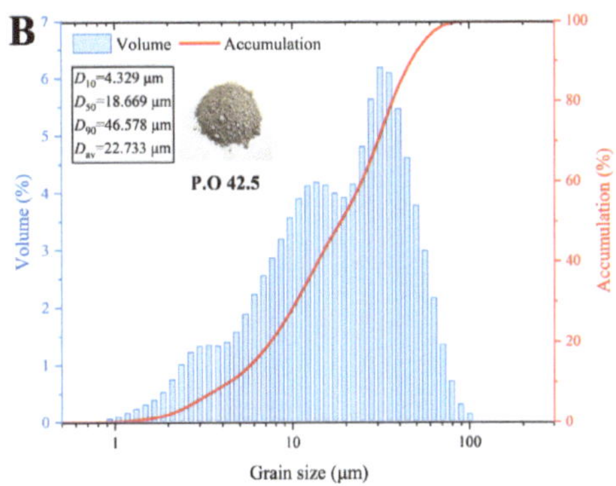

Figure 1. Particle size distribution curve of HIRA (**A**) and P.O 42.5 (**B**).

The relevant performance indicators of HIRA-type materials and P.O 42.5-type materials are shown in Table 3. HIRA-type material proportioning mainly includes main materials and auxiliary materials, where the main materials are sulfur aluminate cement, early strength silicate cement, fine sand, calcium sulfate, calcium carbonate, etc., and auxiliary materials are magnesium oxide, silica micronized powder, Cai system water reducing agent, boric acid, etc. P.O 42.5-type material is mainly composed of tricalcium silicate, dicalcium silicate, tricalcium aluminate, gypsum, etc.

Table 3. Performance indicators for HIRA type materials and P.O 42.5 type materials.

Category	Technical Requirements	Item	Unit	Measured Value
HIRA	Coagulation time	Initial condensation	min	55
		Final condensation	min	219
	Compressive strength	1 d	MPa	31.2
		3 d	MPa	49.3
P.O 42.5	Coagulation time	Initial condensation	min	131
		Final condensation	min	314
	Compressive strength	1 d	MPa	16.7
		3 d	MPa	25.4

2.2. Testing Instruments

The test loading device is a WAW-1000B microcomputer-controlled electro-hydraulic servo universal testing machine with a maximum range of 1000 kN. The test loading method is set to the displacement control with a control value of 5 mm/min, and the broken type judgment is set to be invalid so that the testing machine can continue tensioning after the anchor solid slips and breaks. The strain gauges were made of foil-type non-welded strain gauges with a nominal resistance of 120 Ω, a sensitive grid length of 3 mm, and a sensitive grid structure of AA. The XL2118B static strain gauges were used for data acquisition, and the wiring method was 1/4 bridge. The test machine loading and strain gauge data recording were carried out simultaneously.

2.3. Test Method and Procedure

The test set a total of two variables of maintenance time and material. The water-cement ratio was 0.5, and the specific parameters of the specimens are shown in Table 4. As the water-cement ratio of P.O 42.5 material was 0.5 in previous studies and actual projects, the water-cement ratio of HIRA-type material was finally determined to be 0.5 in order to compare better with P.O 42.5 material and to integrate the results of pre-experiments. The average peak bond strength was calculated by the peak load of the testing machine in the tensioning process, as in Equation (1).

$$\tau_{mA} = \frac{F_m}{\pi dl} \tag{1}$$

where τ_{mA} denotes the average peak bond strength; F_m denotes the ultimate load; d denotes anchor solid diameter; and l denotes the anchor solid length.

Table 4. Sample parameters.

Grouting Materials	Specimen	Maintenance Time (d)	Grouting Length (mm)	Quantity
HIRA	H-1d	1	100	3
	H-3d	3	100	3
	H-7d	7	100	3
	H-14d	14	100	3
P.O 42.5	P-1d	1	100	3
	P-3d	3	100	3
	P-7d	7	100	3
	P-14d	14	100	3

Synthesizing previous studies, it is assumed that the stress–strain relationship in the test is linear elastic, and the axial stress σ_i at each measurement point of the anchor solid under different test loads can be calculated according to Equation (2). The shear stress τ_i is calculated from the equilibrium equation of the two adjacent measurement points of the anchor solid, which is shown in Equation (3).

$$\sigma_i = \varepsilon_i \cdot E \tag{2}$$

$$\tau_i = \frac{F_i - F_{i+1}}{\pi dL} = \frac{(\sigma_i - \sigma_{i+1})\pi \frac{d^2}{4}}{\pi dL} = \frac{(\sigma_i - \sigma_{i+1})d}{4L} \tag{3}$$

where σ_i denotes the axial stress at the measurement point i; ε_i denotes the microstrain at the measurement point i; E denotes the modulus of elasticity of the material; the HIRA material is taken as 28 GPa, and the P.O 42.5 material is taken as 33 GPa; τ_i denotes the average shear stress between the measurement point i and the measurement point $i + 1$; F_i denotes the axial force at the measurement point i; d denotes the diameter of the anchor solid; L denotes the distance between two measurement points.

The correspondences between each measurement point and the axial stress and shear stress are shown in Figure 2. There are 5 measurement points, and each measurement point is spaced 25 mm apart. To avoid damage to the strain gauges by boundary effects, measurement point 1 is located about 5 mm to the right of point 0, and measurement point 5 is located about 95 mm to the left of 100 mm. The axial stress location corresponds to the location of the measurement points, and the shear stress symbol is located at the center of the two adjacent measurement points.

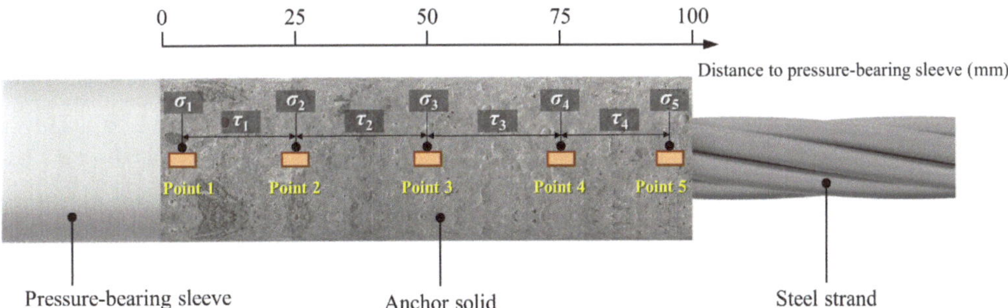

Figure 2. Corresponding relationship between stress and measuring points. The colored square represents the strain gauge, and the "Point" is the measuring point.

3. Results

3.1. Load Displacement Relationship

In order to better describe the changing trend between variables, only typical specimens in the same group are selected in this paper to describe the load–displacement relationship. the load–displacement relationship of HIRA-type anchor solids under different maintenance times is shown in Figure 3A, where load is the value of loading force of the testing machine, and displacement is the displacement of the testing machine collet. From the figure, it can be seen that all the four sets of curves are slowly rising at first, then, with the increase in displacement, the load value rises sharply, reaches the peak, and then falls back to the non-linear change law. The displacement of the 4 sets of curves corresponding to the peak load is 10–15 mm. With the increase in the number of maintenance days, the peak load also increases. The curves grew slowly before reaching the peak load at 1 day of maintenance, and grew rapidly at 3 days and later. The reason for this trend was that the anchor solid strength was not fully formed at 1 day, and the bond between the anchor solid and the rock was not yet complete.

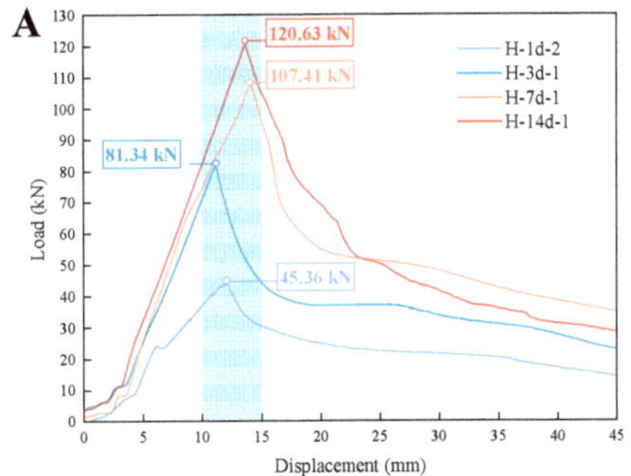

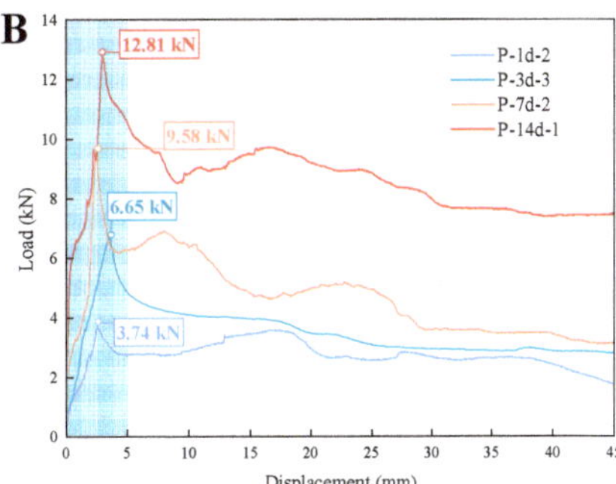

Figure 3. Load–displacement curve of HIRA (**A**) and P.O 42.5 (**B**) anchor solids under different maintenance times.

Figure 3B shows the load displacement curves of P.O 42.5 anchor solids under different maintenance times. From the figure, it can be seen that the four sets of curves all show a non-linear change pattern of rapid rise first, followed by a slow fall after reaching the peak, and the displacement of the four sets of curves corresponding to the peak load is within 5 mm and increases with the increase in maintenance days. The curves were similar in form at 1 and 3 days of maintenance, and more similar at 7 and 14 days. It can be seen that, in the anchor solid formed by cement slurry, the cumulative stage of bond strength development is located before 7 days, and the strength is basically formed after 7 days [2]. At the same time, after reaching the peak load damage, the residual strength of the 7-day and 14-day specimens decreased more slowly with the increase in the displacement of the testing machine, and the strength rebounded. The reason for this phenomenon is that when the bond slip of the anchor solid started to occur, the upper end anchor solid cracked, which directly caused the rapid decrease in bond strength. At this time, the testing machine continued to tension, and, after the cracked part of the anchor solid was slowly pulled out, the uncracked part of the anchor solid started to function, which caused the temporary rebound of strength. However, since the overall anchor solid had already undergone slip damage, the bond strength would eventually decrease with displacement increases and decreases.

3.2. Stress Distribution

Using the strain data obtained from the five measurement points, the shear stresses of the anchor solid under different loads can be obtained according to Equations (2) and (3). The specimens used in this section are the same as those used in the previous section. Since the tension test in this paper is a destructive test, and all strain gauges fail when the anchor solid is damaged by slip, the axial stress and shear stress analyzed in this section are the data before the maximum tension load with strain records.

The distribution curves of shear stress along the length of the HIRA anchor solid at different maintenance times are shown in Figure 4. The value at the dotted line is the average peak bond strength calculated for the corresponding specimen according to Equation (1). The distribution of shear stress in the four sets of curves is not uniform, showing a non-linear form of decreasing shear stress along the length of the anchor solid at lower load levels and a single peak curve at higher load levels [24]. The stress is more concentrated at the location near the bearing sleeve and is generally lower at the location near the hole at the outer anchor end; as the load increases, the shear stress affects the range of the anchor solid, and the peak value no longer increases with the load, but shifts along the anchor solid to the outer anchor end [25–27]. As the load increases, the stress effect range increases, and the peak shear stress appears. The peak no longer increases with the load growth, but shifts along the anchor solid to the outer anchor end. The peak transfer of the shear stress curve along the anchorage section is accompanied by a decrease in the shear stress level at the previously peaked measurement points, which indicates that the effective range of shear stress influence is limited [28,29]. The black dashed line in the figure shows the average ultimate bond strength calculated from the peak loads of the corresponding specimens by Equation (1) (similarly hereinafter), and it can be seen that the peak shear stresses in all four groups of specimens are greater than their average ultimate bond strengths.

After the load grew to 40 kN at 1 day of maintenance, the peak shear stress appeared to be flattened backward, while the peak shear stress stopped growing and was transferred from the shear stress τ_1 between measurement points 1 and 2 to the shear stress τ_2 between measurement points 2 and 3, and the value of τ_1 started to decrease from the peak while the peak shear stress was transferred. At 3 days, when the load value was 75 kN, the peak shear stress at this time was lower than the peak at 70 kN because the shear stress near measurement point 3 at this time cannot be calculated to obtain a specific value. However, according to the phenomenon that τ_1 continues to decrease, τ_2 starts to fall back from the peak at 70 kN, and τ_3 and τ_4 continue to increase during the increase in the load from 70 kN to 75 kN. It can be judged that the peak shear stress at 75 kN is located near measurement point 3. At 3 days, when the anchor solid was pulled out, τ_3 did not reach its peak, but, at 7 days, τ_3 reached its peak before the anchor solid slipped and broke, which indicates that the peak shear stress kept increasing with the growth of maintenance time, while the peak shear stress kept moving toward the outer anchor section of the anchor body with the increase in the load level. The trend of peak shear stress transfer at 14 days of maintenance was similar to that at 3 days.

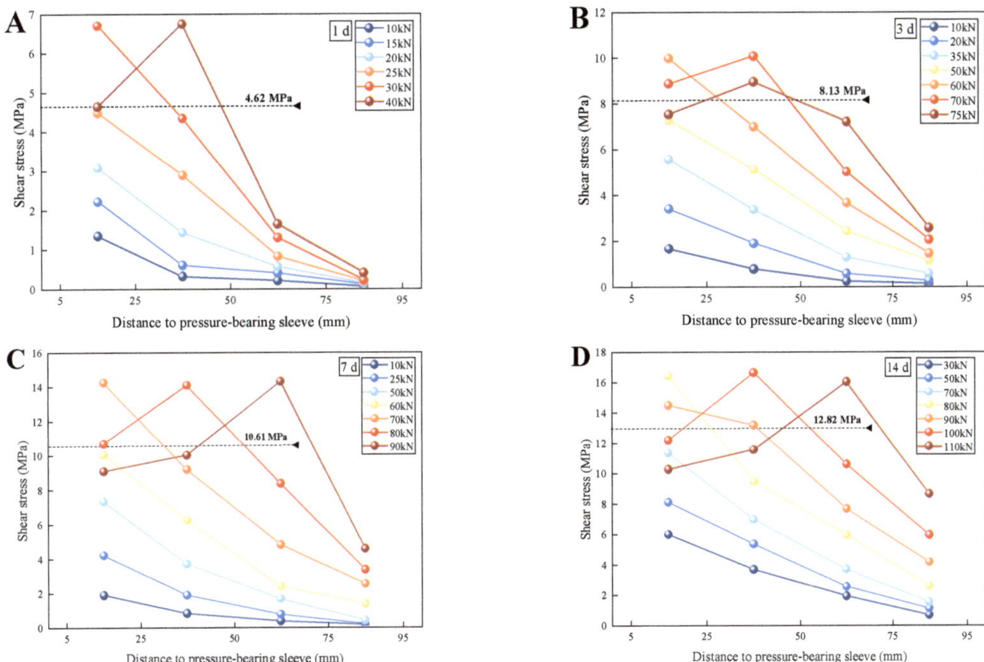

Figure 4. Distribution curve of shear stress along the length of HIRA type anchor solids after 1 (**A**), 3 (**B**), 7 (**C**), and 14 (**D**) days of maintenance time.

The distribution curve of shear stress along the length of the P.O 42.5 anchor solid under different maintenance times is shown in Figure 5, from which it can be seen that the distribution characteristics of shear stress are basically similar to those of the HIRA type, but the difference is that the shear stress presents a non-linear form of decreasing along the length of anchor solid in the early maintenance period, and the single-peak curve appears only after 7 days of maintenance, i.e., the peak shift occurs [30]. As with the HIRA material, the value at the dotted line is the average ultimate bond strength calculated for the corresponding specimen according to Equation (1). When the maintenance time increased to 7 days, the peak shear stress shifted along the length of the anchor solid with the increase in the load level, while the value of τ_1 started to decrease from the peak; when the maintenance was 14 days, the peak shear stress shifted after the load increased to 10 kN, and, in addition, the overall τ_3 was lower at 14 days compared with 7 days, and τ_4 was basically unchanged, which was due to the fact that the second half of the anchor section of the 14-day specimens were cracked and dropped in different degrees, which was the direct cause of the low values of τ_3 and τ_4 at 14 days.

3.3. Failure Mode

The failure mode in this study is mainly reflected in the evolutionary pattern of the type and number of cracks on the anchor solid. To facilitate analysis and in conjunction with existing studies [31–33], typical cracks in anchor solids in the tests were defined as six types: primary tensile crack T_p, secondary tensile crack T_s, primary shear crack S_p, secondary shear crack S_s, far-field shear crack F_s, and spalling region S_a. Figure 6 shows the typical damage patterns of HIRA and P.O 42.5 anchor solids after pull-out for different maintenance times, and the redrawing of each anchor solid crack is also shown in Figure 6. The gray area in the redrawing is the pressure-bearing sleeve. A clear trend in the figure is that with increasing maintenance time, the damage of the HIRA-type anchor solid after pull-out gradually increases, while the P.O 42.5-type gradually becomes smaller, as shown below.

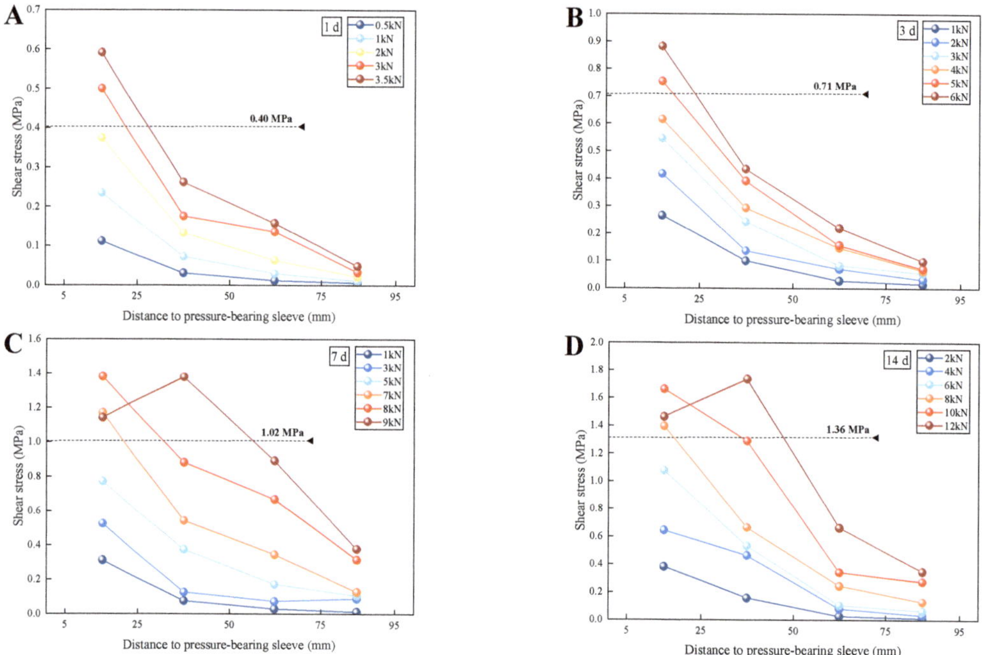

Figure 5. Distribution curve of shear stress along the length of P.O 42.5 anchor solids after 1 (**A**), 3 (**B**), 7 (**C**), and 14 (**D**) days of maintenance time.

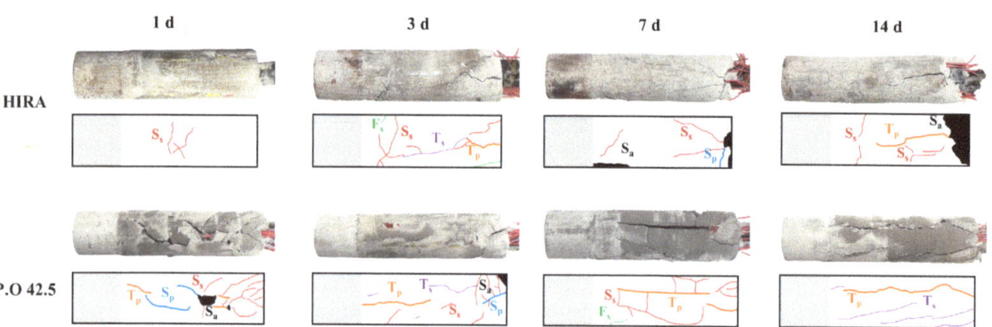

Figure 6. Schematic diagram of typical failure form after anchor solids are pulled out under different maintenance times.

At 1 day of maintenance, the cracks of the HIRA anchor solid were mainly secondary shear cracks S_s (hereinafter referred to as S_s cracks and other cracks as well), with a small number of cracks. When the maintenance time increased to 3 days, it evolved into a combination of multiple cracks, and the number of cracks increased. At 7 days, it changed to a combination of two kinds of cracks, and, at the same time, spalling started to appear at the end of the anchor solid. At 14 days, it was still a combination of both types of cracks, but the spalling area increased further and the damage intensified. It can be seen that the most types and numbers of cracks in the HIRA-type anchor solids after pulling out occurred at 3 days of maintenance, after which the spalling area starts to appear and the area gradually expands as the maintenance time increases.

The most types and numbers of cracks were found in P.O 42.5 anchor solids at 1 day of maintenance. There were mainly three crack combinations of T_p cracks, S_s cracks, and S_p cracks, with a high number of cracks and spalling areas. At 3 days of maintenance, there were still three crack combinations, but the spalling areas were reduced compared with those at 1 day. At 7 days of maintenance, it changed to a combination of two cracks, with no spalling areas. At 14 days, there were still two crack combinations, but the spalling areas were reduced compared with those at 1 day. The number of cracks was further reduced at 14 days, but still in the form of two crack combinations. Unlike HIRA, the type and number of cracks in the P.O 42.5 anchor solids after pull-out gradually decreased with increasing maintenance time.

4. Discussion

4.1. Bond Strength

The peak load of each group of specimens of the HIRA type was brought into Equation (1), and the average peak bond strength was calculated to obtain its variation curve with the maintenance time, as shown in Figure 7A. The average peak bond strength increases with increasing maintenance time, with 4.26 MPa, 9.57 MPa, 10.37 MPa, and 12.80 MPa. It is noteworthy that the 3-day strength increases by 125% compared to the 1-day strength, the largest increase, which is also reflected in Figure 3A, followed by 8.4% and 23.4% increases in strength, in that order. Using the 14-day strength as the final strength, the anchor solid bond strength at 3 days has reached 74.8% of the 14-day strength, which fully reflects the early strength and rapid setting characteristics of the HIRA-type material. The measured values of bond strength at 1 day are more discrete, probably due to the different degrees of material setting at 1 day or the uneven distribution of large particles, such as fine sand, in the three specimens during grouting, resulting in large differences in strength limit values [34].

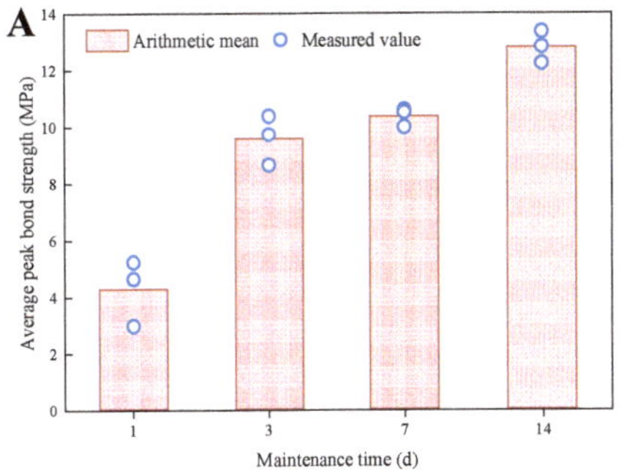

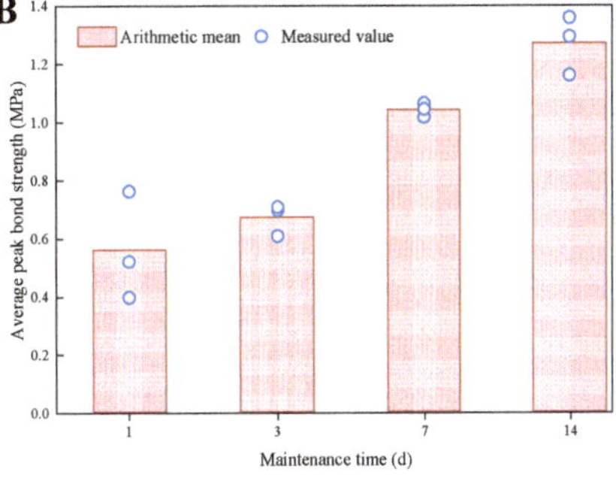

Figure 7. Variation of average peak bond strength of HIRA (**A**) and P.O 42.5 (**B**) anchor solids with maintenance time.

As with the peak load treatment of the HIRA type, the peak load of each group of specimens of P.O 42.5 type was brought into Equation (1), and the average peak bond strength was calculated to obtain its variation curve with the maintenance time, as shown in Figure 7B. The average peak bond strength increases with the increase in the maintenance time [35], to 0.56 MPa, 0.67 MPa, 1.04 MPa, and 1.27 MPa. The increase in the strength from 3 to 7 days is larger, 55.2%, which corresponds to the change of the form of the 3–7 days curve in Figure 4. The strength increased by 19.6% from 1 to 3 days and 22.1% from 7 to 14 days. Using the 14-day strength as the final strength, the 1-day, 3-day, and 7-day anchor solid bond strengths reached 44.1%, 52.8%, and 81.9% of the 14-day strength, respectively.

It can also be found from Figure 7 that the difference between the ultimate average bond strength of HIRA-type material and P.O 42.5-type material is 10.45 times, on average, at each maintenance time point; for example, the ultimate average bond strength of HIRA-type material is 10.37 MPa at 7 days of maintenance, which is 9.97 times higher than that of P.O 42.5-type material at 1.04 MPa.

4.2. Shear Stress Peak Evolution Characteristics

The peak shear stresses of the HIRA and P.O 42.5 anchor solids in Section 3.2 were collapsed to obtain Figure 8. It can be seen from the figure that the peak shear stresses of both increased with time, and the peak shear stresses of HIRA anchor solids increased in decreasing order, that is, 48.7%, 42.1%, and 14.9%. The peak shear stresses of P.O 42.5 anchor solids increased the most in 3–7 days, by 56.8%. This indicates that the bond strength of the HIRA material accumulates fastest when the maintenance time is short, and the strength increase is small in the late maintenance period, which is consistent with the analysis of Figures 3A and 4. Meanwhile, the initial strength of the P.O 42.5 material is available only at 7 days of maintenance, and its strength accumulation is slow, which is consistent with the analysis of Figure 3B, and the analysis of Figure 5C also proves that the peak transfer phenomenon of shear stress begins to appear at this time [36–38]. Under the same maintenance time, the peak shear stress of HIRA type anchor solid is about 10 times higher than that of the P.O 42.5 type. The peak shear stress of the former is 10.2 times higher than that of the latter when maintained for 7 days.

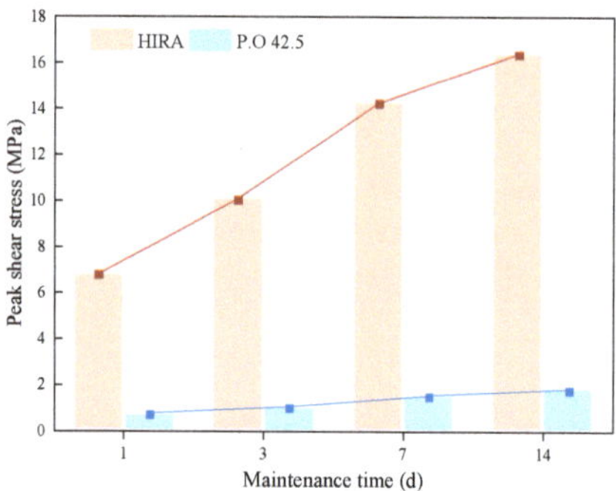

Figure 8. Variation of peak shear stress with maintenance time.

4.3. Association between the Number of Cracks and the Damage Pattern

The number of cracks after pulling out the anchor solids of HIRA and P.O 42.5 were counted, as shown in Figure 9, with three columns corresponding to three parallel samples at each maintenance time. It can be seen from the figure that S_s cracks are predominant for both materials at each maintenance time. After the damage from the anchor solid pullout at the early maintenance time, the HIRA-type material is dominated by S_s cracks, while the P.O 42.5-type material shows a combination of three kinds of cracks, and the number of cracks is more than that of the HIRA type material [39].

With the increase in the maintenance time, the combination of S_s cracks and other cracks became the main damage form of the HIRA-type material, and the overall number of cracks showed a trend of first increasing and then decreasing. According to the aforementioned analysis, the bond strength between the anchor solid and the rock body has reached a high level at 3 days, but, due to the short maintenance time, the anchor solid itself is not fully formed, resulting in more types and numbers of cracks [40]. Later, as the maintenance days become longer, the anchor solid is formed, but, due to the high stress level, the cracks are transformed into spalling areas, resulting in a smaller number of cracks and more severe damage to the anchor solid [41].

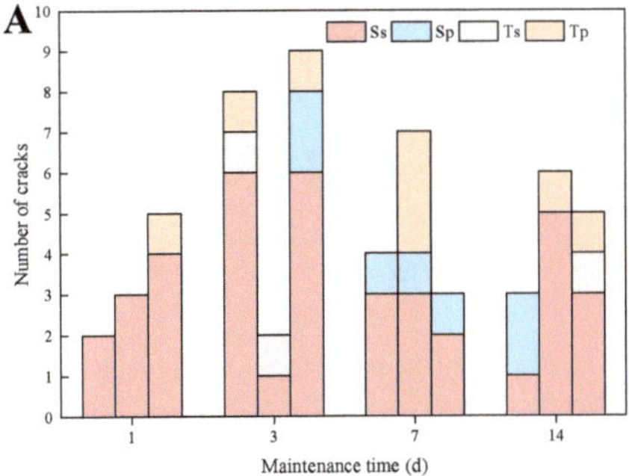
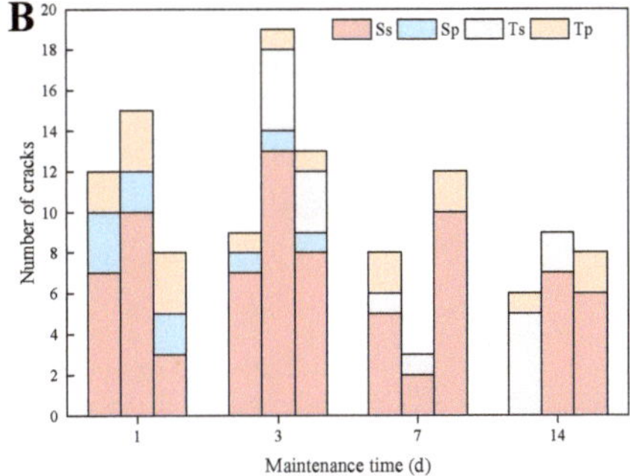

Figure 9. Statistical diagram of crack quantity of HIRA (**A**) and P.O 42.5 (**B**) anchor solids.

The damage form of the P.O 42.5 anchor solid gradually changed to a combination of S_s cracks and T_s cracks as the maintenance time became longer, and its overall number of cracks showed a gradually decreasing trend. Due to the slow solidification of P.O 42.5 material in the early stage, the anchor solid strength was low, and although the stress level was also low at this time. The anchor solid was still severely damaged, and the bond stress increased with the longer maintenance time in the later stage, but the level was always low, and the anchor solid strength was gradually formed, which led to the gradual reduction of the number of cracks in the later stages and the lowest number of cracks at 14 days [42].

Combining the above analysis results, it is easy to find that the characteristics of the HIRA-type material with high strength throughout and fast setting and forming are further verified, while the degree of pull-out damage under four different maintenance times is generally lower than that of the P.O 42.5 material of the same period.

5. Conclusions

(1) The average peak bond strength of anchor solids increases with the increase in maintenance time, and the early strength and rapid setting characteristics of HIRA-type materials are obvious, while cement requires 7 days to reach an approximate proportion of anchor solid bond strength. The difference between the ultimate average bond strength of HIRA-type materials and P.O 42.5-type materials at each maintenance stage is 10.45 times on average. HIRA materials are superior to cement materials commonly used in subsea tunnels in terms of early strength properties.

(2) The shear stress distribution shows a concentration of stress near the bearing sleeve and a lower stress level near the outer anchor end. As the load increases, the range of stress influence increases, and the peak shear stress appears and shifts, accompanied by a decrease in the shear stress level at the previously peaked measurement points. The peak transfer occurs earlier in the HIRA-type anchor solid than in the P.O 42.5 material, which indicates that the overall strength of the HIRA material is higher.

(3) With the increase in maintenance time, the type and number of cracks in HIRA-type anchor solids after extraction first increased and then decreased. The spalling area gradually increased, and the damage degree gradually increased. The type, number and spalling area of P.O 42.5-type anchor solids gradually decreased and the damage degree gradually became smaller. The extraction damage degree of the HIRA-type material was generally lower than that of the P.O 42.5 material in the same period. This proves that the stiffness of HIRA material is lower than that of P.O. 42.5 material.

Author Contributions: Data curation, K.W. and H.P.; funding acquisition, Y.Z. and T.L.; investigation, H.P.; methodology, Q.M. and H.P.; project administration, Q.M., Y.Z. and T.L.; writing—original draft, K.W.; writing—review and editing, Q.M., Y.Z. and T.L. All authors contributed to the article and approved the submitted version. All authors have read and agreed to the published version of the manuscript.

Funding: This work was funded by the National Natural Science Foundation of China (U2006213), the National Natural Science Foundation of China (42277139), the China Postdoctoral Science Foundation (2022M712989), and the Natural Science Foundation of Shandong Province (ZR2022QD103).

Institutional Review Board Statement: Not applicable.

Informed Consent Statement: Not applicable.

Data Availability Statement: The data used to support the findings of this study are included within the article.

Acknowledgments: We appreciate the reviewers for their valuable comments, which are crucial to shaping our manuscript. At the same time, we are also grateful for the financial support provided by the above-mentioned funds.

Conflicts of Interest: The authors declare that the research was conducted in the absence of any commercial or financial relationships that could be construed as a potential conflict of interest.

References

1. China Association of Metros. 2022. Available online: https://www.camet.org.cn/xyxw/11484 (accessed on 6 March 2023).
2. Chen, C.; Liang, G.; Tang, Y.; Xu, Y. Anchoring solid-soil interface behavior using a novel laboratory testing technique. *Chin. J. Geotech. Eng.* **2015**, *37*, 1115–1122.
3. Ding, W.; Liu, J.; Zhang, L. Analysis on interaction of rock-bolts in anchorage support structure of subsea tunnel at different corrosion levels. *J. Cent. South Univ.* **2014**, *45*, 1642–1652.
4. Liu, X.; Li, Z.; Tai, P.; Chen, R.; Fu, W. In-situ Experimental Investigation on Stress Distribution of Grout Body of Tension-type Ground Anchor. *Chin. J. Undergr. Space Eng.* **2021**, *17*, 63–70.
5. Forbes, B.; Vlachopoulos, N.; Diederichs, M.S.; Aubertin, J. Augmenting the in-situ rock bolt pull test with distributed optical fiber strain sensing. *Int. J. Rock Mech. Min. Sci.* **2020**, *126*, 104202. [CrossRef]
6. Martin, L.B.; Tijani, M.; Hadj-Hassen, F.; Noiret, A. Assessment of the bolt-grout interface behaviour of fully grouted rockbolts from laboratory experiments under axial loads. *Int. J. Rock Mech. Min. Sci.* **2013**, *63*, 50–61. [CrossRef]
7. Guo, X.; Stoesser, T.; Zheng, D.; Luo, Q.; Liu, X.; Nian, T. A methodology to predict the run-out distance of submarine landslides. *Comput. Geotech.* **2023**, *153*, 105073. [CrossRef]
8. Shi, M.; Xia, W.; Wang, F.; Liu, H.; Pan, Y. Experimental study on bond performance between polymer anchorage body and silt. *Chin. J. Geotech. Eng.* **2014**, *36*, 724–730.
9. Han, F.; Liu, J.; Liu, J.; Ma, B.; Sha, J.; Wang, X. Study on Anchorage Behavior of Steel Bar in Ultra-high Performance Concrete. *Mater. Rep.* **2019**, *33*, 244–248.
10. Zhang, L.; Wang, R. Research on status quo of anchorage theory of rock and soil. *Rock Soil Mech.* **2002**, *23*, 627–631. [CrossRef]
11. Yang, Q.; Zhu, X.; Ruan, M. Analysis of stress Distribution and Influential Parameters of Anchorage Segment of Pressure-Type Cable. *Chin. J. Rock Mech. Eng.* **2006**, *25*, 4065–4070.
12. Lu, L.; Zhang, Y.; Wu, S. Distribution of stresses on bonded length of compression type rock bolt. *Rock Soil Mech.* **2008**, *29*, 1517–1520. [CrossRef]
13. You, C. Mechanical analysis on anchorage segment of pressure-type cable. *Chin. J. Geotech. Eng.* **2004**, *26*, 828–831.
14. Benmokrane, B.; Chennouf, A.; Mitri, H.S. Laboratory Evaluation of Cement-Based Grouts and Grouted Rock Anchors. *Int. J. Rock Mech. Min. Sci. Geomech. Abstr.* **1995**, *32*, 633–642. [CrossRef]
15. Liao, J.; Tu, B. A Determination Method of Anchorage Length of Compression Anchor. *J. Civ. Archit. Environ. Eng.* **2013**, *35*, 9–14.
16. Zhang, Y.; Lu, L.; Rao, X.; Li, J. Model test research on mechanical behavior of compression type rock bolt. *Rock Soil Mech.* **2010**, *31*, 2045–2050. [CrossRef]
17. Lu, L.; Zhang, Y.; Zhang, S.; Wu, S. Stress Distribution in Fixed Anchor Length of Compression Type Anchor in Soft Rock Mass. *J. Civ. Archit. Environ. Eng.* **2011**, *33*, 69–74.
18. Shen, J.; Gu, J.; Zhang, X.; Chen, A.; Ming, Z. Field Pull-Out Test Research on Tension and Pressrue Unbonded Anchor Cables. *Chin. J. Rock Mech. Eng.* **2012**, *31*, 3291–3297.
19. Lu, L.; Li, L.; Deng, Y.; Hu, T.; Li, X. Failure mechanism of mortar in front of bearing plate of compression type anchor cables. *Chin. J. Geotech. Eng.* **2013**, *35*, 2110–2116.
20. Guo, X.; Liu, Z.; Zheng, J.; Luo, Q.; Liu, X. Bearing capacity factors of T-bar from surficial to stable penetration into deep-sea sediments. *Soil Dyn. Earthq. Eng.* **2023**, *165*, 107671. [CrossRef]
21. Zhang, Y.; Wang, H.; Liu, T.; Liu, H.; Deng, S. Interpretation of pore pressure dissipation of CPTu in intermediate soil considering partial drainage effect. *Ocean Eng.* **2022**, *266*, 112956. [CrossRef]
22. Zhang, Y.; Feng, X.; Ding, C.; Liu, Y.; Liu, T. Study of cone penetration rate effects in the Yellow River Delta silty soils with different clay contents and state parameters. *Ocean Eng.* **2022**, *250*, 110982. [CrossRef]
23. Guo, X.; Nian, T.; Fu, C.; Zheng, D. Numerical Investigation of the Landslide Cover Thickness Effect on the Drag Forces Acting on Submarine Pipelines. *J. Waterw. Port Coast. Ocean Eng.* **2023**, *149*, 04022032. [CrossRef]

24. Bi, D.; You, Z.; Liu, Q.; Wang, C.; Shi, J. Soil anchor solid composite interface element form and mechanical effects. *Rock Soil Mech.* **2017**, *38*, 277–283. [CrossRef]

25. Zhou, B.; Wang, B.; Liang, C.; Wang, Y. Study on load transfer characteristics of wholly grouted bolt. *Chin. J. Rock Mech. Eng.* **2017**, *36*, 3774–3780. [CrossRef]

26. Zhang, X.; Chen, S. Analytical solution for load transfer along anchored section of prestressed anchor cable. *Rock Soil Mech.* **2015**, *36*, 1667–1675. [CrossRef]

27. You, Z.; Fu, H.; You, C.; Zhang, J.; Shao, H.; Bi, D.; Shi, J. Stress transfer mechanism of soil anchor body. *Rock Soil Mech.* **2018**, *39*, 85–92+102. [CrossRef]

28. Huang, M.; Li, J.; Zhao, M.; Chen, C. Nonlinear Analysis on Load Transfer Mechanism of Bolts in Layered Ground. *China J. Highw. Transp.* **2019**, *32*, 12–20+56. [CrossRef]

29. Hoien, A.H.; Li, C.C.; Zhang, N. Pull-out and Critical Embedment Length of Grouted Rebar Rock Bolts-Mechanisms When Approaching and Reaching the Ultimate Load. *Rock Mech. Rock Eng.* **2021**, *54*, 1431–1447. [CrossRef]

30. Chen, J.; Chen, X. Analysis of whole process of bolt pulling based on wavelet function. *Rock Soil Mech.* **2019**, *40*, 4590–4596. [CrossRef]

31. Li, B.; Liang, Q.; Zhou, Y.; Zhao, C.; Wu, F. Research on crack propagation law of granite based on CT-GBM reconstruction method. *Chin. J. Rock Mech. Eng.* **2022**, *41*, 1114–1125. [CrossRef]

32. Li, B.; Zhu, Q.; Zhang, F.; Zhao, C.; Wu, F. Study on crack propagation of heterogeneous rocks with double flaws based on grain based. *Chin. J. Rock Mech. Eng.* **2021**, *40*, 1119–1131. [CrossRef]

33. Hu, X.; Bian, K.; Liu, J.; Li, B.; Chen, M. Discrete element simulation study on the influence of microstructure heterogeneity on the creep characteristics of granite. *Chin. J. Rock Mech. Eng.* **2019**, *38*, 2069–2083. [CrossRef]

34. Salcher, M.; Bertuzzi, R. Results of pull tests of rock bolts and cable bolts in Sydney sandstone and shale. *Tunn. Undergr. Space Technol.* **2018**, *74*, 60–70. [CrossRef]

35. Upadhyaya, P.; Kumar, S. Pull-out capacity of adhesive anchors: An analytical solution. *Int. J. Adhes. Adhes.* **2015**, *60*, 54–62. [CrossRef]

36. Xia, Y.; Ye, H.; Liu, X.; Chen, J. Analysis of shear stress along pressure-type anchorage cable in weathered rock mass. *Rock Soil Mech.* **2010**, *31*, 3861–3866. [CrossRef]

37. Luo, Y.; Shi, S.; Yan, Z. Shear interaction of anchorage body and rock and soil interface under the action of uplift load. *J. China Coal Soc.* **2015**, *40*, 58–64. [CrossRef]

38. Huang, M.; Zhou, Z.; Ou, J. Nonlinear Full-Range Analysis of Load Transfer in Fixed Segment of Tensile Anchors. *Chin. J. Rock Mech. Eng.* **2014**, *33*, 2190–2199. [CrossRef]

39. Li, X.F.; Zhang, Q.B.; Li, H.B.; Zhao, J. Grain-Based Discrete Element Method (GB-DEM) Modelling of Multi-scale Fracturing in Rocks Under Dynamic Loading. *Rock Mech. Rock Eng.* **2018**, *51*, 3785–3817. [CrossRef]

40. Zhang, X.-P.; Ji, P.-Q.; Peng, J.; Wu, S.-C.; Zhang, Q. A grain-based model considering pre-existing cracks for modelling mechanical properties of crystalline rock. *Comput. Geotech.* **2020**, *127*, 103776. [CrossRef]

41. Zhang, Y.; Wong, L.N.Y.; Chan, K.K. An Extended Grain-Based Model Accounting for Microstructures in Rock Deformation. *J. Geophys. Res. Solid Earth* **2019**, *124*, 125–148. [CrossRef]

42. Li, X.F.; Li, H.B.; Liu, L.W.; Liu, Y.Q.; Ju, M.H.; Zhao, J. Investigating the crack initiation and propagation mechanism in brittle rocks using grain-based finite-discrete element method. *Int. J. Rock Mech. Min. Sci.* **2020**, *127*, 104219. [CrossRef]

Journal of
*Marine Science
and Engineering*

Article

Effect of High Temperature on Mechanical Properties and Microstructure of HSFCM

Yanbin Li [1], Qingsheng Meng [1,*], Yan Zhang [2,*], Huadong Peng [3] and Tao Liu [1,4]

[1] Shandong Provincial Key Laboratory of Marine Environment and Geological Engineering, Ocean University of China, Qingdao 266100, China; li117156@163.com (Y.L.)
[2] College of Civil Engineering, Anhui Jianzhu University, Hefei 230601, China
[3] Shanghai Survey, Design and Research Institute (Group) Co., Ltd., Qingdao 266100, China
[4] Qingdao National Laboratory of Marine Science and Technology, Qingdao 266000, China
* Correspondence: qingsheng@ouc.edu.cn (Q.M.); avayan8006@163.com (Y.Z.)

Abstract: A new type of composite cement-based cementing material—high-strength fast cementing material (HSFCM)—will be widely used in marine engineering projects such as submarine tunnels. However, the influence of fire and other high temperature conditions on its material properties have not been explored in previous studies. Mechanical tests and microstructure observations of HSFCM were carried out, and the strength and deformation characteristics, microstructure and composition evolution of HSFCM after high temperature treatment were discussed. After high temperature treatment, the compressive strength of HSFCM deteriorated. The compressive strength of HSFCM decreased by more than half at 400 °C. The peak strain increased at 200 °C with the increase of temperature, and decreased at 400~600 °C with the increase of temperature. High temperature reduces the stiffness of HSFCM, and the elastic modulus decreases with increasing temperature. The influence of high temperature on the microstructure of HSFCM is mainly shown in the increase and enlargement of pores in three-dimensional space, the development of micro-cracks and the thermal decomposition of cementing material into stable oxides without cementing effect. The microscopic changes of HSFCM are in good agreement with the mechanical test results.

Keywords: cement-based material; high temperature; indoor test; mechanical properties; microanalysis

Citation: Li, Y.; Meng, Q.; Zhang, Y.; Peng, H.; Liu, T. Effect of High Temperature on Mechanical Properties and Microstructure of HSFCM. *J. Mar. Sci. Eng.* **2023**, *11*, 721. https://doi.org/10.3390/jmse11040721

Academic Editor: Cristiano Fragassa

Received: 10 March 2023
Revised: 24 March 2023
Accepted: 25 March 2023
Published: 27 March 2023

1. Introduction

Coastal areas are the most economically developed areas in China, with numerous engineering projects [1] and huge uses of cement-based cementing materials [2]. Currently, cement-based materials are widely used in offshore projects, such as undersea tunnel construction [3], offshore oil and gas engineering [4], submarine pipeline construction [5] and offshore bridge construction [6]. Taking Qingdao, a coastal city in northern China, as an example, with the expansion of the construction scope of urban public transport facilities in recent years, the number of undersea tunnels in this area is increasing (as shown in Figure 1A). Cementing materials are widely used in weak strata reinforcement, concrete lining reinforcement, surrounding rock plugging and other conditions (as shown in Figure 1B). The performance of cementing materials directly affects the stability and durability of supporting structures and surrounding rock strata. The increased construction in the area has increased the risk of accidents such as fire. As the undersea tunnel is a narrow and closed structure located in the underwater environment, fire often causes heavy casualties and huge economic losses. At the same time, due to the poor ventilation of undersea tunnels, fire can cause the ambient temperature to reach more than 1000 °C [7]; high temperatures will affect the structural composition of cement-based cementing material, changing the mechanical properties of the material, causing the tunnel structure to crack and affecting the bearing capacity and service life of the undersea tunnel. This can lead to tunnel collapse, threatening the safety and stability of the project [3,8–10].

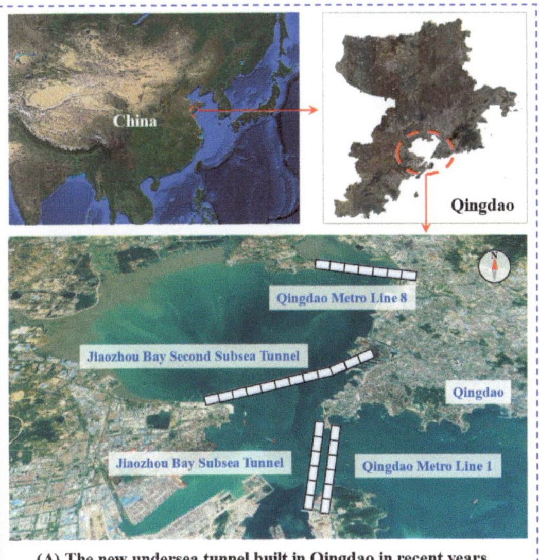

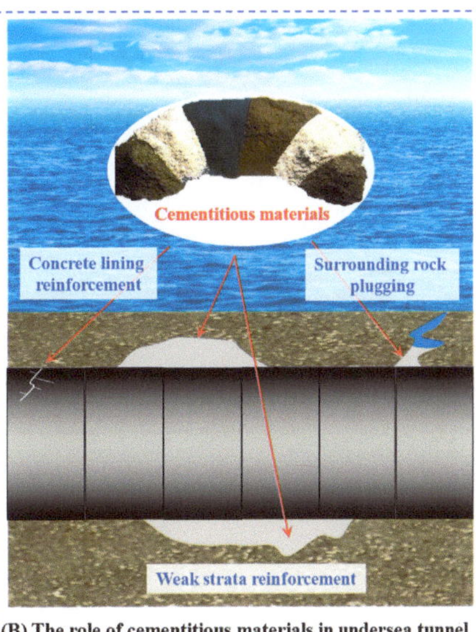

Figure 1. The role of cement-based cementitious materials in growing marine engineering.

It is very important to study the evolution law and mechanism of properties of cement-based cementitious materials at high temperature for accurate evaluation of engineering structure safety and further improvement of material properties. Portland cement is a widely used basic cementing material. When exposed to the influence of a high temperature environment, the hue of its surface color will change, accompanied by volume change and damage [11]. Observation of the microstructure revealed that thermal cracking between different media was the main reason for performance degradation [12]. In some emergency projects, aluminate series cement with shorter curing time is used, and its performance is more significantly affected by high temperature [13–15]. A variety of composite cementing materials have been developed based on cement. The bearing capacity of engineering cementing composites (ECC) is reduced at high temperature [16]. The specimen size does not affect the percentage of strength loss [17]. Fiber can be added to obtain higher toughness [18]; recycled materials such as alkali slag and fly ash are also often used to improve high temperature resistance of cement-based cementitious materials, and obtaining the law of mechanical properties changing with temperature [19,20]. With the development of experimental techniques and equipment, the mechanism of microstructure change of cement-based materials after high temperature has been revealed by various means [21–23].

Due to the particularity and complexity of the location, the construction cost of marine engineering is often very high. The key to saving investment is to use the appropriate construction technology and cementing materials to reduce the construction period [24]. At present, Portland cement is mainly used in marine engineering in China, but its single property, poor durability such as impermeability, corrosion resistance and freeze–thaw resistance, cannot fully meet the special requirements of marine engineering [25]. Different types of cementing materials have different engineering properties. For construction projects in coastal areas that require materials to set quickly and become strong early, our predecessors developed a new high-strength fast cementing material, HSFCM [26], which is a composite cement-based cementitious material. It is often used in grouting engineering to reinforce the weak stratum and lining structure, which can significantly shorten the site construction period [27], and it has a wide application prospect. However, there are few studies on HSFCM at present. Most of these studies were conducted at normal temperature, aiming at the engineering characteristics of the anchorage structure. This makes it is impossible to evaluate the influence of high temperature on this material, which restricts its wider application. Therefore, focusing on the influence of high temperature on HSFCM and studying the evolution law and mechanism of its mechanical properties can provide an important reference for evaluating the properties of materials and further improving materials.

2. Materials and Methods

2.1. Materials and Specimen Preparation

The material used in this study is HSFCM, a composite cement-based cementitious material produced by a company in Shandong Province. The composition of the raw materials is shown in Table 1 [26].

Table 1. The raw materials composition of HSFCM.

Material	Proportion (wt%)
Sulfoaluminate cement	30~50%
Early strength Portland cement	15~25%
Fine sand	25~35%
Calcium sulfate	3~7%
Calcium carbonate	3~7%
Magnesium oxide	0.7~1.3%
Silicon powder	0.7~1.3%
Water reducer	0.7~1.3%
Boric acid	0.07~0.13%

The particle size distribution of cement-based cementitious materials is closely related to their mechanical properties [28]. Particle size analysis of HSFCM was carried out using a Winner 3003 dry laser particle size analyzer, and the particle size distribution curve is obtained as shown in Figure 2. It can be seen that the particle size distribution of HSFCM is uneven, most of the particle size is concentrated in the range of 10~40 μm, particles at the size of about 100 μm are few and the material is poorly graded.

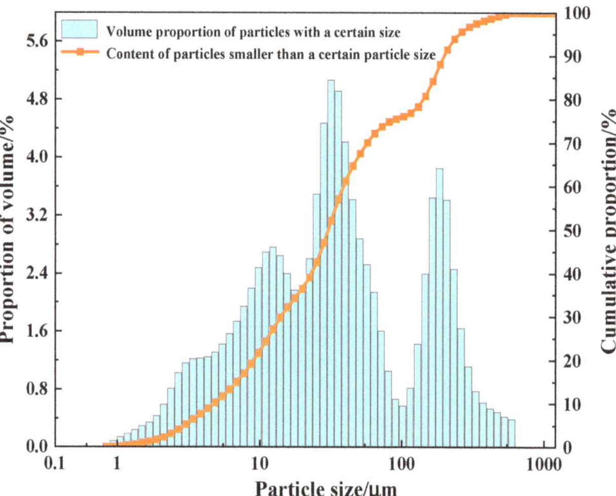

Figure 2. Particle size distribution of HSFCM.

HSFCM is often used as a grouting material in marine engineering. The diffusion capacity of the material can be measured by indicators such as fluidity and setting time. The differences in performance between HSFCM and Portland cement commonly used in marine engineering were compared through laboratory tests. The materials were weighed with the water/material ratio of 0.3 [26,27]. The stirred HSFCM and Portland cement were filled with the cup–cone round mold, respectively, and then the mold was lifted to let the slurry flow freely until it stopped under the condition of no disturbance. The diffusion diameter was measured with calipers to calculate the fluidity. Use Vicat apparatus to measure the setting time, when measuring let the test needle freely sink into the pure pulp, observe the pointer reading and obtain the final setting time. The mobility and setting time data are shown in Table 2. Under the same water–material ratio, HSFCM fluidity is obviously greater than Portland cement, HSFCM water requirement is smaller and the grouting diffusion range is larger. Furthermore, HSFCM setting time is shorter than Portland cement. HSFCM can set and harden faster to achieve strength, which is conducive to the next early process.

Table 2. Comparison of HSFCM and Portland cement grouting properties.

Indicators	HSFCM	Portland Cement
Fluidity (cm)	35	6
Final setting time (min)	80	375

Cylindrical specimens with a diameter of 50 mm and a length-to-diameter ratio of 2.0 were produced for the test according to ASTM C39 recommended dimensional standards [29]. The mixing procedure: HSFCM was poured into a mixing bucket and stirred for 1 min, then mixed with water according to the water/material ratio of 0.3. The mixing was continued for 5 min and then the mixture was injected into the mold prepared in advance, vibrated and then left to stand for 24 h at room temperature before demolding, and then placed in a standard maintenance room at a temperature of (20 ± 3) °C and relative humidity of (95 ± 2)% for 28 d.

2.2. Test Procedure

The GR.AF80/14 atmosphere furnace was used for high temperature treatment of the specimens, and the heating temperature was set to 100 °C, 150 °C, 200 °C, 400 °C and 600 °C for five test groups. The specimens at room temperature (20 °C) were used as the control group. In order to ensure that the specimens were uniformly heated, the heating rate was set to 5 °C/min, and the rated temperature was maintained for 4 h (Figure 3A). Then the heating was stopped and the specimens were taken out after natural cooling in the chamber (Figure 3B).

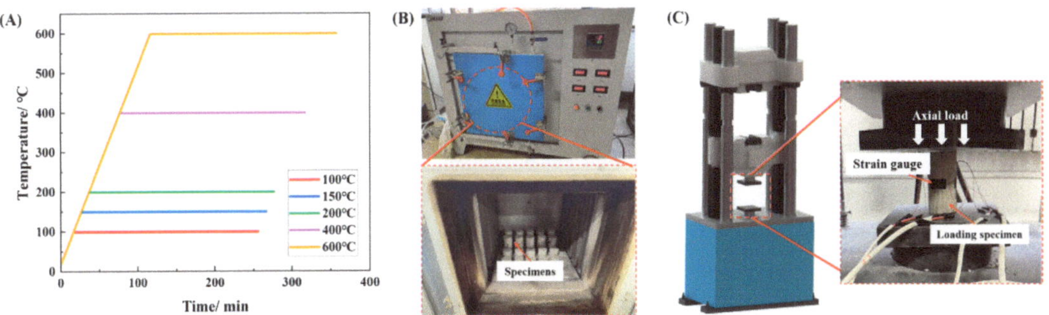

Figure 3. Temperature rise curve (**A**), high temperature treatment equipment (**B**) and test device for uniaxial compression (**C**).

The SHT4106 electro-hydraulic servo-type instrument was used to test the mechanical properties of the specimens after high temperature treatment. The axial displacement-controlled loading mode with a loading rate of 0.5 mm/min was selected. The arrangement of the device is shown in Figure 3C. The microstructure and morphological evolution of

HSFCM were analyzed by a nanoVoxel-2792 X-ray three-dimensional microscopic imaging system and a scanning electron microscope. At the same time, HSFCM specimens were selected for an XRD test, and the diffraction profile was obtained by MDI Jade software to analyze the changes in microscopic composition.

3. Results

3.1. Uniaxial Compression Test

3.1.1. Compressive Strength

Compressive strength, which is closely related to elastic modulus, peak strain and other mechanical parameters, is the most conventional and important index for testing mechanical properties of materials [30]. A uniaxial compression test was conducted on HSFCM specimens treated by atmosphere furnace. Figure 4 shows the compressive strength of HSFCM treated at different temperatures.

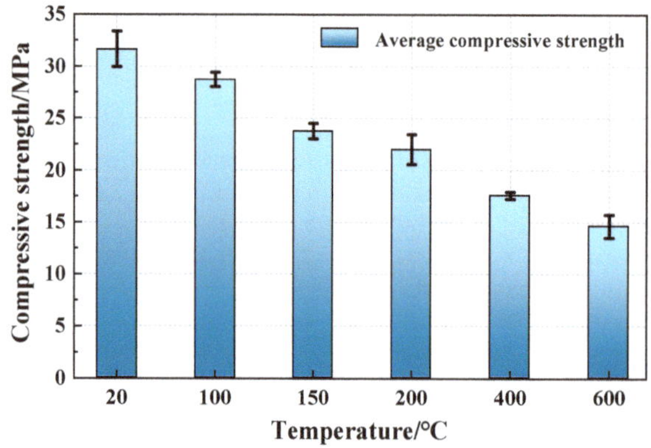

Figure 4. Compressive strength of HSFCM after treatment at different temperatures.

High temperature deteriorates the compressive strength of HSFCM. The average compressive strength of specimens treated at 100 °C, 150 °C, 200 °C, 400 °C and 600 °C was 28.73 MPa, 23.78 MPa, 22.02 MPa, 17.60 MPa and 14.64 MPa, respectively. Compared with 31.65 MPa at 20 °C, the compressive strength decreased by 9.24%, 24.87%, 30.43%, 44.39% and 53.75%, respectively. It can be seen that the compressive strength of HSFCM gradually decreased with the increase in temperature. At 100 °C, the compressive strength decreased less, the specimens were in the stage of free water evaporation in the internal pores and the internal damage was relatively light. At 100~200 °C, the free water in the specimens completely evaporated, the internal damage was aggravated and the strength decreased rapidly. When the temperature rose above 400 °C, a large number of hydration products were dehydrated, the cement matrix was heated and cracked and the internal micro-cracks expanded. At this time, the compressive strength could be reduced by more than half.

3.1.2. Stress–Strain Curve

The stress–strain relationship reflects the most basic mechanical properties of materials and is the main basis for studying mechanical strength and deformation characteristics [31]. Figure 5 shows the compressive stress–strain characteristic curve of HSFCM after high temperature.

After high temperature treatment, the compressive stress–strain curve of HSFCM can be divided into four stages: compression and compact section, elastic growth section, yield section and residual section. At the initial stage of compression, the specimen stiffness increases due to the external force compaction of the internal pores, and the curve presents an upward concave shape. When the strain reaches 20% of the peak strain of the specimen, the approximate linear growth of the curve indicates that the elastic growth stage is entered. As the loading continues, the curve no longer increases linearly; at this time, the internal damage of the specimen intensifies, and the curve becomes concave downward into the yield section until the specimen's brittle failure occurs at the peak, and the curve drops sharply. After that, the compressive stress does not increase and the strain increases, entering the residual section. With the increase of temperature, the peak stress and curve slope decrease, and the growth segment before reaching the peak stress decreases. It is particularly noted that after the specimens were treated at 400~600 °C, the color gradually deepened, the mesh cracks appeared on the surface and gradually expanded and some specimens burst, indicating that the HSFCM specimen had local damage between 400~600 °C. This is also the reason for the change of the peak strain trend of the specimen at 400~600 °C. Due to the already existing failure in the specimen after the high temperature treatment, when a small strain occurs in compression of the specimen, it will be reflected as compression failure. Therefore, at 400~600 °C, the stress–strain curve does not reflect the whole process of specimen failure, but only the process of compression failure.

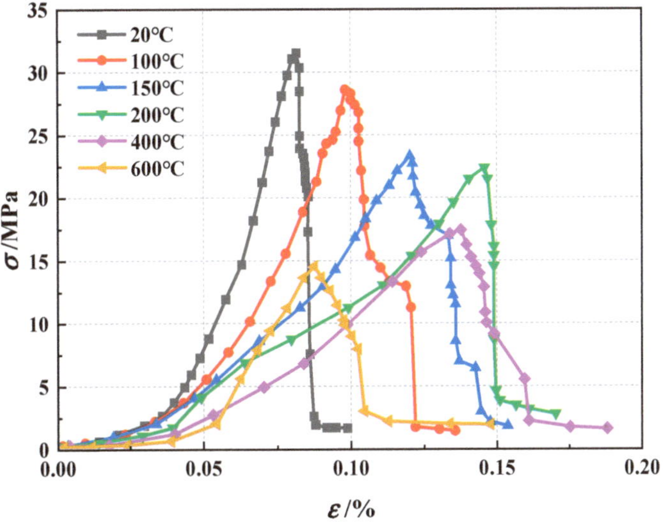

Figure 5. Compressive stress–strain curve of HSFCM after treatment at different temperatures.

3.2. CT Scan of Pore Structure

The macroscopic mechanical properties of cementitious materials depend on the stability and density of the internal microstructure. The thermal decomposition of cementitious materials at high temperature will cause the change of the internal microstructure, resulting in the deterioration of the macroscopic mechanical properties [21,32,33].

The same HSFCM specimen was selected for CT scanning in order to obtain the changes of the microscopic pore structure of the material with temperature. Firstly, a cylindrical specimen with a diameter of 50 mm and a length of 100 mm was prepared, which was drilled and polished into a cylindrical specimen with a diameter of 3 mm and a length of 10 mm for CT scanning. The specimen was scanned in a CT device to obtain fluoroscopic images. After that, the specimen was transferred to the atmosphere furnace for high temperature treatment. After natural cooling, CT scanning was carried out to obtain the images of the specimen treated at 100 °C and 400 °C, successively. After the scanning, the software supporting the instrument was used to reconstruct the overall and internal morphology of the specimen, and the influence of temperature on the microstructure was analyzed by analyzing gray decay and three-dimensional pore structure changes.

Figure 6 shows CT scanning images of different cross sections (n_z = 300 layers, 600 layers, 900 layers), longitudinal sections (n_x = 450 layers) and three-dimensional pore distribution of the specimen treated at 20 °C, 100 °C and 400 °C. The images were rendered—the higher the density, the whiter and brighter the image color—and a threshold segmentation method was used to obtain the distribution of pores within the specimen, marked in red.

By comparison, it can be seen that the influence of high temperature treatment on HSFCM is mainly manifested as the increase in the number of pores and the size of pores in the three-dimensional space. The distribution of pores in the specimen at 100 °C is little different from that at 20 °C, but the number of pores increases, mainly distributed in the top and middle of the specimen, and a large number of coarse particles such as fine sand are distributed at the bottom. At 400 °C, the internal pores of the specimen increased significantly, and with the increase of temperature, the high-density material at the bottom also gradually experienced thermal decomposition. The changes of micro-pores are in good agreement with the evolution of macro-mechanical properties.

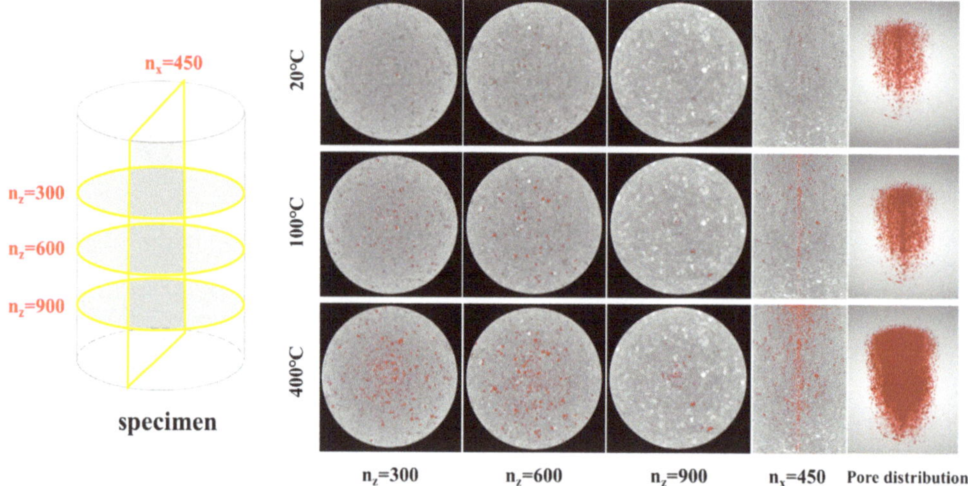

Figure 6. CT images of different sections and internal pores of HSFCM.

3.3. Component Analysis

High temperature deterioration of cement-based cementitious materials is not only a physical process, but also a chemical reaction of composition change [34–36]. Therefore, an X-ray diffraction test (XRD) was used to discuss the composition change under the condition of thermal decomposition of materials, and the XRD pattern as shown in Figure 7 was obtained.

As seen in Figure 7, the XRD pattern of HSFCM changes after high temperature treatment. Multiple ettringite and calcium silicate hydrate diffraction peaks can be identified at 20 °C. The reason why HSFCM can coagulate quickly and become strong early is that it contains a large number of active silica substances, which can react with $Ca(OH)_2$ to form calcium silicate hydrate gels ($CaO \cdot SiO_2 \cdot nH_2O$, C-S-H), etc. Such substances are the products of a hydration reaction. It reflects the main bonding substance in the specimen. When it rises to 100 °C, the diffraction peaks of ettringite decrease obviously, and multiple $CaSO_4$ diffraction peaks appear, indicating that the decomposition of ettringite occurs under heat. The diffraction peaks of calcium silicate hydrate are slightly enhanced, and the diffraction peaks of SiO_2 and $Ca(OH)_2$ are slightly reduced, indicating that the remaining SiO_2 and $Ca(OH)_2$ in the specimen continue to undergo hydration reactions to produce hydration products before the free water evaporates completely. When the temperature exceeds 400 °C, the diffraction peaks of ettringite and calcium silicate hydrate almost disappear, and multiple diffraction peaks of SiO_2 and $CaSO_4$ appear again, indicating that hydration products are further decomposed by heat with the increase in temperature. In addition, multiple diffraction peaks of CaO appear, which may be the CaO generated by the heat dehydration of $Ca(OH)_2$ and the heat decomposition of $CaCO_3$. On the whole, with the increase in treatment temperature, the decomposition of gelling substances in HSFCM becomes more serious, and stable oxides without gelling are eventually generated, resulting in serious deterioration of material properties.

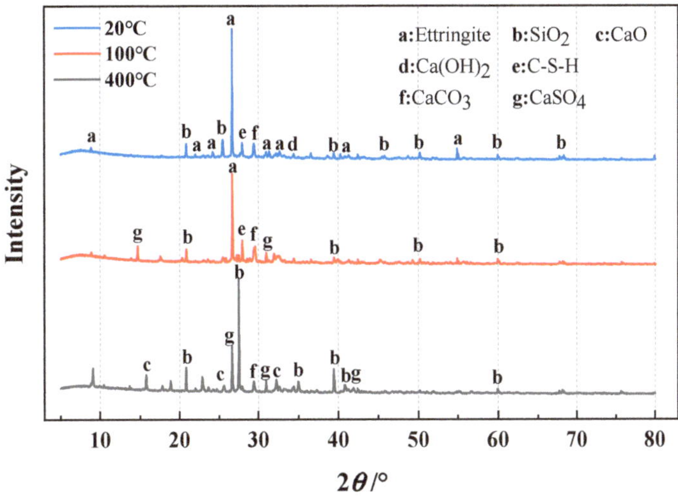

Figure 7. XRD pattern of HSFCM after treatment at different temperatures.

3.4. Surface Morphology Analysis

In order to more directly observe the changes in the surface microstructure of HSFCM, a scanning electron microscope (SEM) was used for observation and analysis, and the scanning images are shown in Figure 8.

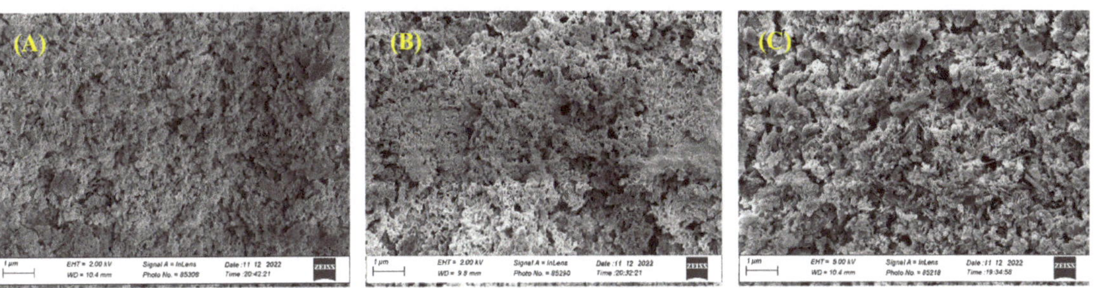

Figure 8. SEM images of HSFCM after treatment at 20 °C (**A**), 100 °C (**B**) and 400 °C (**C**).

The surface of the HSFCM specimen without high temperature treatment is more uniform and flat; the number of pores and micro-cracks is less; the hydration products grow freely; there are $Ca(OH)_2$, ettringite crystals and C-S-H gels, etc.; the contact between particles is close; the bonding degree is higher; and the microstructure is good. When the temperature is 100 °C, the surface is slightly convex; the fold becomes rough; the size and number of pores increase, but no obvious crack development has occurred; and the degree of bonding of the gel becomes low. At 400 °C, a large number of irregular crystals and whiskers appear; the bonding substance between the crystals is reduced; the specimen surface becomes uneven; the honeycomb micro-pores increase; the pore structure increases; and the micro-cracks develop through each other, which is consistent with the results of CT scanning and XRD analysis.

4. Discussion

4.1. Comparison of Compressive Strength with Other Cementitious Materials

The ratio of compressive strength of HSFCM after high temperature treatment f_c^T to compressive strength at room temperature f_c^{20} was selected and compared with silicate cement PO [37], engineering cementitious composites ECC [17], calcium aluminate cement CAC [38], magnesium potassium phosphate cement MKPC [39] and strain-hardening cementitious composites with high-volume of fly ash HVFA-SHCC [20] for comparison (Figure 9).

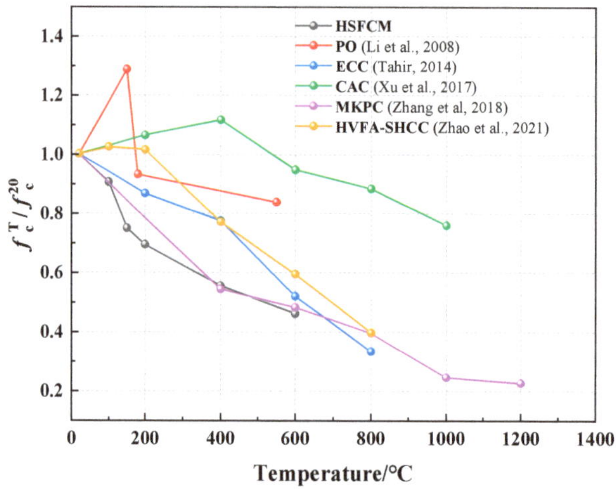

Figure 9. Comparison of compressive strength of several cementitious materials [17,20,37–39].

As can be seen from Figure 9, the influence of high temperature on the compressive strength of different types of cement-based cementitious materials is different. For pure cement cementitious materials such as PO and CAC, the compressive strength tends to rise with the increase of temperature, which may be caused by the continuation of the unfinished hydration reaction due to the increase in temperature. After reaching a certain temperature, the strength begins to decline. For composite cementitious materials composed of a variety of raw materials, the compressive strength is more seriously affected by the deterioration of high temperature, and decreases continuously with the increase of temperature, which is caused by different material components affected by temperature (pyrolysis, chemical reaction, etc.). In general, HSFCM is more deteriorated at high temperature than other materials, because there is more sulfoaluminate cement in the ingredients of HSFCM. Sulfoaluminate cement has a fast setting speed, high early strength and good freezing resistance [40], but it is generally not suitable for high temperature environments [15]. The content of sulfoaluminate cement can be reduced in the follow-up HSFCM high temperature resistance improvement.

4.2. Peak Strain of HSFCM after High Temperature Treatment

Peak strain is an important index reflecting the deformation capacity, which can be used to analyze the deformation law of cement-based cementitious materials [41,42]. The peak strain of HSFCM treated at each temperature is selected from the stress–strain curve, and the ratio of peak strain treated at high temperature ε^T to peak strain at room temperature ε^{20} varies with temperature (Figure 10).

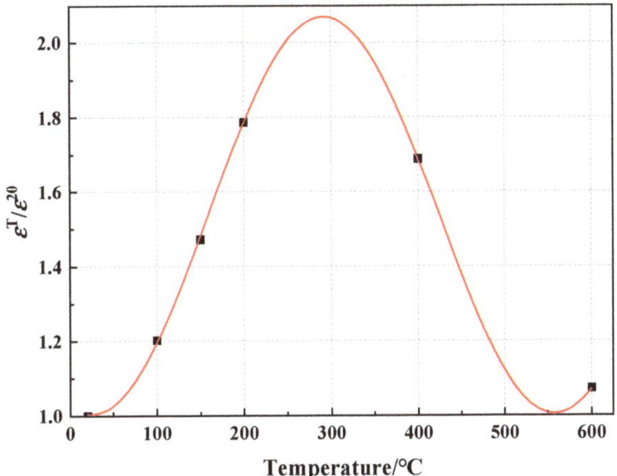

Figure 10. Peak strain ratio changes with temperatures.

The peak strain ratio increases with the increase of temperature within 200 °C. At 400~600 °C, due to severe internal heat damage, the compressive resistance and deformation ability of the specimens deteriorated, and the peak strain showed a downward trend, but all of them were greater than the peak strain at 20 °C. The peak strain ratio obtained from the test was fitted with the temperature, and the relation Equation (1) was obtained from Figure 10:

$$\frac{\varepsilon^T}{\varepsilon^{20}} = 0.53\sin\left(\pi\frac{T-159}{266}\right) + 1.54 \tag{1}$$

where T is the treatment temperature. The correlation coefficient R^2 of peak strain ratio and temperature fitting curve is 0.999, indicating a good degree of fitting.

4.3. Elastic Modulus of HSFCM after High Temperature Treatment

Elastic modulus is an index to measure the difficulty of material deformation [43]. It is calculated by referring to Equation (2) in ASTM C 469/C 469M:

$$E = \frac{\sigma_{0.4} - \sigma_1}{\varepsilon_{0.4} - \varepsilon_1} \tag{2}$$

where $\sigma_{0.4}$ is the stress value at 40% of the peak stress; $\varepsilon_{0.4}$ is the strain value corresponding to $\sigma_{0.4}$; and σ_1 and ε_1 are the stress and strain values, respectively, corresponding to 1 kN load.

The elastic modulus was calculated; the ratio of the elastic modulus after high temperature treatment E^T and the elastic modulus at room temperature E^{20} changed with temperature (Figure 11).

High temperature decreased the stiffness of HSFCM, the elastic modulus decreased with increasing temperature and the deterioration rate was first rapid and then slowed down. Equation (3) is obtained by fitting the relationship between the ratio of elastic modulus and temperature in Figure 11:

$$\frac{E^T}{E^{20}} = \frac{2.56}{T^{0.17}} \tag{3}$$

where T is the treatment temperature. The correlation coefficient R^2 of the elastic modulus ratio of HSFCM to the fitting curve of temperature is 0.991, indicating a good degree of fitting.

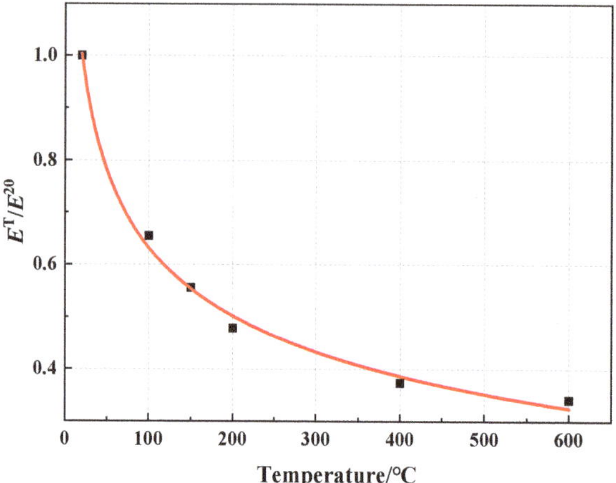

Figure 11. Elastic modulus ratio changes with temperatures.

5. Conclusions

(1) The compressive strength of HSFCM deteriorated after high temperature treatment, and the compressive strength decreased gradually with the increase in temperature. At 100 °C, the compressive strength decreased less; at 100~200 °C, the specimen damage increased and the strength decreased faster; at 400 °C, the compressive strength decreased by more than half. HSFCM is brittle under compression. With the increase of temperature, the peak stress and curve slope decrease, and the growth segment before reaching the peak decreases.

(2) The peak strain of HSFCM increased with increasing temperature at 200 °C, and decreased with increasing temperature at 400~600 °C. High temperature decreased the stiffness of HSFCM, the elastic modulus decreased with increasing temperature and the deterioration rate was first fast and then slow.

(3) A CT scan was used to study the microstructure changes of HSFCM after high temperature treatment, and it was found that the influence of high temperature on HSFCM mainly manifested as the increase of pore number and pore size in three-dimensional space; the high-density material was also gradually decomposed by heat with the increase in temperature.

(4) By comparing XRD patterns, it was found that hydration products such as ettringite and calcium silicate hydrate are gradually decomposed by heat, and multiple diffraction peaks of SiO_2, $CaSO_4$ and CaO appear, indicating that with the increase of temperature, the decomposition of gelling substances in HSFCM becomes more serious, and stable oxides are eventually generated.

(5) Through SEM analysis, it was found that a large number of irregular crystals and whiskers appeared when HSFCM was heated, the cementing material between crystals decreased, the surface bulges and folds gradually became uneven, the honeycomb micro-pores increased, the pores increased and the micro-cracks developed through each other.

In summary, the changes of microstructure and composition of HSFCM are in good agreement with the experimental results of mechanical properties after high temperature, which is the fundamental reason for the deterioration of its macroscopic mechanical properties at high temperature.

Author Contributions: Methodology, Y.L., Q.M. and Y.Z.; formal analysis, H.P.; investigation, H.P.; data curation, Y.L.; writing—original draft preparation, Y.L.; writing—review and editing, Y.Z. and T.L.; supervision, H.P. and T.L.; project administration, Q.M. and Y.Z.; funding acquisition, Q.M., Y.Z. and T.L. All authors have read and agreed to the published version of the manuscript.

Funding: National Natural Science Foundation of China (U2006213, 42277139, 42207172, 42272327); China Postdoctoral Science Foundation (2022M712989); National Natural Science Foundation of Shandong Province (ZR2022QD103).

Institutional Review Board Statement: Not applicable.

Informed Consent Statement: Not applicable.

Data Availability Statement: The data used to support the findings of this study are included within the article.

Conflicts of Interest: The authors declare no conflict of interest.

References

1. Guo, X.; Liu, Z.; Zheng, J.; Luo, Q.; Liu, X. Bearing capacity factors of T-bar from surficial to stable penetration into deep-sea sediments. *Soil Dyn. Earthq. Eng.* **2023**, *165*, 107671. [CrossRef]
2. Li, C. China cement industry structural adjustment and development report. *China Cem.* **2021**, *2022*, 10–17.
3. Lin, J.; Dong, Y.; Duan, J.; Zhang, D.; Zheng, W. Experiment on single-tunnel fire in concrete immersed tunnels. *Tunn. Undergr. Space Technol.* **2021**, *116*, 104059. [CrossRef]
4. El-Khoury, M.; Roziere, E.; Grondin, F.; Cortas, R.; Chehade, F.H. Experimental evaluation of the effect of cement type and seawater salinity on concrete offshore structures. *Constr. Build. Mater.* **2022**, *322*, 126471. [CrossRef]
5. Guo, X.; Nian, T.; Fu, C.; Zheng, D. Numerical Investigation of the Landslide Cover Thickness Effect on the Drag Forces Acting on Submarine Pipelines. *J. Waterw. Port Coast. Ocean. Eng.* **2023**, *149*, 04022032. [CrossRef]
6. Peng, R.X.; Qiu, W.L.; Teng, F. Research on performance degradation analysis method of offshore concrete piers in cold regions. *Ocean. Eng.* **2022**, *263*, 112304. [CrossRef]
7. Zhou, Y.; Xin, Y.; Zhang, S. Review on recent grave accidents in long tunnels of highway around the world for safety facility. *Chin. J. Rock Mech. Eng.* **2004**, *23*, 4882–4887.
8. Zhang, Z. Research of Mechanics Behavior for Immersed Tuunel under Fire Load. Master's Thesis, Chongqing Jiaotong University, Chongqing, China, 2013.
9. Guo, Q. A Study on Temperature Distribution of Tunnel and Damage of the Lining Structure under High Temperature of Fire. Master's Thesis, Taiyuan University Of Technology, Taiyuan, China, 2015.
10. Hu, X. Study on Structural Fire Safety of Long and Large Subsea Railway Shield Tunnel. Master's Thesis, Southwest Jiaotong University, Chengdu, China, 2021.
11. Yuzer, N.; Akoz, F.; Ozturk, L.D. Compressive strength-color change relation in mortars at high temperature. *Cem. Concr. Res.* **2004**, *34*, 1803–1807. [CrossRef]
12. Feng, J.; Fu, Y.; Chen, Z.; Zhang, R. Effect of high temperatures on microstructure of cement-based composite material. *J. Build. Mater.* **2009**, *12*, 318–322.
13. Zhou, W.; Xu, Z.; Deng, M. High temperature stability of cement containing 20% gypsum sulfoaluminate. *China Concr. Cem. Prod.* **2000**, *5*, 50–53. [CrossRef]
14. Xu, Z.; Zhou, W.; Deng, M. Stability of hardened sulph-aluminate cement paste treated at high temperature. *J. Chin. Ceram. Soc.* **2001**, *29*, 104–108.
15. Xiao, Z.; Lou, Y.; Guo, J. Effect of high temperature drying treatment on properties of sulfoaluminate cement mortar and its mechanism. *Cement* **2018**, *7*, 7–10. [CrossRef]
16. Chowdary, M.; Asadi, S.S.; Poluraju, P. Impact of materials on characteristics of engineered cementitious composite at elevated temperatures: An integrated approach. In Proceedings of the 1st International Conference on Advanced Lightweight Materials and Structures (ICALMS), Hyderabad, India, 6–7 March 2020; pp. 1389–1393.
17. Erdem, T.K. Specimen size effect on the residual properties of engineered cementitious composites subjected to high temperatures. *Cem. Concr. Compos.* **2014**, *45*, 1–8. [CrossRef]
18. Wang, Z.-B.; Han, S.; Sun, P.; Liu, W.-K.; Wang, Q. Mechanical properties of polyvinyl alcohol-basalt hybrid fiber engineered cementitious composites with impact of elevated temperatures. *J. Cent. South Univ.* **2021**, *28*, 1459–1475. [CrossRef]
19. Zhu, J. Basic Research on High Temperature Resistance of Alkali-Activated Slag Cementitious Material and Its Application in Engineering. Ph.D. Thesis, Harbin Institute of Technology, Harbin, China, 2014.
20. Zhao, J.; Niu, M.; Zhou, J.; Wang, Z. Uniaxial compressive behavior and constitutive relationship of HVFA-SHCC after exposure to high temperature. *J. Build. Mater.* **2021**, *24*, 22–30.
21. Su, X.-G.; Du, X.-J.; Yuan, H.-H.; Li, B.-K. Research of the thermal stability of structure of resin anchoring material based on 3D CT. *Int. J. Adhes. Adhes.* **2016**, *68*, 161–168. [CrossRef]

22. Peng, Y.; Zhao, X.; Xu, S.; Li, Q. Microstructure characteristics of ultra-high toughness cementitious composites after exposure to high temperature. *J. Chin. Electron Microsc. Soc.* **2019**, *38*, 236–244.
23. Takahashi, H.; Sugiyama, T. Application of non-destructive integrated CT-XRD method to investigate alteration of cementitious materials subjected to high temperature and pure water. *Constr. Build. Mater.* **2019**, *203*, 579–588. [CrossRef]
24. Guo, X.; Stoesser, T.; Zheng, D.; Luo, Q.; Liu, X.; Nian, T. A methodology to predict the run-out distance of submarine landslides. *Comput. Geotech.* **2023**, *153*, 105073. [CrossRef]
25. Gao, Y. Preparation of Multi-Source Solid Waste Based Marine Grouting Material and Research on Its Performance. Master's Thesis, Shandong University, Shandong, China, 2021.
26. Yang, Z.; Liu, Q.; Huang, X.; Ling, X. Preparation method of high strength fast anchorage agent and its grout. China Patent CN201810739147, 23 November 2018.
27. Liu, Q.; Huang, C.; Liu, L.; Ye, S.; Zang, G.; Gao, W.; Zhang, J. Reinforcement-Mud-Rock bonding test based on FAST-1 new high strength anchoring agent. *J. Jilin Univ.* **2021**, *51*, 1570–1577. [CrossRef]
28. Wang, Y.; Xu, L. Research progress of the influence of particle size distribution on the properties of cement. *Mater. Rep.* **2010**, *24*, 68–71+80.
29. Yi, S.T.; Kim, J.K.; Oh, T.K. Effect of strength and age on the stress-strain curves of concrete specimens. *Cem. Concr. Res.* **2003**, *33*, 1235–1244. [CrossRef]
30. Jia, B. Static and Dynamic Mechanical Behavior of Concrete at Elevated Temperature. Ph.D. Thesis, Chongqing University, Chongqing, China, 2011.
31. Duan, A. Research on Constitutive Relationship of Frozen-Thawed Concrete and Mathematical Modeling of Freeze-Thaw Process. Ph.D. Thesis, Tsinghua University, Beijing, China, 2009.
32. Su, X.; Du, X.; Su, L.; Wang, M.; Yuan, H.; Li, B. Experimental analysis on the micro-structure and the mechanical-properties of resin anchor material at high temperature. *J. China Coal Soc.* **2015**, *40*, 2408–2413. [CrossRef]
33. Su, X.; Du, X.; Zhang, S.; Yang, Z.; Guan, J. Anchoring properties and CT analysis affected by the pyrolysis of the resin anchoring material at high temperature. *Chin. J. Rock Mech. Eng.* **2016**, *35*, 964–970. [CrossRef]
34. Ibrahim, R.K.; Hamid, R.; Taha, M.R. Fire resistance of high-volume fly ash mortars with nanosilica addition. *Constr. Build. Mater.* **2012**, *36*, 779–786. [CrossRef]
35. Wang, G.; Zhang, C.; Zhang, B.; Li, Q.; Shui, Z. Study on the high-temperature behavior and rehydration characteristics of hardened cement paste. *Fire Mater.* **2015**, *39*, 741–750. [CrossRef]
36. Wu, X.; Dai, S.; Li, Z.; Li, W.; Wang, J. Compressive strength and microstructure of cement-based materials after high temperature. *Bull. Chin. Ceram. Soc.* **2019**, *38*, 1755–1758. [CrossRef]
37. Li, Q.; Yao, Y.; Sun, B.; Li, Z. Mechanism of effect of elevated temperature on compressive strength of cement mortar. *J. Build. Mater.* **2008**, *11*, 699–703.
38. Xu, W.; Dai, J.-G.; Ding, Z.; Wang, Y. Polyphosphate-modified calcium aluminate cement under normal and elevated temperatures: Phase evolution, microstructure, and mechanical properties. *Ceram. Int.* **2017**, *43*, 15525–15536. [CrossRef]
39. Zhang, X.; Li, G.; Niu, M.; Song, Z. Effect of calcium aluminate cement on water resistance and high-temperature resistance of magnesium-potassium phosphate cement. *Constr. Build. Mater.* **2018**, *175*, 768–776. [CrossRef]
40. Chen, J. Research of Properties Modification and Applications of Sulphoaluminate Cement. Master's Thesis, Wuhan University, Wuhan, China, 2005.
41. Qin, L. Study on the Strength and Deformation of Concrete under Multiaxial Stress after High-Temperature or Freeze-Thaw Cycling. Ph.D. Thesis, Dalian University of technology, Dalian, China, 2003.
42. Zhang, Z. Experimental Study on the Multi-Axial Mechanical Behavior of Concrete after Freeze-Thaw Cycling or High Temperature. Ph.D. Thesis, Dalian University of Technology, Dalian, China, 2007.
43. Tang, J.; Ma, W.; Pang, Y.; Fan, J.; Liu, D.; Zhao, L.; Sheikh, S.A. Uniaxial compression performance and stress-strain constitutive model of the aluminate cement-based UHPC after high temperature. *Constr. Build. Mater.* **2021**, *309*, 125173. [CrossRef]

Journal of
Marine Science and Engineering

MDPI

Article

Stabilizing and Destabilizing Breaching Flow Slides

Said Alhaddad [1,*], Dave Weij [2], Cees van Rhee [1] and Geert Keetels [1]

[1] Section of Offshore and Dredging Engineering, Faculty of Mechanical, Maritime and Materials Engineering, Delft University of Technology, 2628 CN Delft, The Netherlands
[2] Plaxis BV, Computerlaan 14, 2628 XK Delft, The Netherlands
* Correspondence: s.m.s.alhaddad@tudelft.nl

Abstract: As a result of the dilation of soil matrix, dense submarine sand slopes can temporarily be steeper than the natural angle of repose. These slopes gradually fail by the detachment of individual grains and intermittent collapses of small coherent sand wedges. The key question is whether steep disturbances in a submarine slope grow in size (destabilizing breaching) or gradually diminish (stabilizing breaching) and thereby limit the overall slope failure and resulting damage. The ability to predict whether the breaching failure is stabilizing or destabilizing is also crucial for the assessment of safety of submarine infrastructure and hydraulic structures located along rivers, lakes, and coasts. Through a set of large-scale laboratory experiments, we investigate the validity of an existing criterion to determine the failure mode of breaching (i.e., stabilizing or destabilizing). Both modes were observed in these experiments, providing a unique set of data for analysis. It is concluded that the existing method has limited forecasting power. This was quantified using the mean absolute percentage error, which was found to be 92%. The reasons behind this large discrepancy are discussed. Given the complexity of the underlying geotechnical and hydraulic processes, more advanced methodologies are required.

Keywords: stabilizing breaching; destabilizing breaching; flow slides; underwater slope failure; dilative slope failure; sand erosion

Citation: Alhaddad, S.; Weij, D.; van Rhee, C.; Keetels, G. Stabilizing and Destabilizing Breaching Flow Slides. *J. Mar. Sci. Eng.* **2023**, *11*, 560. https://doi.org/10.3390/jmse11030560

Academic Editors: Xiaolei Liu, Thorsten Stoesser and Xingsen Guo

Received: 30 January 2023
Revised: 28 February 2023
Accepted: 3 March 2023
Published: 6 March 2023

1. Introduction

Subaqueous slope failure is a common problem in the fields of geotechnical, hydraulic, and dredging engineering, posing a serious threat to underwater infrastructure and flood defence structures along rivers, lakes, and coasts. One of the possible failure mechanisms is 'flow slide', which occurs when the sediment deposit loses its stability and runs downslope, forming a gentler slope than the initial one [1]. During flow slides, the sediment is transported as a sediment–water mixture, which behaves as a viscous fluid [2]. Two end members of flow slides are distinguished in the literature: liquefaction flow slides and breaching flow slides. The former occurs in loosely-packed sand, which shows a contractive behaviour under shear forces; the soil structure collapses abruptly and a large amount of the soil body flows downslope. The latter, on the other hand, does not occur as an abrupt collapse. Rather, sand grains peel off particle by particle, generating a turbidity current propagating over the slope surface (or 'breach face') [3–5].

Unlike static liquefaction, breaching is mostly encountered in densely-packed sand, which dilates under shear forces [3,6]. Dilative sand undergoes an increase in pore volume under shear deformation, leading to the generation of a negative pore pressure, which substantially retards the sand erosion process [7]. An inward hydraulic gradient is generated, as a result of the pressure difference, forcing the ambient water to seep into the pores, dissipating the negative pressure. Consequently, the sand grains located at the sand–water interface destabilize and gradually peel off, almost grain by grain [5,8]. These grains mix and interact with the ambient water, forming a sustained turbidity current travelling along the breach face and subsequently down the slope toe [9].

Breaching-generated turbidity currents are self-accelerating [5]. They induce an additional shear stress on the breach face, thereby picking up more sediment, which makes the turbidity current denser. As a consequence, the current expedites downslope, boosting its erosive capacity and leading to higher erosion rates in the downstream direction. This implies that sediment entrainment and acceleration of the turbidity current are coupled in a positive feedback loop. Next to the grain-by-grain erosion, intermittent collapses of coherent sand wedges, termed surficial slides, have been observed in breaching lab experiments [3,5,10]. The current understanding of the occurrence of surficial slides remains poor [8].

In dredging engineering, breaching is triggered deliberately for sand mining purposes [3]. It should be noted, however, that natural breaching events can also occur. The mechanisms that trigger these events are not yet fully understood. For example, at Amity Point in Queensland, Australia, breaching events continue to be observed without a clear understanding of their triggers [11].

Breaching, both naturally occurring and anthropogenic, can last for several hours, travelling towards nearby or remote shorelines or river banks, posing a major risk [11,12]. Additionally, breaching could cause instabilities during the construction of underwater slopes [6]. Such risky situations could be avoided by developing a good understanding of the spatio-temporal evolution of breaching. In this regard, prior experimental observations showed that two failure modes of breaching can be distinguished [3]: stabilizing and destabilizing (Figure 1). Breaching is regarded as destabilizing when the breach face increases in height over time, resulting in an uncontrolled retrogressive failure of the slope. In contrast, during stabilizing breaching, the height of the breach face decreases over time until it disappears completely.

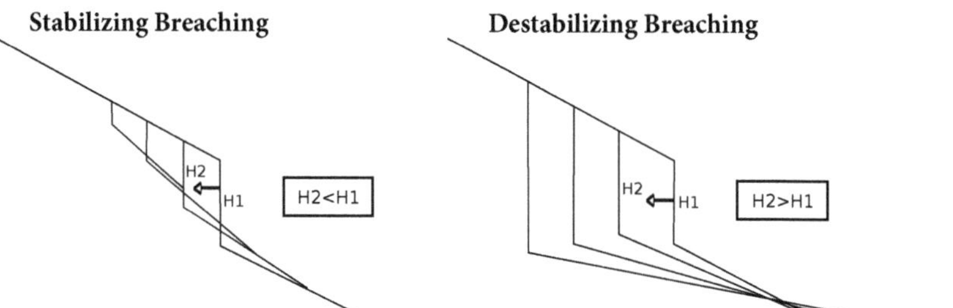

Figure 1. Diagrammatic illustration of the breaching modes, stabilizing breaching (**left**), and destabilizing breaching (**right**); H is the breach height.

On the modelling side, a few numerical models for breaching-generated turbidity currents have been proposed in the literature. Most of these numerical investigations were mostly restricted to layer-averaged, one-dimensional models, where several empirical closure relations are required (e.g., [1,4,9]). However, Alhaddad et al. [8] presented 3D large-eddy-resolving numerical simulations, providing deeper insights into the flow structure and hydrodynamics of breaching-generated turbidity currents. Additionally, Van Rhee [6,13] simulated breaching by a two-dimensional computational fluid dynamics code coupled with a bed boundary condition to model sediment erosion and as well as an empirical approach to assess the failure mode of breaching. The latter will be discussed in detail in this study.

This paper presents results of laboratory experiments where we investigate the failure mode of breaching. Several initial conditions were tested, allowing us to observe both stabilizing and destabilizing breaching. Following that, we utilize our experimental measurements to provide the first insights into the validity of the approach used in the literature to assess the failure mode of breaching. Lastly, key future research directions are defined to improve the predictability of breaching mode.

2. Existing Method for Failure Mode Assessment

The mode of a 2D breaching failure can be assessed by a simple geometric argument. Using the schematized slope profile depicted in Figure 2, the following relation can be derived for the spatial change of the breach height, H:

$$\frac{dH}{dx} = \frac{1}{\Delta x}(\Delta y_2 - \Delta y_1) = \frac{\Delta x \tan \beta_{top} - \Delta x \tan \beta_{toe}}{\Delta x} = \tan \beta_{top} - \tan \beta_{toe}, \tag{1}$$

where β_{toe} is the angle of the slope at the toe of the breach face and β_{top} is the angle of the slope at the top of the breach face. It follows that

$$\text{breaching mode is} \begin{cases} \text{stabilizing if} & \frac{dH}{dx} < 0 \\[2mm] \text{destabilizing if} & \frac{dH}{dx} > 0. \end{cases} \tag{2}$$

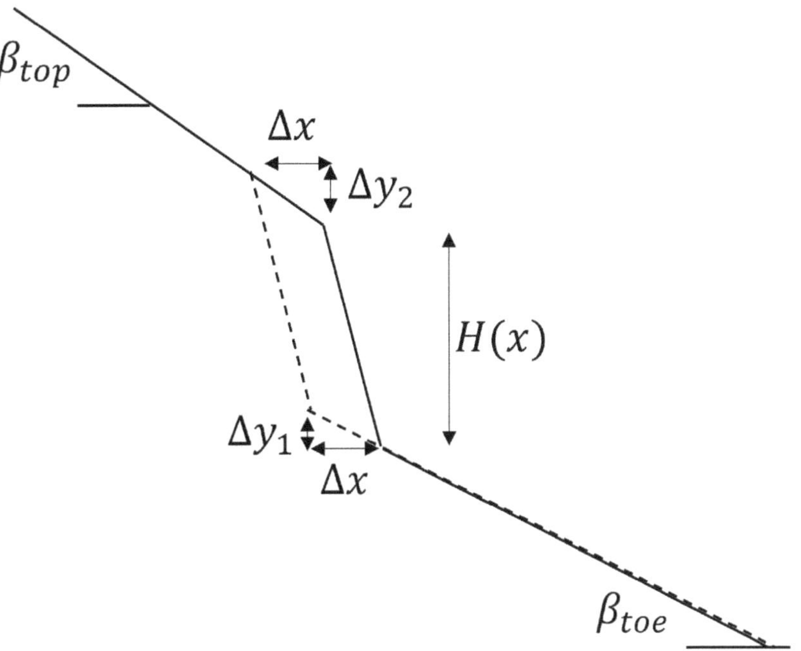

Figure 2. Schematic overview of a 2D breaching failure. During a time interval Δt the breach face moves over a distance Δx and the slope at the toe progresses by sedimentation over the same distance. It is assumed that during this interval the angles β_{top} and β_{toe} remain constant.

This means that to predict the sign of dH/dx it is essential to estimate the angle of the slope at the toe, β_{toe}. Van Rhee [6] suggests that a first estimate for this slope can be obtained following the findings of the sand fill experiments of Mastbergen et al. [14]. In their study, a constant sand flux, s, was applied via a pipe at the beginning of a flume (32 m $\times$ 2.5 m $\times$ 0.5 m). At the start of the experiment, the flume was filled with water only. Initially, the sand pile was formed below the pipe until the free surface was reached. In the next phase, the sand body gently migrated in a horizontal direction. After some time, the produced slope angle became constant, while the slope kept migrating horizontally; this situation is shown in Figure 3.

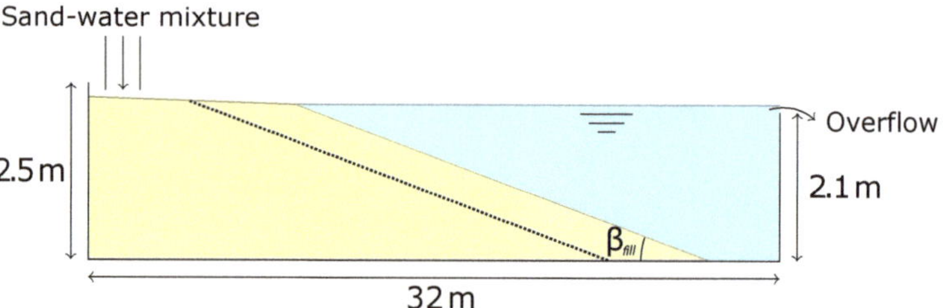

Figure 3. Steady slope angle, β_{fill}, produced in an experimental run by Mastbergen et al. [14].

The following empirical relation was proposed for the eventual steady slope angle, β_{fill} (°):

$$\tan \beta_{fill} = \frac{1623 D_{50}^{0.92}}{s^{0.39}}, \tag{3}$$

where D_{50} (m) is the median sand grain size and s (kg/m/s) is the sand flux. It should be realized that the calibration coefficient (1623) is not dimensionless. Assuming that the slope at the toe has enough time to adapt to the sand flux provided by the breach face, it follows that

$$\beta_{toe} \approx \beta_{fill}. \tag{4}$$

For breaching, the sand flux, s, at the beginning of the toe slope reads [13]:

$$s = v_{breach} H \rho_s (1 - n_0), \tag{5}$$

where v_{breach} is the erosion velocity of the breach face, ρ_s is the density of the sand particles and n_0 is the in situ porosity. The critical value between stabilizing breaching ($dH/dx < 0$) and destabilizing breaching ($dH/dx > 0$) can be found by setting,

$$0 = \frac{dH}{dx} = \tan \beta_{top} - \tan \beta_{toe}. \tag{6}$$

Now it follows that

$$\tan \beta_{toe} = \tan \beta_{fill} = \frac{1623 D_{50}^{0.92}}{s^{0.39}} = \tan \beta_{top}. \tag{7}$$

Substitution of Equation (5) gives the following expression for the critical breach height:

$$H_{crit} = \left(\frac{1623 D_{50}^{0.92}}{\tan \beta_{top}} \right)^{2.6} \frac{1}{v_{breach} \rho_s (1 - n_0)}. \tag{8}$$

If desired, the erosion velocity of the breach face v_{breach} can be calculated [6] as follows:

$$v_{breach} = \frac{1 - n_l}{n_l - n_0} k_l (1 - n_0) \frac{\rho_s - \rho_w}{\rho_w} \frac{\sin(\beta_{breach} - \phi)}{\sin \phi}, \tag{9}$$

where ϕ (°) is the internal friction angle, β_{breach} (°) is the slope angle of the breach face, n_0 (-) is the in situ porosity of the sand, n_l (-) is the maximum porosity of the sand, k_l (m/s) is the sand hydraulic conductivity at maximum porosity, ρ_s (kg/m³) is the density of the particles, and ρ_w (kg/m³) is the density of water.

If $H > H_{crit}$, we have destabilizing breaching and if $H < H_{crit}$, we have stabilizing breaching. Figure 4 visualizes the criterion expressed by Equation (8). This visualization would be very helpful for practical assessment of the risk of breaching. Unfortunately, as shown above, the derivation of Equation (8) requires several assumptions and, therefore, it needs to be validated by experiments.

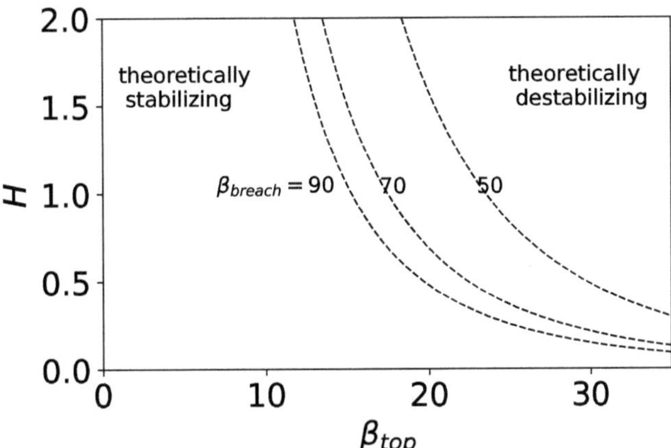

Figure 4. Illustration of the criterion expressed by Equation (8). For a given sand type and in situ porosity, the breaching mode is determined by three parameters: the breach height H, the angle of the breach face β_{breach} and the the slope angle at the top of the breach face β_{top}. The criterion curve is plotted for three angles of the breach face: $90°$, $70°$, and $50°$. For a given breach face angle, the corresponding contour line separates the regions in the (H, β_{top}) plane that result in $dH/dx < 0$ (left side of dashed line, stabilizing) and $dH/dx > 0$ (right side of dashed line, destabilizing). Sand parameters correspond with GEBA type as defined in Table 1 and $n_1 = 0.455$.

3. Laboratory Experiments

The present set of experiments aims at investigating the mode of the breaching phenomenon. Unlike laboratory experiments reported in the literature (e.g., [3,5,15,16]), the sand part above the breach face in some of our experiments is oblique. It is expected that this allows for a transition from stabilizing to destabilizing breaching or vice versa.

This section describes the experimental setup, characterization of sands, test procedure, and data analysis, respectively.

3.1. Experimental Setup

The experimental setup is composed of several components: a breaching tank, impermeable removable gate, false floor, and sedimentation tank (see Figure 5). The breaching tank is 5.1 m long, 0.5 m wide, and 2 m high. The front side of this tank is made of glass to facilitate failure tracking and flow visualization (see Figure 6). Since the current knowledge about the triggering mechanisms of breaching flow slides is very limited [1], the experiments were carried out for over-steepened vertical slopes, which initially fail due to the gravitational force. An impermeable removable gate is used to prevent the failure of sand during the construction process of the sand deposit and to shape the vertical breach face.

A false floor of a height of 0.3 m and a length of 4.8 m is placed at the bottom of the tank to create a 0.3 m high pump sump at the right end of the tank. A centrifugal pump is mounted there to prevent the reflection of the turbidity current back upstream; the sand–water mixture is pumped out from the pump sump to a sedimentation tank, which is 4.5 m long, 1.25 m wide, and 1.25 m high. On the other side of the sedimentation tank, a second submersible pump is placed behind a 1 m high dividing wall, which pumps clean water back into the breaching tank, so as to maintain a constant water level therein.

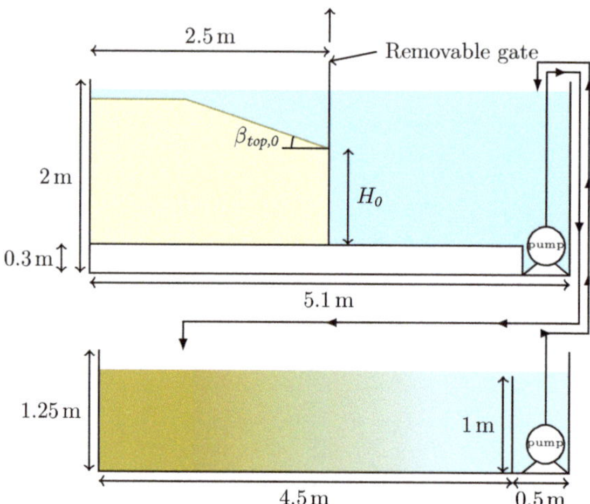

Figure 5. Front view of the experimental setup illustrating all components: breaching tank (**Top**) and sedimentation tank (**Bottom**).

Figure 6. Front side of the breaching tank just before Experiment 8. The view is partially obstructed by the vertical steel bars in the middle of the tank.

3.2. Characterization of Sands

In the experiments, two types of sand are used, namely GEBA and D9, the properties of which are summarized in Table 1. The initial porosity, n_0, of each sand type was determined by taking samples before the start of the experiments for two experiments with GEBA sand, and two experiments with D9 sand. The median and 15th percentile grain size, D_{50} and D_{15}, were determined using sieve analysis tests. The internal friction angle, ϕ, was determined using direct shear tests.

Table 1. Properties of the sands used in the experiments.

	D_{50} (μm)	D_{15} (μm)	n_0 (-)	ϕ (°)
GEBA	120	80	0.415	35.8
D9	330	225	0.430	40.1

3.3. Test Procedure

Each experiment was executed following the next sequence of steps:

- The false floor is placed at the bottom of the breaching tank.
- The breaching tank is filled with clean water.
- The removable gate is lowered down until it reaches the bottom of the breaching tank.
- A layer of sand is placed into the breaching tank and compacted by a vibrator needle.

- The previous step is repeated until reaching the target breach height, H_0, which was up to 1.47 m.
- For some experimental runs, a slope at the crest of the breach face is formed. The angle of this slope, $\beta_{top,0}$, was up to 30°.
- The pumps are switched on and then the gate is automatically removed, which takes about 10–17 s depending on the initial breach height.

Table 2 summarises the initial conditions of the experiments conducted within this study.

Table 2. A summary of the experiments conducted within this study. H_0 is the breach height and $\beta_{top,0}$ is the angle of the slope at the top of the breach face, both at the start of the experiment (see Figure 5).

Test #	H_0 (m)	$\beta_{top,0}$ (°)	Sand Type
1	0.66	0	GEBA
2	0.66	0	GEBA
3	0.66	0	GEBA
4	1.17	0	GEBA
5	1.17	0	GEBA
6	1.17	0	GEBA
7	0.8	20	GEBA
8	1.47	0	GEBA
9	0.8	30	GEBA
10	1.47	0	GEBA
11	0.66	30	GEBA
12	0.66	20	GEBA
13	0.66	0	D9
14	1.17	0	D9
15	0.8	30	D9
16	1.47	0	D9

3.4. Data Analysis

After the execution of experiments, deposit profiles demonstrating the temporal evolution of the sand failure are extracted from videos recorded by a GoPro Hero 3 camera (Figure 7). From these deposit profiles, the parameters v_{breach}, H, and β_{breach} are retrieved. This allows for robust analysis and comparison between experiments.

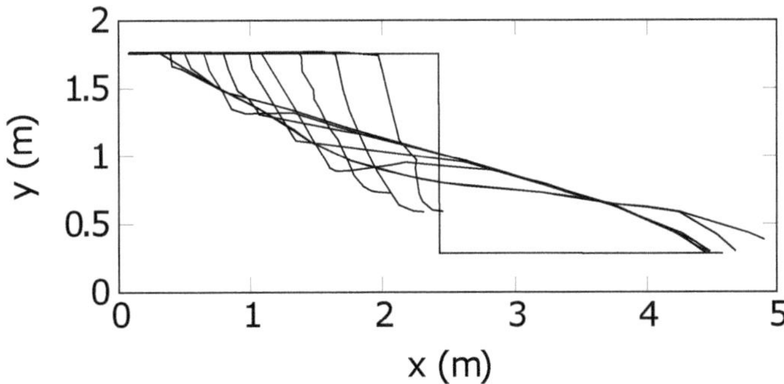

Figure 7. Deposit profiles plotted every 90 s for Experiment 8. Linear interpolation was used for obscured parts by the vertical steel bars.

The breach face is defined as the portion of the slope profile where the angle is steeper than the internal friction angle. The height of the breach face, H, is the vertical distance between the start and end points of the breach face. The breach face angle, β_{breach}, is defined as the angle of the straight line between the start and end points of the breach wall (see Figure 8). Since the sand erosion velocity along the breach face is not uniform (due to the self-acceleration of the turbidity currents), the average velocity is considered the erosion velocity of the breach face, v_{breach}.

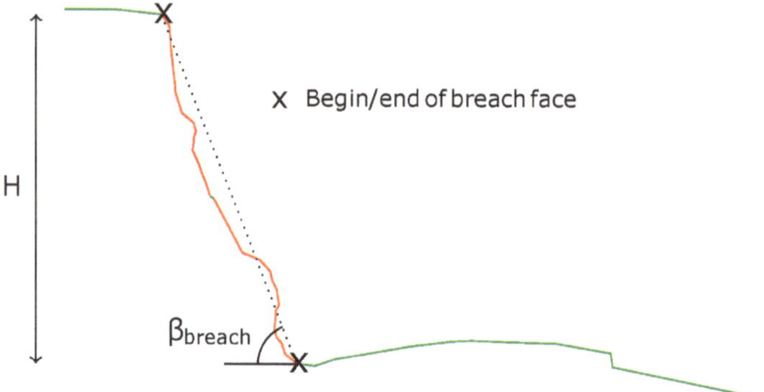

Figure 8. Definitions used for the breach height, H, and breach face angle, β_{breach}.

To check the reproducibility of our experiments, we have repeated some experimental runs one time or two times. For example, Experiments 1–3 have the same initial conditions (GEBA sand, $H_0 = 0.66$ m, $\beta_{top,0} = 0°$), and so have Experiments 8 and 10 (GEBA sand, $H_0 = 1.47$ m, $\beta_{top,0} = 0°$). The largest variations between similar experiments are v_{breach} and the size of the surficial slides occurring from time to time. This is attributed to the difficulty of constructing sand deposits of the same in situ porosity, which plays a major role in the failure progression [8]. On the other hand, the total distance travelled by the breach face for experiments having identical initial conditions is found to be very similar. The differences between breach heights, breach face angles are also found to be relatively small (see, e.g., Figure 9), while analysing the data, the variance between experiments having the same initial conditions is taken into account by considering the average values.

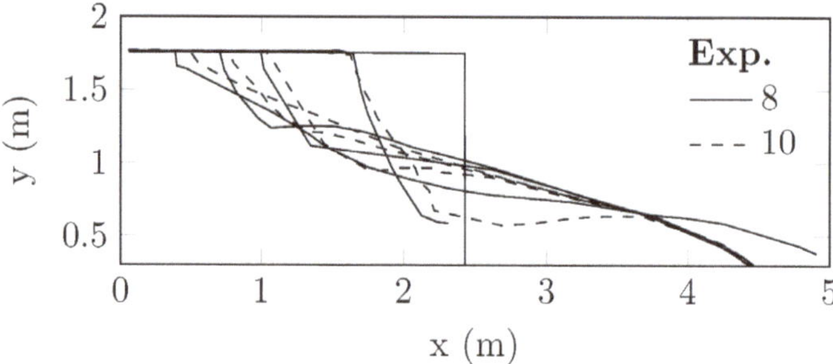

Figure 9. Profiles plotted at 0, 180, 410, 600, and 810 s for Experiments 8 and 10.

4. Experimental Results

4.1. General Description of the Deposit Failure

Upon the removal of the gate, the underwater sand deposit starts to fail as particles peel off almost one by one under the influence of gravity, creating a shower of sand which drags the ambient water along to generate a downward turbidity current. This current makes a turn at the breach face toe and travels as a net-depositional current over the downstream region. From time to time, an internally coherent sand wedge slides down the slope (surficial slide). The size of these slides correlates with the breach height; larger breach heights result in larger slides. It was also observed that the contribution of these surficial slides to erosion is significantly larger in the case of the courser sand (D9), compared with the finer sand (GEBA). Additionally, a lower breach face angle results in less frequent surficial slides, which is in line with the observations of Alhaddad et al. [5].

4.2. Temporal Evolution of Breaching

The temporal change of the breach height was constructed for all experimental runs. Figure 10 compares experiments with two different initial top angles, $\beta_{top,0}$. It is manifestly seen that the breach height remains significantly larger for the experimental runs with $\beta_{top,0} = 30°$ compared to the experiments with an initial top angle of $\beta_{top,0} = 20°$. This observation relates to the first term on the right-hand side of Equation (1) that describes the growth of the breach height, which is proportional to the tangent of β_{top}.

On the other hand, Figure 11 compares experiments with two different sand types (GEBA and D9). The breach height in the experiments with the finest sand type, GEBA, remains significantly larger compared to the experiments with the coarser sand type, D9. This observation can be attributed to both the differences in the sand hydraulic conductivities and grain sizes. The sand hydraulic conductivity largely affects the erosion velocity of the breach face, see Equation (9), which controls the time scale of the breaching process, since $dH/dt = v_{breach}dH/dx$. The grain size affects the sedimentation process at the toe of the breach face and thus the loss of breach height according to the second term on the right-hand side of Equation (1).

The breaching mode can be determined by looking at the temporal change of the breach height. For instance, results of Experiments 8 and 16 show that both breaching failures were invariably stabilizing throughout the experiments (see Figure 11 top). In addition, transition from stabilizing to destabilizing breaching is observed during Experiment 11 at time $\simeq 60$ s, and vice versa at time $\simeq 230$ s (see Figure 10 top).

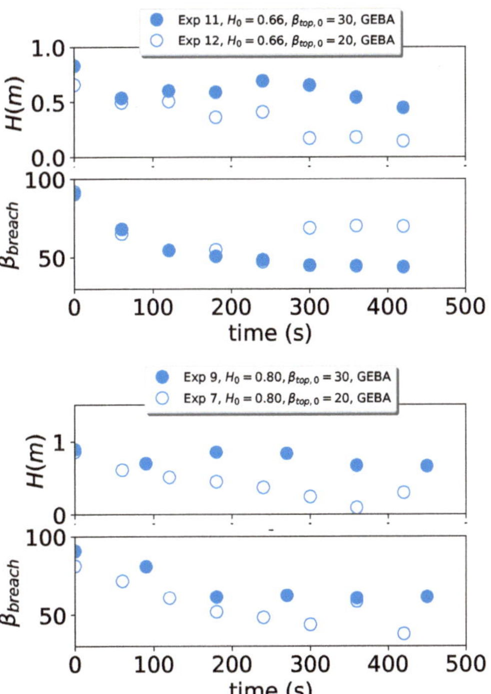

Figure 10. Effect of the initial angle of the slope at the top of the breach face, $\beta_{top,0}$, on the temporal evolution of the breach height, H, and breach face angle, β_{breach}.

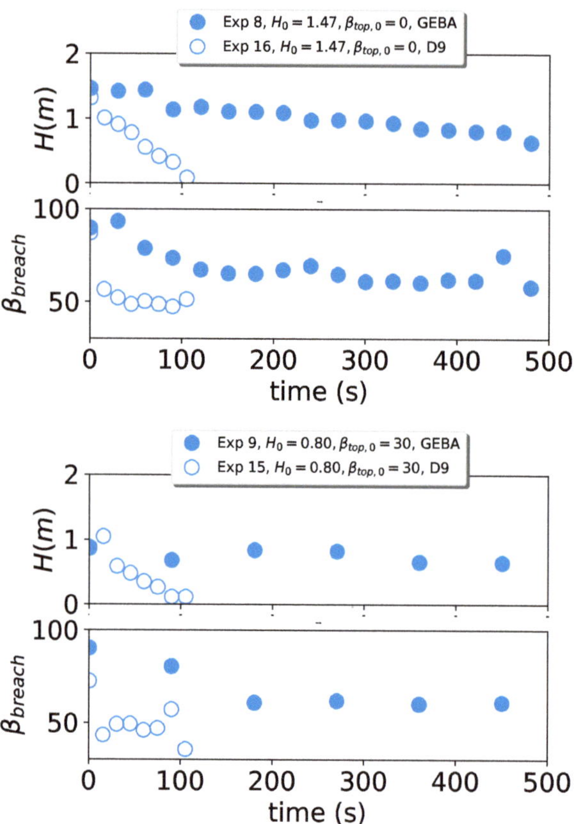

Figure 11. Effect of sand type on the temporal evolution of the breach height, H, and breach face angle, β_{breach}.

5. Validity Investigation of Existing Assessment Method

In this section, we examine the validity of the assessment method presented in Section 2 based on our experimental measurements. Following that, we raise discussion points and define research directions to allow for a robust assessment of the failure mode of breaching.

5.1. Comparison of Results

As a result of the finite size of our experimental setup, the number of observations with a clear destabilizing mode $dH/dx > 0$ and thus $dH/dt > 0$ is very limited. Alternatively, we here investigate the validity of the equation for H_{crit} (Equation (8)) by examining the underlying relations for dH/dx.

Figure 12 demonstrates that the combination of Equations (1) and (3)–(5), which yields the criterion of Equation (8), is not supported by the present experimental data. This can be further quantified by using the mean absolute percentage error (MAPE) defined as,

$$\text{MAPE} = \frac{100}{N} \sum_{1}^{N} \left| \frac{(dH/dx)_o - (dH/dx)_c}{(dH/dx)_o} \right|, \tag{10}$$

where $(dH/dx)_o$ denotes experimental observations of dH/dx, $(dH/dx)_c$ denotes the corresponding computed values, and the summation runs over all N observations shown in Figure 12. The MAPE is 92%, which demonstrates a substantial deviation. The observed dH/dx is significantly smaller than the computed dH/dx in most cases. This large discrepancy could be attributed to several factors. Firstly, the deviation of the geometrical simplification of the breaching process shown in Figure 2. A large number of the momentary profiles presented in Figures 7 and 9 typically show the development of a plateau at the toe of breach face, after which a slope is developed. The presence of the observed plateau is not taken into account in the assessment method. Secondly, the sedimentation process at the region down the toe is complex. The turbidity current that develops at the breach face impinges at the toe and subsequently continues along the toe slope. The formation of the bed in these conditions can substantially differ from the sand fill study of Mastbergen et al. [14]. This raises questions about the applicability of Equation (3) for breaching. Thirdly, the time variation of the breach face height and toe angle, as shown in Figures 10 and 11, is not taken into account in the derivation of Equation (8). In other words, the steady slope angle in the experiments of Mastbergen et al. [14] was observed after some time within which the sand flux was constant. This time is not available during breaching due to the dynamic nature of the breach face; the breach height varies over time and the breach face moves continuously backward as a result of sediment erosion. Lastly, the effect of surficial slides is not considered in the assessment method.

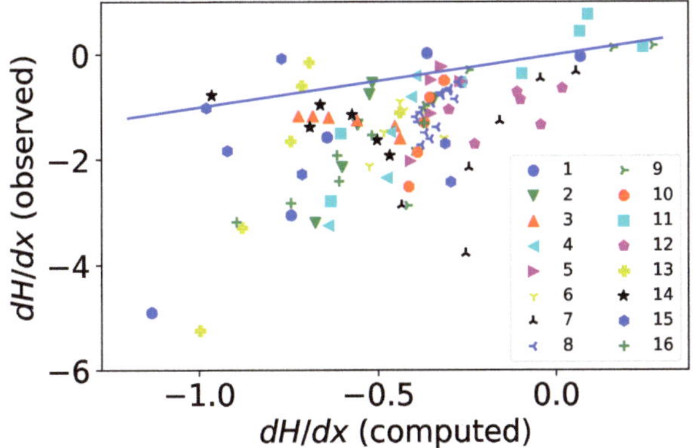

Figure 12. Observed change of the breach height dH/dx in Experiments 1-16 versus the computed dH/dx using Equations (1) and (3)–(5). The solid line represents perfect agreement.

5.2. Outlook

In the previous subsection, we revealed that the existing assessment method of the breaching mode is unreliable. The process is geometrically more complicated than suggested by Equation (1). In fact, the sedimentation process at the toe is not in equilibrium with the instantaneous sand flux, resulting in Equation (3) being unsuitable here. It is not straightforward to correct these issues in an analytical way. Therefore, to advance our predictive capability of the breaching mode, all physical processes discussed in the previous subsection should be taken into account in a robust numerical model. This implies that numerical simulations must be capable of reproducing the hydrodynamics and sediment transport of breaching-generated turbidity currents [5]. This is because the influence of the turbidity current largely determines whether the breaching process is stabilizing or destabilizing [6]. When the turbidity current deposits sediment at the breach face toe, the breach height will gradually decrease and eventually disappear. In contrast, if the turbidity current erodes sediment at the breach face toe, the breach height will increase over time, resulting in destabilizing breaching.

Moreover, the simulation of the turbidity current should be coupled to an advanced soil model that incorporates the dilation of sand and associated pore pressure feedback. Recently it was found that soil dilatancy and pore pressure feedback can be incorporated in a fluid mechanical model of tilted sand layers with a variable in situ relative density [17,18].

This gives an interesting perspective to model the complete soil and hydromechanical interaction underlying the breaching process in the foreseeable future. Given the three dimensionality of the breaching problem [5], 3D numerical simulations would also be necessary for accurate prediction of the breaching mode.

6. Conclusions

The ability to assess whether the failure during breaching will gradually grow in size (destabilizing breaching) or diminish (stabilizing breaching) is critical for the safety of nearby and remote underwater infrastructure as well as flood defence structures. To this end, an empirical assessment method has been developed and applied in previous studies of breaching. In this study, we evaluated the performance of this method using our unique data of large-scale experiments, where we observed both failure modes. A large discrepancy is found between the experimental data and the outcome of the underlying formula of the existing assessment method. The mean absolute percentage error was found to be 92%. Although the method is useful to acquire some qualitative understanding of the progression of breaching, it is unreliable for quantitative assessment. A more fundamental 3D approach is essential to capture breach face dynamics, breaching-generated turbidity currents, occurrence of surficial slides, and sedimentation processes at the toe of the breach face.

Author Contributions: Conceptualization and methodology: All authors; execution of experiments and data curation: D.W.; writing of original draft: S.A. and G.K.; funding acquisition: G.K. and C.v.R.; reviewing: C.v.R. All authors have read and agreed to the published version of the manuscript.

Funding: This study was partly supported by Stichting Speurwerk Baggertechniek (SSB) and Rijkswaterstaat.

Institutional Review Board Statement: Not applicable.

Informed Consent Statement: Not applicable.

Data Availability Statement: Data will be made available on request.

Conflicts of Interest: The authors declare no conflict of interest.

References

1. Alhaddad, S.; Labeur, R.J.; Uijttewaal, W. Breaching Flow Slides and the Associated Turbidity Current. *J. Mar. Sci. Eng.* **2020**, *8*, 67. [CrossRef]
2. Alhaddad, S.; Labeur, R.J.; Uijttewaal, W. The need for experimental studies on breaching flow slides. In Proceedings of the Second International Conference on the Material Point Method for Modelling Soil-Water-Structure Interaction, Cambridge, UK, 8–10 January 2019; pp. 166–172.
3. Van Rhee, C.; Bezuijen, A. The breaching of sand investigated in large-scale model tests. In Proceedings of the 26th International Conference on Coastal Engineering, Copenhagen, Denmark, 22–26 June 1998; pp. 2509–2519.
4. Mastbergen, D.R.; Van Den Berg, J.H. Breaching in fine sands and the generation of sustained turbidity currents in submarine canyons. *Sedimentology* **2003**, *50*, 625–637. [CrossRef]
5. Alhaddad, S.; Labeur, R.J.; Uijttewaal, W. Large-scale Experiments on Breaching Flow Slides and the Associated Turbidity Current. *J. Geophys. Res. Earth Surf.* **2020**, *125*, e2020JF005582. [CrossRef]
6. Van Rhee, C. Slope failure by unstable breaching. In *Proceedings of the Institution of Civil Engineers-Maritime Engineering*; Thomas Telford Ltd.: London, UK, 2015; Volume 168, pp. 84–92.
7. Weij, D.; Keetels, G.; Goeree, J.; Van Rhee, C. An approach to research of the breaching process. In Proceedings of the WODCON XXI, Miami, FL, USA, 13–17 June 2016; pp. 13–17.
8. Alhaddad, S.; de Wit, L.; Labeur, R.J.; Uijttewaal, W. Modeling of Breaching-Generated Turbidity Currents Using Large Eddy Simulation. *J. Mar. Sci. Eng.* **2020**, *8*, 728. [CrossRef]
9. Eke, E.; Viparelli, E.; Parker, G. Field-scale numerical modeling of breaching as a mechanism for generating continuous turbidity currents. *Geosphere* **2011**, *7*, 1063–1076. [CrossRef]
10. You, Y.; Flemings, P.; Mohrig, D. Mechanics of dual-mode dilative failure in subaqueous sediment deposits. *Earth Planet. Sci. Lett.* **2014**, *397*, 10–18. [CrossRef]
11. Mastbergen, D.R.; Beinssen, K.; Nédélec, Y. Watching the Beach Steadily Disappearing: The Evolution of Understanding of Retrogressive Breach Failures. *J. Mar. Sci. Eng.* **2019**, *7*, 368. [CrossRef]
12. Alhaddad, S.; Labeur, R.; Uijttewaal, W. Preliminary Evaluation of Existing Breaching Erosion Models. In Proceedings of the 10th International Conference on Scour and Erosion, ISSMGE, Arlington, VA, USA, 18–20 October 2021; pp. 619–627.

13. Van Rhee, C. Simulation of the breaching process—Experimental validation. In Proceedings of the 22nd World Dredging Conference, Changhai, China, 25–29 April 2019.
14. Mastbergen, D.; Winterwerp, J.; Bezuijen, A. On the construction of sand fill dams. Part 1: Hydraulic aspects. In *Modelling Soil-Water Structure Interaction*; Kolkman, P.A., Lindenberg, J., Pilarczyk, K.W., Eds.; CRC Press: Delft, The Netherlands, 1988; pp. 353–362.
15. Eke, E.; Parker, G.; Wang, R. Breaching as a mechanism for generating sustained turbidity currents. In *Proceedings of the 33rd Internation Association of Hydraulic Engineering & Research Congress: Water Engineering for a Sustainable Environment*; IAHR: Vancouver, BC, Canada, 2009.
16. You, Y.; Flemings, P.; Mohrig, D. Dynamics of dilative slope failure. *Geology* **2012**, *40*, 663–666. [CrossRef]
17. Montellà, E.; Chauchat, J.; Chareyre, B.; Bonamy, C.; Hsu, T. A two-fluid model for immersed granular avalanches with dilatancy effects. *J. Fluid Mech.* **2021**, *925*. [CrossRef]
18. Lee, C.H.; Chen, J.Y. Multiphase simulations and experiments of subaqueous granular collapse on an inclined plane in densely packed conditions: Effects of particle size and initial concentration. *Phys. Rev. Fluids* **2022**, *7*, 044301. [CrossRef]

Journal of
*Marine Science
and Engineering*

Article

Experimental Investigation on the Cyclic Shear Mechanical Characteristics and Dynamic Response of a Steel–Silt Interface in the Yellow River Delta

Peng Yu [1,2], Jie Dong [1,2], Yong Guan [1,2], Qing Wang [1,2], Shixiang Jia [1,2], Meijun Xu [1,2], Hongjun Liu [3] and Qi Yang [4,*]

[1] Key Laboratory of Geological Safety of Coastal Urban Underground Space, Ministry of Natural Resources, Qingdao 266100, China
[2] Qingdao Geo-Engineering Surveying Institute (Qingdao Geological Exploration Development Bureau), Qingdao 266100, China
[3] Department of Environmental Science and Engineering, Ocean University of China, Qingdao 266100, China
[4] State Key Laboratory of Hydroscience and Engineering, Tsinghua University, Beijing 100084, China
* Correspondence: qi.yang@polimi.it

Abstract: The shear behavior and dynamic response of a steel–silt interface are significant for the safety and stability of offshore structures in the Yellow River Delta. A series of steel–silt interface cyclic shear tests under constant normal load conditions (CNL) were carried out to explore the effects of normal stress, shear amplitude, roughness, and water content on the interface shear strength, shear stiffness, and damping ratio using a large interface shear apparatus. The preliminary results showed that the amplitude of normal stress and shear amplitude affected the interface's shear strength, stiffness, and damping ratio in a dominant manner. The roughness and water content were also crucial factors impacting the rule of shear strength, shear stiffness, and damping ratio, changing with the number of cycles. Under various scenarios, the steel–silt interface weakened distinctively, and the energy dissipation tended to be asymptotic with the cyclic shear.

Keywords: steel–silt interface; mechanical behavior; dynamic response; cyclic shear test

Citation: Yu, P.; Dong, J.; Guan, Y.; Wang, Q.; Jia, S.; Xu, M.; Liu, H.; Yang, Q. Experimental Investigation on the Cyclic Shear Mechanical Characteristics and Dynamic Response of a Steel–Silt Interface in the Yellow River Delta. *J. Mar. Sci. Eng.* **2023**, *11*, 223. https://doi.org/10.3390/jmse11010223

Academic Editor: Constantine Michailides

Received: 12 December 2022
Revised: 5 January 2023
Accepted: 13 January 2023
Published: 15 January 2023

1. Introduction

Soil–structure interactions have gradually become a worldwide hot issue in offshore engineering since numerous offshore structure failures have been reported due to the reduction in the bearing capacity associated with the soil–structure interaction. The shear behavior of the soil–structure interface is often regarded as an essential factor affecting the bearing capacity of deep structures. Offshore structures, such as offshore oil platforms and offshore wind turbines (OWT), are often subjected to vertical cyclic loads under storm loads. Resistance degradation may occur at the soil–structure interface, thus reducing the vertical bearing capacity of the foundation [1]. In this process, shear stiffness and damping ratio are crucial indicators to characterize its dynamic response [2]. Therefore, it is of paramount importance for offshore engineering design and safety evaluation to clarify the mechanical properties, shear failure mode, and dynamic response law of the soil–structure interface and further reasonably determine its side friction.

The shear behavior and dynamic response at the soil–structure interface vary with the distinctive mechanical properties of different soil types. Some scholars have carried out cyclic shear research on different types of soil, such as sand [3], coarse-grained soil [4], frozen soil [5], bentonite sand mixtures (BSM) [6], and calcareous sand [7]. However, only some studies have focused on the shear behavior and dynamic response of the silt–structure interface. As is known, silt possesses the composite mechanical characteristics of sand and clay and has a unique engineering property. Expressly, it should be noted that the silt in the Yellow River Delta is typically formed under rapid deposition, and the pore water inside the soil cannot be discharged in time [8,9]. In this regard, the sediment on the surface has no time for consolidation, which, in turn, induces a relatively large pore pressure, high water content, and compressibility—but low shear strength [10] (Figure 1).

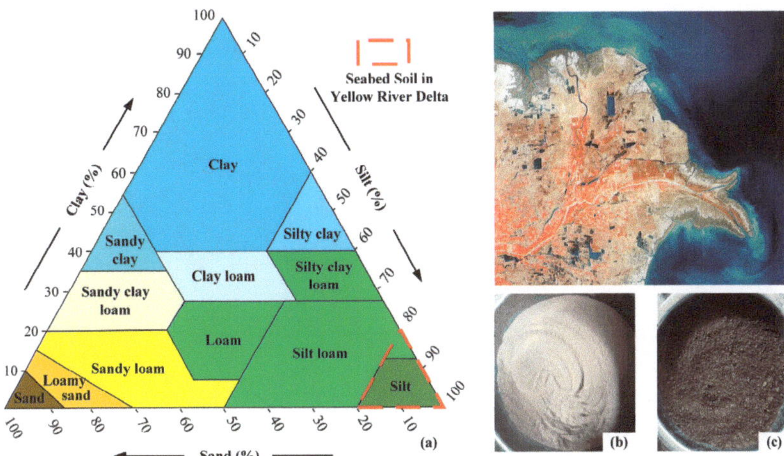

Figure 1. Silt in the Yellow River Delta. (**a**) Concept map of the soil classification adapted from Ren et al., 2020 [11]. (**b**) Dry condition. (**c**) Wet condition.

Until now, there have been few studies on the silt–structure interface in the Yellow River Delta, and we reported most of these studies. In previous studies, the direct shear characteristics [12] and volumetric strain [13] were investigated, and this paper is the follow-up of these precedent ones. In the present work, the effect of five factors, namely, the number of cycles, shear amplitude, normal stress, roughness, and water content, were clarified for the cycle shear mechanical characteristics and dynamic response law of the pile–soil interface under cyclic loads that were analogous to a storm. This paper aimed to provide a theoretical and technical reference for designing and evaluating offshore structures in the area (Figure 2).

Figure 2. The offshore structures in the Yellow River Delta. (**a**) The Chengdao offshore oil platform. (**b**) The Hekou jack-up marine ranching platform.

2. Large-Scale CNL Cyclic Shear Tests

2.1. Testing Apparatus

The tests were carried out under constant normal load conditions (CNL) with a JAW-500 large shear apparatus (Figure 3). The shear instrument adopted computer-controlled hydraulic loading. It was equipped with a relevant cooling system to ensure efficient and constant loading during the test. Twenty springs were used to connect the shear box with a lower sliding plate; thus, it allowed for free movement in a vertical direction, which enabled the efficient measuring of the volume change during the shear process [14,15]. The apparatus collected real-time data during the test through the matched monitoring instrument and data acquisition system (Figure 4).

Figure 3. Large-scale shear instrument. (**a**) 3D schematic diagram of the shear box. (**b**) Displacement measuring device. (**c**) Spring under the shear box.

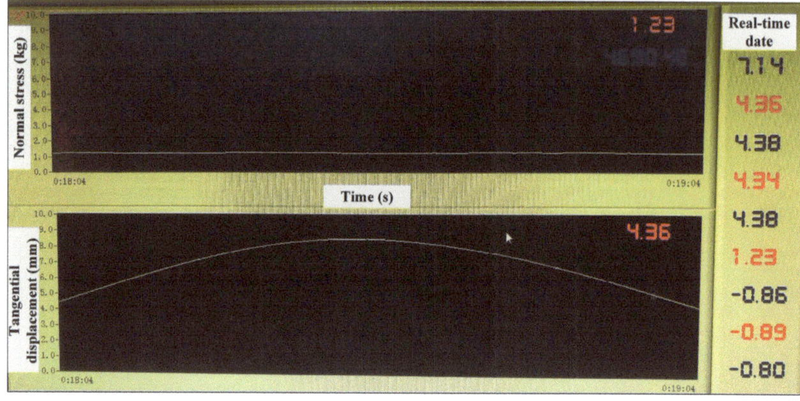

Figure 4. Data acquisition system.

2.2. Testing Materials

The soil used in the test was taken from the Yellow River silt of the Chendao Oilfield. Its physical characteristics included high roundness and porosity, poor permeability, and inter-particle occlusion. The particle size and basic physical properties of the soil sample are shown in Figure 5. Considering the disturbance during the construction of the structure, the undisturbed soil was remolded according to Standard GB/T50123-2019 for soil testing. The process of running a series of interface tests must adopt the large instruments mentioned above, which consume many materials. Therefore, three soil samples with different saturations of 63.5%, 79.3%, and 95.2% (with water contents of 16%, 20%, and 24%, respectively) were prepared by controlling the dry density (1.61 g/cm^3).

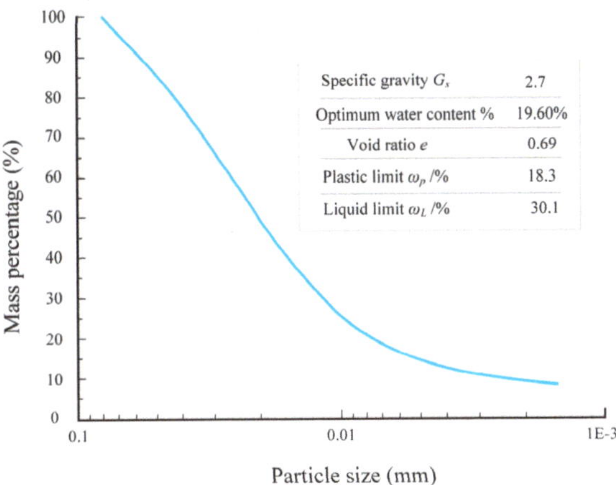

Figure 5. Silt particle size curve and physical properties.

With consideration of the fact that the foundation surface was not entirely smooth in the actual project due to rust or erosion, the grooving process was used to process three different roughness levels of a steel plate: R_0 (smooth), R_1 (slightly rough), and R_2 (rough) (Figure 6). The sand filling method, average profile height, fractal dimension method, etc., typically determine the surface roughness. In this test, three parameters (the peak valley distance H, groove cross-section width L_1, and platform spacing L_2) were defined to determine the surface roughness according to the modified sand-filling method [16]. The steel plate had no material loss during the tests.

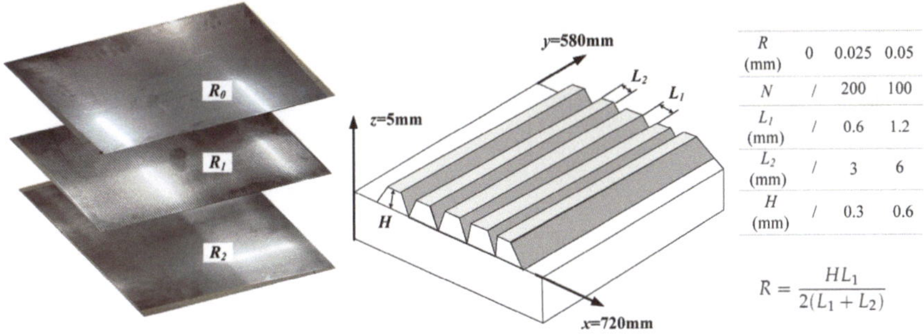

Figure 6. Steel plates with different roughness levels.

2.3. Testing Scheme

The test scheme used was the same as that of our previous study on interfacial shear properties, and so it will not be repeated here [13]. This test conducted nine undrained shear tests, which included three types of normal stress, shear amplitude, roughness, and water content (Table 1). The maximum saturation of the sample was 95.2%, which met the requirements of the indoor shear test [17]. Relevant studies have found that the influence of shear rate on interface shear properties under low-stress conditions can be ignored, and the test rate was 1 mm/s [18]. The interface after the test was completed i9s shown in Figure 7.

S

Table 1. Testing scheme.

Number	Normal Stress (kPa)	Shear Amplitude (mm)	Roughness (mm)	Water Content (%)	Others
1	100	5	0.05	20	Number of cycles = 30 Frequency = 0.01 HZ Shear rate = 1 mm/s
2					
3		15			
4			0		
5	200	35	0.025		
6			0.05	16	
7				24	
8					20
9	300	5			

Figure 7. Shear interface after the test was completed.

3. Results and Discussion

Since the data are too many to be displayed individually, the results of the 1st, 5th, 10th, 15th, 20th, 25th, and 30th cycle tests were selected as representative datasets for the analysis. In this section, the shear strength and dynamic response of the steel–silt interface will be studied in detail.

3.1. Interface Shear Strength

To further explore the shear mechanics under different cycles, the shear strength in each cycle was set as the peak shear stress τ^{max}. The peak shear stress was equal to the average magnitude of the positive and negative maximum shear stress. Furthermore, the definition of the stress ratio D_τ was introduced to characterize the relationship between the peak shear stress and the cycle number and between the stress ratio and the cycle number.

$$\tau^{max} = \frac{\tau_1 + |\tau_2|}{2} \tag{1}$$

$$D_\tau = \frac{\tau_1^{max}}{\tau_n^{max}} \tag{2}$$

where τ_1 is the maximum positive shear stress of one cycle and the maximum negative shear stress of one cycle, and τ_1^{max} is the peak shear stress of the first cycle while τ_n^{max} is the peak shear stress of the *n*-th cycle.

3.1.1. The Influence of Normal Stress

Figure 8 illustrates the shear strength-cycle number curve of test groups 1, 2, and 9 (where the normal stress levels were 100 kPa, 200 kPa, and 300 kPa, respectively). The test results showed that:

(1) The shear strength decreased with the increase in the number of cycles. At the beginning of shear, the soil particles occluded tightly under normal stress, and the cohesion between the steel soil interface was strong, resulting in the soil interface having a more remarkable ability to resist shear. With the increase in the shear cycle, the soil was compacted and the corners of the soil particles were rounded. After the first few shear cycles, the soil particles near the interface under the same repeated shear stress had a particular historical stress path, which led to the shear strength decreasing continuously and finally becoming stable.

(2) With the increase in normal stress, the shear strength and residual shear strength (the peak shear stress of the 30th cycle) also increased.

(3) The first two cycles showed hardening under the low-stress conditions of 100 kPa and 200 kPa. The results were inconsistent with the findings of Shang (2016), Li (2018), and others [19,20]. One possible reason could be attributed to the constant stress conditions used in the tests and the normal stress remaining unchanged during the whole test, resulting in a discontinuous reduction in normal stress under the low-stress conditions. Furthermore, combined with the direct test results, under the low-stress condition, the shear stress–shear displacement curve appeared to develop a hardening behavior, and so when the overall shear displacement was small (the first two cycles), the shear stress increased. The softening feature seemed to align with the rise in the overall shear displacement (after two cycles) [12].

(4) Similar to the shear stress–shear displacement curve in the direct shear test, the peak shear stress–total shear displacement curve in the cyclic test could be used as the shear strength–cycle number curve. Three stages could be recognized, i.e., the elastic stage of the rapid change in shear stress (the first five cycles), the plastic stage of the slow evolution of the shear stress (5–10 cycles), and the shear failure stage of equilibrium of the shear stress (10–30 cycles).

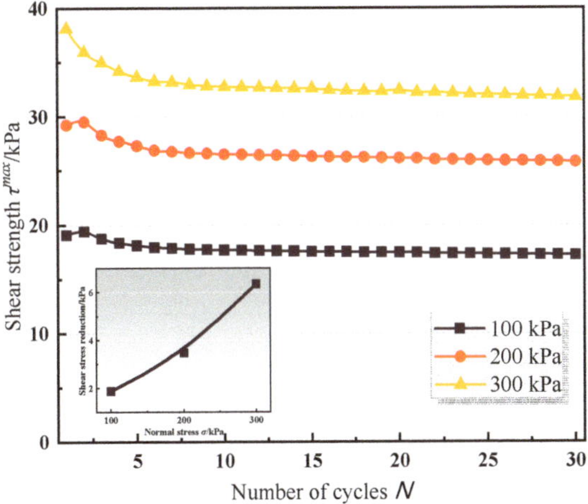

Figure 8. Shear strength–number of cycles curve under different normal stress levels.

The stress ratio–cycle number curve in Figure 9 shows that normal stress significantly impacted the interface cyclic weakening profile. With the increase in normal stress, the weakening degree increased. The trend agreed with those reported by Oumarou et al. (2005) and Feng et al. (2018) [21,22].

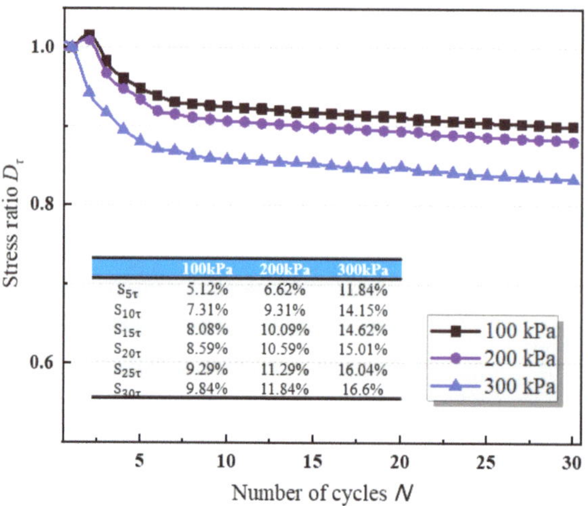

Figure 9. Stress ratio–number of cycles curve under different normal stress levels.

To more clearly capture the relationship between this weakening phenomenon and the number of cycles, the shear strength weakening coefficient $S_{n\tau}$ was further adopted based on the stress ratio D_τ to characterize the cumulative weakening degree of the first n cycles. The representative weakening coefficients $S_{5\tau}$, $S_{10\tau}$, $S_{15\tau}$, $S_{20\tau}$, $S_{25\tau}$, and $S_{30\tau}$ under the three normal stresses were calculated. The results are shown in Figure 9. The maximum weakening coefficients $S_{30\tau}$ under the three normal stresses were 9.84%, 11.84%, and 16.6%, respectively.

$$S_{n\tau} = \frac{\tau_1^{max} - \tau_n^{max}}{\tau_1^{max}} \times 100\%$$

(3)

3.1.2. The Influence of Shear Amplitude

Figure 10 illustrates the shear strength–cycle number curve of test groups 2, 3, and 8 (where the shear amplitudes were 5 mm, 15 mm, and 35 mm, respectively). It was found from the test results that:

(1) With the increase in the shear amplitude, the peak and residual shear strength increased. However, with the increased shear amplitude, its increasing rate decreased and the residual shear strength was similar to those when the amplitude was 15 mm and 35 mm.

(2) The test group with a shear amplitude of 5 mm showed hardening in the first two cycles. With the increase in the number of cycles, the shear strength decreased. The results were similar to the findings of Vieira et al. (2013), indicating that the cyclic loading of the interface under a small displacement would not reduce the shear strength, while the cyclic shear strength decreased significantly under a large displacement [23]. The explanation is the same as the previous one. In the case of large shear amplitude, the interface has reached the direct shear strength under the same conditions in the first few cycles (approximately 2–5), which affected the development of the subsequent cycle shear strength.

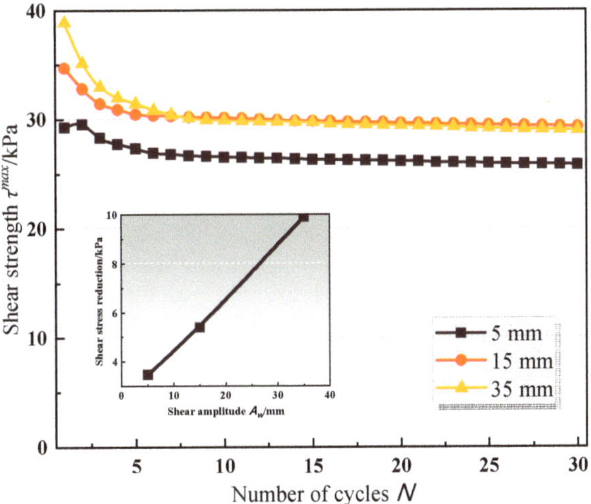

Figure 10. Shear strength–number of cycles curve under different shear amplitude levels.

Similarly, the stress ratio–cycle number curve demonstrated in Figure 11 implies that the shear strength decreased significantly with the increase in the shear amplitude corresponding to the shear displacement. Liu et al. (2021) reached the same conclusion [24]. To explain this phenomenon, the total shear displacement was the product of the shear amplitude and the number of cycles. With the amplitude increase, the number of cycles required for achieving shear failure decreased, and the total shear displacement increased under the same number of cycles. Therefore, the weakening degree became intense until it reached equilibrium. The representative weakening coefficient $S_{n\tau}$ under the three shear amplitudes was calculated, as shown in Figure 11. The maximum weakening coefficients $S_{30\tau}$ under the three shear amplitudes were 11.84%, 15.59%, and 25.51%, respectively.

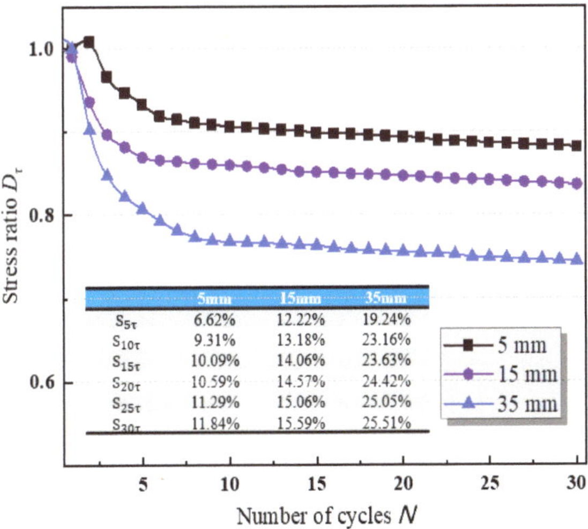

Figure 11. Stress ratio–number of cycles curve under different shear amplitude.

3.1.3. The Influence of Roughness

Figure 12 illustrates the shear strength–cycle number curve of test groups 4, 5, and 8 (where the roughness R was 0 mm, 0.025 mm, 0.05 mm, respectively). The test results show that:

(1)   With the roughness increase, the shear strength of the first cycle increased, but the difference was insignificant. Referring to the conclusion of the direct test, the influence of roughness on shear stress decreased under a larger normal stress, which could reasonably explain the above phenomenon [12].

(2)   Smaller roughness levels reached the shear failure stage earlier, but the residual shear strength decreased with the roughness increase. The stress ratio–cycle number curve presented in Figure 12 demonstrates that the weakening of the shear strength became apparent with the increase in roughness. Taha and Fall (2014) drew a similar conclusion for their cyclic shear test on steel marine clay [25]. With the roughness increase, the dislocation and rearrangement process of the soil particles near the interface became intense. With the progressive number of cycles, the strength degradation became more significant.

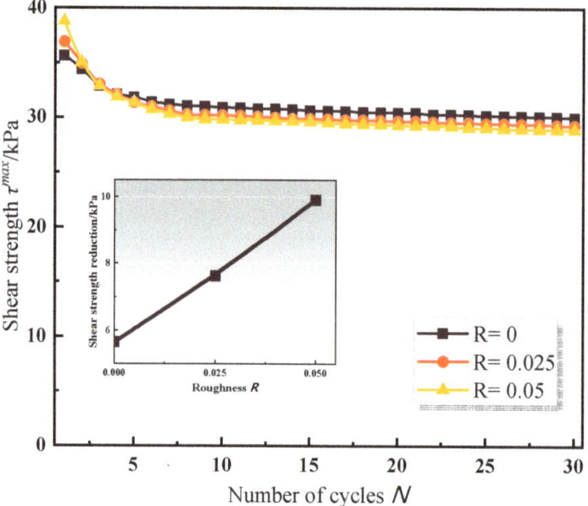

Figure 12. Shear strength–number of cycles curve under different roughness levels.

The representative weakening coefficients $S_{n\tau}$ under the three shear amplitudes were calculated, as shown in Figure 13. The maximum weakening coefficients $S_{30\tau}$ under the three roughness conditions were 15.84%, 20.65%, and 25.51%, respectively, which verified the above conclusions.

3.1.4. The Influence of Water Content

Figure 14 illustrates the shear strength–cycle number curve of test groups 6, 7, and 8 (where the water contents were 16%, 24%, and 20%, respectively). The results suggested that shear strength increased with the decrease in water content. However, the difference between the test results for the 16% and 20% water content levels was smaller than those of 20% and 24%. The shear behavior in the positive direction of the first cycle of the cyclic test was equivalent to that of the direct shear test. According to the conclusion of the direct shear test, the distinction mentioned above was induced by the optimal water content (19.6%) [12]. With the increase in water content, the cohesion first increased and then decreased. The cohesion peak occurred when the water content reached the plastic limit or when the optimal water content was used.

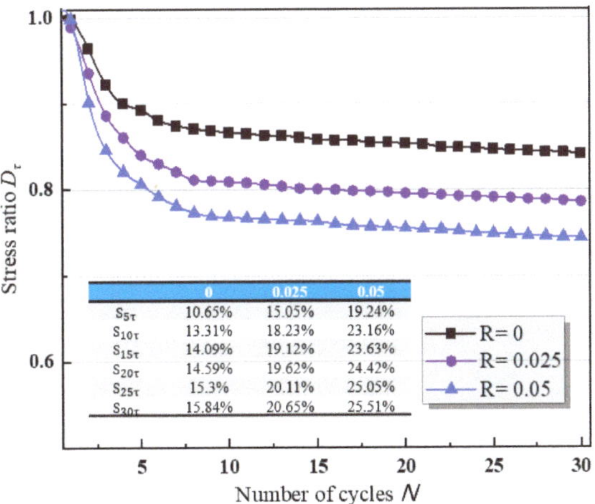

Figure 13. Stress ratio–number of cycles curve under different roughness levels.

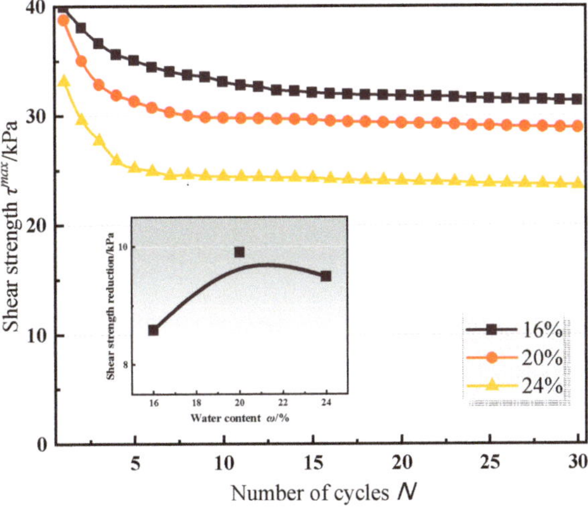

Figure 14. Shear strength–number of cycles curve under different water content levels.

The curves of the stress ratio–cycle number presented in Figure 15 show that the number of cycles required to reach the shear failure stage decreased with the increase in water content. With the increased water content, the thickness of the interface water film increased and the shear work was reduced due to the lubrication during the reciprocating shear process such that the shear failure stage could be reached earlier. In addition, the correlation between the shear strength reduction and the water content shown in Figure 15 implied that the weakening degree of the interface did not increase further with the increase in water content, but rather, it reached the maximum value of 20% near the optimal water content. The reason for this is that the cohesion first increased and then decreased. Liu (2017) and Chen (2018) reached similar conclusions [26,27]. The maximum weakening coefficient $S_{30\tau}$ were 21.5%, 25.51%, and 28.6% under the three water content conditions, respectively.

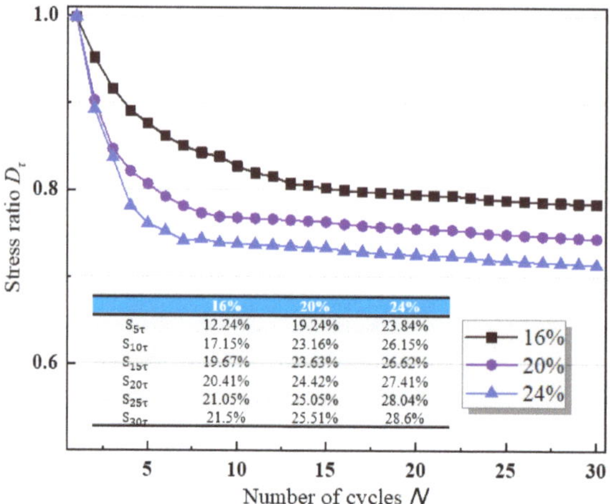

	16%	20%	24%
$S_{5\tau}$	12.24%	19.24%	23.84%
$S_{10\tau}$	17.15%	23.16%	26.15%
$S_{15\tau}$	19.67%	23.63%	26.62%
$S_{20\tau}$	20.41%	24.42%	27.41%
$S_{25\tau}$	21.05%	25.05%	28.04%
$S_{30\tau}$	21.5%	25.51%	28.6%

Figure 15. Stress ratio–number of cycles curve under different water content conditions.

3.2. Shear Stiffness and Damping Ratio

Compared with the direct shear test, the number of cycles and shear amplitude were considered in the cyclic shear tests. Similar to Nye et al. (2007), Fox et al. (2011), and others, the shear stiffness and damping ratio were introduced to describe the dynamic characteristics and response analysis of soil mass and to analyze the weakening behavior of the steel–silt interface under the cyclic shear scenario [28,29]. The schematic diagram is shown in Figure 16, and the formula used is as follows:

$$K = \frac{K_1 + K_2}{2} = \frac{\tau_1^{Max} + \tau_2^{Max}}{2A_w} \tag{4}$$

where K_1 and K_2 represent the shear stiffness in the positive and negative shear directions, respectively, and τ_2^{Max} is the peak shear stress, while A_w is the shear amplitude.

$$D = \frac{D_1 + D_2}{2} = \frac{1}{2}\left(\frac{S}{4\pi S_1} + \frac{S}{4\pi S_2}\right) = \frac{S}{4\pi A_w}\left(\frac{1}{\tau_1^{Max}} + \frac{1}{\tau_2^{Max}}\right) \tag{5}$$

where D_1 and D_2 represent the damping ratio in the positive and negative shear directions, S is the hysteresis loop area, and S_1 and S_2 are shadow areas.

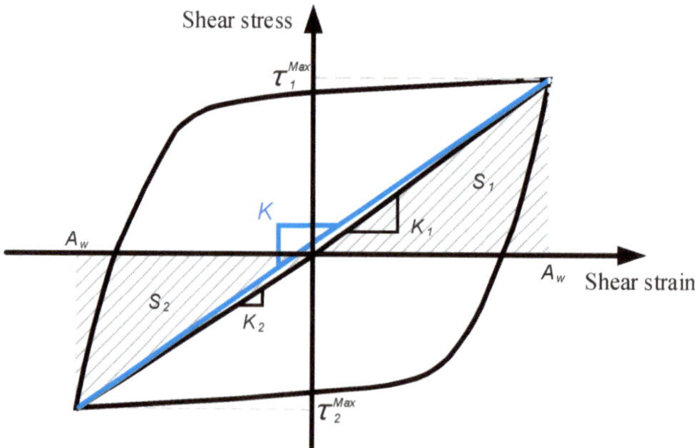

Figure 16. Stiffness and damping ratio.

Due to a large number of data points, the cycle data for cycles 1, 5, 10, 15, 25, and 30 were selected as representative points for the following analysis. The area of each cycle hysteresis loop was calculated using the Origin commercial software, and the curves of the shear stiffness–cycle number and damping ratio–cycle number were plotted, respectively.

3.2.1. The Influence of Normal Stress

It has been reported that different normal stresses lead to an additional cyclic shear stiffness and damping ratio of the interface [30,31]. As shown in Figure 17, in the K–N (shear stiffness–cycle number) curve, a weakening feature tends to appear under all normal stress conditions. For the same number of cycles, with the increase in normal stress, shear stiffness increased, and its softening trend became apparent with the number of cycles, which was consistent with the influence of normal stress on shear strength.

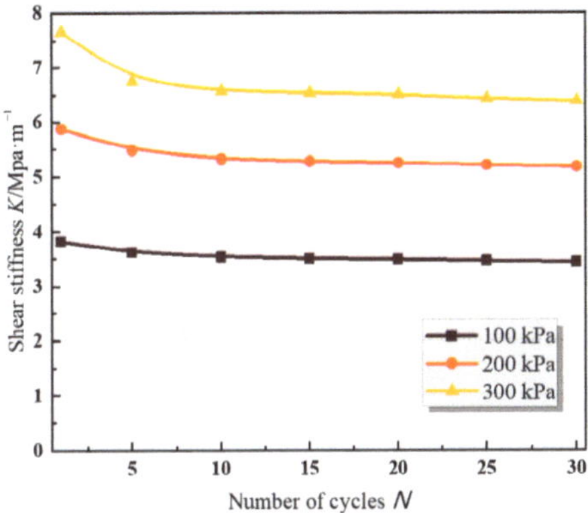

Figure 17. Shear stiffness–cycle number curve under different normal stress conditions.

The D–N (damping ratio–cycle number) curve in Figure 18 shows that with the progress in cyclic shear, the D–N curve also tended to present a weakening trend, indicating that the cyclic shear had an adverse impact on the safety and stability of the interface. The damping ratio and the reduction rate decreased with the increase in normal stress. In addition, with the progress in cyclic shear, the difference in the damping ratio under different normal stress conditions decreased. Similar conclusions emerged in the research conducted by Ying et al. (2020), i.e., the energy dissipation tended to converge with the progress in cyclic shear [32].

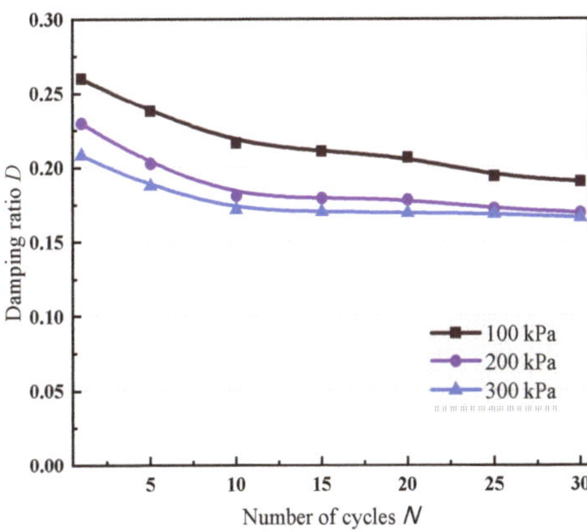

Figure 18. Damping ratio–cycle number curve under different normal stress conditions.

3.2.2. The Influence of Shear Amplitude

Figures 19 and 20 show the *K–N* and *D–N* curves of the interface under different shear amplitudes conditions (test groups 2, 3, and 8). The *K–N* curve in Figure 19 indicates that with the increase in cyclic shear amplitude, the shear stiffness corresponding to the same number of cycles decreased. Vieira et al. (2013) obtained the same test results [23]. The *D–N* curve in Figure 20 implies that with the increase in shear amplitude, the energy dissipation and the damping ratio increased. However, the group with a shear amplitude of 35 mm had an opposite rule to the other two groups. With the rise in shear, the damping ratio increased first and then appeared asymptotic.

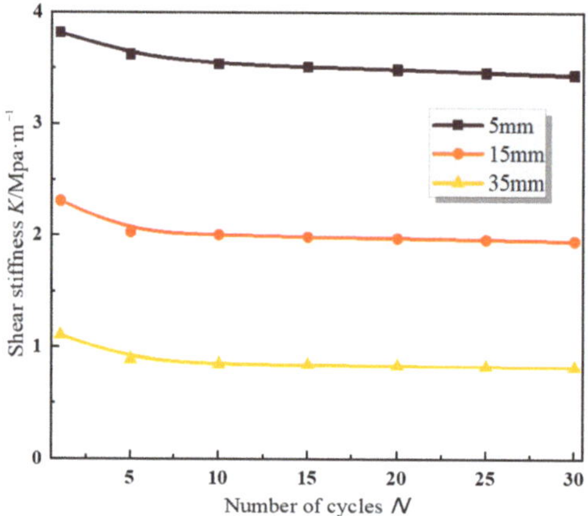

Figure 19. Shear stiffness–cycle number curve under different shear amplitude levels.

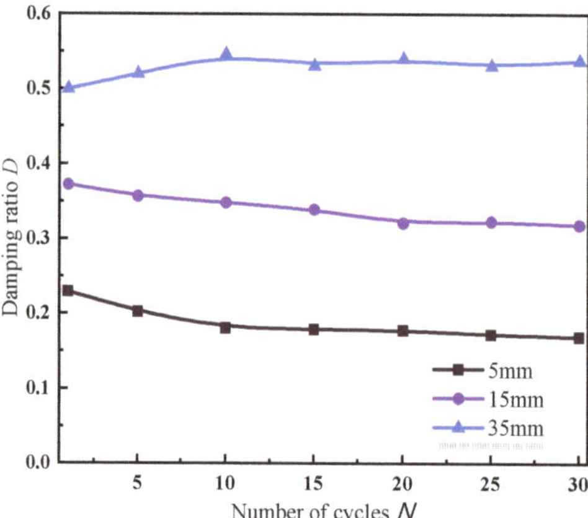

Figure 20. Damping ratio–cycle number curve under different shear amplitude levels.

3.2.3. The Influence of Roughness

Figures 21 and 22 demonstrate the *K–N* and *D–N* curves of the interface under different roughness conditions (test groups 4, 5, and 8). The *K–N* curve in Figure 21 shows that with the progress in cyclic shear, the shear stiffness passed through three stages: rapid decline, slow decline, and equilibrium. The relationship between shear stiffness and roughness was the same as the relationship between shear strength and roughness in the previous section. The rough interface made the frictional occlusion effect apparent at the beginning of cyclic shear, and then the shear stiffness increased. With the roughness increase, the strength degradation of the interface became remarkable, and the decrease in the shear stiffness became significant.

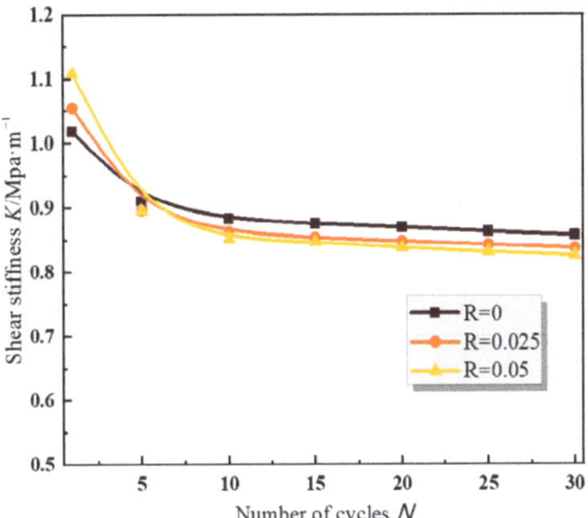

Figure 21. Shear stiffness–cycle number curve under different roughness levels.

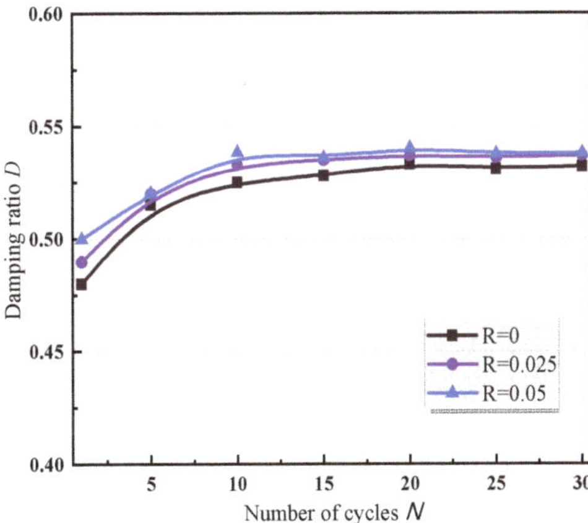

Figure 22. Damping ratio–cycle number curve under different roughness levels.

The *D–N* curve in Figure 22 implies that with the roughness increase, the interface's energy dissipation caused by cyclic shear became fast, and the damping ratio increased [33]. In addition, the damping ratios of the interfaces with different roughness levels varied significantly in the first few cycles. As the number of cycles increased, the damping ratio difference decreased. A possible reason for this may be that during the shear process, the grooves having various roughness levels were filled with soil particles to different extents, and their contact surfaces tended to be smooth. Thus, the differences in the volume and shape between the grooves were gradually reduced, which ultimately weakened the impact of the roughness on the damping ratio.

3.2.4. The Influence of Water Content

Figures 23 and 24 present the interface *K–N* and *D–N* curves under different water content conditions (test groups 6, 7, and 8). The *K–N* curve shows that the relationship between the water content and the shear stiffness was the same as the relationship between the water content and the shear strength.

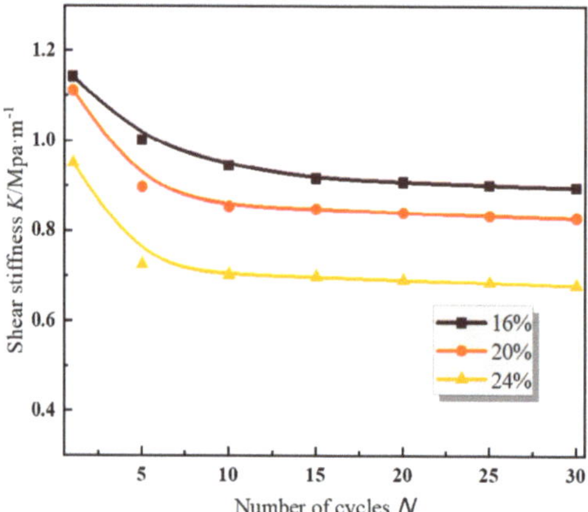

Figure 23. Shear stiffness–cycle number curve under different water content conditions.

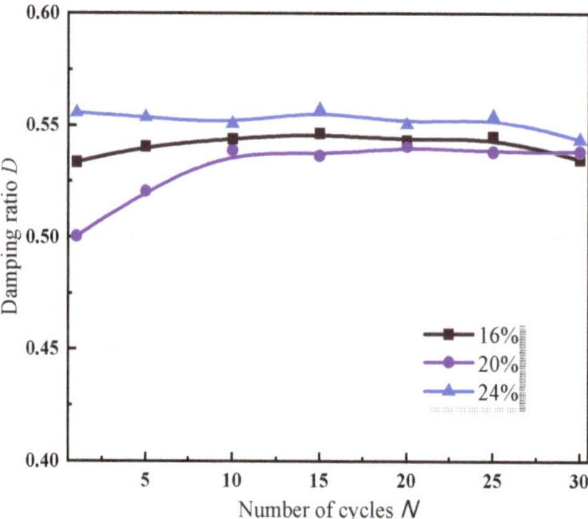

Figure 24. Damping ratio–cycle number curve under different water content conditions.

The *D–N* curve in Figure 24 shows that the relationship between water content and damping ratio was complex and did not show a general rule. For the soil sample with a high water content, its damping ratio was relatively large, indicating that the energy dissipation of the interface of the soil sample with a high water content under the action of cyclic shear was fast because the interface was near a saturation state. As the number of cycles increased, some water may have flowed out during the shear process, which decreased the water content, leading to a downward trend in the damping ratio. The damping ratio of 20% water content, which was closer to the optimal moisture content, was greater than that of the 16% condition, which may have been related to many factors, such as the compactness and suction of the unsaturated soil matrix [34]. The reasons will be further explored and analyzed. In addition, the damping ratio under the three water contents tended to reach a specific value with the increase in the cycle number. The results indicated that the difference in damping ratio caused by the water content gradually was reduced, resulting in the convergence of energy dissipation at the interface in the late period of cyclic shear.

4. Conclusions

In this work, using a JAW-500 apparatus, a series of cyclic shear tests were carried out on the steel–silt interfaces concerning a background of the Yellow River Delta. The results indicated that the normal stress and shear amplitude can affect the steel–silt interface's shear strength, shear stiffness, and damping ratio. The roughness and water content can significantly influence the variations in shear strength, shear stiffness, and damping ratio with the number of cycles. The following conclusions may be drawn based on the current results:

(1) Generally, three distinctive stages appeared in the curve of the peak shear stress: the elastic stage, in which the shear stress changed rapidly (the first five cycles); the plastic stage, in which the shear stress changed slowly (5–10 cycles), and the shear failure stage, in which shear stress equilibrium was achieved (10–30 cycles). The shear strength, shear stiffness, and damping ratio decreased, and the energy dissipation tended to be asymptotic with the increase in the cycle number.

(2) The normal stress conditions significantly influenced the action of the cyclic interface weakening. As the normal stress increased, the degree of interface weakening increased. At the same time, the shear strength and shear stiffness of the interface increased while the damping ratio decreased.

(3) With the increase in the shear amplitude, the degree of the interface weakening, the shear strength, the damping ratio, and the energy dissipation increased, while the shear stiffness also decreased.

(4) As the roughness increased, the shear strength's weakening became apparent and the energy dissipation became fast.

(5) The shear strength and stiffness increased when the water content decreased. The weakening degree of the interface reached its maximum near the optimal water content.

Author Contributions: Conceptualization, Q.Y. and P.Y.; formal analysis, Y.G. and Q.W.; data curation, S.J. and P.Y.; writing—original draft preparation, P.Y. and H.L.; writing—review and editing, Q.Y. and P.Y.; visualization, M.X. and P.Y.; project administration, J.D.; funding acquisition, J.D. and P.Y. All authors have read and agreed to the published version of the manuscript.

Funding: This research was funded by the Scientific Research Fund Project of the Qingdao Geo-Engineering Surveying Institute (grant no. 2022-QDDZYKY06) and the Shandong Provincial Bureau of Geology and Mineral Resources (grant no. KY202223).

Institutional Review Board Statement: Not applicable.

Informed Consent Statement: Not applicable.

Data Availability Statement: The data presented in this study are available on request from the corresponding author.

References

1. Yang, Q.; Pan, G.; Liu, H.J.; Wang, Q. Bearing capacity of offshore umbrella suction anchor foundation in silty soil under varying loading modes. *Mar. Georesources Geotechnol.* **2018**, *36*, 781–794. [CrossRef]
2. Carneiro, D.; White, D.J.; Danziger, F.A.B.; Ellwanger, G. A novel approach for time-dependent axial soil resistance in the analysis of subsea pipelines. *Comput. Geotech.* **2015**, *69*, 641–651. [CrossRef]
3. Pra-Ai, S.; Boulon, M. Soil–structure cyclic direct shear tests: A new interpretation of the direct shear experiment and its application to a series of cyclic tests. *Acta Geotech.* **2017**, *12*, 107–127. [CrossRef]
4. Rehman, Z.; Zhang, G. Three-dimensional elasto-plastic damage model for gravelly soil-structure interface considering the shear coupling effect. *Comput. Geotech.* **2021**, *129*, 103868. [CrossRef]
5. Wang, T.L.; Wang, H.H.; Hu, T.F.; Song, H.-F. Experimental study on the mechanical properties of soil-structure interface under frozen conditions using an improved roughness algorithm. *Cold Reg. Sci. Technol.* **2019**, *158*, 62–68. [CrossRef]
6. Li, L.; Fall, M.; Fang, K. Shear behavior at interface between compacted clay liner–geomembrane under freeze-thaw cycles. *Cold Reg. Sci. Technol.* **2020**, *172*, 103006. [CrossRef]
7. Li, Y.J.; Guo, Z.; Wang, L.Z.; Li, Y.; Liu, Z. Shear resistance of MICP cementing material at the interface between calcareous sand and steel. *Mater. Lett.* **2020**, *274*, 128009. [CrossRef]
8. Liu, X.L.; Zhang, H.; Zheng, J.W.; Guo, L.; Jia, Y.; Bian, C.; Li, M.; Ma, L.; Zhang, S. Critical role of wave–seabed interactions in the extensive erosion of Yellow River estuarine sediments. *Mar. Geol.* **2020**, *426*, 106208. [CrossRef]

9. Liu, X.L.; Zheng, J.W.; Zhang, H.; Zhang, S.-T.; Liu, B.-H.; Shan, H.-X.; Jia, Y.-G. Sediment Critical Shear Stress and Geotechnical Properties along the Modern Yellow River Delta, China. *Mar. Georesources Geotechnol.* **2018**, *36*, 875–882. [CrossRef]
10. Zhang, H.; Liu, X.L.; Jia, Y.G.; Du, Q.; Sun, Y.; Yin, P.; Shan, H. Rapid consolidation characteristics of Yellow River–derivedsediment: Geotechnical characterization and its implications for the deltaicgeomorphic evolution. *Eng. Geol.* **2020**, *270*, 105578. [CrossRef]
11. Ren, Y.; Xu, G.; Xu, X.; Zhao, T.; Wang, X. The initial wave induced failure of silty seabed: Liquefaction or shear failure. *Ocean Eng.* **2020**, 106990. [CrossRef]
12. Yu, P.; Liu, C.; Liu, H.J. Large–scale direct shear test study on the silt–steel interface in the Yellow River Delta. *Period. Ocean Univ. China* **2021**, *51*, 71–79. (In Chinese)
13. Yu, P.; Dong, J.; Liu, H.S.; Xu, R.; Wang, R.; Xu, M.; Liu, H. Analysis of cyclic shear stress–displacement mechanical properties of silt–steel interface in the Yellow River Delta. *J. Mar. Sci. Eng.* **2022**, *10*, 1704. [CrossRef]
14. Guo, X.S.; Nian, T.K.; Zhao, W.; Gu, Z.; Liu, C.; Liu, X.; Jia, Y. Centrifuge experiment on the penetration test for evaluating undrained strength of deep-sea surface soils. *Int. J. Min. Sci. Technol.* **2022**, *32*, 363–373. [CrossRef]
15. Guo, X.S.; Nian, T.K.; Wang, D.; Gu, Z. Evaluation of undrained shear strength of surficial marine clays using ball penetration-based CFD modeling. *Acta Geotech.* **2022**, *17*, 1627–1643. [CrossRef]
16. Liu, J.W.; Cui, L.; Zhu, N.; Han, B. Investigation of cyclic pile-sand interface weakening mechanism based on large-scale CNS cyclic direct shear tests. *Ocean Eng.* **2019**, *194*, 106650. [CrossRef]
17. Lu, Z.H.; Qi, C.Z.; Jin, T.W. Experimental study on the mechanical characteristics of saturated sandy soil under different loading stress paths. *Sci. Technol. Eng.* **2019**, *19*, 294–301. (In Chinese)
18. Wang, Y.H.; Zhang, M.Y.; Bai, X.Y. Experimental research on effect of shear rate on shear strength of clayey soil-concrete interface. *J. Civ. Environ. Eng.* **2019**, *41*, 48–54. (In Chinese)
19. Shang, W.C. Study on the Degenerate Mechanism of Pile–Soil Interface under Cyclic Loading. Master's Thesis, Qingdao Technological University, Qingdao, China, 2016. (In Chinese)
20. Li, X.L. Study on the Degenerate Mechanism of Pile–Soil Interface under Cyclic Loading. Master's Thesis, Qingdao Technological University, Qingdao, China, 2018. (In Chinese)
21. Oumarou, T.A.; Evgin, E. Cyclic behavior of a sand–steel plate interface. *Can. Geotech. J.* **2005**, *42*, 1695–1704. [CrossRef]
22. Feng, D.K.; Zhang, J.M.; Deng, L.J. Three–dimensional direct and cyclic behavior of a gravel–steel interface from large–scale simple shear tests. *Can. Geotech. J.* **2018**, *55*, 1657–1667. [CrossRef]
23. Vieira, C.S.; Lopes, M.L.; Caldeira, L.M. Sand geotextile interface characterization through direct and cyclic direct shear tests. *Geosynth. Int.* **2013**, *20*, 26–38. [CrossRef]
24. Liu, F.Y.; Ying, M.J.; Yuan, G.H.; Wang, J.; Gao, Z.Y.; Ni, J.F. Particle shape effects on the cyclic shear behaviour of the soil–geogrid interface. *Geotext. Geomembr.* **2021**, *49*, 991–1003. [CrossRef]
25. Taha, A.; Fall, M. Shear behavior of sensitive marine clay–steel interfaces. *Acta Geotech.* **2014**, *9*, 969–980. [CrossRef]
26. Liu, Z.Y. Reciprocating Shear Behavior and Strength Weakening Mechanism of Steel–Soil Interface in Rock–Socketed Filling Pile with Steel Tube. Master's Thesis, Chongqing Jiaotong University, Chongqing, China, 2017. (In Chinese)
27. Chen, C. Reciprocating Shear Behavior and Strength Weakening Mechanism of Coarse–Grained Soil–Steel Interface. Master's Thesis, Chongqing Jiaotong University, Chongqing, China, 2018. (In Chinese)
28. Nye, C.J.; Fox, P.J. Dynamic shear behavior of a needle–punched geosynthetic clay liner. *J. Geotech. Geoenvironmental Eng.* **2007**, *133*, 973–983. [CrossRef]
29. Fox, P.J.; Ross, J.D.; Sura, J.M. Geomembrane damage due to static and cyclic shear over compacted gravelly sand. *Geosynth. Int.* **2011**, *18*, 272–279. [CrossRef]
30. Lee, K.M.; Manjunath, V.R. Soil–geotextile interface friction by direct shear tests. *Can. Geotech. J.* **2000**, *37*, 238–252. [CrossRef]
31. Li, M.J.; Fang, H.Y.; Du, M.; Zhang, C.; Su, Z.; Wang, F. The behavior of polymer-bentonite interface under shear stress. *Constr. Build. Mater.* **2020**, *248*, 118680. [CrossRef]
32. Ying, M.J.; Liu, F.Y.; Wang, J.; Wang, C.; Li, M. Coupling effects of particle shape and cyclic shear history on shear properties of coarse-grained soil–geogrid interface. *Transp. Geotech.* **2020**, *27*, 100504. [CrossRef]
33. Jotisankasa, A.; Rurgchaisri, N. Shear strength of interfaces between unsaturated soils and composite geotextile with polyester yarn reinforcement. *Geotext. Geomembr.* **2018**, *46*, 338–353. [CrossRef]
34. Shahrour, I.; Rezaie, F. An elastoplastic constitutive relation for the soil-structure interface under cyclic loading. *Comput. Geotech.* **1997**, *21*, 21–39. [CrossRef]

Journal of
*Marine Science
and Engineering*

Article

A Case Study Assessing the Liquefaction Hazards of Silt Sediments Based on the Horizontal-to-Vertical Spectral Ratio Method

Qingsheng Meng [1,2,3,*], Yang Li [1,*], Wenjing Wang [1], Yuhong Chen [1] and Shilin Wang [1]

[1] College of Environmental Science and Engineering, Ocean University of China, Qingdao 266100, China
[2] Key Laboratory of Marine Environment Science and Ecology, Ministry of Education, Qingdao 266100, China
[3] Key Laboratory of Marine Environmental Geological Engineering, Qingdao 266100, China
*   Correspondence: qingsheng@ouc.edu.cn (Q.M.); liyang4009@stu.ouc.edu.cn (Y.L.);
    Tel.: +86-138-5327-4653 (Q.M.); +86-176-6090-8189 (Y.L.)

Abstract: Silt liquefaction can occur due to the rapid cyclic loading of sediments. This can result in the loss of the bearing capacity of the underlying sediments and damage to the foundations and infrastructure. Therefore, assessing liquefaction hazards is an important aspect of disaster prevention and risk assessment in geologically unstable areas. The purpose of this study is to assess the liquefaction hazards of silt sediments by using the horizontal-to-vertical spectral ratio method. Single-station noise recording was carried out in the northern plain of the Yellow River Delta, and a new method was adopted to identify the fundamental frequency. The dynamic parameters of the silt, such as the fundamental frequency, amplification, and vulnerability index, were used as indicators to assess the liquefaction potential. The results show that the silty soils in different areas have different stable ranges of values of the fundamental frequency. Moreover, the distribution of the observations is in good agreement with the geological conditions in the area, which indicates the potential applicability and reliability of the new method for identifying fundamental frequency. The vulnerability index is inversely related to the fundamental frequency, with the southwestern part of the study area having a lower fundamental frequency and a higher vulnerability index, meaning a greater liquefaction risk compared to other areas. The horizontal-to-vertical spectral ratio method has great advantages in characterizing subsurface dynamic parameters and can be applied to liquefaction hazard assessments of silt sediments in large areas, which is critically important in terms of providing information and guidance for urban construction and planning.

Keywords: HVSR; liquefaction potential; fundamental frequency; amplification; vulnerability index

Citation: Meng, Q.; Li, Y.; Wang, W.; Chen, Y.; Wang, S. A Case Study Assessing the Liquefaction Hazards of Silt Sediments Based on the Horizontal-to-Vertical Spectral Ratio Method. *J. Mar. Sci. Eng.* **2023**, *11*, 104. https://doi.org/10.3390/jmse11010104

Academic Editor: Zhen-Gang Ji

Received: 10 November 2022
Revised: 24 December 2022
Accepted: 28 December 2022
Published: 4 January 2023

1. Introduction

Continuous vibrations, such as from earthquakes or strong storms, can cause loosely deposited sediments on or near the surface to undergo liquefaction. Liquefaction is a common and harmful form of unstable failure of soil and is a scientific issue of widespread interest among the domestic and international geotechnical and earthquake engineering communities. When liquefaction occurs, the stability of the soil skeletal is disrupted, and the soil particles take on a free state. This causes the soil to exhibit fluid-like properties, which loses its bearing capacity and results in damage to overlying structures. For example, the 1964 Niigata earthquake in Japan [1], the 1994 Northridge earthquake in California [2], the 1999 Chi-Chi earthquake in China [3], and the 2011 Christchurch earthquake in New Zealand [4] all caused severe damage to buildings, embankments, roads, and underground facilities due to soil liquefaction.

There are many factors that affect soil liquefaction, and real liquefaction events are rare and difficult to reproduce. As a result, liquefaction hazard assessments are critical for preventing these hazards. At present, assessing liquefaction potential primarily adopts laboratory tests and field tests [5]. The dynamic triaxial test [6] uses stress and strain of soil as the criteria to assess liquefaction. The standard penetration test (SPT) [7], the piezocone penetration test (PCPT) [8,9], and geophysical tests [10,11] are used to directly characterize the liquefaction potential of soils based on data from field investigations. Due to their safety and reliability, the above methods have been successfully applied to the study of sand liquefaction [12–15]. However, SPT and CPT are not suitable for large areas because of their complex implementation and high cost. Furthermore, laboratory tests cannot guarantee the structural characteristics of in situ soils; thus, they cannot truly reflect the dynamic response of sediments. Moreover, most studies have only focused on sandy soils, and no consensus has been reached on liquefaction assessments of fine-grained sediments containing silt, silty sand, and silty clay. Unlike sandy soils, silty soils are dominated by fine particles such as silt, fine sand, and small amounts of interspersed clay. With weaker permeability and poorer water stability, these sediments are prone to uneven settlement. Therefore, it is of great practical importance to assess the liquefaction potential of silty soil during local engineering and construction projects, and to ensure the safety of life and property.

Ambient noise is a kind of vibration caused by natural or cultural origins such as waves, tides, the effects of wind, industrial machinery, and so on. The horizontal-to-vertical spectral ratio (HVSR) method is based on ambient noise recordings and is a key technique proposed by Nakamura [16] to determine the dynamic response of soft sediment [17,18]. This method only requires a three-component geophone to quickly investigate the dynamic characteristics of undisturbed soil over a large area by measuring properties such as fundamental frequency (f_0) and amplification (A). Due to these advantages, HVSR has been rapidly developing in the fields of site effect evaluation [19,20], seismic microzonation [21,22], and evaluation of sedimentary cover thickness [23,24]. In the past, the frequency that corresponded to the peak amplitude of the HVSR curve was considered to be the fundamental frequency. However, the spectrum of the actual recording has multiple peaks that Nakamura [25,26] described as being caused by multiple reflections of the vertically incident SH waves. Moreover, the energy of the fundamental Rayleigh waves in the noise wavefield generates additional peak amplitudes on the HVSR curve [27], which interferes with the identification of the fundamental frequency. Therefore, it is necessary to find a more suitable method for identifying the fundamental frequency.

The purpose of this paper is to assess silt liquefaction hazards using the HVSR method. We established a method to identify the fundamental frequency based on the variable differences of the amplitude spectrum between the horizontal and vertical ambient noise components and conducted single-station noise recording tests at multiple sites in the Yellow River Delta plain. Then, the HVSR method was used to process the data to obtain the fundamental frequency and amplification for each station, and the vulnerability index (K_g) was calculated to estimate the potential liquefaction sites. The liquefaction hazards in the Yellow River Delta are evaluated based on geological information and the distribution of the fundamental frequency and amplification and the vulnerability index.

2. Methodology

The HVSR method is a passive source seismic technique that requires only a single three-component geophone to record ambient noise. Fourier spectrum analysis is then performed for each component of the ambient noise, converting the time-domain data to frequency-domain data. The method analyzes the spectral ratio of the Fourier amplitudes of the horizontal and vertical noise components, from which the HVSR curves and fundamental frequencies of the stations are then estimated [28].

$$\text{HVSR} = \frac{\sqrt{(H_{EW}(f)^2 + H_{EW}(f)^2)/2}}{H_V(f)} \tag{1}$$

where $H_{EW}(f)$ is the Fourier amplitude spectrum of the east–west component of the noise signal, $H_{NS}(f)$ is the Fourier amplitude spectrum of the north–south component and $H_V(f)$ is the Fourier amplitude spectrum of the vertical component. Fourier amplitude spectra can understand the frequency components of ambient noise and be used for hazard analysis. In addition, several ground motion parameters, such as fundamental frequency and amplification can be obtained from the Fourier amplitude spectrum.

The HVSR method can estimate the characteristics of ground motion by recording ambient noise at the surface. The HVSR curve is plotted based on the ratio of the Fourier amplitude spectra of the horizontal and vertical components of the ambient noise, which is usually related to the wave impedance. By determining the fundamental frequency of the site and the corresponding spectral amplitude, the amplification characteristics of the actual soil can be estimated. Based on this premise, Nakamura [29] proposed a vulnerability index, to estimate the risk of soil damage according to the HVSR results.

$$K_g = \frac{A^2}{f_0} \tag{2}$$

where K_g means the vulnerability indices of ground, and g means ground. A is the amplification and f_0 is the fundamental frequency. The vulnerability index (K_g) value is derived from the ground motion parameters and is related to the strain on the ground and the basement. Therefore, K_g can be used to characterize the easiness of deformation at the measurement point, which can be used to detect ground weak points.

3. Application in the Yellow River Delta

3.1. Geologic Setting

As shown in Figure 1, the study area is located in the northern part of the Yellow River Delta plain, which has an overall gentle terrain. The Yellow River Delta is a typical fast-depositing delta, with a characteristic dual sedimentary structure consisting of river alluvium covering the marine layer [30,31]. The overlying sediments were formed by rapid scouring and siltation driven by the hydrodynamic forces and coastal sedimentation processes. These sediments are mainly composed of silt and silty clay, and become thinner from southwest to northeast. Due to the high sand content, high water content, high liquidity index, and poor grading, the silt has low strength and a low bearing capacity. Additionally, during the continuous deposition of the deltaic strata, soft soils have been generally developed and are unevenly distributed. These conditions may lead to local geohazards in the soil when subjected to dynamic loadings, such as by uneven soil settlement or liquefaction.

3.2. Instruments and Field Investigation

In the field test, the equipment used for ambient noise recording during the field investigation included a WZG-6A engineering seismograph for data acquisition, a PS-2B three-component velocity geophone with a dominant frequency of 2 Hz, and a compass, as shown in Figure 2.

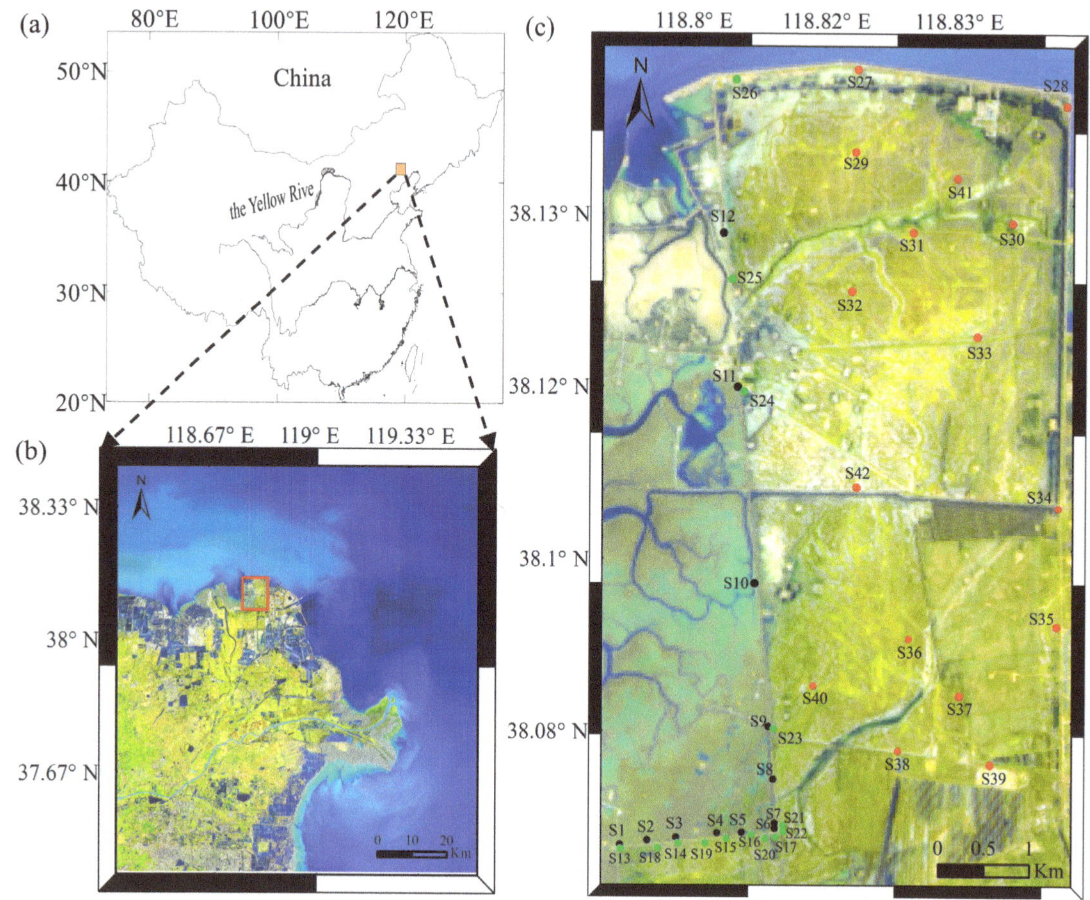

Figure 1. (**a**) Site of the Yellow River Delta (orange rectangle). (**b**) Map of the study area (red rectangle). (**c**) Distribution of measurement stations within the study area.

The study area is divided into three zones based on the location of the stations in Figure 1, of which S1–S12 are located in the beach area in front of the embankment (BFE), S13–S26 are located in the area behind the embankment (ABE), and S27–S42 are located in the wetland in the subaerial delta (WISD). The first step of the test was to determine the location of the observation station by GPS and the north direction by compass. The second step was to set up the three-component geophone. The geophones were installed directly on the cleared ground or in a hole (approximately 40 cm deep), oriented so that the X-axis pointed east and the Y-axis pointed north, and then leveled. The main purpose of installing the geophone in a hole was to enhance the coupling between the geophone and the soil and to minimize the effects of bad weather. Finally, the status of the device connection was checked, and the recording at each station had a duration of approximately 3 min, with a sampling frequency of 100 Hz. During acquisition, it is essential to avoid interference from transient strong signals caused by vehicles and footsteps, and so on, and to ensure that the geophone remains sufficiently coupled to the soil [32,33].

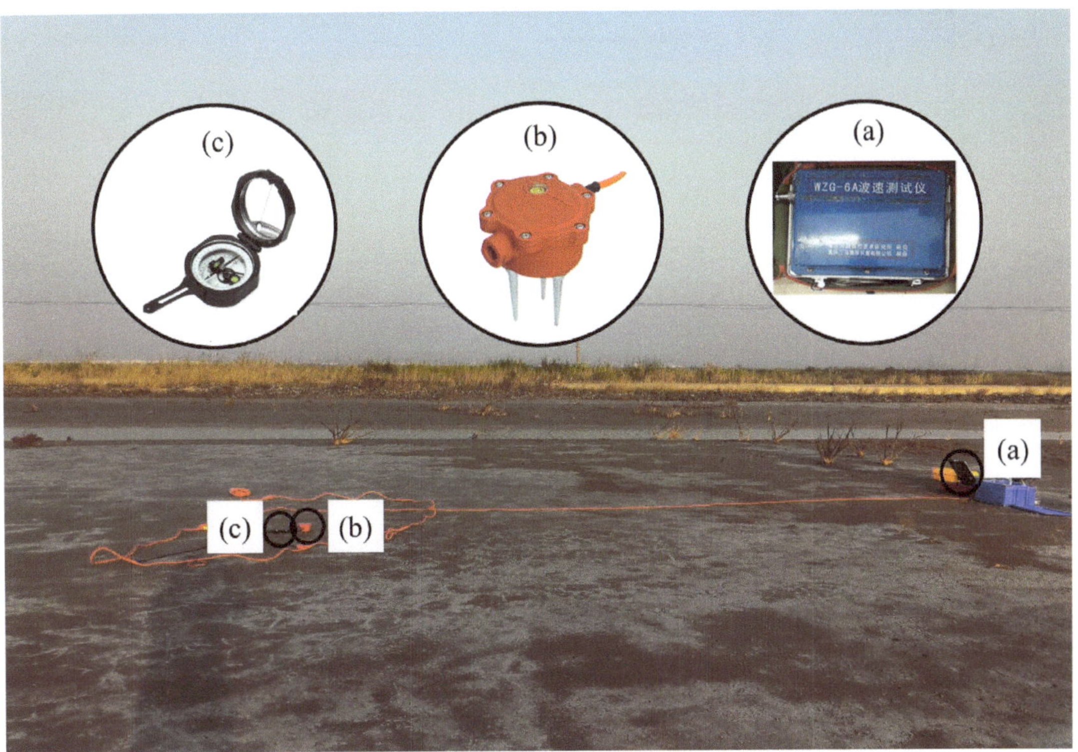

Figure 2. Pictures of the field test. (**a**) Data acquisition instrument. (**b**) Three-component geophone. (**c**) Compass.

3.3. Data Processing

In this study, we used the "OpenHVSR-Processing toolkit" computer program to process the recorded signals. This program is a new, interactive visualization tool for the HVSR processing of large noise datasets, which can highlight the spatial correlation of noise [34]. The calculation steps of the HVSR were shown in Figure 3. In the first data processing step, the continuously recorded three-component noise data are filtered and split into a stationary window of 10 s, with STA/LTA of 2.0 and overlap of 50%. This is to eliminate the interference of noise deviating from the normal distribution caused by transient impulse noise (e.g., vehicles passing near the geophone). In the second step, the Fourier amplitude spectrum for each time window is calculated and smoothed using the Konno and Ohmachi [35] method with a bandwidth coefficient (b) of 40, and the window's cosine tapering value of 5%. The subsequent step is to average the Fourier spectra of the two horizontal components and calculate the H/V of each window. Finally, the average HVSR of all windows is calculated.

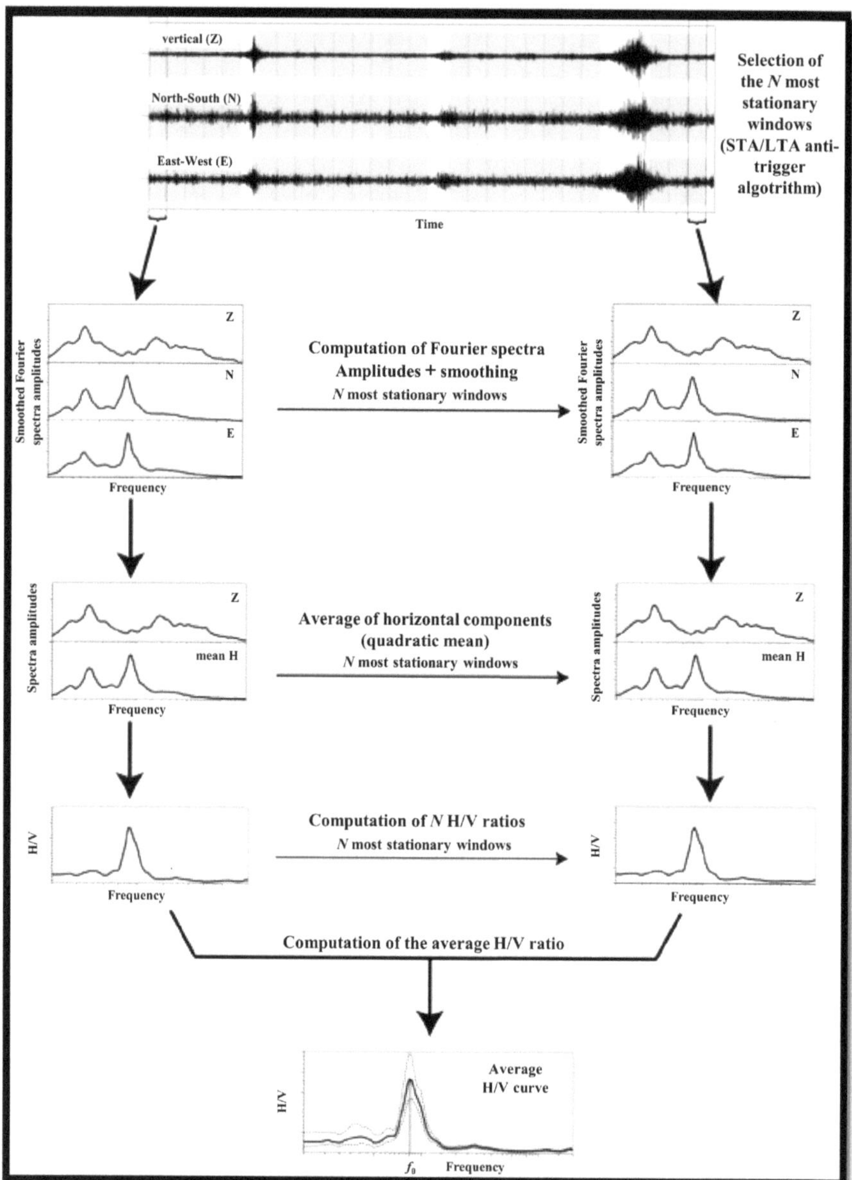

Figure 3. Flow chart shows the calculation steps of the HVSR. (http://www.geopsy.org/documentation/geopsy/hv.html) "URL (accessed on 10 November 2022)".

3.4. Identify Fundamental Frequency

It was shown that the fundamental frequency identified by HVSR coincides with the resonant frequency of the shear waves in the surface sediments, irrespective of the nature of the noise wavefield [36]; thus, it is a crucial step to correctly identify the fundamental frequency. Nakamura [25–27] argued that when seismic waves propagate through firm and homogeneous strata, the vibrations are uniform in each frequency range and each direction and do not amplify the amplitude. However, when seismic waves are vertically incident to the firm strata and propagate into the overlying loose sediment strata, the horizontal motion of the seismic waves is amplified but the vertical motion is barely amplified.

The amplification of seismic waves is related to factors such as density variations between strata, P-wave velocity, S-wave velocity, and layer thickness. The HVSR method is based on monitoring the amplification of shear waves on the surface. Therefore, I propose a new method to identify the fundamental frequency based on the difference in amplification of vertically incident S-waves. This method determines the fundamental frequency by comparing the frequency bands where the horizontal component signal is enhanced but the vertical component is not. When noise propagates in the ground, vibrations in the noise wavefield with the same frequency as the resonant band of the ground will cause a resonant response in the soil. The corresponding ground motion with the strongest energy of seismic waves can be detected. The amplitude of the horizontal component of seismic waves is amplified when the seismic waves are incident vertically from the hard strata into the soft strata. Therefore, the fundamental frequency is accurately identified by monitoring the horizontal and vertical components of ground motion at the surface and considering the energy distribution of the Fourier spectrum.

Figure 4 shows the processing results for the S1 stations. Figure 4a shows the record of the V, E–W, and N–S noise components after filtering and windowing. The white windows are those that were filtered or cleaned because they contained transient signals. It shows the tiled views of the spectral ratios for H/V, E/V, and N/V, respectively, in Figure 4b, and the tiled views of the Fourier spectra for each window of the three components in Figure 4c. As shown in Figure 4d, the mean HVSR and 95% confidence intervals for all windows are shown in the left panel, the comparison of the mean curves labeled H/V, E/V, and N/V are presented in the middle, and the smoothed amplitude spectral curves for the E, N, and V components are shown in the right panel. Finally, Figure 4e shows the spectral ratios for H/V, E/V, and N/V respectively. It can be seen from Figure 4, the amplitude spectrum of each component shows that the E–W and N–S components begin to increase from 1.0 Hz, while the vertical component barely increases in the 1.0–1.85 Hz range. Furthermore, according to the Fourier spectrum of the N–S component, the vertical component starts to increase at approximately 1.8 Hz, and the tiled views of the spectral ratio are at their maximum at that frequency. Therefore, 1.8 Hz was identified as the fundamental frequency of the soil at this station.

The results of the identification of fundamental frequency using the method of this paper and another method, as shown in Figure 5. The previous method finds the fundamental frequency by determining the peak amplitude, as the green line in Figure 5a. Nevertheless, there are obviously multiple peaks in the HVSR curve. Combined with the observation data and local geological conditions, the site is located in the beach area in front of the embankment, which could not have such a high value of fundamental frequency. The previous method may have some errors when applied in the actual complex surroundings. Thus, we use the method proposed in this paper for identification. As can be seen in Figure 5c, the vertical component of the noise is barely amplified in the range of 0.8–1.6 Hz, while the horizontal component of the noise is amplified. The HVSR curve has a peak at 1.4 Hz, as shown in the red line in Figure 5a. Moreover, the H/V, E/V, and N/V ratios have peaks here in Figure 5b. Therefore, we conclude that the value of the fundamental frequency at this station is 1.4 Hz.

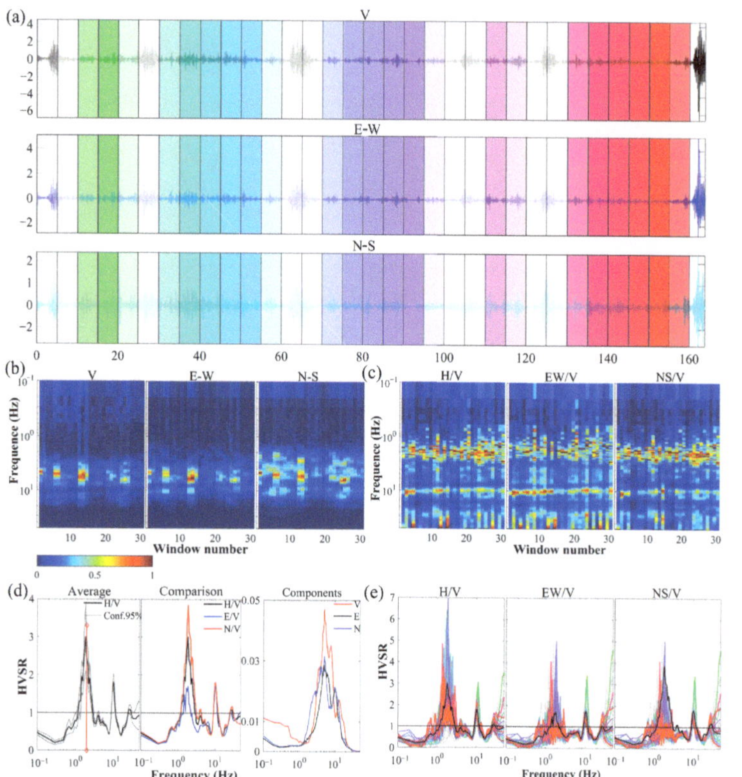

Figure 4. Processing result for S1. (**a**) Windowing. (**b**) Tiled view of Fourier spectra. (**c**) Tiled view of spectral ratios. (**d**) Average HVSR curve (left), H/V, E/V, and N/V ratios (center), the smoothed E, N, and V spectra (right). (**e**) HVSR curves for all windows, H/V, E/V, and N/V are shown from left to right.

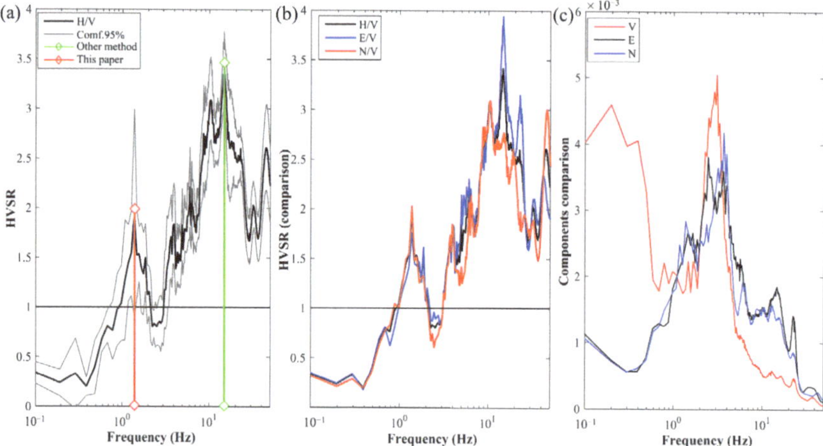

Figure 5. The identification of the fundamental frequency of S9 using the method of this paper and other method. (**a**) Average HVSR curve. (**b**) The H/V, E/V, and N/V ratios. (**c**) The smoothed E, N, and V spectra.

4. Discussion

To evaluate the liquefaction hazard in the Yellow River Delta, the ambient noise recordings from 42 stations were processed to calculate the HVSR, of which 31 results met the criteria for a reliable HVSR curve as defined by SESAME [37]. The results are shown in Table 1. The poor results from the other 11 stations may have been caused by poor coupling of the instrument to the soil or by severe weather conditions.

Table 1. The results of ambient noise recording from the Yellow River Delta.

ID	Longitude	Latitude	Altitude	f_0 (Hz)	A	K_g	Number of Stationary Window
S1	4,226,376	920,357	0.2	1.8	3.0	5.0	23
S2	4,226,434	920,654	−0.1	1.5	1.8	2.2	19
S3	4,226,474	920,970	0.0	3.7	2.1	1.2	20
S4	4,216,312	394,829	0.0	1.4	2.5	4.5	21
S5	4,226,555	921,692	0.0	1.5	2.1	2.9	22
S6	4,226,608	922,055	0.3	1.4	2.1	3.2	25
S7	4,216,401	395,467	0.0	1.3	2.0	3.1	32
S8	4,227,136	922,019	0.6	0.8	3.5	15.3	28
S9	4,227,697	921,944	0.3	1.4	1.9	2.6	23
S10	4,229,229	921,736	0.0	N/A	N/A	N/A	19
S11	4,231,335	921,477	0.3	N/A	N/A	N/A	8
S12	4,232,986	921,262	0.4	1.5	1.8	2.2	22
S13	4,226,335	920,364	1.2	N/A	N/A	N/A	20
S14	4,226,417	920,992	1.1	3.7	2.0	1.1	28
S15	4,226,485	921,521	1.3	3.8	2.8	2.1	26
S16	4,226,524	921,809	1.1	3.7	3.2	2.8	26
S17	4,226,518	922,066	1.0	3.7	2.5	1.7	24
S18	4,216,175	394,172	1.3	3.8	2.5	1.6	22
S19	4,216,212	394,689	1.2	3.8	2.3	1.4	28
S20	4,216,244	395,354	1.1	4.0	2.5	1.6	22
S21	4,216,373	395,557	1.2	3.7	3.0	2.4	23
S22	4,226,579	922,147	0.8	3.7	3.0	2.4	29
S23	4,227,677	922,001	0.9	1.4	3.0	6.4	23
S24	4,232,495	921,369	0.9	N/A	N/A	N/A	10
S25	4,234,639	921,337	0.6	3.3	1.9	1.1	27
S26	4,234,362	922,710	1.5	3.3	1.9	1.1	20
S27	4,231,288	921,546	1.6	6.3	2.5	1.0	20
S28	4,234,481	924,999	1.9	5.6	3.0	1.6	26
S29	4,233,906	922,691	1.9	4.0	2.2	1.2	28
S30	4,233,200	924,451	0.0	N/A	N/A	N/A	24
S31	4,233,063	923,358	0.9	3.6	2.1	1.2	27
S32	4,232,411	922,703	1.8	N/A	N/A	N/A	27
S33	4,231,969	924,106	1.7	N/A	N/A	N/A	31
S34	4,230,155	925,059	0.8	5.4	3.2	1.9	23
S35	4,228,883	925,090	1.0	N/A	N/A	N/A	7
S36	4,228,696	923,460	1.6	N/A	N/A	N/A	9
S37	4,228,105	924,045	2.1	N/A	N/A	N/A	11
S38	4,227,490	923,386	1.9	4.0	3.3	2.7	28
S39	4,227,375	924,408	1.6	N/A	N/A	N/A	29
S40	4,228,228	922,430	1.0	4.0	2.5	1.6	30
S41	4,233,658	923,826	0.3	9.8	2.4	0.6	26
S42	4,219,971	396,474	1.3	6.1	3.5	2.0	29

Note: N/A is unavailable data.

4.1. Fundamental Frequency

It has commonly been assumed that the distribution of f_0 is closely related to geological conditions and depends mainly on shear wave velocity (V_s) and sediment thickness [38,39], which is related to the average vs. and inversely related to the sediment thickness. Figure 6 shows the distribution of the fundamental frequency of all stations, with f_0 being distributed between 0.8 and 9.8 Hz and gradually increasing from southwest to northeast. The sediment thickness in the area gradually becomes thinner from southwest to northeast, and the distribution of fundamental frequencies is inverse to the thickness layer. This indicates that the fundamental frequency values identified by the new method are consistent with the geological conditions, which indicates the potential applicability and reliability of the new method.

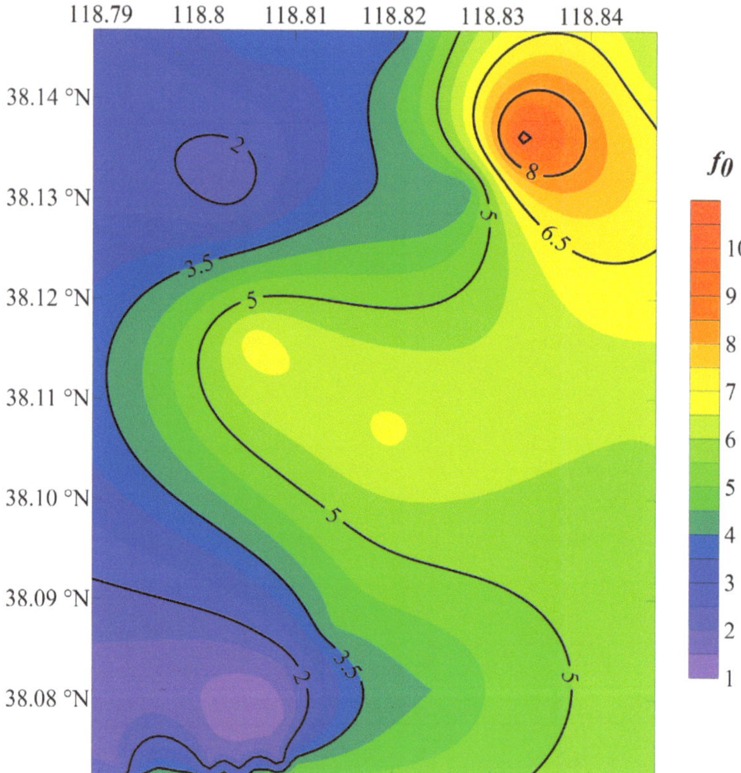

Figure 6. Distribution map of the fundamental frequency.

According to Figure 7, the stations in the BFE, ABE, and WISD areas have a stable distribution of the fundamental frequency values. It can be seen that the f_0 values of stations in BFE are lower than the others and basically stay between 1.3 and 1.8 Hz, while S3 has a higher f_0 of 3.7 Hz. This inconsistency may be related to the fact that this station is located on the back side of the seaward extending road, where the soils are less affected by wave-driven erosion and are more naturally consolidated than in other places. The f_0 values of stations in ABE are basically centered around 3.8 Hz, while the f_0 value at S23 is anomalous at 1.4 Hz. It is speculated that this f_0 value is low because this station had been inundated with water before the test, resulting in loose soils and shallow groundwater levels; thus, the wave velocity and amplification are weakened when the S-wave reaches the surface. The distribution of f_0 values in WISD ranges from 3.6 to 9.8 Hz, which ranges from middle to high levels.

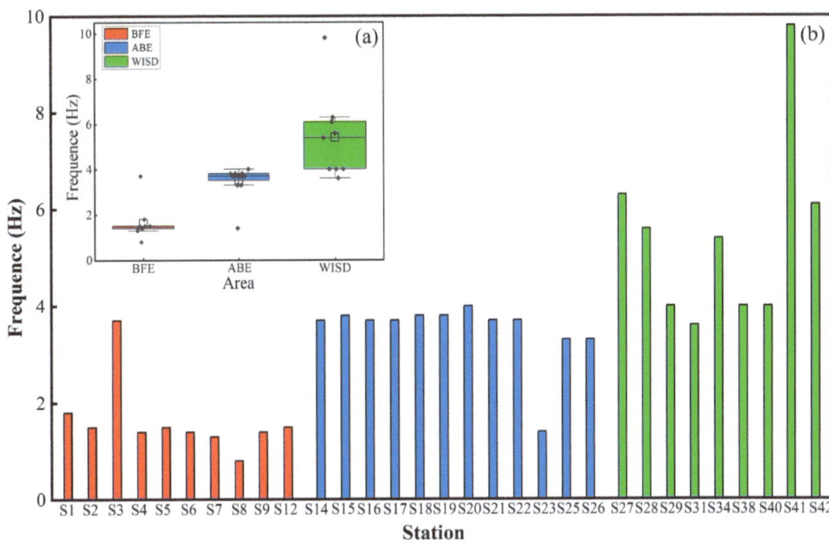

Figure 7. Fundamental frequency statistics for each station. (**a**) The distribution of fundamental frequency for stations in different study areas. (**b**) The fundamental frequency for all stations.

According to geological data, the study area is a flood plain established by the successive oscillations, migrations, and alluvial deposits of the Yellow River. Due to the effects of coastal currents and waves, thick deposits of loose sediments are widely distributed within the study area, with a gradual decrease in thickness from southwest to northeast. In addition, the stability of the embankment is undermined by the scouring and erosive actions of the waves and tides. These areas are more susceptible to ground motion amplification due to energy aggregation and the multiple reflections of seismic waves when an earthquake occurs, which will potentially cause substantial damage and possibly even liquefaction. The terrain behind the embankment slopes gently contains shallow groundwater, and the soil is basically saturated due to the poor drainage of groundwater and surface water. The stability of the soil behind the embankment is also low due to the frequent events of wave overtopping and scouring. The f_0 values in BFE and ABE are lower compared to those in WISD, or more precisely, to the soil on both sides of the embankment. Therefore, it is necessary to take precautionary measures regarding foundation treatments to prevent the embankment from being damaged by dynamic loads caused by strong storms or earthquakes.

4.2. Amplification

The peak amplitude, *A*, gives a rough indication of the difference in density between the surface sediments and the bedrock. The amplification of the seismic wave could occur due to significant differences in density between layers. In other words, when a seismic wave propagates from a high-density medium to a low-density medium, the amplitude of the seismic wave will increase. As shown in Figure 8, the amplification values are in the range of 1.8–3.5, which is in the medium to high amplification range. This indicates that the amplitude of the seismic wave will increase when it propagates through the overlying loose sediments in the area. From the HVSR curves of all stations, the peak amplitudes of the HVSR are mainly concentrated around three frequency bands, which are consistent with the distribution of fundamental frequencies in the BFE, ABE, and WISD regions. Additionally, "False peaks" are found in some curves in the lower frequency bands (0.2–0.4 Hz), which were mainly due to the disturbance caused by the turbulence of the wind near the geophone during observation. It has been reported that buildings can be damaged when amplification exceeds a factor of 3 [28], and a total of 10 measurement stations recorded this amplification. Therefore, based on the distribution of the amplification, not only are both sides of the embankment potentially hazardous areas but there are also areas of vulnerability within the wetland.

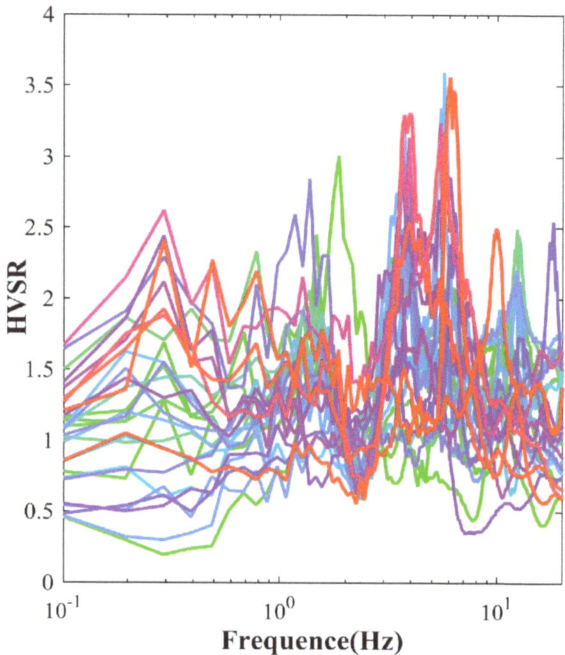

Figure 8. HVSR curves for all stations. Different colors represent different HVSR curves for different stations.

In addition, the multiple peaks in the HVSR curves also reflect the presence of multiple density interfaces in the stratum, indicating different impedance ratios at different depths [40,41]. These facts are consistent with the binary structure of the sediments in the Upper Delta Plain and the intercalated unevenly distributed soft soil layers. The different shapes of the HVSR curve reflect the lateral heterogeneity of the formation [42].

4.3. Vulnerability Index

The vulnerability index value can be considered an indicator of the degree of deformation of the soil [43], and it will help to detect weak points of the ground by comparing the K_g values at different stations. Figure 9 shows that the distribution of K_g ranges from 0.6 to 15.3, gradually decreasing from southwest to northeast, which is closely related to the local geological conditions. The K_g values on both sides of the embankment are in the range of 1.1–15.3, while the K_g values of the stations located on the wetlands are in the range of 0.6–2.7. Comparing Figures 6 and 9, it can be seen that the distribution of K_g strongly coincides with f_0, for which a high vulnerability index is usually found in the areas with low fundamental frequencies. The stations in BFE have a higher vulnerability index due to the damage caused by scouring and hollowing or the impacts of siltation on the stability of the soil.

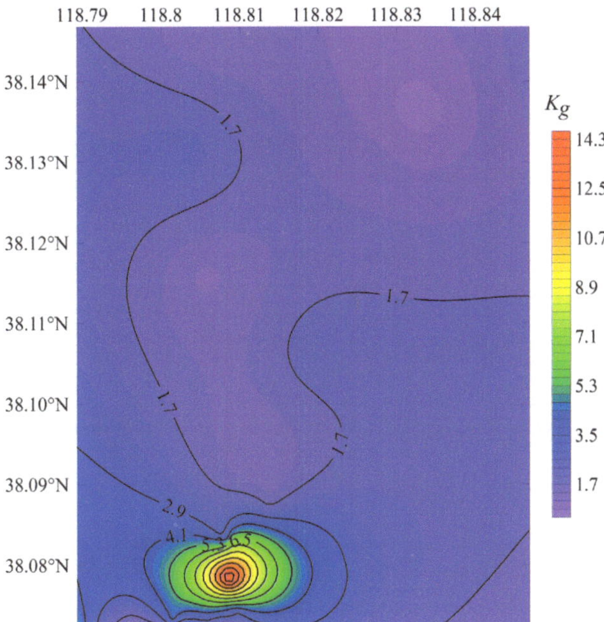

Figure 9. Distribution map of the vulnerability index.

The K_g value is a qualitative description [44]. The higher the K_g value is, the higher the liquefaction potential [45]. In the southwestern part of the study area, the stations on the beach area in front of the corner of the two embankments have the largest K_g values, which are the weakest areas. It is more likely that the amplitude of the seismic motion will be amplified here due to the aggregation and multiple reflections of the energy of seismic waves; therefore, unstable soil is more likely to fail, and the liquefaction hazard is the highest. Therefore, it is necessary to adopt methods for treating foundations, such as replacement method and dynamic consolidation, to prevent the loose soil in front of the embankment from undergoing liquefaction or failures driven by uneven settlement during strong storms or earthquakes, which would result in damage to the embankment.

5. Conclusions

This work presents a case study evaluating the liquefaction potential of silt sediments based on the HVSR method. The ambient noise was recorded at multiple stations in the Yellow River Delta, and a new method was used to identify the fundamental frequency and amplification of the silt sediment. The results showed that f_0 was in the range of 0.8–9.8 Hz and A was in the range of 1.8–3.5. According to the fundamental frequency and amplification, the vulnerability index was calculated to evaluate the liquefaction potential of the silt foundation. The following conclusions were drawn from this analysis:

1. This study presents a new method to identify the fundamental frequency from ambient noise recording in the Yellow River Delta. The results show that the soils in the areas with different geological conditions have a stable range of the fundamental frequency respectively, and the results are consistent with the inverse relationship between the fundamental frequency and the thickness of the sediment layer. Thus, the new method is potentially applicable and reliable for identifying the fundamental frequency;

2. The vulnerability indexes of the Yellow River Delta range from 0.6 to 15.3 and are dependent on the variability of the local geological conditions. In the southwestern part of the study area, the maximum value of K_g occurs on the beach in front of the corner of the two embankments, where the liquefaction hazard is greatest; therefore,

the construction and maintenance of buildings should be carefully considered here. However, we suggest that combining the HVSR method with geotechnical investigation methods, such as field tests or indoor geotechnical tests, can further investigate the extent of liquefiable layers, which is essential for a more reliable assessment of the liquefaction potential of silty soils;

3. The results show that the HVSR method, when based on single station noise recordings, is suitable for rapidly assessing liquefaction potential. Compared with other field or laboratory tests, the HVSR method is a convenient, practical, faster, and non-destructive tool for assessing the liquefaction potential of silty sedimentary, such as the Mississippi River Delta. This will help to carry out liquefaction hazard assessment of silty soil in a new perspective, which can provide services for urban disaster prevention and mitigation and can also provide a reference basis for establishing regional seismic codes with good practical application value.

Author Contributions: Conceptualization and methodology, Q.M., Y.L., and W.W.; formal analysis, S.W. and W.W.; data curation, Y.L. and Y.C.; writing—original draft preparation, Y.L. and Q.M.; writing—review and editing, Y.L. and Q.M.; visualization, Y.C. and S.W.; project administration, Q.M.; funding acquisition, Q.M. All authors have read and agreed to the published version of the manuscript.

Funding: This research was funded by The National Natural Science Foundation of China (grant number 42272327) and the Social and livelihood project of Shandong Province (grant number 2021, 202131001).

Institutional Review Board Statement: Not applicable.

Informed Consent Statement: Not applicable.

Data Availability Statement: The data presented in this study are available on request from the corresponding author.

Acknowledgments: The authors would like to thank Zhiyuan Chen who participated in recording the ambient noise in the field and Yupeng Ren who participated in revising the manuscript.

Conflicts of Interest: The authors declare no conflict of interest.

References

1. Seed, H.B.; Idriss, I.M. Analysis of Soil Liquefaction: Niigata Earthquake. *J. Soil Mech. Found. Div.* **1967**, *93*, 83–108. [CrossRef]
2. Holzer, T.L.; Bennett, M.J.; Ponti, D.J.; Tinsley, J.C., III. Liquefaction and Soil Failure During 1994 Northridge Earthquake. *J. Geotech. Geoenviron. Eng.* **1999**, *125*, 438–452. [CrossRef]
3. Yuan, H.; Hui Yang, S.; Andrus, R.D.; Hsein Juang, C. Liquefaction-Induced Ground Failure: A Study of the Chi-Chi Earthquake Cases. *Eng. Geol.* **2004**, *71*, 141–155. [CrossRef]
4. Chen, L.; Yuan, X.; Sun, R. Review of Liquefaction Phenomena and Geotechnical Damage in the 2011 New Zealand Mw6.3 Earthquake. *World Earthq. Eng.* **2013**, *29*, 1–9. (In Chinese) [CrossRef]
5. Youd, T.L.; Idriss, I.M. Liquefaction Resistance of Soils: Summary Report from the 1996 NCEER and 1998 NCEER/NSF Workshops on Evaluation of Liquefaction Resistance of Soils. *J. Geotech. Geoenviron. Eng.* **2001**, *127*, 297–313. [CrossRef]
6. Kumar, S.S.; Krishna, A.M.; Dey, A. Assessment of Dynamic Response of Cohesionless Soil Using Strain-Controlled and Stress-Controlled Cyclic Triaxial Tests. *Geotech. Geol. Eng.* **2020**, *38*, 1431–1450. [CrossRef]
7. Seed, H.B.; Idriss, I.M.; Arango, I. Evaluation of Liquefaction Potential Using Field Performance Data. *J. Geotech. Eng.* **1983**, *109*, 458–482. [CrossRef]
8. Boumpoulis, V.; Depountis, N.; Pelekis, P.; Sabatakakis, N. SPT and CPT Application for Liquefaction Evaluation in Greece. *Arab. J. Geosci.* **2021**, *14*, 1631. [CrossRef]
9. Geyin, M.; Maurer, B.W. Evaluation of a Cone Penetration Test Thin-Layer Correction Procedure in the Context of Global Liquefaction Model Performance. *Eng. Geol.* **2021**, *291*, 106221. [CrossRef]
10. Qin, L.; Ben-Zion, Y.; Bonilla, L.F.; Steidl, J.H. Imaging and Monitoring Temporal Changes of Shallow Seismic Velocities at the Garner Valley Near Anza, California, Following the M7.2 2010 El Mayor-Cucapah Earthquake. *J. Geophys. Res. Solid Earth* **2020**, *125*, e2019JB018070. [CrossRef]
11. Lin, A.; Wotherspoon, L.; Bradley, B.; Motha, J. Evaluation and Modification of Geospatial Liquefaction Models Using Land Damage Observational Data from the 2010–2011 Canterbury Earthquake Sequence. *Eng. Geol.* **2021**, *287*, 106099. [CrossRef]
12. Amini, F.; Qi, G.Z. Liquefaction Testing of Stratified Silty Sands. *J. Geotech. Geoenviron. Eng.* **2000**, *126*, 208–217. [CrossRef]

13. Hatzor, Y.H.; Gvirtzman, H.; Wainshtein, I.; Orian, I. Induced Liquefaction Experiment in Relatively Dense, Clay-Rich Sand Deposits. *J. Geophys. Res. Solid Earth* **2009**, *114*, B02311. [CrossRef]

14. Dobry, R.; Abdoun, T. Recent Findings on Liquefaction Triggering in Clean and Silty Sands during Earthquakes. *J. Geotech. Geoenviron. Eng.* **2017**, *143*, 04017077. [CrossRef]

15. Ye, B.; Hu, H.; Bao, X.; Lu, P. Reliquefaction Behavior of Sand and Its Mesoscopic Mechanism. *Soil Dyn. Earthq. Eng.* **2018**, *114*, 12–21. [CrossRef]

16. Nakamura, Y. A method for Dynamic Characteristics Estimation of Subsurface Using Microtremor on the Ground Surface. *Railw. Tech. Res. Inst. Q. Rep.* **1989**, *30*, 25–33.

17. Lunedei, E.; Albarello, D. Theoretical HVSR Curves from Full Wavefield Modelling of Ambient Vibrations in a Weakly Dissipative Layered Earth. *Geophys. J. Int.* **2010**, *181*, 1093–1108. [CrossRef]

18. Sánchez-Sesma, F.J.; Rodríguez, M.; Iturrarán-Viveros, U.; Luzón, F.; Campillo, M.; Margerin, L.; García-Jerez, A.; Suarez, M.; Santoyo, M.A.; Rodríguez-Castellanos, A. A Theory for Microtremor H/V Spectral Ratio: Application for a Layered Medium. *Geophys. J. Int.* **2011**, *186*, 221–225. [CrossRef]

19. Bonnefoy-Claudet, S.; Cotton, F.; Bard, P.Y. The Nature of Noise Wavefield and Its Applications for Site Effects Studies: A Literature Review. *Earth-Sci. Rev.* **2006**, *79*, 205–227. [CrossRef]

20. Maghami, S.; Sohrabi-Bidar, A.; Bignardi, S.; Zarean, A.; Kamalian, M. Extracting the shear wave velocity structure of deep alluviums of "Qom" Basin (Iran) employing HVSR inversion of microtremor recordings. *J. Appl. Geophys.* **2021**, *185*, 104246. [CrossRef]

21. Panou, A.A.; Theodulidis, N.; Hatzidimitriou, P.; Stylianidis, K.; Papazachos, C.B. Ambient Noise Horizontal-to-Vertical Spectral Ratio in Site Effects Estimation and Correlation with Seismic Damage Distribution in Urban Environment: The Case of the City of Thessaloniki (Northern Greece). *Soil Dyn. Earthq. Eng.* **2005**, *25*, 261–274. [CrossRef]

22. Leyton, F.; Ruiz, S.; Sepúlveda, S.A.; Contreras, J.P.; Rebolledo, S.; Astroza, M. Microtremors' HVSR and Its Correlation with Surface Geology and Damage Observed after the 2010 Maule Earthquake (Mw 8.8) at Talca and Curicó, Central Chile. *Eng. Geol.* **2013**, *161*, 26–33. [CrossRef]

23. Ullah, I.; Prado, R.L. Soft Sediment Thickness and Shear-Wave Velocity Estimation from the H/V Technique up to the Bedrock at Meteorite Impact Crater Site, Sao Paulo City, Brazil. *Soil Dyn. Earthq. Eng.* **2017**, *94*, 215–222. [CrossRef]

24. Moon, S.W.; Subramaniam, P.; Zhang, Y.; Vinoth, G.; Ku, T. Bedrock Depth Evaluation Using Microtremor Measurement: Empirical Guidelines at Weathered Granite Formation in Singapore. *J. Appl. Geophys.* **2019**, *171*, 103866. [CrossRef]

25. Nakamura, Y. Basic Structure of QTS (HVSR) and Examples of Applications. In Proceedings of the Increasing Seismic Safety by Combining Engineering Technologies and Seismological Data, Dubrovnik, Croatia, 19–21 September 2007; pp. 33–51.

26. Nakamura, Y. What Is the Nakamura Method? *Seismol. Res. Lett.* **2019**, *90*, 1437–1443. [CrossRef]

27. Nakamura, Y. On the H/V Spectrum. In Proceedings of the 14th Word Conference on the Earthquake Engineering, Beijing, China, 12–17 October 2008; pp. 12–17.

28. Nakamura, Y. Clear Identification of Fundamental Idea of Nakamura's Technique and Its Applications. In Proceedings of the 12th World Conference on Earthquake Engineering, Auckland, New Zealand, 30 January–4 February 2000; pp. 1–8.

29. Nakamura, Y. Seismic Vulnerability Indices for Ground and Structures Using Microtremor. In Proceedings of the World Congress on Railway Research, Florence, Italy, 16–19 November 1997; pp. 1–7.

30. Zhao, G.; Ye, Q.; Ye, S.; Ding, X.; Yuan, H.; Wang, J. Holocene Stratigraphy and Paleoenvironmental Evolution of the Northern Yellow River Delta. *Mar. Geol. Quat. Geol.* **2014**, *34*, 25–32. (In Chinese)

31. Wang, K.; Li, N.; Wang, W. Characteristics of Coastline Change and Multiyear Evolution of the Yellow River Delta. *J. Appl. Oceanogr.* **2018**, *37*, 330–338. (In Chinese)

32. Chatelain, J.L.; Guillier, B.; Cara, F.; Duval, A.M.; Atakan, K.; Bard, P.Y.; The WP02 SESAME Team. Evaluation of the influence of experimental conditions on H/V results from ambient noise recordings. *Bull. Earthq. Eng.* **2008**, *6*, 33–74. [CrossRef]

33. Guillier, B.; Atakan, K.; Chatelain, J.L.; Havskov, J.; Ohrnberger, M.; Cara, F.; Duval, A.M.; Zacharopoulos, S.; Teves-Costa, P.; The SESAME Team. Influence of Instruments on the H/V Spectral Ratios of Ambient Vibrations. *Bull. Earthq. Eng.* **2008**, *6*, 3–31. [CrossRef]

34. Bignardi, S.; Yezzi, A.J.; Fiussello, S.; Comelli, A. OpenHVSR—Processing Toolkit: Enhanced HVSR Processing of Distributed Microtremor Measurements and Spatial Variation of Their Informative Content. *Comput. Geosci.* **2018**, *120*, 10–20. [CrossRef]

35. Konno, K.; Ohmachi, T. Ground-Motion Characteristics Estimated from Spectral Ratio between Horizontal and Vertical Components of Microtremor. *Bull. Seismol. Soc. Am.* **1998**, *88*, 228–241. [CrossRef]

36. Field, E.; Jacob, K. The Theoretical Response of Sedimentary Layers to Ambient Seismic Noise. *Geophys. Res. Lett.* **1993**, *20*, 2925–2928. [CrossRef]

37. Bard, P.Y.; Acerra, C.; Aguacil, G.; Anastasiadis, A.; Atakan, K.; Azzara, R.; Basili, R.; Bertrand, E.; Bettig, B.; Blarel, F.; et al. Guidelines for the Implementation of the H/V Spectral Ratio Technique on Ambient Vibrations Measurements, Processing and Interpretation. *Bull. Earthq. Eng.* **2008**, *6*, 1–2. [CrossRef]

38. Parolai, S.; Bormann, P.; Milkereit, C. New Relationships between Vs, Thickness of Sediments, and Resonance Frequency Calculated by the H/V Ratio of Seismic Noise for the Cologne Area (Germany). *Bull. Seismol. Soc. Am.* **2002**, *92*, 2521–2527. [CrossRef]

39. Gosar, A.; Rošer, J.; Šket Motnikar, B.; Zupančič, P. Microtremor Study of Site Effects and Soil-Structure Resonance in the City of Ljubljana (Central Slovenia). *Bull. Earthq. Eng.* **2010**, *8*, 571–592. [CrossRef]
40. Guéguen, P.; Chatelain, J.L.; Guillier, B.; Yepes, H.; Egred, J. Site Effect and Damage Distribution in Pujili (Ecuador) after the 28 March 1996 Earthquake. *Soil Dyn. Earthq. Eng.* **1998**, *17*, 329–334. [CrossRef]
41. D'Amico, V.; Picozzi, M.; Baliva, F.; Albarello, D. Ambient Noise Measurements for Preliminary Site-Effects Characterization in the Urban Area of Florence, Italy. *Bull. Seismol. Soc. Am.* **2008**, *98*, 1373–1388. [CrossRef]
42. Matsushima, S.; Hirokawa, T.; De Martin, F.; Kawase, H.; Sánchez-Sesma, F.J. The Effect of Lateral Heterogeneity on Horizontal-to-Vertical Spectral Ratio of Microtremors Inferred from Observation and Synthetics. *Bull. Seismol. Soc. Am.* **2014**, *104*, 381–393. [CrossRef]
43. Khan, M.Y.; Turab, S.A.; Riaz, M.S.; Atekwana, E.A.; Muhammad, S.; Butt, N.A.; Abbas, S.M.; Zafar, W.A.; Ohenhen, L.O. Investigation of Coseismic Liquefaction-Induced Ground Deformation Associated with the 2019 Mw 5.8 Mirpur, Pakistan, Earthquake Using near-Surface Electrical Resistivity Tomography and Geological Data. *Near Surf. Geophys.* **2021**, *19*, 169–182. [CrossRef]
44. Beroya, M.A.A.; Aydin, A.; Tiglao, R.; Lasala, M. Use of Microtremor in Liquefaction Hazard Mapping. *Eng. Geol.* **2009**, *107*, 140–153. [CrossRef]
45. Akkaya, İ. Availability of Seismic Vulnerability Index (Kg) in the Assessment of Building Damage in Van, Eastern Turkey. *Earthq. Eng. Eng. Vib.* **2020**, *19*, 189–204. [CrossRef]

Journal of
Marine Science and Engineering

Article

Stability Characteristics of Horizontal Wells in the Exploitation of Hydrate-Bearing Clayey-Silt Sediments

Xiaofeng Sun [1], Qiaobo Hu [1,2,3,*], Yanlong Li [2,3,*], Mingtao Chen [2,3,4] and Yajuan Zhang [2,3]

[1] Sanya Offshore Oil and Gas Research Institute, Northeast Petroleum University, Sanya 572025, China
[2] Key Laboratory of Gas Hydrate, Ministry of Natural Resources, Qingdao Institute of Marine Geology, Qingdao 266237, China
[3] Laboratory for Marine Mineral Resources, Pilot National Laboratory for Marine Science and Technology, Qingdao 266237, China
[4] College of Oceanography, Hohai University, Nanjing 210098, China
* Correspondence: hqbnepu@stu.nepu.cn (Q.H.); ylli@qnlm.ac (Y.L.)

Abstract: The mechanical properties of hydrate-bearing strata in clayey-silt sediments are significantly different from those of either conventional reservoirs or hydrate-bearing sandy sediments, which poses great challenges for wellbore stability analyses. The stability characteristics of a deviated borehole during drilling in hydrate-bearing clayey-silt sediments (HBS-CS) remain to be studied. In this paper, an analysis of the wellbore stability characteristics of a deviated borehole using the Mohr–Coulomb (M-C) criterion and Drucker–Prager (D-P) criterion was carried out based on the elastic stress distribution model of the surrounding strata of the wellbore and the triaxial shear tests of the HBS-CS. The results imply that the collapse pressure and safety density window are symmetrically distributed with deviation angle and azimuth. Considering the effect of hydrate decomposition, the collapse pressure gradient could become higher and the instability risks would be amplified. Considering the combined effects of collapse, fracture pressure gradient, and the safety density window, it is suggested that the borehole be arranged along an azimuth of 60–120°, which could greatly reduce the risk in a drilling operation.

Keywords: natural gas hydrate; clayey-silt; deviated borehole; stability characteristics; hydrate saturation; collapse pressure; fracture pressure; safety density window

Citation: Sun, X.; Hu, Q.; Li, Y.; Chen, M.; Zhang, Y. Stability Characteristics of Horizontal Wells in the Exploitation of Hydrate-Bearing Clayey-Silt Sediments. *J. Mar. Sci. Eng.* **2022**, *10*, 1935. https://doi.org/10.3390/jmse10121935

Academic Editors: Xiaolei Liu, Thorsten Stoesser and Xingsen Guo

Received: 17 November 2022
Accepted: 2 December 2022
Published: 7 December 2022

Publisher's Note: MDPI stays neutral with regard to jurisdictional claims in published maps and institutional affiliations.

1. Introduction

Natural gas hydrate (NGH) is a kind of low-carbon fossil energy, and vigorously developing NGH is of great significance [1–3]. At present, more than 30 countries and/or regions have been actively involved in the development of NGH. In 1976, Russia began to exploit NGH from the permafrost of the Messoyakha [4,5]. In 2002, Japex jointly carried out the NGH pilot production in the Mackenzie delta of Canada in 2002 and 2007–2008 [6,7]. In 2008, the United States successfully exploited NGH on the northern slopes of Alaska by using CO_2 replacement [8,9]. In 2013, Japan completed the world's first offshore NGH development in the Nankai Trough [10,11]. In 2017 [12] and 2020 [13], China completed two rounds of NGH production trials from the clayey-silt sediment in the northern South China Sea.

To improve the gas productivity from HBS-CS, China's second marine NGH production trial was carried out via a horizontal wellbore. In the process of wellbore construction, the whipstock section passes through the overlaying strata and extends horizontally in the NGH-bearing layer. One of the most crucial challenges during horizontal well drilling is to control the stability of the borehole [14–16], which is highly coupled with fluid circulation and fluid losses [17]. For the study of NGH production wellbore stability, Freij-Ayoub et al. (2016) established a wellbore stability analytic model for NGH production [18]. Zhang et al. (2018) analyzed the effect of temperature on the stability of the NGH production borehole [19]. Sun et al. (2018) analyzed the effects of drilling fluid, and the initial reservoir conditions based on the geological backgrounds of the first exploration in the Shenhu area, northern South China Sea [20]. Yuan et al. (2020) simulated the wellbore stability of a vertical wellbore during depressurization [21]. These studies mainly focus on the wellbore stability of vertical or horizontal wellbores. However, the stability of the wellbore of highly deviated sections was rarely reported. Additionally, the M-C criterion was mostly adopted in the failure evaluation, but the influence of intermediate principal stress was ignored, which might affect the calculation accuracy of wellbore stability.

In this paper, the stability characteristics of an inclined wellbore during drilling are analyzed. The failure criteria of the M-C and the D-P are deployed, respectively. Taking the collapse pressure, fracture pressure, and the safety density window as the main evaluation indices, the effects of hydrate saturation (S_h), the deviation angle, and the azimuth on the stability of the borehole are clarified. The results may have some significance for drilling design in the clayey-silt HBS in the northern South China Sea.

2. Mechanical Parameters of HBS-CS

The basic mechanical parameters of the HBS-CS were obtained from triaxial shear tests. A detailed description of the experimental devices can be found in our previous publication [22,23]. The steps of the triaxial shear tests are as follows: (1) HBS-CS skeleton prefabrication; (2) triaxial testing system installation; (3) apply confining pressure; and (4) hydrate formation, which has real-time monitoring of the hydrate saturation by time domain reflectometry, and finally the triaxial shear tests.

The sediment in the Shenhu area, northern South China Sea is mainly composed of argillaceous siltstone with poor cementation strength and a high content of silt and clay [23,24]. The particle size distribution of the sediment is shown in Figure 1, in which the medium grain size is about 7 μm. The actual S_h range is between 20% and 60% [14], and hence, the predetermined S_h is 15%, 30%, 45%, and 60% in the experiments. The effective confining pressure was set as 1 MPa, 2 MPa, and 4 MPa in the triaxial shear tests, respectively.

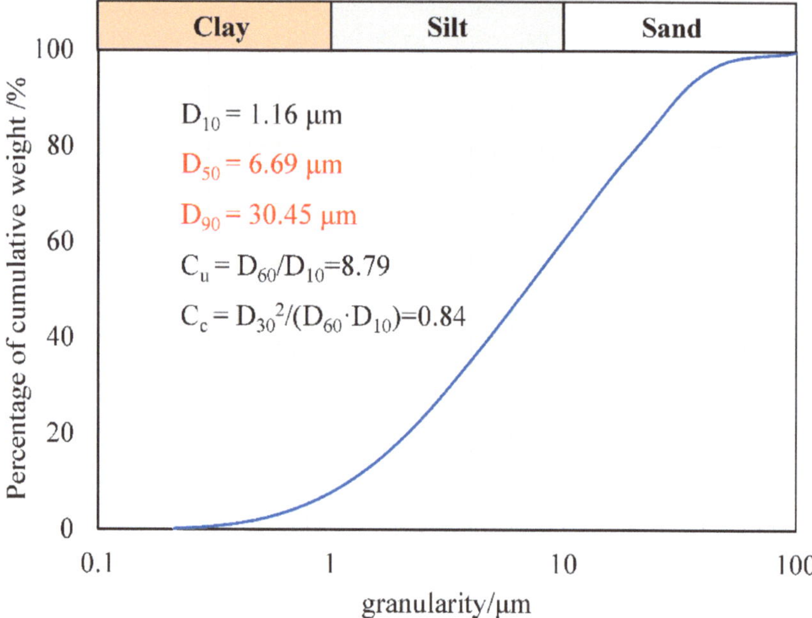

Figure 1. Particle size distribution of the filler.

The basic mechanical parameters of HBS-CS were obtained, as shown in Table 1. It could be seen that the cohesion and the internal friction angle of the HBS-CS are much lower than that of the conventional natural gas reservoirs (Li et al., 2020), and the cohesion and friction angle increased with the increase in S_h. For a detailed analysis of mechanical properties, refer to the literature [25,26].

Table 1. Mechanical parameters of the hydrate.

S_h	σ_3/MPa	σ_1/MPa	Cohesion C/MPa	Friction Angle ϕ/°
15%	1	1.62	0.0953	10.340
	2	3.17		
	4	5.95		
30%	1	2.48	0.426	11.612
	2	4.16		
	4	7.03		
45%	1	3.54	0.811	12.275
	2	5.12		
	4	8.17		
60%	1	4.29	1.071	13.568
	2	6.00		
	4	9.15		

3. Model Description

3.1. Stress Distribution around the Wellbore

Currently, scholars have mainly investigated borehole stability in NGH-bearing sediments by referring to similar problems in conventional formations. NGH-bearing sediments are always regarded as elastic or elastic–plastic materials when doing borehole stability analysis. In the analysis of collapse pressure, the elastic model was often used [18,20,27,28]. Therefore, in this study, the HBS-CS formation is assumed to be an elastic material, and an elastic model is established to analyze the wellbore stability. The elastic model has few parameters and clear physical meaning. The calculation results of the model are analytical solutions rather than numerical solutions, and the calculation speed is extremely fast. It can be used for the rapid analysis of the situation in a well. It has great advantages in dealing with emergencies.

3.1.1. Conversion of Axis Coordinates of Wellbore

In the process of calculating the stress surrounding a deviated wellbore, the coordinate transformation should be carried out first. The principal in situ stress coordinate (x_2, y_2, z_2) shall be converted into a wellbore axis coordinate system (x, y, z). The in situ stress coordinate system, corresponding to the Ox_2 axis, Oy_2 axis, and Oz_2 axis is consistent with the principal ground stresses σ_H, σ_h, and σ_v directions, respectively. In the wellbore axis coordinate system, axis Oz corresponds to the deviated wellbore axis and axes Ox and Oy are located in a plane perpendicular to the well axis, and ψ, and φ are the deviation angle and azimuth, respectively (Figure 2).

The principal in situ stress coordinate system rotates with the wellbore axis, and the coordinate system transformation relationship can be derived from direction cosine, $M_{(\varphi,\psi)}$, for which the expressions are listed as follows:

$$M_{(\varphi)} = \begin{pmatrix} \cos\varphi & -\sin\varphi & 0 \\ \sin\varphi & \cos\varphi & 0 \\ 0 & 0 & 1 \end{pmatrix} \quad M_{(\psi)} = \begin{pmatrix} \cos\psi & 0 & \sin\psi \\ 0 & 1 & 0 \\ -\sin\psi & 0 & \cos\psi \end{pmatrix} \tag{1}$$

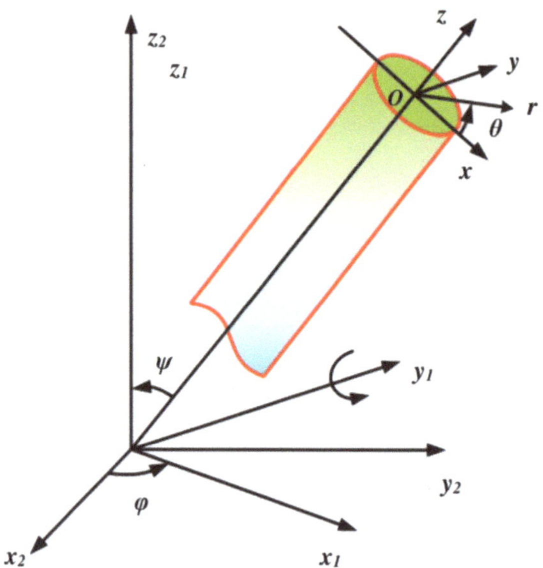

Figure 2. Coordinate transformation diagram of a deviated well axis.

$$M_{(\varphi,\psi)} = M_{(\varphi)} \cdot M_{(\psi)} = \begin{pmatrix} \cos\varphi\cos\psi & -\sin\varphi & \cos\varphi\sin\psi \\ \sin\varphi\cos\psi & \cos\varphi & \sin\varphi\sin\psi \\ -\sin\psi & 0 & \cos\psi \end{pmatrix} \quad (2)$$

where φ is the deviation angle of the wellbore in $°$; ψ is the azimuth in $°$.

In the coordinate system (x, y, z), each stress components calculation formula is as follows:

$$\begin{pmatrix} \sigma_{xx} & \tau_{xy} & \tau_{xz} \\ \tau_{yx} & \sigma_{yy} & \tau_{yz} \\ \tau_{zx} & \tau_{zy} & \sigma_{zz} \end{pmatrix} = M_{(\varphi,\psi)} \cdot \begin{pmatrix} \sigma_H & 0 & 0 \\ 0 & \sigma_h & 0 \\ 0 & 0 & \sigma_v \end{pmatrix} \cdot M_{(\varphi,\psi)}^T \quad (3)$$

where σ_H, σ_h, and σ_v represent the in situ stresses in MPa; σ_{xx}, σ_{yy}, and σ_{zz} are the principal stresses in MPa in directions, x, y, and z, respectively; and τ is the shear stress in MPa.

Since the wellbore section is circular, polar coordinates can be used for simplicity. The cartesian coordinates to polar coordinates transformation is performed as follows:

$$P_{(\theta)} = \begin{pmatrix} \cos\theta & -\sin\theta & 0 \\ \sin\theta & \cos\theta & 0 \\ 0 & 0 & 1 \end{pmatrix} \quad (4)$$

where θ is the wellbore rounded angle in $°$.

Referring to Equation (3), transform to polar coordinates (r, θ, z), and each stress component can be expressed as:

$$\begin{cases} \sigma_r = \sigma_{xx}\cos^2\theta + \sigma_{yy}\sin^2\theta + 2\tau_{xy}\sin\theta\cos\theta \\ \sigma_\theta = \sigma_{xx}\sin^2\theta + \sigma_{yy}\cos^2\theta - 2\tau_{xy}\sin\theta\cos\theta \\ \sigma_z = \sigma_{zz} \\ \tau_{\theta r} = (\sigma_{yy} - \sigma_{xx})\sin\theta\cos\theta + \tau_{xy}(\cos^2\theta - \sin^2\theta) \\ \tau_{zr} = \tau_{xz}\cos\theta + \tau_{yz}\sin\theta \\ \tau_{z\theta} = -\tau_{xz}\sin\theta + \tau_{yz}\cos\theta \end{cases} \quad (5)$$

where σ_r, σ_θ, and σ_z are the radial stresses of the reservoir, the tangential stress of the reservoir, and the vertical stress of the reservoir, respectively, in MPa.

3.1.2. The Elastic Solution of the Stress around the Wellbore

In polar coordinates (r, θ, z), the problem of the redistribution of stresses generated by the principal in situ stress σ_H, σ_h, and σ_v on the surrounding strata of the wellbore is analyzed. It is assumed that the surrounding strata of the wellbore located in the HBS-CS are homogeneous, isotropic, linear–elastic, without creep, and independent from viscosity behavior. The stress of the original strata is in an isotropic state. The analytical solution is solved according to the principle of elasticity. The stress on the surrounding strata meets the plane-stress mechanics' equilibrium and consistent equation [29], in which the plane-stress equilibrium equation can be expressed as:

$$\begin{cases} \frac{\partial \sigma_r}{\partial r} + \frac{\partial \tau_{\theta r}}{r \partial \theta} + \frac{\sigma_r - \sigma_\theta}{r} = 0 \\ \frac{\partial \tau_{\theta r}}{\partial r} + \frac{\partial \sigma_\theta}{r \partial \theta} + \frac{2\tau_{\theta r}}{r} = 0 \end{cases} \tag{6}$$

The consistent equation is expressed as:

$$\left(\frac{\partial^2}{\partial r^2} + \frac{\partial}{r \partial r} + \frac{\partial^2}{r^2 \partial \theta^2} \right)(\sigma_r + \sigma_\theta) = 0 \tag{7}$$

It is assumed that the pore pressure is independent of time and the borehole radius is at any position of the borehole. According to Equation (5), the stress components and boundary conditions could be obtained by the independent action of liquid column pressure, p_i, and principal stress, $\sigma_{xx}, \sigma_{yy}, \tau_{xy}, \tau_{xz}$, and τ_{yz}. Equation (5) is substituted into Equations (6) and (7) to obtain the surrounding strata stresses of the wellbore (Tables 2 and 3). In the vertical well, each component in Table 3 is 0.

Table 2. Stress distribution expression for the surrounding strata caused by p_i, σ_{xx}, σ_{yy}, τ_{xy}, τ_{xz}, and τ_{yz}.

σ_r	σ_θ	$\tau_{\theta r}$
$\frac{R^2}{r^2} p_i$	$-\frac{R^2}{r^2} p_i$	/
$\left(1 - \frac{R^2}{r^2}\right)\frac{\sigma_{xx}}{2} + \left(1 - \frac{4R^2}{r^2} + \frac{3R^4}{r^4}\right)\frac{\sigma_{xx}}{2}\cos 2\theta$	$\left(1 + \frac{R^2}{r^2}\right)\frac{\sigma_{xx}}{2} - \left(1 + \frac{3R^4}{r^4}\right)\frac{\sigma_{xx}}{2}\cos 2\theta$	$-\left(1 + \frac{2R^2}{r^2} - \frac{3R^4}{r^4}\right)\frac{\sigma_{xx}}{2}\sin 2\theta$
$\left(1 - \frac{R^2}{r^2}\right)\frac{\sigma_{yy}}{2} - \left(1 - \frac{4R^2}{r^2} + \frac{3R^4}{r^4}\right)\frac{\sigma_{yy}}{2}\cos 2\theta$	$\left(1 + \frac{R^2}{r^2}\right)\frac{\sigma_{yy}}{2} + \left(1 + \frac{3R^4}{r^4}\right)\frac{\sigma_{yy}}{2}\cos 2\theta$	$\left(1 + \frac{2R^2}{r^2} - \frac{3R^4}{r^4}\right)\frac{\sigma_{yy}}{2}\sin 2\theta$
$\left(1 - \frac{4R^2}{r^2} + \frac{3R^4}{r^4}\right)\tau_{xy}\sin 2\theta$	$-\left(1 + \frac{3R^4}{r^4}\right)\tau_{xy}\sin 2\theta$	$\left(1 + \frac{2R^2}{r^2} - \frac{3R^4}{r^4}\right)\tau_{xy}\cos 2\theta$

Table 3. Stress distribution expression for the surrounding strata caused by τ_{xy}, and τ_{yz}.

τ_{rz}	$\tau_{\theta z}$
$\tau_{xz}\left(1 - \frac{R^2}{r^2}\right)\cos \theta$	$-\tau_{xz}\left(1 + \frac{R^2}{r^2}\right)\sin \theta$
$\tau_{yz}\left(1 - \frac{R^2}{r^2}\right)\sin \theta$	$\tau_{yz}\left(1 + \frac{R^2}{r^2}\right)\cos \theta$

The stresses of the surrounding strata can be solved according to Hooke's law. The expression of Hooke's law in direction z is as follows:

$$\varepsilon_z = \frac{1}{E}\left[\sigma_{zz} - v\left(\sigma_{xx} + \sigma_{yy}\right)\right] \tag{8}$$

where ε_z and v are the strain in direction z, and the Poisson's ratio, respectively, and are dimensionless; E is the elastic modulus in MPa.

Neglecting the deformation of the formation in the vertical direction z, and substituting $\varepsilon_z = 0$ into Equation (8), the equation is transformed into:

$$\sigma_{zz} = v\left(\sigma_{xx} + \sigma_{yy}\right) \tag{9}$$

Following this step, the component stresses of directions r and θ (Table 2) are summed to obtain the expression of the wellbore surrounding stress caused by the overburden pressure.

$$\sigma_z = \sigma_{zz} - v\left[2(\sigma_{xx} - \sigma_{yy})\frac{R^2}{r^2}\cos 2\theta + 4\tau_{xy}\frac{R^2}{r^2}\sin 2\theta\right] \tag{10}$$

The linear superposition of all the stress components obtains the stress component at the shaft wall ($r = R$). The principal in situ stresses σ_H, σ_h, and σ_v are obtained by substituting the stress component of the wellbore axis coordinate system. These expressions are as follows:

$$\begin{cases} \sigma_r = p_i \\ \sigma_\theta = A\sigma_H + B\sigma_h + C\sigma_v - p_i \\ \sigma_z = D\sigma_H + E\sigma_h + F\sigma_v \\ \sigma_{\theta z} = G\sigma_H + H\sigma_h + J\sigma_v \\ \sigma_{r\theta} = \sigma_{rz} = 0 \end{cases} \tag{11}$$

where

$$\begin{cases} A = \cos\psi\left[\cos\psi\cos^2\varphi(1 - 2\cos 2\theta) - 2\sin 2\varphi\sin 2\theta\right] + \cos^2\varphi(1 + 2\cos 2\theta) \\ B = \cos\psi\left[\cos\psi\sin^2\varphi(1 - 2\cos 2\theta) + 2\sin 2\varphi\sin 2\theta\right] + \cos^2\varphi(1 + 2\cos 2\theta) \\ C = \sin^2\psi(1 - 2\cos 2\theta) \\ D = \sin^2\psi\cos^2\varphi - 2v\left[\cos\psi\sin 2\varphi\sin 2\theta - (\sin^2\varphi - \cos^2\psi\cos^2\varphi)\cos 2\theta\right] \\ E = \sin^2\psi\sin^2\varphi + 2v\left[\cos\psi\sin 2\varphi\sin 2\theta + (\cos^2\varphi - \cos^2\psi\sin^2\varphi)\cos 2\theta\right] \\ F = \cos^2\psi - 2v\sin^2\psi\cos 2\theta \\ G = \sin\psi\sin 2\varphi\cos\theta - \sin 2\psi\cos^2\varphi\sin\theta \\ H = -(\sin\psi\sin 2\varphi\cos\theta + \sin 2\psi\sin^2\varphi\sin\theta) \\ J = \sin 2\psi\sin\theta \end{cases} \tag{12}$$

The rock unit on the deviated wellbore is shown in Figure 3. Since σ_r is the principal stress, the deviated shaft wall is still a principal stress surface. To judge the location of a rock failure, the other two principal stress planes must be solved first.

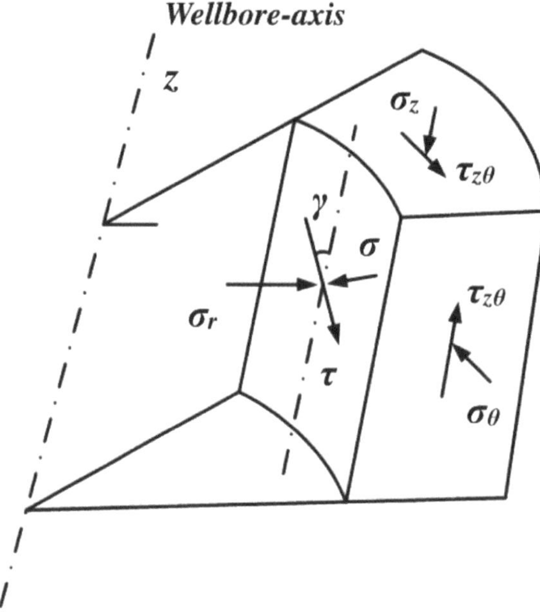

Figure 3. Schematic diagram of the stress distribution of the rock element on a deviated wellbore.

According to stress analysis, the relationship between the normal stress, σ_r, shear stress, τ, and each component is as follows:

$$\begin{cases} \sigma = \sigma_\theta \cos^2 \gamma + \sigma_z \sin^2 \gamma + 2\tau_{\theta z} \sin \gamma \cos \gamma \\ \tau = \frac{1}{2}(\sigma_z - \sigma_\theta) \sin 2\gamma + \tau_{\theta z} \cos 2\gamma \end{cases} \tag{13}$$

by substituting $d\sigma/d\gamma = 0$ into Equation (13), the two principal stress expressions can be obtained. Considering the influence of the pore pressure, the expression of the stress distribution of the hydrate-deviated wellbore can be obtained as follows:

$$\begin{cases} \sigma_i = \sigma_r = p_i - \alpha p_p \\ \sigma_j = \frac{1}{2}(\sigma_\theta + \sigma_z) + \frac{1}{2}\sqrt{(\sigma_\theta - \sigma_z)^2 + 4\tau_{\theta z}{}^2} - \alpha p_p \\ \sigma_k = \frac{1}{2}(\sigma_\theta + \sigma_z) - \frac{1}{2}\sqrt{(\sigma_\theta - \sigma_z)^2 + 4\tau_{\theta z}{}^2} - \alpha p_p \end{cases} \tag{14}$$

where σ_θ and σ_z are the expressions as in Equation (11); p_p is the pore pressure in MPa; α is the effective stress coefficient and is dimensionless.

3.2. Failure Criteria

In this study, two failure criteria were used, namely the M-C failure criterion and the D-P failure criterion [30,31]. The M-C failure criterion is the most widely used criterion when analyzing the stability of the boreholes during drilling [32,33]. The criterion, however, only considers the effect of maximum and minimum principal stresses and ignores the influence of the intermediate principal stress. On the other hand, the D-P failure criterion also takes into account the influence of the intermediate principal stress, and adds hydrostatic pressure, thereby overcoming the main weakness of the M-C failure criterion. The D-P failure criterion was also applied in the numerical analysis of formation stability at home and abroad.

3.2.1. The M-C Failure Criterion

Coulomb proposed that the failure of rock was mainly caused by shear failure. The strength of the rock, namely the frictional strength, was supposed to be equal to the adhesion force of the rock, itself against the friction imposed by the shear, and the friction force generated by the normal force on the shear surface. The general form is as follows [34]:

$$\tau = C + \sigma \tan \phi \tag{15}$$

where τ, σ, and C, all in MPa, are the shear stress, the normal stress on the shear plane, and the cohesive force, respectively; ϕ is the angle of internal friction in $^\circ$.

To facilitate the calculation, the principal stress form is rewritten as:

$$\sigma_1 = \sigma_3 K^2 + 2CK \tag{16}$$

where K^2 is the influence coefficient of confining pressure on axial bearing capacity, $K = \cot\left(\frac{\pi}{4} - \frac{\phi}{2}\right)$.

3.2.2. The D-P Failure Criterion

The D-P failure criterion is an extension of the M-C failure criterion and the von Mises failure criterion [35]. The difference from the M-C failure criterion is that the corner of the yield surface for the D-P failure criterion is smooth and conical in the principal stress space, eliminating the singular point caused by sharp angles, thereby facilitating numerical calculation of the stability of the wellbore (Figure 4).

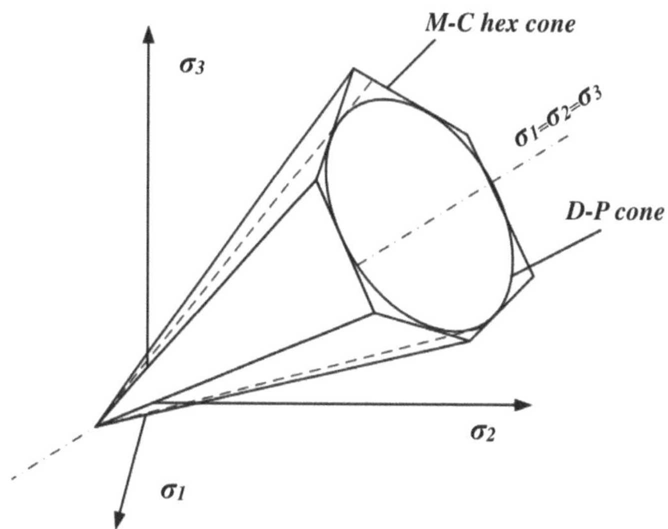

Figure 4. A comparison between the three-dimensional failure surface derived from the M-C criterion and the D-P criterion.

The standard form of the D-P failure criterion is as follows [36]:

$$f = Q_f I_1(\sigma_{ij}) + \sqrt{J_2(S_{ij})} + K_f = 0 \tag{17}$$

where f, $I_1(\sigma_{ij})$, and $J_2(S_{ij})$ are the plastic potential function, the first invariant of the stress tensor, and the second invariant of the stress partial tensor, respectively. Q_f and K_f are a function of cohesion C, and internal friction angle ϕ, respectively, as follows:

$$Q_f = \frac{\sqrt{3}\sin\phi}{3\sqrt{3+\sin^2\phi}}, K_f = -\frac{\sqrt{3}C\cos\phi}{\sqrt{3+\sin^2\phi}}.$$

3.3. Collapse Pressure and Fracture Pressure

Since the M-C failure criterion only considers the influence of the maximum and the minimum principal stresses and ignores the influence of the intermediate principal stress, it is necessary to establish the principal stress Equation (14) to obtain the maximum and the minimum principal stresses. Then, by substituting them into the M-C failure criterion Equation (16), the value of p_i that makes the equation hold was obtained, which is the collapse pressure of the wellbore under the M-C failure criterion.

Similarly, by substituting these three principal stresses into the D-P failure criterion Equation (17), the collapse pressure of the hydrate wellbore can be obtained by the D-P failure criterion.

For the calculation of the fracture pressure, since only σ_k can be negative values in Equation (16), the fracture pressure calculation expression is as follows:

$$f(p_b) = \frac{1}{2}(\sigma_\theta + \sigma_z) - \frac{1}{2}\sqrt{(\sigma_\theta - \sigma_z)^2 + 4\tau_{\theta z}^2} - \alpha p_p + S_t = 0 \tag{18}$$

where S_t is the uniaxial tensile strength of the HBS-CS in MPa.

The above equations were substituted into Matlab to solve for the collapse and fracture pressure of the HBS-CS-deviated wellbore.

4. Wellbore Stability Analyses

The basic formation parameters are shown in Table 4. The collapse and fracture pressure of the deviated wellbore located in the HBS-CS are obtained.

Table 4. Basic parameter values [37,38].

Parameter	Value
Maximum principal stress, g/cm^3	1.203
Minimum principal stress, g/cm^3	1.159
Overburden rock stress, g/cm^3	1.282
Poisson's ratio	0.45
Effective stress factor	0.6

4.1. Effect of S_h on Collapse Pressure of Wellbore

To describe the stability characteristics of the deviated wellbore located in HBS-CS, the lower hemisphere projection method proposed by Zoback was adopted [39] to show the collapse pressure gradient distribution cloud diagram with different saturation at different azimuth and deviation angles. The cloud diagram of the collapse pressure gradient distribution around the deviated wellbore with different saturation by the M-C and D-P failure criteria are shown in Figures 5 and 6, respectively.

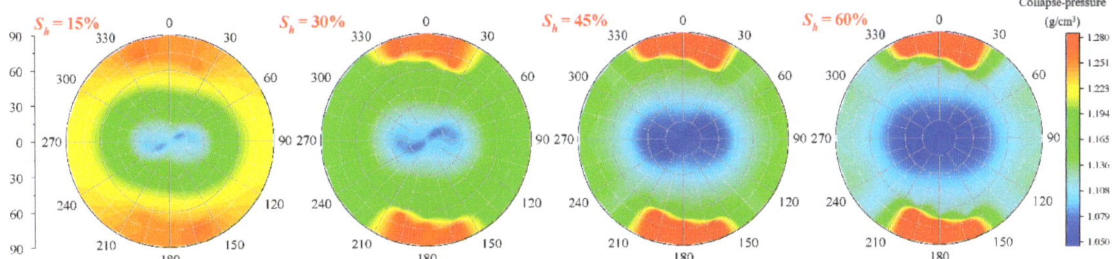

Figure 5. Cloud diagram of the collapse pressure gradient distribution under M-C failure criterion.

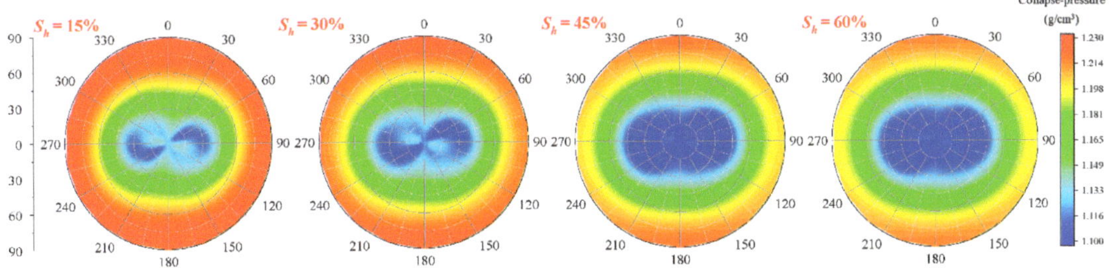

Figure 6. Cloud diagrams for the collapse pressure gradient distribution under the D-P failure criterion.

A brief comparison between Figures 5 and 6 implies that the collapse pressure gradient varies with the azimuth in a centrosymmetric distribution at different saturations. This is independent of the chosen failure criteria. The low collapse pressure gradient area is concentrated in the middle "8" region. However, there are differences in the distribution locations of the high collapse pressure gradient area for the different criteria. In the M-C failure criterion, the area where the azimuth ranges over 330°-30° and the deviation angle is greater than 75° has the largest collapse pressure gradient. On the other hand, the collapse pressure gradient is the highest in the annular region (the maximum principal stress direction, deviation angle > 60°; the minimum principal stress direction, deviation angle > 75°) by the D-P failure criterion.

The relationship between S_h and the collapse pressure gradient by the two criteria is shown in Figure 7. The collapse pressure gradient increased with decreasing S_h. When the saturation decreases from 60% to 15%, the collapse pressure gradient increases by

7–10%. Li et al. (2020) and Sun et al. (2018) obtained the same discovery from the analysis of stratum mechanical properties [17,20]. When the S_h decreases, the strength of HBS decreases, the plastic zone increases, and the wellbore stability becomes worse. Therefore, considering the influence of the hydrate decomposition on wellbore stability, the equivalent density of drilling fluid should be appropriately increased to meet the needs of the wellbore stability after hydrate decomposition.

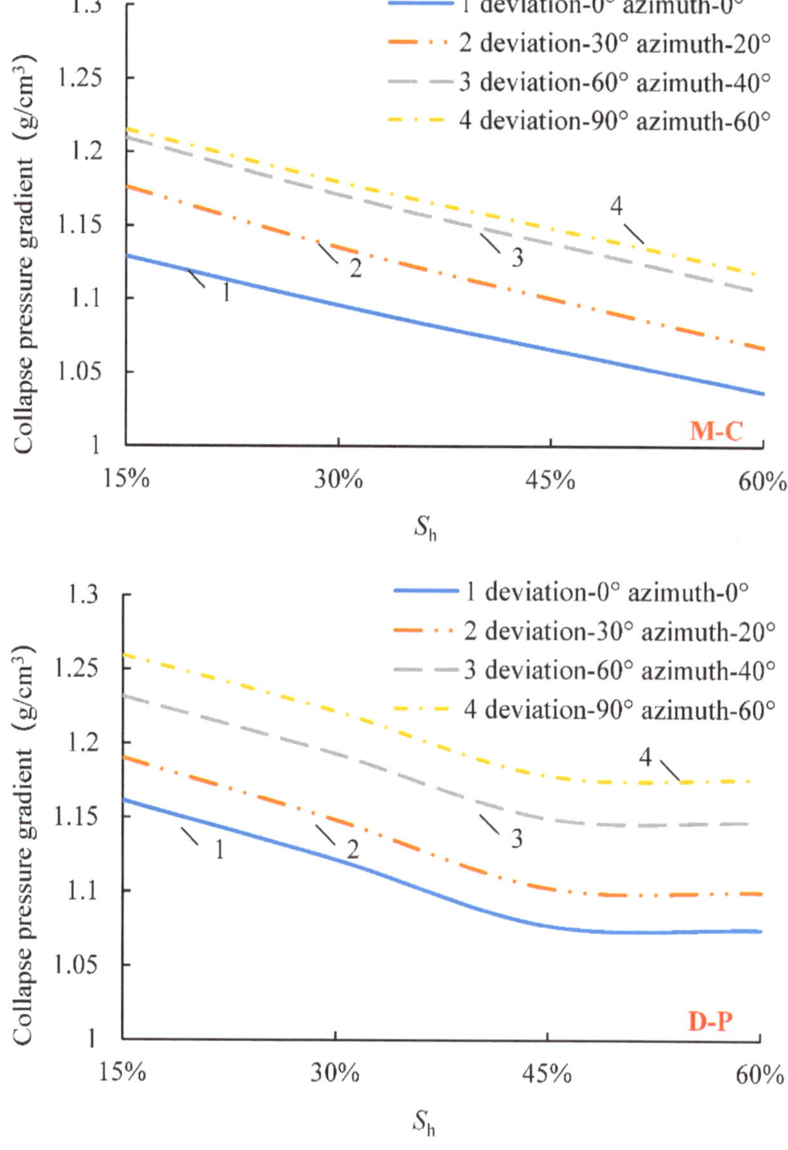

Figure 7. Hydrate saturation and collapse pressure gradient relationship curves by the two criteria.

By comparing the curves of the two criteria, the collapse pressure gradient decreases linearly with an increase in S_h under low saturation (<45%) by the two criteria. The reason is that with decreasing S_h, the formation pore–space increases, and the cementation ability of hydrate to the formation weakens, which results in a continuous increase in the collapse pressure gradient and the deterioration of the stability of the wellbore. At high saturation (>45%), the M-C failure criterion curve trend is the same as is seen at low saturation, while the D-P failure criterion curve is flat. At the same position, the calculated value of the D-P failure criterion is greater than that of the M-C failure criterion at low saturation, and the opposite is true at high saturation. It showed that the formation stability in high saturation is higher when the influence of the intermediate principal stress is fully considered. However, higher drilling fluid densities are required to maintain formation stability in low saturations.

4.2. Influence of the Deviation Angle on the Collapse Pressure Gradient of the Wellbore

According to the formation basic parameters and the hydrate mechanics parameters, the influence of the deviation angle on the collapse pressure gradient was calculated. Taking 15% saturation as an example, the relationship curves between the deviation angle and the collapse pressure gradient by the two criteria were obtained (Figure 8). Within a quarter of the cycle (azimuth 0~90°), the trends of the two curves are the same, and the collapse pressure gradient rises, on the whole, with increasing deviation angle. When the azimuth is less than 60° (curves 1–3), the collapse pressure gradient continues to increase with increasing deviation angle. When the azimuth is higher than 60 ° (curves 4 and 5), the collapse pressure gradient decreases first and then increases with changing deviation angle, and the position of the inflection point is gradually delayed with increasing azimuth. The collapse pressure gradient increases by 7.2–9.2% from the vertical wellbore section (deviation angle 0°) to the horizontal wellbore section (deviation angle 90°). In other words, the safe density that satisfies the horizontal section can also ensure the wellbore stability of other sections, and will not collapse.

By comparing the two curves obtained from the two criteria, it is found that when the azimuth is higher than 60°, the inflection point of the collapse pressure gradient by the M-C failure criterion appears earlier than that of the D-P failure criterion. When the azimuth is 60° and 80°, the inflection points of the M-C failure criterion curve are 10° and 20°, while considering that the inflection point of the D-P failure criterion curve, under the influence of the intermediate principal stress, is 15° and 30°. The M-C failure criterion curve increases abnormally near 65°, while the D-P curve is smooth and without distortion. The reason for this is that the yield surface of the D-P failure criterion is smooth and without sharp corners, while the yield surface of the M-C failure criterion is hexagonal and with sharp corners (Figure 4). Hence, the D-P criterion is more suitable for the wellbore stability calculation.

4.3. Effect of the Azimuth on the Collapse Pressure Gradient of the Wellbore

According to the formation of the basic parameters and the hydrate mechanics parameters, the influence of the azimuth on the collapse pressure gradient was calculated. Taking 15% S_h as an example, the relationship between the azimuth and the collapse pressure gradient by the two criteria was obtained (see Figure 9). It was found that the collapse pressure gradient decreases and then increases with an increasing azimuth on the high inclination well sections (curves 2–6). The collapse pressure gradient reaches a minimum value in the direction of the minimum principal stress (azimuth of 90°). The same was reported in the literature [40]. With an increase in the deviation angle, the fluctuation of the curve is lower. That is, the difference in the collapse pressure gradient is not significant during drilling in any direction. When the deviation angle is 30° and 90°, the differences are 0.038 and 0.018 g/cm^3, respectively. In the near-vertical sections (curve 1), the minimum collapse pressure gradient was obtained at an azimuth of 60°. In summary, the wellbore should be arranged along the azimuth between 60° and 120° for high wellbore stability.

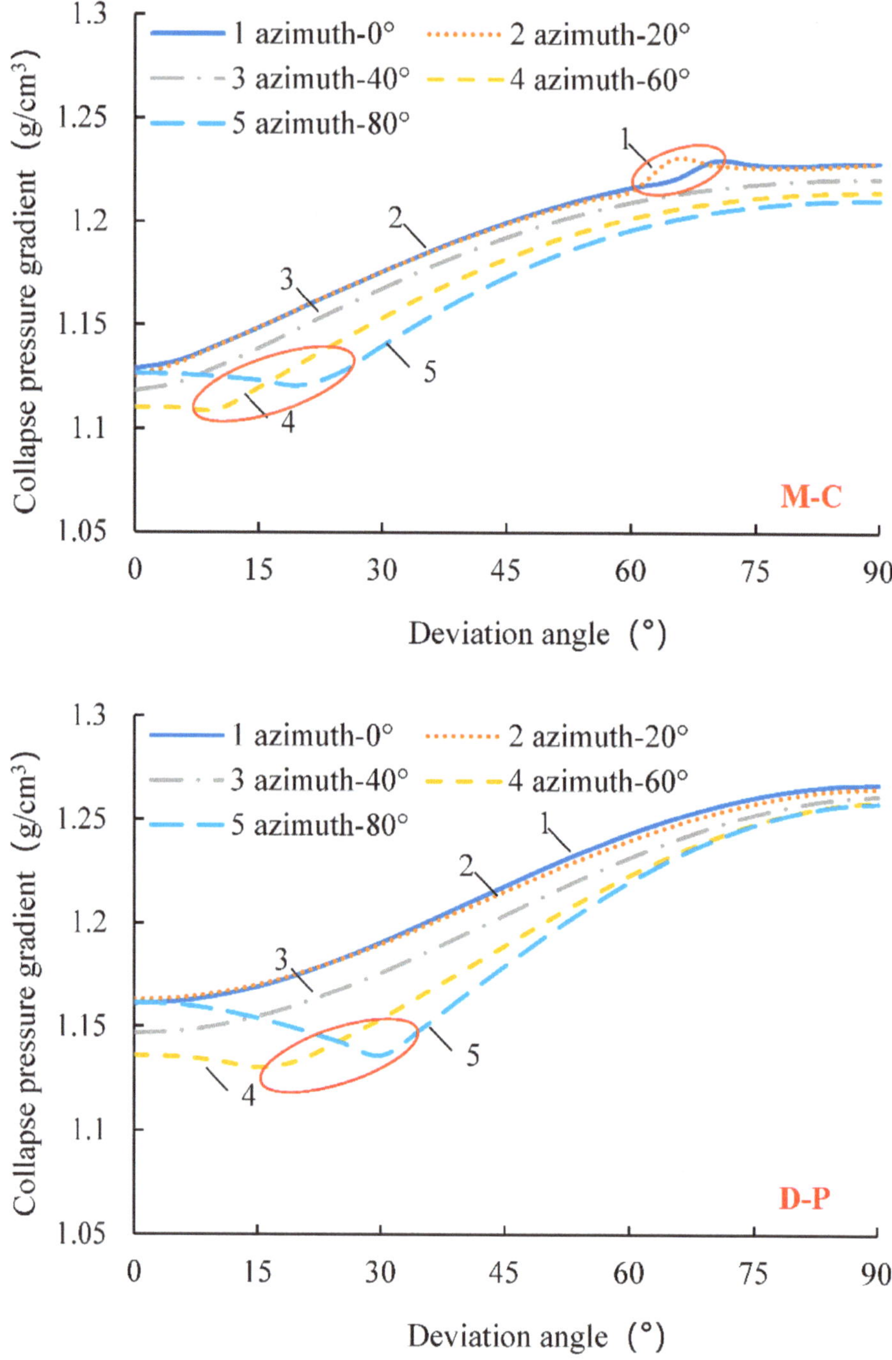

Figure 8. The deviation angle and the collapse pressure gradient relationship curves by the two criteria.

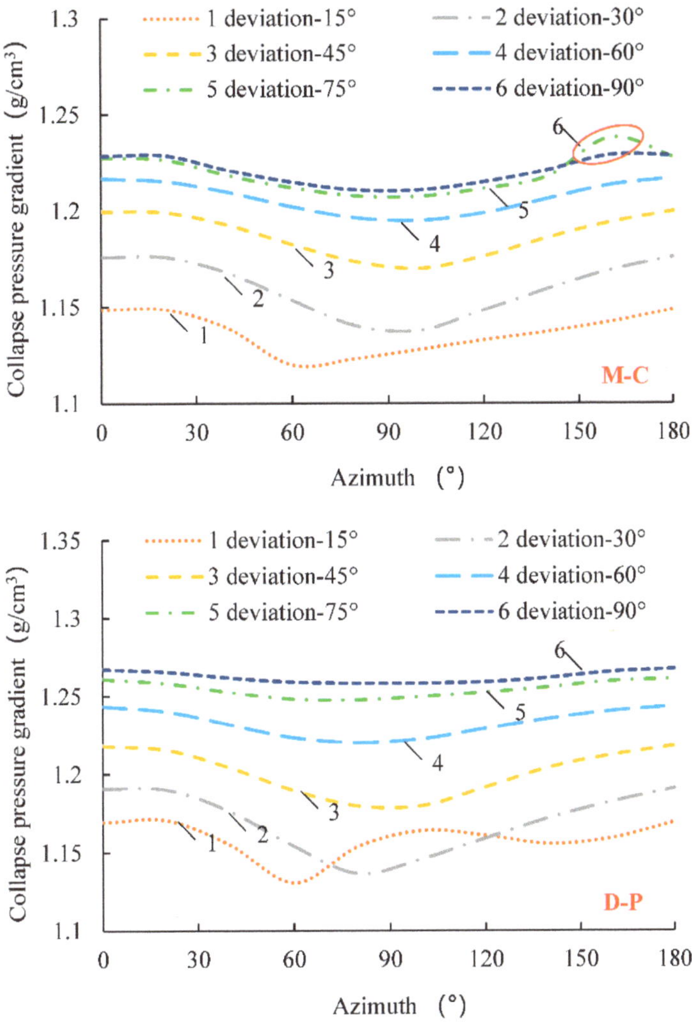

Figure 9. The azimuth and the collapse pressure gradient relationship curves by the two criteria.

4.4. Fracture Pressure Gradient and the Safe Drilling Fluid Density Window

The safe drilling fluid density window is defined as the difference between the collapse pressure gradient and the fracture pressure gradient. It is safer for drilling with a relatively wide safe drilling fluid density window. Therefore, the variation pattern of fracture pressure was analyzed first. According to the formation of the basic parameters and the hydrate mechanics parameters, and taking the 15% S_h as an example, the relationship curve between the azimuth and fracture pressure gradient was simulated, as shown in Figure 10. When the deviation angle was higher than 45° (curves 4–7), the fracture pressure gradient increased and then decreased with increasing azimuth, and the maximum value is in the direction of the minimum principal stress. When the deviation angle is less than 45° (curves 1–3), the fracture pressure gradient generally increases and then decreases with increasing azimuth, but the peak value exists between 120° and 150°. The azimuth corresponding to the peak decreases gradually with increasing deviation angle, closer to the direction of the minimum principal stress. During the drilling of horizontal wellbores, the drilling fluid density needs to meet the stability of the wellbore in all sections. Therefore, a drilling fluid density is selected in the yellow zone formed below the vertical and horizontal wellbore curves (curves 1 and 7), which can be safely drilled in any direction.

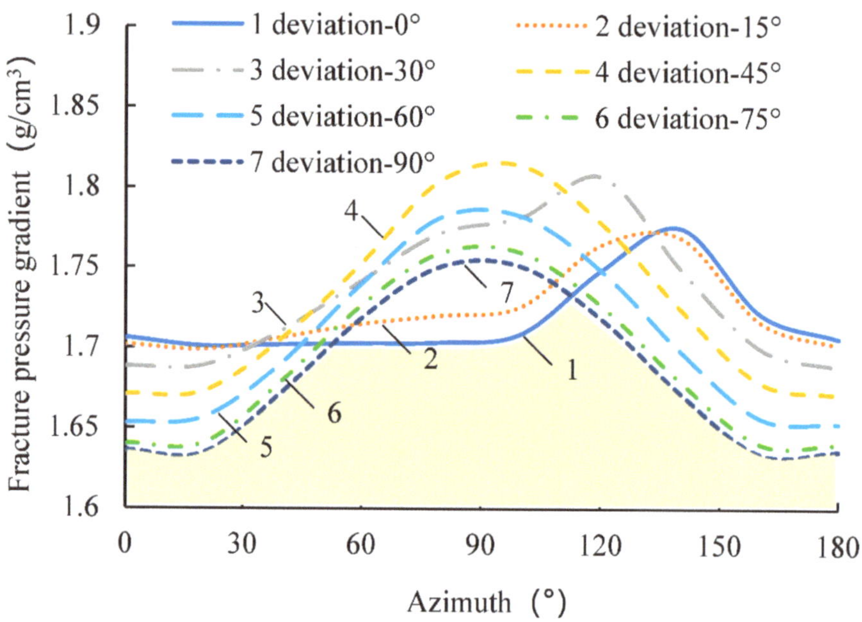

Figure 10. The azimuth and the fracture pressure gradient curves of the wellbore.

The cloud diagram of the safe density window is drawn by the lower hemisphere projection method, as shown in Figure 11. The safe density window has a center-symmetric strip distribution in the range of 0.368–0.660 g/cm^3. The density window of the upper and lower belt regions is narrow, while the density window of the middle region is large. The maximum safe density window is the "crescent" region in the direction of the minimum principal stress. Drilling operations near this zone are most conducive to wellbore stability.

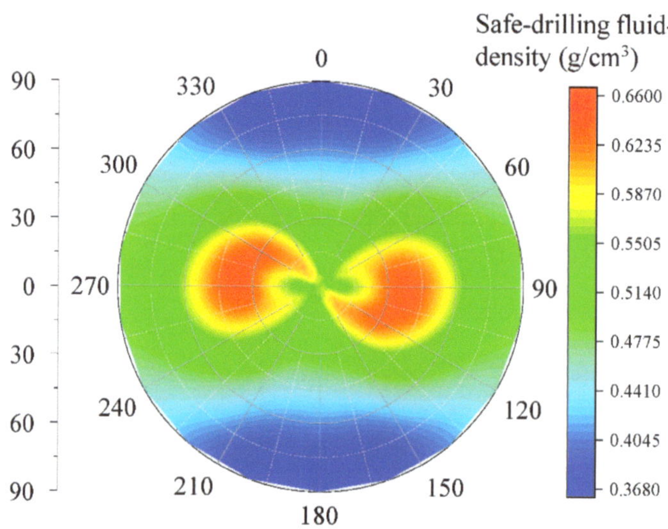

Figure 11. Window distribution cloud diagram for safe drilling fluid density.

5. Conclusions and Suggestions

1. The window distribution cloud chart of the collapse pressure gradient and the safe drilling fluid density of HBS-CS has a centrosymmetric distribution with a deviation angle and azimuth.

2. Hydrate decomposition will lead to a higher collapse pressure and poorer stability of the formation. Therefore, the drilling fluid density should be appropriately increased by 7~10% during drilling to ensure the stability of formation after hydrate decomposition.

3. The collapse pressure gradient increases by 7.2–9.2% from the vertical wellbore to the horizontal wellbore. From the perspective of preventing wellbore collapse, the safe density that satisfies the horizontal section can also ensure the wellbore stability of other sections. To prevent the wellbore fracture, a safe density that satisfies both the horizontal and vertical sections is necessary to ensure wellbore stability in all sections.

4. Considering the combined effects of collapse, fracture pressure gradient, and safety density window, it is suggested that the borehole be arranged along the azimuth of 60–120°, which could greatly reduce the risk of the drilling operation.

The model can provide a safe density window in horizontal drilling design and guide the selection of drilling fluid for construction. In case of an emergency, such as a kick and overflow, the risk of wellbore instability can be calculated quickly by this model to guide the formulation of corresponding disposal measures. However, the parameters involved in the model are not comprehensive enough, and the calculation accuracy is slightly insufficient. Therefore, the elastic model is just the beginning of the HBS-CS wellbore stability study. Next, we will establish the plastic model and damage model and strive to improve the prediction accuracy of wellbore stability, helping NGH mining.

Author Contributions: Conceptualization, Q.H.; data curation, Y.L. and Y.Z.; methodology, X.S. and Y.L.; software, M.C. and Y.Z.; validation, M.C.; writing—original draft, X.S. and Q.H.; writing—review and editing, Q.H. All authors have read and agreed to the published version of the manuscript.

Funding: This study was supported by the Key R&D Projects of Hainan Province (Grant no. ZDYF2022SHFZ063) and the Shandong Special Fund of Pilot National Laboratory for Marine Science and Technology (Qingdao) (Grant no. 2021QNLM020002).

Institutional Review Board Statement: Not applicable.

Informed Consent Statement: Not applicable.

Data Availability Statement: Not applicable.

Conflicts of Interest: The authors declare no conflict of interest.

References

1. Wan, Z.; Luo, J.; Yang, X.; Zhang, W.; Liang, J.; Zuo, L.; Sun, Y. The Thermal Effect of Submarine Mud Volcano Fluid and Its Influence on the Occurrence of Gas Hydrates. *J. Mar. Sci. Eng.* **2022**, *10*, 832. [CrossRef]
2. Hu, G.; Bu, Q.; Lyu, W.; Wang, J.; Chen, J.; Li, Q.; Gong, J.; Sun, J.; Wu, N. A comparative study on natural gas hydrate accumulation models at active and passive continental margins. *Nat. Gas Ind. B* **2021**, *8*, 115–127. [CrossRef]
3. Qin, X.-W.; Lu, J.-A.; Lu, H.-L.; Qiu, H.-J.; Liang, J.-Q.; Kang, D.-J.; Zhan, L.-S.; Lu, H.-F.; Kuang, Z.-G. Coexistence of natural gas hydrate, free gas and water in the gas hydrate system in the Shenhu Area, South China Sea. *China Geol.* **2020**, *3*, 210–220. [CrossRef]
4. Makogon, Y.F. Natural gas hydrates—A promising source of energy. *J. Nat. Gas Sci. Eng.* **2010**, *2*, 49–59. [CrossRef]
5. Makogon, Y.; Omelchenko, R. Commercial gas production from Messoyakha deposit in hydrate conditions. *J. Nat. Gas Sci. Eng.* **2013**, *11*, 1–6. [CrossRef]
6. Kurihara, M.; Sato, A.; Funatsu, K.; Ouchi, H.; Yamamoto, K.; Numasawa, M.; Ebinuma, T.; Narita, H.; Masuda, Y.; Dallimore, S.R.; et al. Analysis of Production Data for 2007/2008 Mallik Gas Hydrate Production Tests in Canada. In Proceedings of the International Oil and Gas Conference and Exhibition in China, Beijing, China, 8–10 June 2010; p. SPE-132155-MS. [CrossRef]
7. Dubreuil-Boisclair, C.; Gloaguen, E.; Bellefleur, G.; Marcotte, D. Non-Gaussian gas hydrate grade simulation at the Mallik site, Mackenzie Delta, Canada. *Mar. Pet. Geol.* **2012**, *35*, 20–27. [CrossRef]
8. Boswell, R.; Yamamoto, K.; Collett, T.S.; Okinaka, N. Virtual Special Issue of Recent Advances on Gas Hydrates Scientific Drilling in Alaska. *Energy Fuels* **2022**, *36*, 7921–7924. [CrossRef]

9. Li, X.-S.; Xu, C.-G.; Zhang, Y.; Ruan, X.-K.; Li, G.; Wang, Y. Investigation into gas production from natural gas hydrate: A review. *Appl. Energy* **2016**, *172*, 286–322. [CrossRef]
10. Yu, T.; Guan, G.; Abudula, A.; Yoshida, A.; Wang, D.; Song, Y. Gas recovery enhancement from methane hydrate reservoir in the Nankai Trough using vertical wells. *Energy* **2018**, *166*, 834–844. [CrossRef]
11. Liu, Z.; Wang, Z.; Sun, J.; Chen, L.; Wang, J.; Sun, B. Risk and preventive strategies of hydrate reformation in offshore gas hydrate production trials: A case study in the Eastern Nankai Trough. *J. Nat. Gas Sci. Eng.* **2022**, *103*, 104602. [CrossRef]
12. Li, J.-F.; Ye, J.-L.; Qin, X.-W.; Qiu, H.-J.; Wu, N.-Y.; Lu, H.-L.; Xie, W.-W.; Lu, J.-A.; Peng, F.; Xu, Z.-Q.; et al. The first offshore natural gas hydrate production test in South China Sea. *China Geol.* **2018**, *1*, 5–16. [CrossRef]
13. Ye, J.; Qin, X.; Xie, W.; Lu, H.; Ma, B.; Qin, H.; Liang, J.; Lu, J.; Kuang, Z.; Lu, C.; et al. Main progress of the second gas hydrate trial production in the South China Sea. *Geol. China* **2020**, *47*, 557–568. [CrossRef]
14. Wei, N.; Zhao, J.; Liu, A.; Zhou, S.; Zhang, L.; Jiang, L. Evaluation of Physical Parameters and Construction of the Classification System of Natural Gas Hydrate in the Northern South China Sea. *Energy Fuels* **2021**, *35*, 7637–7645. [CrossRef]
15. Rodger, P.M. Stability of gas hydrates. *J. Phys. Chem.* **1990**, *94*, 6080–6089. [CrossRef]
16. Liu, C.; Li, Y.; Liu, L.; Hu, G.; Chen, Q.; Wu, N.; Meng, Q. An integrated experimental system for gas hydrate drilling and production and a preliminary experiment of the depressurization method. *Nat. Gas Ind. B* **2020**, *7*, 56–63. [CrossRef]
17. Li, Y.; Cheng, Y.; Yan, C.; Song, L.; Liu, H.; Tian, W.; Ren, X. Mechanical study on the wellbore stability of horizontal wells in natural gas hydrate reservoirs. *J. Nat. Gas Sci. Eng.* **2020**, *79*, 103359. [CrossRef]
18. Freij-Ayoub, R.; Tan, C.; Clennell, B.; Tohidi, B.; Yang, J. A wellbore stability model for hydrate bearing sediments. *J. Pet. Sci. Eng.* **2006**, *57*, 209–220. [CrossRef]
19. Zhang, H.; Cheng, Y.; Li, Q.; Yan, C.; Han, X. Numerical analysis of wellbore instability in gas hydrate formation during deep-water drilling. *J. Ocean Univ. China* **2018**, *17*, 8–16. [CrossRef]
20. Sun, J.; Ning, F.; Lei, H.; Gai, X.; Sánchez, M.; Lu, J.; Li, Y.; Liu, L.; Liu, C.; Wu, N.; et al. Wellbore stability analysis during drilling through marine gas hydrate-bearing sediments in Shenhu area: A case study. *J. Pet. Sci. Eng.* **2018**, *170*, 345–367. [CrossRef]
21. Yuan, Y.; Xu, T.; Xin, X.; Xia, Y.; Li, B. Mechanical stability analysis of strata and wellbore associated with gas production from oceanic hydrate-bearing sediments by depressurization. *Chin. J. Theor. Appl. Mech.* **2020**, *52*, 544–555. [CrossRef]
22. Dong, L.; Li, Y.; Liu, C.; Liao, H.; Chen, G.; Chen, Q.; Liu, L.; Hu, G. Mechanical Properties of Methane Hydrate-Bearing Interlayered Sediments. *J. Ocean Univ. China* **2019**, *18*, 1344–1350. [CrossRef]
23. Jin, Y.; Wu, N.; Li, Y.; Yang, D. Characterization of Sand Production for Clayey-Silt Sediments Conditioned to Hydraulic Slotting and Gravel Packing: Experimental Observations, Theoretical Formulations, and Modeling. *SPE J.* **2022**, 1–20. [CrossRef]
24. Lu, J.; Lin, D.; Li, D.; Liang, D.; Wen, L.; Wu, S.; Zhang, Y.; He, Y.; Shi, L.; Xiong, Y. Microcosmic Characteristics of Hydrate Formation and Decomposition in the Different Particle Size Sediments Captured by Cryo-SEM. *J. Mar. Sci. Eng.* **2022**, *10*, 769. [CrossRef]
25. Li, Y.; Liu, C.; Liao, H.; Lin, D.; Bu, Q.; Liu, Z. Mechanical properties of the clayey-silty sediment-natural gas hydrate mixed system. *Nat. Gas Ind. B* **2021**, *8*, 154–162. [CrossRef]
26. Dong, L.; Liao, H.; Li, Y.; Meng, Q.; Hu, G.; Wang, J.; Wu, N. Analysis of the Mechanical Properties of the Reconstituted Hydrate-Bearing Clayey-Silt Samples from the South China Sea. *J. Mar. Sci. Eng.* **2022**, *10*, 831. [CrossRef]
27. Tan, C.P.; Clennell, M.B.; Freij-Ayoub, R.; Tohidi, B.; Yang, J. Mechanical and Petrophysical Characterisation and Wellbore Stability Management in Gas Hydrate-Bearing Sediments. In Proceedings of the Alaska Rocks 2005, the 40th US Symposium on Rock Mechanics (USRMS), Anchorage, Alaska, 25–29 June 2005; OnePetro: Richardson, TX, USA, 2005.
28. Birchwood, R.; Noeth, S.; Hooyman, P.; Winters, W.; Jones, E. Wellbore Stability Model for Marine Sediments Containing Gas Hydrates. In Proceedings of the American Association of Drilling Engineers National Conference and Exhibition, Houston, TX, USA, 5–7 April 2005.
29. Risnes, R.; Bratli, R.K.; Horsrud, P. Sand Stresses Around a Wellbore. *Soc. Pet. Eng. J.* **1982**, *22*, 883–898. [CrossRef]
30. Yan, C.; Ren, X.; Cheng, Y.; Song, B.; Li, Y.; Tian, W. Geomechanical issues in the exploitation of natural gas hydrate. *Gondwana Res.* **2020**, *81*, 403–422. [CrossRef]
31. Liu, X.; Sun, Y.; Guo, T.; Rabiei, M.; Qu, Z.; Hou, J. Numerical simulations of hydraulic fracturing in methane hydrate reservoirs based on the coupled thermo-hydrologic-mechanical-damage (THMD) model. *Energy* **2021**, *238*, 122054. [CrossRef]
32. Sujatono, S. Determination of cohesion and friction angle on sedimentary rock based on geophysical log. *Géoméch. Geophys. Geo-Energy Geo-Resour.* **2022**, *8*, 1–10. [CrossRef]
33. Jiang, H.; Xie, Y. A note on the Mohr–Coulomb and Drucker–Prager strength criteria. *Mech. Res. Commun.* **2011**, *38*, 309–314. [CrossRef]
34. Labuz, J.F.; Zang, A. Mohr–Coulomb Failure Criterion. *Rock Mech. Rock Eng.* **2012**, *45*, 975–979. [CrossRef]
35. Kesarev, A.G.; Vlasova, A.M. Generalized von Mises Criterion as a Tool for Determining the Strength Properties of Hexagonal Materials. *Phys. Met. Met.* **2022**, *123*, 186–192. [CrossRef]
36. Zhao, Y.; Wang, Y.; Tang, L. The compressive-shear fracture strength of rock containing water based on Druker-Prager failure criterion. *Arab. J. Geosci.* **2019**, *12*, 542. [CrossRef]
37. Chen, L.; Feng, Y.; Okajima, J.; Komiya, A.; Maruyama, S. Production behavior and numerical analysis for 2017 methane hydrate extraction test of Shenhu, South China Sea. *J. Nat. Gas Sci. Eng.* **2018**, *53*, 55–66. [CrossRef]

38. Li, G.; Moridis, G.J.; Zhang, K.; Li, X.-S. Evaluation of the Gas Production Potential of Marine Hydrate Deposits in the Shenhu Area of the South China Sea. In Proceedings of the Offshore Technology Conference, Houston, TX, USA, 3–6 May 2010; p. OTC-20548-MS. [CrossRef]
39. Zoback, M.D. *Reservoir Geomechanics*; Cambridge University Press: Cambridge, UK, 2010.
40. Li, W.; Gao, D.; Yang, J. Study of mud weight window of horizontal wells drilled into offshore natural gas hydrate sediments. *J. Nat. Gas Sci. Eng.* **2020**, *83*, 103575. [CrossRef]

Journal of
*Marine Science
and Engineering*

Review

Seabed Dynamic Responses Induced by Nonlinear Internal Waves: New Insights and Future Directions

Tian Chen [1,2], Zhenghui Li [1], Hui Nai [3], Hanlu Liu [1], Hongxian Shan [1,4] and Yonggang Jia [1,4,*]

[1] Shandong Provincial Key Laboratory of Marine Environment and Geological Engineering, Ocean University of China, Qingdao 266100, China
[2] Key Laboratory of Coastal Science and Integrated Management, Ministry of Natural Resources, Qingdao 266061, China
[3] School of Earth System Science, Tianjin University, Tianjin 300072, China
[4] Laboratory for Marine Geology, Qingdao National Laboratory for Marine Science and Technology, Qingdao 266061, China
* Correspondence: yonggang@ouc.edu.cn

Abstract: Strong nonlinear internal waves generate a significant pressure force on the seafloor and induce a pore-pressure response penetrated in the seabed and are thus an important driver of sediment resuspension and a potential trigger of seabed failure. The following provides an overview of the seabed responses induced by nonlinear internal waves and the theory, models, and limited observations that have provided our present knowledge. The pressure disturbance is generated by the combined effect of interface displacement and near-bottom acceleration by the nonlinear internal waves. Recent observations in the South China Sea have shown that the pressure magnitudes up to 4 kPa, which is the largest known disturbance. Intense pore-pressure changes in roughly the top 1 m of the weakly conductive seabed are expected during the shoaling and breaking of the nonlinear internal waves and lead to 2 cm sediments of the local seabed appearing in transient liquefaction. Since the fluid seepage reduces the specific weight of the bed, results show that the contribution of vertical seepage on sediment resuspension is estimated at 11% for a seabed saturation of 0.97. Finally, in situ observations are needed to confirm theoretical knowledge and to help improve our ability to model the multiscale interaction process between the seabed and internal waves in the future.

Keywords: internal solitary waves; seabed stability; pore pressure; dynamic response; seepage

Citation: Chen, T.; Li, Z.; Nai, H.; Liu, H.; Shan, H.; Jia, Y. Seabed Dynamic Responses Induced by Nonlinear Internal Waves: New Insights and Future Directions. *J. Mar. Sci. Eng.* **2023**, *11*, 395. https://doi.org/10.3390/jmse11020395

Academic Editor: George Kontakiotis

Received: 14 December 2022
Revised: 29 January 2023
Accepted: 3 February 2023
Published: 10 February 2023

1. Introduction

Nonlinear internal waves (NLIWs) are a large-amplitude short-period fluctuation that occur in the interior of the stratification ocean [1,2]. Internal solitary waves (ISWs) are a particular type of NLIW that appear frequently, in which the wave structure is isolated, propagation is unidirectional, and the velocity and waveform remain essentially constant during propagation [3,4]. One type of NLIW that is found in the upper ocean depresses the isopycnals and hence is referred to as a depression wave. Another type of NLIW elevates the isopycnals that are located in the bottom half of the ocean and hence is referred to as an elevation wave [5–7]. NLIWs are widely observed in the global oceans, such as the South China Sea [8,9], the Andaman Sea [10], the Massachusetts Bay [11], and the New Jersey's Shelf [12], among many others (Figure 1a).

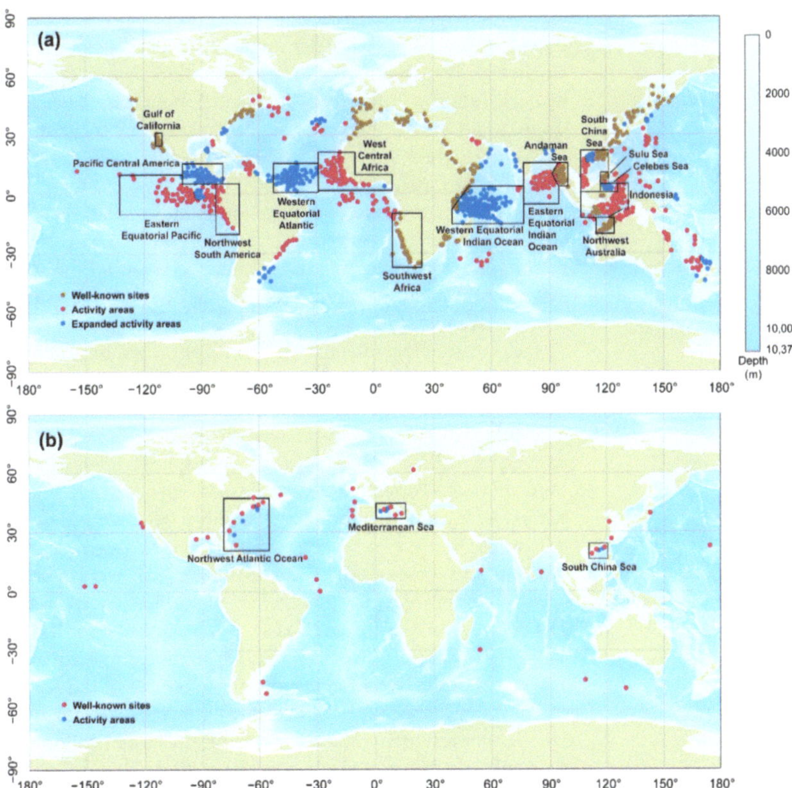

Figure 1. Nonlinear internal wave and nepheloid layer distribution around the world. (**a**) Global internal waves observed in MODIS imagery from 2002 through 2004, adapted from Jackson [13]. (**b**) Global nepheloid layers observed from 1953 through 2019, adapted from Tian et al. [14]. The well-known sites are regions where internal waves or nepheloid layers have been previously reported and their characteristics well-documented. The activity areas are regions where internal wave or nepheloid layer activity has been reported but their characteristics have not been determined. The expanded activity areas are regions in which the MODIS survey found a significant amount of internal wave activity where little or no activity had been previously reported.

Typically, NLIWs have wavelengths more than tens of kilometers long, amplitudes often exceed 200 m, and periods last from tens of minutes to several hours [2]. The strongest ISWs in the world were reported by Huang et al. [15] in the South China Sea, with amplitudes exceeding 240 m and a westward current reaching 2.55 m/s. On the one hand, NLIWs induce bottom shear stress [16] and benthic boundary layer instability [2], which may become sufficient for particle initiation and sediment resuspension, especially during ISW shoals in the continental shelf and slope [17,18], where the powerful nepheloid layers were often observed (Figure 1b). On the other hand, NLIWs are capable of inducing significant pressure (positive or negative) upon the seafloor [19,20] and excess pore pressure [21–23] in the sediment during their passage, whether they are depression waves or elevation waves. The pore pressure is the fluid pressure (liquid pressure and gas pressure) in the seabed pore space. In addition, pore fluid circulation and seepage appear in the strong pore-pressure gradient zone beneath the seafloor, the enhanced sediment resuspension, and the driving bottom nepheloid layer burst events [21,22]. Understanding this relationship is necessary, since changes in seabed stability and sediment resuspension potentially modify the submarine topography [19,24], the solute transport across the sediment-water interface [25], and the biogeochemical cycles [26].

Sediment resuspension and transport by NLIWs receive the most attention and benthic boundary layer instability driven by NLIWs is believed to be the principal mechanism [2,17]. During NLIW propagation, the observation data at 2000 m depth showed that the acoustic backscattering intensity of the benthic boundary layer was enhanced, velocities exceeded 0.6 m/s and the suspended particulate matter concentration increased by two orders of magnitude and reached 14.3 mg/L above 7.9 m of the bottom [27]. Much of what has been inferred about bottom shear stress generated by NLIWs and their effects in eroding and transporting sediment is taken from extensive compilation results of field observations, theoretical inference, numerical simulation, and laboratory experiments [16,28–30]. Unfortunately, individual research data provide little or no information about the seabed dynamic response of NLIW events and they have rarely been coupled with direct measurements of seafloor pressure or seabed pore pressure.

Conductivity, stiffness, and saturation are important mechanical properties of sediments [21,31]. The stiffness and saturation dictate how the pressure that waves exert on the seafloor is transmitted through the sediments with a fine material of weak conductivity and eventually cause the seabed dynamic responses [32], even the seabed instability. Surface waves, including solitary surface waves, have been identified as the occurrence of wave-induced seabed effective stress attenuation so far as to liquefaction [33]. For an unsaturated seabed with a small amount of gas, only a deep trough wave that induces strongly negative bottom pressure can be generating powerful seepages vertically from the seabed, then eroding and resuspending the sediment [34]. However, in deep-sea environments with depths greater than 200 m, the effect of surface waves on the seabed is extremely weak [21]. The NLIWs, whose amplitudes are 20–30 times larger than that of surface waves under the same energy conditions, can generate significant perturbations in the deep sea bottom [35]. The study of the dynamic response mechanism of the seabed under the action of NLIWs is of great significance for engineering safety and geohazard forecasts [36,37].

It is noteworthy to note that pore pressure has long been recognized by researchers as a crucial aspect of marine geohazards [38]. The earliest known attempt to measure pore pressures in submarine sediments was made by Lai et al. [39]. Since then, Richards et al. [40] first discussed the effect of pore pressures caused by the NLIW-cyclical loading of the seafloor. NLIW shoaling and breaking could cause cyclical wave pressure [41,42], which may cause changes in soil behavior. These ideas about patterns of seabed dynamic responses induced by NLIWs have been supported by subsequent numerical models and filed measurements [21–23]. The first numerical investigation work to explore the interaction between the internal wave and the elastic seabed of a porous medium was by Chen and Hsu [23]. For the highly nonlinear ISWs of the depression type, the pore-pressure field has been examined using numerical simulations [21]. Not until the late 2000s and early 2010s were time-series measurements of bottom high-accuracy pressure transducers made simultaneously with a moored acoustic doppler current profiler (ADCP) [12,20,43]. Episodic response in NLIW-induced pressure in the ocean was first documented in long-time series measurements (1.5-month duration) using a Paroscientific Model 6000-200A pressure transducer moored on the seafloor on the New Jersey shelf [43]. The first known used submarine piezometer to acquire ISW-induced pore pressure was by Ocean University of China (OUC). These authors describe a differential FBG-Piezometer that was designed to operate in water depths of up to 3500 m [44]. Richards et al. [40] hypothesized that the NLIWs induced seabed dynamic responses; since then numerous studies have contributed to understanding their effect and mechanism. Nonetheless, major questions remain: How intense, variable, or persistent are NLIW-induced seabed dynamic responses? How do pore fluid circulation and seepage affect sediment resuspension? Are high-frequency internal waves required to both initiate and maintain the buildup of pore pressure or have NLIWs generated seabed strength weakening and triggered geohazards?

In this paper, we first review known characteristics of NLIW-induced pressure structure (theoretical and numerical results, as well as the field observation data) in Section 2. The time-series bottom pressure measurements were obtained in months' long deployments of a pressure transducer and other moorings. We compare the different bottom pressure records of existing studies to gain a perception on major questions. We then investigate how the pore pressure (Section 3) and fluid seepage (Section 4) variations of the seabed relate to NLIWs. Section 5 provides a conclusion and future outlook on these issues. Our results provide important new insights into the intense, variable, and persistent NLIW-induced seabed dynamic responses. Because pore fluid circulation carries nutrients and trace elements into the benthic boundary layer [45], our results also provide key information that will be useful to submarine biogeochemical processes where pore pressure is critical for understanding NLIW–seabed interactions.

2. The Pressure Disturbance of the Seafloor

During the NLIW propagation, the seafloor pressure disturbance is a significant cause of seabed pore-pressure dynamic response, which is also reflected in the magnitude of perturbation of the NLIW–seabed interaction [21]. The pressure of NLIWs on the seafloor has been the subject of gradually intensive investigations, through both theoretical analysis and numerical simulations, supplemented in more recent years by in situ observations based on high-precision pressure sensors. All these methods are complementary and are illustrated in the following subsections.

2.1. Theoretical and Numerical Results

2.1.1. Wave-Pressure Structure of the Bottom

Theoretically, bottom pressure disturbance caused by NLIWs has been investigated previously as a problem in density and velocity changes of the water column [23,43,46,47]. As mapped by Moum and Smyth [46], the wave pressure structure of the seafloor consists of three components (as in Figure 2):

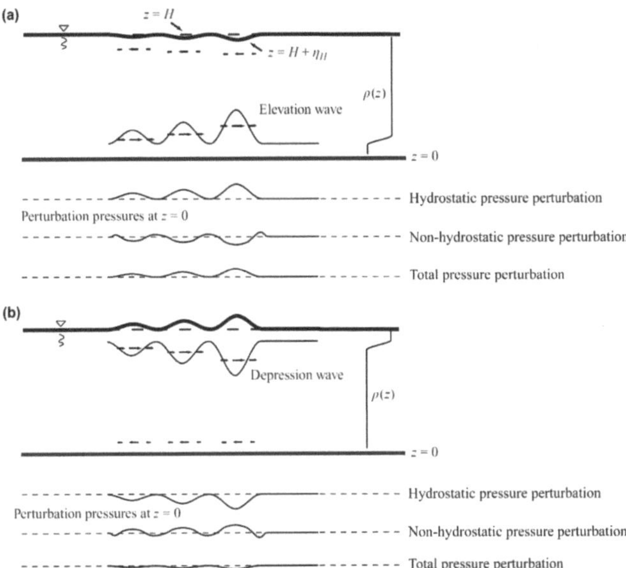

Figure 2. Schematic bottom pressure structure induced by NLIWs. (**a**) Comparable between surface wave displacement, hydrostatic pressure, non-hydrostatic pressure, and total pressure in an NLIW train of elevation propagating to the right. The black arrows show the relative velocity of the wavefronts and induced near-surface velocity. (**b**) A similar structure for NLIW train of depression propagating to the right. Panels (**a,b**) adapted from Moum and Smyth [46], with permission from Cambridge University Press.

a. An internal hydrostatic pressure perturbation as a result of the NLIW-driven isopycnals displacement, which for depression or elevation waves are negative or positive pressure, respectively.

b. An external hydrostatic pressure perturbation as a result of the free surface fluctuation, related to near-surface velocity convergence and divergence that is driven by NLIWs, which for depression or elevation waves are positive or negative pressure, respectively.

c. A non-hydrostatic pressure perturbation as a result of NLIW-driven near-seabed accelerations in the vertical, which for depression or elevation waves are dominated by positive or negative pressure, respectively.

2.1.2. Governing Equations for Wave Forcing

In the pioneering work of Chen and Hsu [23], the two-layer interfacial wave system (as in Figure 3) with inviscid and irrotational assumption was considered for solving the NLIW–seabed interaction problem. In this study, the upper layer is assumed to be a free surface and the lower layer is assumed to be a rigid boundary (i.e., as a seabed). Moreover, the NLIW motion $\eta_2(x, t)$ and free surface fluctuation $\eta_1(x, t)$ is specified as simple harmonic motion. Chen and Hsu [23] considered the independent velocity potential ϕ in both layers. Moreover, the internal wave pressure equation is deduced based on the superposition principle which was previously used to solve the boundary value problem of the complex conditions of the interaction between the surface wave and the seabed [48].

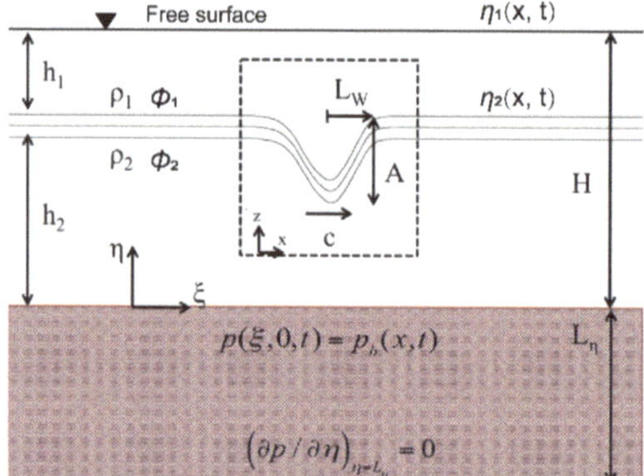

Figure 3. Typical definition sketch of a two-layer fluid system for a depression ISW propagating over a flat seabed. The seabed thickness L_η, the ISW amplitude A, half-wavelength L_w, and a phase speed c are also shown. The ISW lies in a reference frame, x–z, that moves with the wave. The bed lies in a fixed reference frame, ξ–η. The boundary conditions of the porous medium come from Rivera-Rosario et al. [21]. Figure adapted from Rivera-Rosario et al. [21], with permission from Wiley.

According to Chen and Hsu [23], the internal wave pressures $P_{Chen\ \&\ Hsu}$ at the seafloor ($z = 0$) is expressed as

$$P_{Chen\ \&\ Hsu} = \frac{1}{2} p_0 \mathrm{Re}\left[\left(e^{i(kx-\sigma t)} + e^{i(kx+\sigma t)} \right) \right], \tag{1}$$

where Re is the function real part and $i = \sqrt{-1}$ stands for the imaginary part of a complex variable. p_0 is related to the bottom pressure to the first order of the wave theory, which is given by

$$p_0 = -\rho_w \left(\sum \dot{\phi}_j \right), \tag{2}$$

where ρ_w is the fluid reference density, $j = (1, 2)$ represents the upper layer and the lower layer, and $(\cdot)$ stands for the time differentiation. The final form for the p_0 can be expressed as

$$p_{0,1} = -\rho_1\dot{\phi}_1 = -\frac{\rho_w b\sigma^2}{k}\left\{\frac{gk\cosh kd - \sigma^2 \sin h\,kd}{gk\sinh kh_1 - \sigma^2\cosh kh_1}\right\}, \tag{3}$$

$$p_{0,2} = -\rho_2\dot{\phi}_2 = \frac{b\rho_w}{k\rho_2}\left\{\frac{[-gk\rho_2\sigma^2\cosh kh_1 + (g^2k^2(\rho_2 - \rho_1) + \rho_1\sigma^4)\sinh kh_1]}{(gk\sinh kh_1 - \sigma^2\cosh kh_1)}\cosh kh_2\right.$$
$$\left. + \frac{(\rho_2\sigma^4\cosh kh_1 - gk\rho_2\sigma^2\sinh kh_1)}{(gk\sinh kh_1 - \sigma^2\cosh kh_1)}\sinh kh_2\right\}, \tag{4}$$

where $p_{0,1}$ is the bottom pressure induced by free surface fluctuation $\eta_1(x,t)$ in the upper layer, $p_{0,2}$ is the bottom pressure induced by internal wave motion $\eta_2(x,t)$ in the lower layer, h_1 and h_2 are the water depth for the upper layer and lower layer, respectively, b is the amplitudes of the internal wave, ρ_1 and ρ_2 are the water density in the upper layer and lower layer, respectively, k is the internal wave number that is solved by a new wave dispersion relationship, and ϕ_1 and ϕ_2 are the velocity potential in the upper layer and lower layer, respectively (Figure 3).

In the equation of the internal wave pressures proposed by Chen and Hsu [23], the internal wave and the free surface perturbation have been assumed to be a linear wave (i.e., the wave amplitudes are small compared with the wavelength). In the same assumption, Williams and Jeng [47] showed the attenuation of internal wave propagation in a porous elastic seabed. However, unlike Chen and Hsu [23], Williams and Jeng [47] suggested that the velocity potential ϕ_1 in the upper layer does not affect the seafloor and only the velocity potential ϕ_2 of the lower layer is required to address the wave pressure problems. Furthermore, they extended the solution to the second-order nonlinear terms of the wave theory to understand the influence of the internal wave with finite amplitude on the porous seabed. According to Williams and Jeng [47], the internal wave pressures $P_{Williams\,\&\,Jeng}$ at the seafloor ($z = 0$) are expressed as

$$P_{Williams\,\&\,Jeng} = \sum_{n=\mathrm{I}}^{\mathrm{II}} A^{(n)}e^{in(kx - wt)}, \tag{5}$$

where $A^{(n)}$ stands for the high-order pressure amplitude of the internal wave motion and the superscripts I and II represent the order of approximation.

The above model regarding internal wave pressure derived by both Chen and Hsu [23] and Williams and Jeng [47] has not been supported by laboratory scale or field scale measurements and must be reviewed in more detail. In fact, the assumption of small amplitudes is not an accurate description of NLIWs with large amplitude features [49–51] and would give rise to a significant underestimation of seabed dynamic responses by NLIWs [21]. Moum and Smyth [46] revisited the problem using elevation NLIW forms modeled by the Korteweg de Vries (KdV) equation, despite the KdV equation being in accordance with the weakly nonlinear theory. In this study, the pressure model is derived from the vertical and horizontal components of the momentum equations, associated with a two-dimensional nonlinear non-hydrostatic internal wave, based on a non-rotating inviscid Boussinesq fluid assumption.

According to Moum and Smyth [46], the NLIW pressures $P_{Moum\,\&\,Smyth}$ is expressed as

$$P_{Moum\,\&\,Smyth} = p_u + p_{Wh} + \rho_0 g\eta_H + p_{nh}, \tag{6}$$

where p_U is the ambient profile pressure, p_{Wh} is the NLIW-driven disturbance pressure, $\rho_0 g \eta_H$ is the external hydrostatic pressure, and p_{nh} is the non-hydrostatic pressure. The p_U, p_{Wh}, η_H, and p_{nh} are written as

$$p_U = \int_z^H (\rho_0 + \rho_U) g \, dz, \tag{7}$$

$$p_{Wh} = \int_z^H \rho_w g \, dz, \tag{8}$$

$$\eta_H = -\frac{1}{g} \int_{-\infty}^x \frac{Du_H}{Dt} dx, \tag{9}$$

$$p_{nh} = \rho_0 \int_z^H \frac{Dw}{Dt} dz, \tag{10}$$

respectively, where ρ_0 is the reference density, $\rho_0 + \rho_U$ represents the ambient density profile, and where ρ_w is the NLIIW-driven perturbation density, η_H is the free surface displacement, and u_H represents the wave propagation velocity. These models about the pressure of NLIWs have been supported by subsequent in situ measurements [46] and are validated in detail by Moum and Nash [43]. The highly sensitive pressure sensors provide details of wave pressure dynamics and subsequently filter raw time series of pressure required to extract the pressure signal of NLIWs. The computation of the bottom pressure signal of an NLIW from measurements of density and velocity is compared with the in situ observed pressure data to verify the correctness of the equation [43]. While it is widely known that the wave amplitude based on the KdV equation compared with the large amplitude observed in the field is too narrow [3,52], the fundamental pressure mechanism suggested by Moum and Smyth [46] is trusted to hold. Following the numerical methods given by Moum and Smyth [46], Rivera-Rosario et al. [21] revisited the issue by employing large-amplitude ISW described by the fully nonlinear Dubriel-Jacotin–Long (DJL) equation; both depression and elevation wave- types were considered.

Theoretical pressure models are useful for understanding the impact of NLIWs on the seabed and further pore-pressure response can be calculated. Compared to $P_{Moum~\&~Smyth}$, both $P_{Chen~\&~Hsu}$ and $P_{Williams~\&~Jeng}$ take a small amplitude linear internal wave assumption in with a two-layer simplified within the water column. In this way, a two-layer system simplifies the governing equations to accelerate the calculation time but introduces a greater uncertainty in the quantification of the NLIW pressures. In contrast, $P_{Moum~\&~Smyth}$ is more suitable for simulating large-amplitude NLIW pressures in the continuously stratified water column of real marine environments. However, for $P_{Moum~\&~Smyth}$, the assumption that without a change of NLIW form limits the estimation of pressure during wave breaking or wave mode transformation, as well as waveform inversion and for NLIW propagation with mode-2. Although theoretical and numerical results may guide seabed dynamic responses, direct application to the ocean to predict changes in the seafloor must be conducted with caution.

2.1.3. Characteristics of Wave-Pressure Disturbance

Based on the pressure structure of NLIWs shown in Figure 2, the wave pressure is dominated by internal hydrostatic pressure perturbation, whether it is a depression-wave type or an elevation-wave type [22,46]. Furthermore, the combination of external hydrostatic pressure perturbation and non-hydrostatic pressure perturbation always creates a negative contribution to the wave pressure, reducing the amplitude of the total pressure [43,46] and weakening the disturbance effect of NLIWs on the seafloor [21]. For depression NLIWs, these signs are reversed (Figure 2b). The wave pressure structures of depression NLIWs and elevation NLIWs have opposite mirror features [43,46].

As mapped by Moum and Smyth [46], the pressure disturbance on the seafloor is controlled by internal hydrostatic pressure perturbation, that is to say, the greater the NLIW amplitudes, the greater the wave pressure generated (Figure 4a). However, in other numerical studies, simple linear correlations between NLIW amplitudes and seafloor pressures were not as expected. Rivera-Rosario et al. [21] investigated the dependency between the maximum ISW amplitude that is permissible and the wave pressure to determine which environmental parameter scheme gains the largest response. Numerical parameterized results show a significant pressure decrease when the ISW amplitude increases (i.e., changing $h_1/h_2 = 1/7$ to $h_1/h_2 = 1/10$, h_1/h_2 represents the ratio of upper and lower water depths) considering the different ratios of pycnocline thickness δ_{pyc} (the pycnocline thickness is defined as the thickness of the water column in which the density transition occurs) and water depth H (as shown in Figure 4b). They argued that higher amplitude ISWs also produced a simultaneously higher non-hydrostatic pressure which was associated with variations of fluid vertical motions. The non-hydrostatic pressure reduced the total ISW pressure on the bottom.

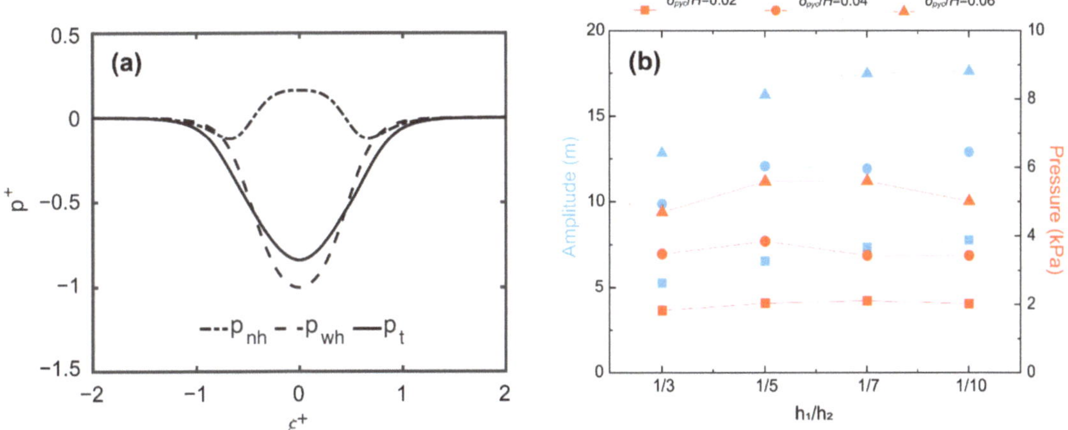

Figure 4. Bottom pressure induced by internal solitary waves for different environmental parameter schemes. (**a**) ISW-induced bottom pressure with a thickness ratio of $h_1/h_2 = 1/7$. The dash–dash line, dash–dot line, and solid line show the hydrostatic pressure, the non-hydrostatic pressure, and the summation of the above, respectively. (**b**) Amplitude and pressure changes of ISWs for the different ratios of pycnocline thickness δ_{pyc} and the ratio of upper and lower water depths. Panel (**a**) adapted from Rivera-Rosario et al. [21], with permission from Wiley. And the data of panel (**b**) come from Table 2 of Rivera-Rosario et al. [21].

2.2. Observation Results

Published observations of pressure disturbance by NLIWs provide the most realistic seabed response, which can be compared with the numerical results to examine the pressure model. The in situ NLIW pressure results from each of the observation regions are reviewed below (as shown in Table 1).

Table 1. Summary of past 15 years on the observation details of seafloor pressure disturbance induced by internal waves.

Observation Area	Range	Water Depth (m)	Transducer	Full Scale (MPa)	Resolution (Pa)	Sampling (Hz)	Height (mab)	Maximum Pressure (kPa)	Wave Event	Wave Amplitude (m)
New Jersey Shelf [43]	38.80–39.2° N 72.75–73.50° W	70–110	Paroscientific Model 6000-200A	1.4	1.4	1	0	−0.765	Depression NLIWs	/
Massachusetts Bay [11]	41.78–42.68° N 69.99–71.09° W	10–80	Paroscientific Model 6000-200A	1.4	1.4	1/0.5	0	±0.2	High-frequency NLIWs	~20

Table 1. *Cont.*

Observation Area	Range	Water Depth (m)	Transducer	Full Scale (MPa)	Resolution (Pa)	Sampling (Hz)	Height (mab)	Maximum Pressure (kPa)	Wave Event	Wave Amplitude (m)
Great Meteor Seamount [53]	30.00° N 28.30° W	549	Sea-Bird Scientific SBE 53	/	/	0.33	1.7	0.05	High-frequency internal waves	/
Marsdiep Strait [20]	52.98° N 4.77° E	23	Sea-Bird Scientific SBE 26	/	/	4	0.08	0.1	High-frequency internal waves	/
South China Sea [19]	20.3° N 115.4° E	481	/	/	/	/	10	20	Obliquely incident ISWs	~80
South China Sea [54]	/	/	/	/	/	/	/	4	Depression ISWs	/
Aogashima Island Slope [55]	32–33° N 140–141° E	1470– 2240	Paroscientific 8B7000-I-005	68.95	/	4/0.7	0	0.05	Internal tide wave	/

Note: mab—meters above the bottom; SBE—Sea-Bird Electronics; NLIW—nonlinear internal waves; ISW—internal solitary waves.

2.2.1. North American Atlantic Coast

In the Shallow Water 2006 experiment on the New Jersey shelf, Moum and Nash [43] deployed three Paroscientific Model 6000-200A pressure transducers (sampled at 1 Hz) mounted in three landers, respectively, to capture NLIWs that propagate onto the shelf. These seafloor NLIW pressure measurements provide adequate experimental verification to confirm the NLIW pressure model derived by Moum and Smyth [46]. Furthermore, observed NLIW trains supply a detailed understanding of the NLIW pressure field which has not previously been quantified. The seafloor NLIW pressure disturbance signals become visible by a low-pass bandpass filter at 1/20 Hz and a high-pass bandpass filter at 1/2000 Hz. A train of negative bottom pressure signal relates to the typical signatures of horizontal and vertical components of NLIWs velocity and the maximum pressure magnitude over 765 Pa (as shown in Figure 5). In addition, it does not appear that NLIW packets have a powerful pressure signal and near seabed current velocity [43].

Most importantly, the correlation between pressure and velocity provided an important understanding of the spatial distribution and structure of NLIW energy transport flux [7,11]. Thirteen high-frequency NLIW packets were observed within 6.5 days in Massachusetts Bay [11], with obvious pressure properties in along-shore and cross-shore spatial variability. The bottom pressure data were collected from an array of 14 Paroscientific Model 6000-200A pressure transducers at 1 Hz or 2 Hz and band-pass filtered low-pass at 1/120 Hz and high-pass at 1/1800 Hz to distinguish the NLIW pressure signal. The pressure amplitude oscillations weaken from 580 Pa to 36 Pa with NLIW propagation cross-shore to the shallow sea (as shown in Figure 6), accompanied by synchronized decay of the high-frequency NLIW energy. Thomas et al. [11] argued that the strong connection between kinetic energy and seabed pressure variation allows qualitatively the kinetic energy variability of high-frequency NLIWs using the bottom pressure transducer array. Moreover, because of the strong dependence on background shear currents and water depth, an attempt to quantify the NLIW energy flux is impracticable.

Because of the significant seafloor pressure response by NLIWs, the ocean bottom pressure (OBP) direct measurement allows real-time automated detection and alert of NLIWs [12,55] without on-duty personnel or expensive instrumentation. Stöber and Moum [12] demonstrate an NLIW detection algorithm using seafloor pressure data from the New Jersey shelf, which could be employed for warning wave events that pressure amplitude over 250 Pa. Before this, OBP was commonly used for the detection of tsunamis [56], seismic waves [57], seafloor deformation [58], and many others.

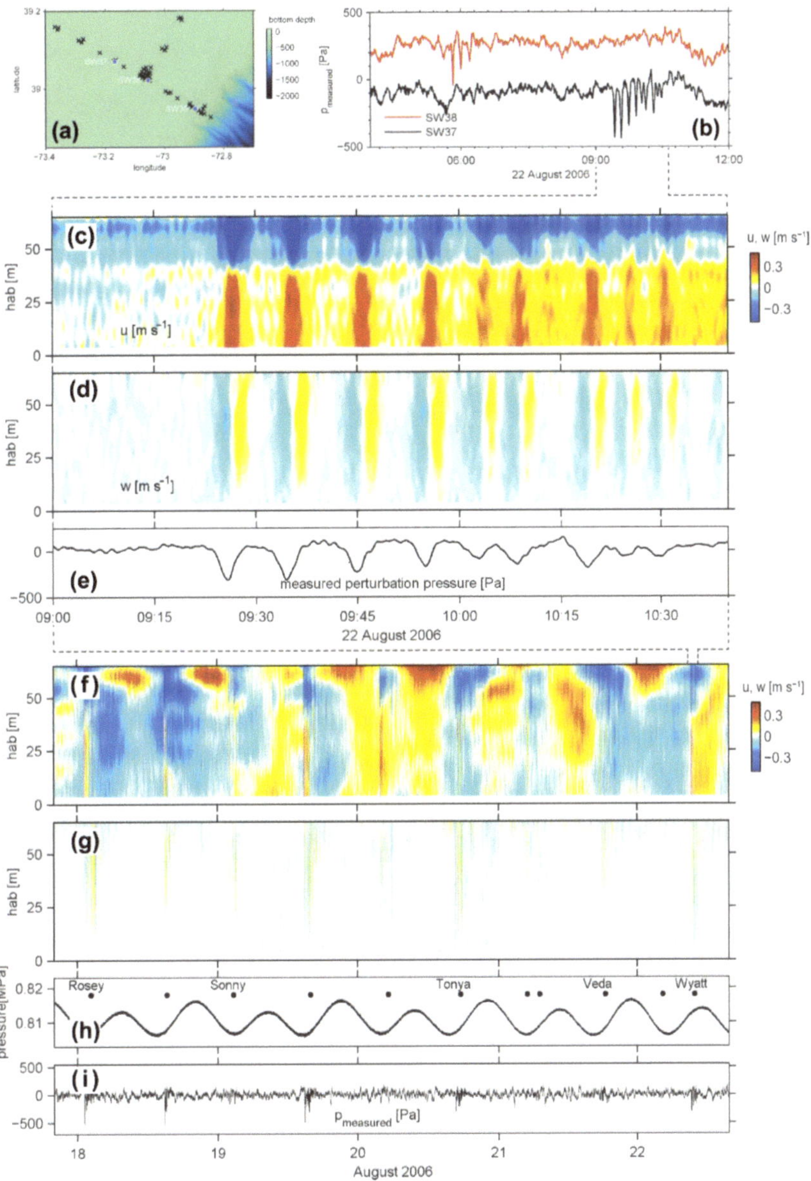

Figure 5. Seafloor pressure measurements of NLIWs on the New Jersey shelf. (**a**) Mooring locations of SW37, SW38, and SW39 where bottom pressure observations were conducted. (**b**) The NLIW named Wyatt induces bottom pressures at SW37 and SW38, respectively. Note the bottom pressure signals have been offset for clarity. For NLIW Wyatt, comparison between horizontal velocity (**c**), vertical velocity (**d**), and bottom pressure (**e**) at SW37 on 22 August 2006. For all NLIW observed at SW37, comparison between time series of (**f**) horizontal velocity, (**g**) vertical velocity, (**h**) bottom pressure, and (**i**) filtered bottom pressure from 18 to 23 of August 2006. Dots in (**h**) show the named NLIWs, including NLIW Wyatt. The hab is expressed as height above the bottom. Panels (**a**–**i**) adapted from Moum and Nash [43], with permission from the American Meteorological Society.

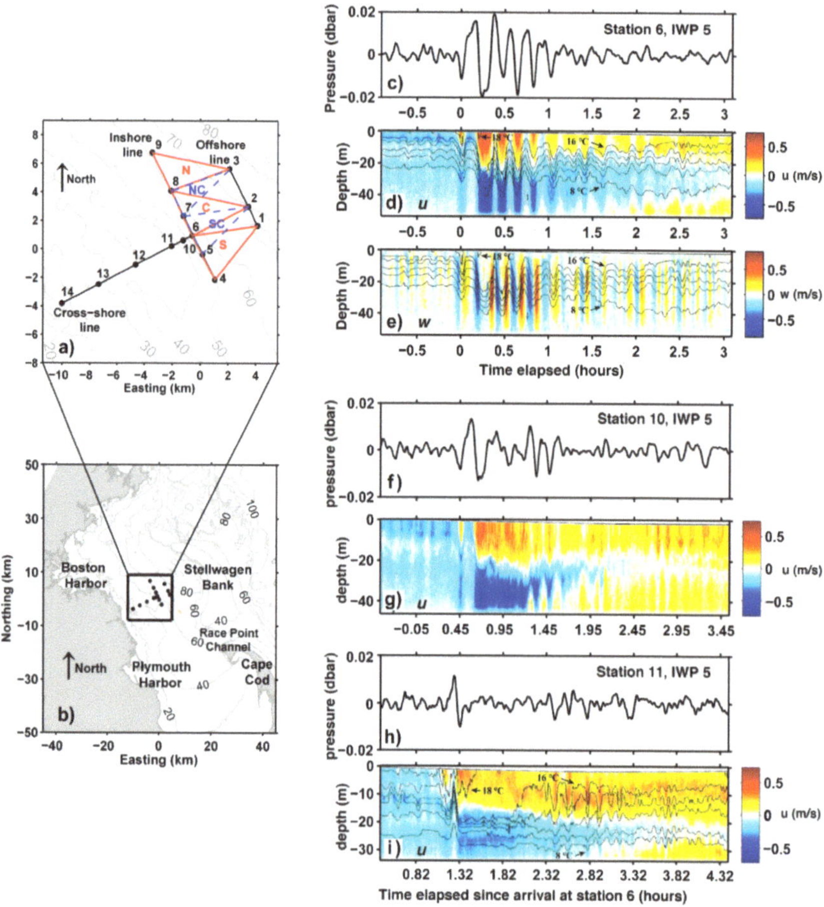

Figure 6. Seafloor pressure variability of the internal wave packet (IWP) in Massachusetts Bay detected by a sensor array. (**a**) Station locations (black dots) of bottom pressure sensors. (**b**) Map of the investigation area in Massachusetts Bay, the black box is (**a**). For IWP 5, comparison between (**c**) filtered bottom pressure, (**d**) horizontal velocity u and temperature, and (**e**) vertical velocity w and temperature at station 6. Comparison between (**f**) filtered bottom pressure and (**g**) horizontal velocity u at station 10. Comparison between (**h**) filtered bottom pressure and (**i**) horizontal velocity u and temperature at station 11. The temperatures in (**d,e,i**) represent black contour lines at 2 °C intervals from 8 °C to 18 °C. Time 0 h is the timing of IWP 5 reaching station 6. Panels (**a–i**) adapted from Thomas et al. [11], with permission from Wiley.

2.2.2. North American Pacific Coast

In the Oregon shelf, the properties of ISWs, internal bores, and gravity currents were observed concurrently with high-intensity acoustic scattering layers induced by sediments resuspending having a vertical extent over 30 m above the seafloor [7]. A SeaBird conductivity and temperature recorder that was located 1 m above the bottom captured intense pressure pulses due to NLIW propagation. Moum et al. [7] found that the pressure–velocity energy flux had significant contributions from the effects of non-hydrostatic and free surface displacement, which implied the substantial impact of non-hydrostatic pressure and external hydrostatic pressure perturbation. The vertical component of non-hydrostatic effects and free surface displacement dominates the NLIW energy transporting near the seabed.

2.2.3. South China Sea

Energetic ISWs were consistently observed in the South China Sea, causing strong near-seabed oscillatory currents over 0.8 m/s [19]. A pressure dramatic fluctuation has been found over 2 kPa using a pressure transducer mounted on a mooring at about 10 m above the seabed, which was accompanied by an ISW with an amplitude exceeding 80 m crossing the slope (as shown in Figure 7). The largest velocity of the westward and northward components are both over 0.4 m/s. Yang et al. [54] observed large amplitude depression ISWs of mode-1 with large amplitude produced near bottom pressure perturbations more than 4 kPa in the South China Sea. A 2.5 kPa bottom pressure event is induced by ISWs, which corresponds to approximately 1.5 m/s velocities in the horizontal component. The bottom pressure perturbation caused by ISWs can be described well by the Bernoulli balance.

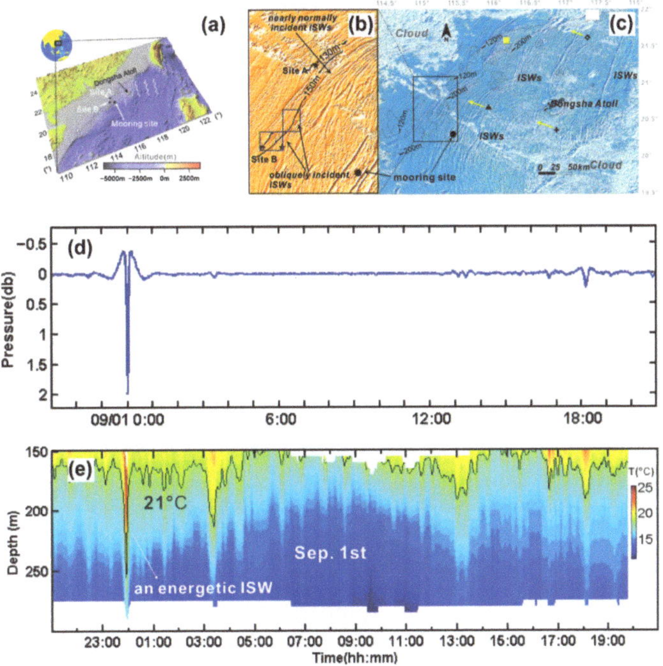

Figure 7. Seafloor pressure of obliquely incident ISWs in the northern South China Sea. (**a**) Location of the mooring site. (**b**) Expansion of boxed region in (**c**). The obliquely incident ISWs (deep brown imprints) near the mooring site are shown in the amplified MODIS image. (**c**) The MODIS image illustrates the ISWs' direction. Approximately 2.5 kPa of pressure changes at 10 m above the bottom at the mooring site (**d**), when the largest fluctuation of 21 °C isotherms is down about 80 m (**e**), reflecting a crossing energetic ISW propagation. Panels (**a–e**) adapted from Ma et al. [19], with permission from Wiley.

2.2.4. European North Atlantic Coast

In the Marsdiep strait between Texel Island and the Netherlands mainland, van Haren [20] observed a pressure oscillation corresponding to high-frequency internal wave propagation using a SeaBird SBE 26 wave and tide recorder with a pressure transducer (4 Hz sampling rate). The spectra analysis of bottom pressure was used to distinguish the surface waves, near-surface turbulence and waves by wind and ferry, turbulent overturns, and high-frequency internal waves. van Haren [59] synthesized bottom pressure data collected in the range from a shallow Marsdiep strait via the Baltic Sea and the seamount in the open Atlantic Ocean to the deep Gulf of Mexico and assessed contributions of internal waves and turbulence in different bottom pressure data with the support of additional temperature and current data. All observations are in sloping bottom topography, where

NLIW propagation becomes near critical and shows internal wave–turbulence transition. In the Great Meteor Seamount of the Atlantic Ocean, van Haren [53] observed that the pressure changes were equivalent to the fluctuation of hydrostatic pressure due to high-frequency internal waves propagating up to 100 m above the bottom and non-hydrostatic pressure (reduced by 50 Pa) following turbulence due to breaking of the internal wave in the range of 50 m above the bottom.

2.2.5. Japan Western Pacific Coast

On the eastern slope of Aogashima Island, Fukao et al. [55] observed the power spectral density of the pressure changes of the M2 internal tide, using a pressure transducer array of 10 stations with a 10 km distance that formed equilateral triangles. The results show that the pressure power spectral density of the M2 internal tide was about 10^8 Pa^2/Hz, equivalent to a bottom pressure on the order of 50 Pa.

3. The Pore-Pressure Variation of the Seabed

The pore-pressure variation of the seabed occupies a central position in seabed dynamic responses during the NLIW-action for conceptual and practical reasons. Conceptually, pore pressures are necessary to evaluate seabed seepage from calculated vertical gradients and thus to allow the rational explanation of the sediment resuspension and benthic boundary layer instability. Previously, from a theoretical perspective, sediment resuspension caused by internal waves was recognized as an issue with hydrodynamic stability [2]. Practically, the pore pressure in marine sediments is often an easier measure than other aspects of seabed-response behavior, because it is omnidirectional [60]. At present, there are theoretical and numerical results in the study of pore-pressure variation by NLIWs. The following subsections set forth the basic theoretical formulations and characteristics of pore-pressure response and some of the experimental evidence for their existence.

3.1. Governing Equations for Pore-Pressure Response

The fundamental theory for porous seabed response was first developed in surface waves and further considered in internal waves by Chen and Hsu [23]. Biot's consolidation equation has been widely used for wave–seabed interactions since 1978 [32] and was discussed in depth in Jeng [61,62]. The description of the seabed is, in general, based on the equations of compressible fluids in an elastic and infinite-thickness porous media. The fluid flow in the pore space of the porous seabed obeys Darcy's law.

Using a combined model of Biot's consolidation equation in the two-dimensional space [63] and the storage equation [64], Chen and Hsu [23] presented a numerical method for the pore pressure and sediment displacements in an isotropic poroelastic seabed induced by internal waves. The governing equation of excess pore pressure P in the seabed is written as

$$K_x \frac{\partial^2 P}{\partial x^2} + K_z \frac{\partial^2 P}{\partial z^2} - r_w n \beta \frac{\partial P}{\partial t} = r_w \frac{\partial \varepsilon}{\partial t}, \tag{11}$$

where K_x and K_z are the permeability coefficients in the x and z directions, respectively, r_w is the unit weight of the pore water in the seabed, n is the porosity of sediment, and β is the compressibility of the pore fluid in the seabed. The ε is the volume strain obtained by

$$\varepsilon = \frac{\partial \zeta}{\partial x} + \frac{\partial \chi}{\partial z}, \tag{12}$$

where ζ and χ are the sediment displacements in the x and z directions, respectively. Moreover, further consideration of the parameterization of saturation S_r, the compressibility β is given by

$$\beta = \frac{1}{k_w} + \frac{1 - S_r}{P_0}, \tag{13}$$

where k_w is the bulk modulus of elasticity of water and P_0 is the absolute pore pressure. The excess pore pressure P, volume strain ε, and sediment displacements that ξ and χ can be related by equilibrium equations that are written as

$$G\nabla^2\xi + \frac{G}{(1-2\mu)}\frac{\partial\varepsilon}{\partial x} = \frac{\partial P}{\partial x},\tag{14}$$

$$G\nabla^2\chi + \frac{G}{(1-2\mu)}\frac{\partial\varepsilon}{\partial z} = \frac{\partial P}{\partial z},\tag{15}$$

where μ is the Poisson ratio and G is the shear modulus and can be related by

$$G = \frac{E}{2(1+\mu)},\tag{16}$$

where E is the Young modulus. The governing Equations (11)–(16) incorporated within the boundary conditions constitute the components needed from which the pore pressure and sediment displacements in the seabed can be solved [23]. On the basis of Biot's consolidation equation, the seabed response theories proposed by Chen and Hsu [23] have been used widely with some success to model the seabed interaction with internal waves [35,47,65].

Pore-pressure response inside the porous seabed has been explored with numerical solutions of the mass conservation equation [66]. More recent examples are the numerical examining of Rivera-Rosario et al. [21], who sought solutions for a pore-pressure response for large-amplitude NLIWs. In this method, the seabed horizontal deformation was neglected. The pore pressure p is expressed as

$$\frac{k}{\rho_0 g}\nabla^2 p = (\alpha(z)+n\beta)\frac{\partial p}{\partial t},\tag{17}$$

$$\alpha(z) = \frac{1}{2\overline{u}(z)+\overline{\lambda}(z)},\tag{18}$$

where k is the permeability of the seabed, ρ_0 is the reference density of the seawater, $\alpha(z)$ is the vertical compressibility, n is the porosity, β is the compressibility and is determined by Equation (13), $\overline{u}$ is the effective shear modulus, and $\overline{\lambda}$ is Lamé's effective first parameter. The numerical method of Fourier–Legendre collocation is used to solve Equation (17).

Although the seabed is an elastoplastic material, the elastic models have played the primary role in elucidating the transient features of the pore pressure and deformation, if not considering the precise details of permanent deformation and accumulated pore pressures. They have the advantage of permitting modeling with reduced equations wherein the parameters are fewer and easy to obtain. For elastoplastic models, accumulated pore pressures and deformation can be studied, but the ease of generalization is often lost. Furthermore, internal waves are restricted to slower phase velocity and longer periods compared with surface waves and were consequently considered unable to generate pore-pressure build-up. Thus, the elastic models have been adopted as the wide application of choice. However, elastoplastic models are needed to accurately describe the properties of seabed dynamic responses.

3.2. Vertical Profile of the Pore-Pressure Changes

The changes in internal waves inducing pore pressure of the transient state were considered firstly by Chen and Hsu [23] and the instance of NLIWs is further considered by Rivera-Rosario et al. [23]. Along with the propagation of depression NLIWs, a negative pore-pressure enhancement occurred inside the superficial seabed under the trough of the wave [23,35], when a strong countercurrent against the wave direction was present as positive pressure imposed from the bottom boundary layer down into the seafloor [43,46]. Since the total pressure load by NLIWs is negative, the wave propagation produces an insufficiency in the pre-existent pore pressure and this insufficiency is more obvious under the wave trough [21]. Before the arrival of the wave trough, the pore pressure inside the superficial seabed shows a slight increase, roughly estimates one order of magnitude lower than that under the wave trough, and decays rapidly in the deeper seabed (as shown in Figure 8). This is corresponding to the horizontal pattern of the bottom pressure disturbance by NLIWs.

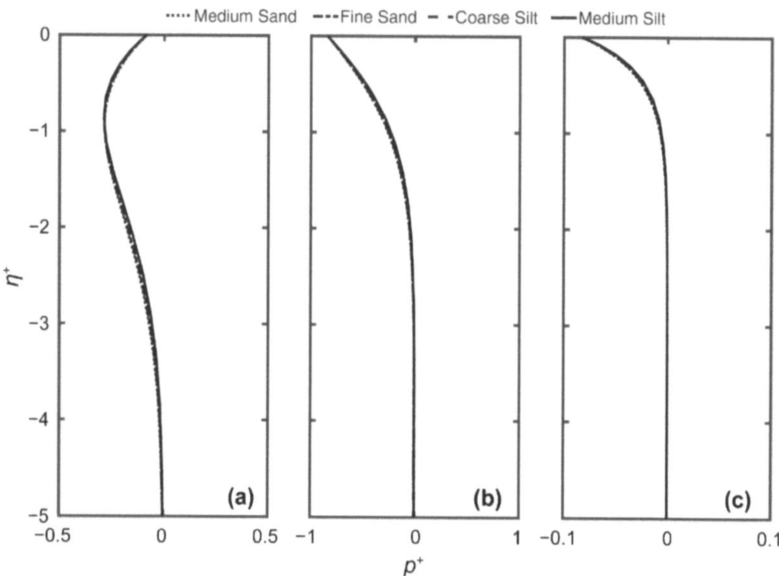

Figure 8. Pore-pressure changes in the vertical direction by ISWs. For different seabed materials, the changes of pore pressure, where (**a**) is after the ISW trough, (**b**) is under the ISW trough, and (**c**) is before the ISW trough. Panels (**a–c**) adapted from Rivera-Rosario et al. [21], with permission from Wiley.

After the passage of the wave trough, the superficial pore pressure is attenuated to the degree before the wave trough, but the pore-pressure insufficiency persists under the seafloor after the wave has passed by. Rivera-Rosario et al. [21] employed an effective diffusion coefficient κ based on Equation (17) to simulate the phenomenon of pore-pressure lag. This phenomenon of lag was first observed in surface wave induced pore-pressure response [32]. The effective diffusion coefficient κ is expressed as

$$\kappa = \frac{k}{\rho_0 g(\alpha(z) + n\beta)} \tag{19}$$

with units of length squared over time. Since the pressure diffusive time scale, L_w^2/κ, is larger than the time scale of the ISW propagation, L_w/c, a pore-pressure phase lag inside the seabed will occur after the passage of the wave trough (as shown in Figure 8).

The numerical results depicted that vertical distributions of the negative pore-pressure by internal waves decrease as the soil depth increases, but the rate of pore-pressure decay depends on the wave and seabed parameters [22,23]. A parametric method is applied to examine seabed dynamic responses induced by NLIWs and has been considered in detail in numerous studies [22,23,47]. Chen and Hsu [23] compared vertical pore-pressure decay by varying relevant parameters of internal wave characteristics and seabed properties. The results demonstrate that the rate of pore-pressure decay occurs faster when accompanied by a large water depth and an internal wave period, as well as the sediment permeability and saturation, or small sediment shear modulus, hydraulic anisotropy, and porosity.

High-accuracy numerical simulations on the NLIWs with large amplitude reported the variations in pore pressure to penetrate deeper into the seabed by a depression wave in the presence of a small amount of pore gas in the seabed [21]. The content of trapped-pore-gas increase will strengthen the pore-pressure build-up and provide steeper vertical gradients in the seabed. For a high permeability of sediment, the negative pore pressure will penetrate deeper into the seabed and approaches a length comparable to the amplitude of ISWs [21], but also means that the pore pressure will dissipate faster compared with low permeability of sediment. For NLIWs in a lake environment, Olsthoorn et al. [22] demonstrated that the long wavelength induced the penetration depth of pore-pressure increase and expanded the range of seepage through the porous domain. Tian et al. [35] modeled seabed pore pressure and experimentally visualized pore-pressure penetration depth from ISWs over a sloping bottom. The results presented that the perturbations approximated tens-to-hundreds of meters in extent, one order of magnitude smaller than the wavelength, which perhaps was deeper than that of surface wave induced pore-pressure changes. Nevertheless, in other studies, the quantification of penetration depth from ISWs was not as clear. Rivera-Rosario et al. [21] found the depth in a pore-pressure impact approximately 1 m below the water–sediment interface, for a seabed with lower permeability. They argued that the higher phase velocity of NLIWs implies an enhancement of the pore-pressure build-up near the superficial sediments but will not permit deeper in the seabed (Figure 9).

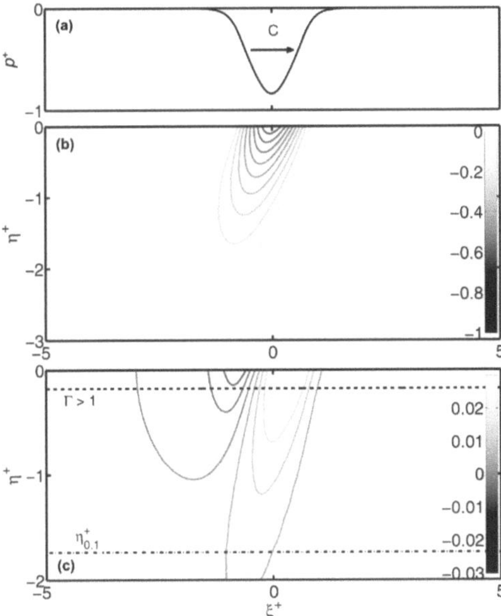

Figure 9. Contour of the pore-pressure changes by ISWs. The ISW traveling to the right with phase speed c, where (**a**) is the ISW-induced pressure forcing and (**b**) is the pore-pressure changes for an unsaturated seabed composed of medium sand. (**c**) Vertical pore-pressure gradient with an unsaturation seabed composed of medium silt. The dashed-dot line marked as $\eta_{0.1}^{+}$ shows the penetration depth, where the changes of pore pressure are 10% of the bottom pressure under the wave trough. The dashed line marked as $\Gamma > 1$ shows the failure criterion, where the buoyant unit weight of sediment is less than the vertical pore-pressure gradient. Panels (**a**–**c**) adapted from Rivera-Rosario et al. [21], with permission from Wiley.

The experimental investigation of Tian et al. [35] provided qualitative confirmation of the theoretical predictions of the pore-pressure changes caused by NLIWs. The experiments were conducted with a 9° slope and the negative pore-pressure inversion accompanied by depression ISW shoaled and transformed into an elevation wave was observed. In the experimental work by Qiao et al. [67], a similar negative pore-pressure depression was observed along with sediment resuspension during the ISW breaking as the depression wave shoaled, indicating that the seabed response is sufficient to produce a significant contribution to sediment resuspension. The phenomenon of polarity changes and the breaking of ISW has been observed in continental slopes [68–70]. Li et al. [71] experimentally visualized the pore-pressure response from ISWs over a sloping bottom. The results demonstrated that the accumulation and release of pore pressure occurred simultaneously before the seabed failure, but the increase of the wave amplitude will enhance the release process. Although experiments may guide first-order pore-pressure dynamic responses by NLIWs, direct application to the in situ observation must be conducted with circumspection.

3.3. Potential Failure Due to Seabed Instability

The above review motivates further consideration of potential failure due to seabed instability under the presence of pore-pressure changes by NLIWs. The motivation is to emphasize the contribution of the seabed dynamic response induced by NLIWs during seabed instability and to motivate subsequent in situ measurements that will support the theoretical findings. The transient liquefaction of the seabed induced by depression NLIWs was considered by Rivera-Rosario et al. [21]. The vertical pore-pressure gradient can be compared to the buoyant unit weight of sediment as a measure of possible transient liquefaction [66]. The buoyant unit weight γ_s is expressed as

$$\gamma_s = (1 - n)(\rho_s - \rho_0)g,\tag{20}$$

where n is the porosity of the seabed, ρ_s is the sediment density, and ρ_0 is the fluid density in the bottom. Therefore, if the vertical pore-pressure gradient is greater than the buoyant unit weight of sediment, transient liquefaction will occur, or

$$-\frac{\partial p}{\partial \eta} \geq \gamma_s\tag{21}$$

for a seabed in a fixed reference frame $\xi - \eta$. Thus, yields the normalizing failure criteria Γ of transient liquefaction by NLIWs as

$$\Gamma = \frac{\partial p}{\partial \eta}\left[\frac{(\rho_s - \rho_0)gA}{\gamma_s\eta_0}\right] \geq 1,\tag{22}$$

where A is the amplitude of depression ISW and η_0 is the vertical dimension that is scaled by the diffusive depth as

$$\eta_0 = \sqrt{\frac{\kappa L_w}{c}},\tag{23}$$

where κ is the effective diffusion coefficient, L_w and c are the half-wavelength and phase speed of depression ISW, respectively. With the early exploration of the fully nonlinear model of NLIWs with numerical solutions of the DJL equation, Rivera-Rosario et al. [21] recognized that the pore-pressure changes influence roughly the top 1 m of the thickness of the low-permeable seabed, with only 2 cm sediments appearing in transient liquefaction under the superficial seabed. When the saturation of the seabed is decreased, the vertical pore-pressure gradient is powerful enough to exceed the buoyant unit weight and induce sediment transient liquefaction observed in the simulation [21]. In the experimental work by Li et al. [71], elevated pore-pressure response with a decreased gradient of the slope was seen as the ISW shoaled during the drawdown period before breaking, indicating that the gently sloped seabed response is more sensitive to the wave amplitude changes.

For the stratified water column, the elevation NLIWs can form when the thickness of the upper layer is larger than that of the lower layer [69,72,73]. The mechanisms of seabed failure induced by the elevation NLIWs were argued by Rivera-Rosario et al. [21]. Due to the positive bottom pressure load dominated by hydrostatic pressure [46], related to the pycnocline displacement, positive pore pressure penetrated the seabed as a result of the elevation NLIWs. The vertical pore-pressure gradient was also generated by the elevation NLIWs, but was insufficient to exceed the buoyant unit weight of the sediment [21]. Thus, during the propagation of elevation NLIWs, the wavefront is steep enough to induce a horizontal pore-pressure gradient that is powerful enough to overcome the mobilized internal friction of the seabed and capable of resulting in seabed failure [74]. Based on Equation (17), for a seabed in a fixed reference frame $\xi - \eta$, the dimensional seabed failure criterion can be expressed as

$$\frac{\partial p}{\partial \xi} > \gamma_s \tan \phi, \tag{24}$$

where ϕ is the angle of internal friction ϕ of the seabed. Equation (24) has been used to examine seabed instability by a surface solitary wave [75].

While there is no in situ measurement of the pore-pressure changes for NLIWs, field observation, including surface wave-induced seabed failure, has reported referable results. Torum [76] reported that pore-pressure changes induced by surface waves penetrated more than 10 m depths in the unsaturated seabed. Another fieldwork by Cross et al. [77], reported pore pressure reaching 37 m below the seafloor by a surface wave height of at least 3 m at a water depth of 5 m. Bennett and Faris [78] found a penetration depth of more than 6 m at the Mississippi Delta at a water depth of 13 m. However, as noted earlier, a similar study implies that the depth below the seafloor of NLIWs inducing seabed instability would be smaller than that of surface waves, but has not been explored in a detailed examination with strong NLIWs in the field.

4. The Seepage and Fluid Circulation in Sediment

The vertical gradient of pore pressure within the seabed can be a measure of seepage in the vertical direction and, therefore, of the possible impact on sediment resuspension [21]. Seepage of pore fluid in a permeable seabed was accompanied by fluid circulation through the sediment–water interface [22], which can have far-reaching implications on heat and chemical exchanges between the seabed and the water column [45,79]. The primary manner of generating seepage and fluid circulation in sediment is through the spatial variability of pressure fluctuations at the bottom [80]. In recent years, few studies have examined the seepage and fluid circulation induced by the passage of an NLIW in any systematized way; rare numerical results in existence are presented below.

4.1. Governing Equations for Seepage of Pore Fluid

Darcy's law is a classical relationship that links pore-pressure gradients and the seepage velocity in the seabed, although it is experimentally determined. Once the wave pressure distribution is determined at the bottom, the distribution of fluid seepage in the seabed can be acquired by Darcy's law [81]. A useful extension of the fluid seepage due to internal waves was proposed by Olsthoorn et al. [22]. They calculated that NLIWs caused seepage through the lakebeds of deep regions, where surface gravity waves had little impact, using the Dubreil-Jacotin–Long (DJL) equation and Darcy's law. The lakebed is assumed to be rigid and saturated, thus the seepage velocity $\vec{w}$ based on Darcy's law is expressed as

$$\vec{w} = -\frac{k}{\mu}\vec{\nabla}(P + \Pi), \tag{25}$$

where k is the permeability of the lakebed, μ is the dynamic viscosity which is assumed to be a constant for a given fluid, P is the pressure, and Π is the gravitational potential energy per unit volume. $P + \Pi$ in Equation (25) is defined as the gauge pressure. For an incompressible fluid, the condition is expressed as

$$-\vec{\nabla} \cdot \left(\frac{k}{u}\vec{\nabla}P\right) = 0. \tag{26}$$

Furthermore, the stratification of permeability k in the porous lakebed was considered and was taken as a hyperbolic tangent function. The permeability k is written as

$$k_{Inc} = 1 + \frac{1}{2}\left[1 + \tanh\left(\frac{z - 0.5}{0.1}\right)\right], \tag{27}$$

$$k_{Dec} = 2 - \frac{1}{2}\left[1 + \tanh\left(\frac{z - 0.5}{0.1}\right)\right], \tag{28}$$

where k_{Inc} and k_{Dec} are the permeability k for increasing and decreasing in the vertical direction of the seabed, respectively.

The impressive point of the seepage model presented above was to employ the bottom pressure extracted from the solution of the DJL equation to drive seepage in a porous lakebed. However, this model has a potential limitation as a seepage tool by NLIWs. As opposed to previous numerical results [21,35], the seepage model proposed by Olsthoorn et al. [22] showed that the seepage velocity magnitude distributions are symmetrical across the crest of NLIWs (Figure 9). Rivera-Rosario et al. [21] found that the seepage is asymmetrical about the wave trough since the bottom pressure of NLIWs translates along the seabed. Thus, the NLIWs leave a diffusive imprint of pore pressure during their passage, which is similar to that described under surface solitary waves [75].

4.2. Seepage Variations Due to Internal Wave Propagation

During internal wave passage, the seepage of pore fluid variations inside the porous seabed have been explored with numerical solutions of the DJL equation. More recent instances are the research of Olsthoorn et al. [22] and Rivera-Rosario et al. [21], who sought seabed solutions for seepage and potential failure, respectively. Olsthoorn et al. [22] investigated the seepage field due to wind-generated internal waves in the lake and found the geometrical field of seepage is powerfully governed by both the topography and the ratio of seabed thickness to the bottom pressure broad (Figure 10). The numerical results show that the depression NLIWs act essentially as a vacuum cleaner because of their intrinsic negative bottom pressure to extract the pore fluid from the seabed. For depression NLIWs, the pore fluid is drawn out under the wave trough where negative pressure is highest and invades in the seabed regions along the wave flanks where negative pressure vanishes. A similar phenomenon was previously found by Huettel et al. [82] in the laboratory context. They demonstrated that the traced particles suspended in the bottom boundary layer intruded quickly into the upper layer of sandy sediments, driven by interfacial flows related to topography. Simultaneously, the increased bottom pressure at the upstream and downstream of a small dune drove water access to the interior and the decreased bottom pressure downstream of the slope drew pore fluid from the interior to the water column.

In a numerical study of the generation of seepage in the porous seabed driven by trapped internal waves, Olsthoorn et al. [22] showed the evolution of the vorticity due to trapped internal waves inducing the currents, which indicates the expansion of a highly turbulent bottom boundary layer. Smaller particle matter will then be resuspended by the instability bottom boundary layer and enhanced benthic turbulence, breaking the seal of the surface pores at the seafloor. On the downstream of the trapped internal waves, the consequent permeable changes significantly enhance exchanges between the bottom water and the pore fluid. Rivera-Rosario et al. [21] used the vertical pore-pressure gradient to evaluate the vertical seepage within the seabed. The results demonstrated that the penetration depth of the vertical seepage was significantly sensitive to the seabed permeability and saturation. For an unsaturated seabed with higher permeability, the enhanced negative pore-pressure build-up was accompanied by reduced saturation and penetrated deeper, thus generating steeper vertical pore-pressure gradients, that is, strong vertical seepage within the seabed.

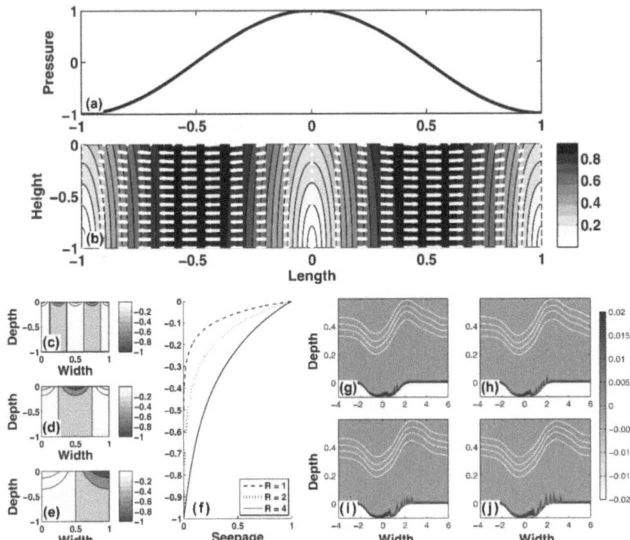

Figure 10. Seepage of pore fluid in lake sediment due to NLIWs propagating. (**a**) Normalized bottom pressure induced by NLIWs. (**b**) The contour of the normalized seepage. White arrows indicate the seepage direction. Normalized pore-pressure changes due to sinusoidal pressure forcing for wavelengths (**c**) R = 1, (**d**) R = 2, and (**e**) R = 4. Note the R gives the proportion of the horizontal scale, determined by the bottom pressure perturbation, to the depth of the seabed. (**f**) The seepage profiles for wavelengths R = 1, R = 2, and R = 4 under the wave trough. Evolution of the normalized vorticity in four nondimensional times of (**a**) 11.2, (**b**) 12.6, (**c**) 14, and (**d**) 15.4, caused by a development of a highly turbulent boundary layer of the trapped NLIWs. Four white lines expressed as the isopycnic fluctuation of NLIWs. Panels (**a**–**j**) adapted from Olsthoorn et al. [22], with permission from Wiley.

4.3. Possible Impact on Sediment Resuspension

As in a previous study, sediment resuspension by NLIWs has traditionally been parameterized according to the bottom shear stress or Shields parameter [2]. High bottom shear stress induced by NLIWs, which may trigger sediment movement sufficiently, resuspended sediments by bottom boundary layer instability [7,83–85]. Nevertheless, in some studies, the results indicated that resuspension events occurred in durations of low bottom shear stress [83,86,87], which indicated a vague relationship between sediment resuspension and bottom shear stress (Figure 11). It is confirmed by oceanic observations that only reaching a critical bottom shear stress does not sufficiently lead to sediment resuspension [83], and that enhanced vertical fluid motions that boost sediments off the seabed are required [88,89].

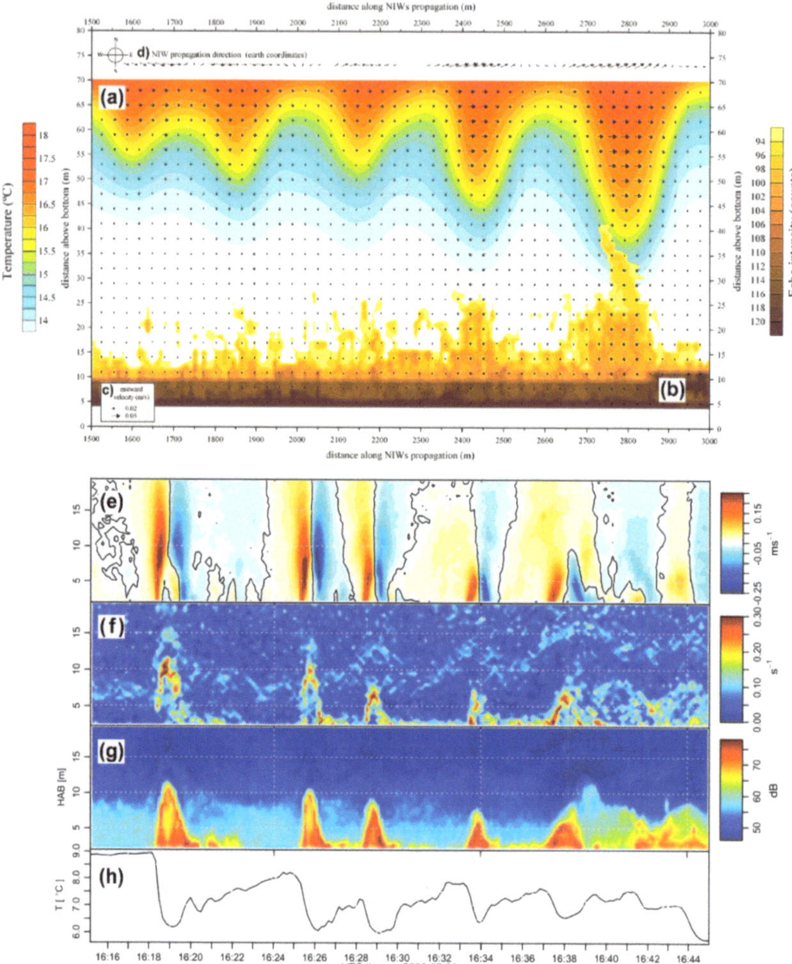

Figure 11. Sediment resuspension due to NLIWs propagating. Comparison between (**a**) temperature profile, (**b**) echo intensity, (**c**) eastward velocity, and (**d**) NLIW propagation direction during an NLIW train propagation. A 30 min segment of an NLIW shoaling event, comparison between (**e**) vertical velocity, (**f**) vertical shear of the horizontal current, (**g**) acoustic backscatter, and (**h**) bottom temperature. The black line in (**e**) shows zero. Panels (**a–d**) adapted from Quaresma et al. [84], with permission from Elsevier, and panels (**e–h**) adapted from Richards et al. [90], with permission from Wiley.

The known mechanisms leading to sediment resuspension by NLIWs discussed above fall into the problem of hydrodynamic stability. Aghsaee et al. [91] parameterized sediment resuspension to the Shields parameter θ_{ISW}, which is written as

$$\theta_{ISW} = \frac{\tau_{ISW}}{(\rho_s - \rho_0)gd},\tag{29}$$

where τ_{ISW} is the shear stress in the bottom, ρ_s is the sediment density, ρ_0 is the fluid density in the bottom, and d is the particle size. However, for the particle incipient motion, the vertical seepage in the seabed was an additional force to be reckoned with since it works to reduce the buoyant weight of the sediments, thus easily inducing suspension. Therefore, Chen and Hsu [23] revisited the issue using a linear internal wave model; the soil response was solved by the equation equilibrium and the storage equation. The numerical results show that the maximum sediment displacements on the seafloor are 38 m in the horizontal direction and 15 m in the vertical direction. Although it is widely known that the internal waves for large amplitude described by the linear wave model are inaccurate, the mechanism of sediment resuspension considered as the soil displacements induced by internal waves [23] is believed to hold. In addition, Rivera-Rosario et al. [21] considered the vertical seepage force and derived a modified Shields parameter to address the problem of particle incipient motion by NLIWs. For a seabed in a fixed reference frame $\xi - \eta$, the modified dimensionless Shields parameter θ_{ISW}^{+} is written as

$$\theta_{ISW}^{+} = \frac{\tau_{ISW}}{(\rho_s - \rho_0)gd - \left[-\partial\left(p/\gamma_f\right)/\partial\eta\right]d},$$

(30)

where p is the pore pressure in the bottom and γ_f is the specific weight of fluid. Equation (30) considered NLIW-induced vertical pore-pressure gradient as the vertical seepage force to the modified buoyant force of particles compared with Equation (29). Based on the observation results of Quaresma et al. [84] and the expression of shear stress τ_{ISW} derived by Aghsaee and Boegman [89], the contribution of vertical seepage which represents the magnitude of the reduction in the buoyant weight of the particles is estimated as 11% and 8% for a saturation of 0.97 and 0.99, respectively [21]. Accordingly, the NLIW train will generate an adverse pore-pressure gradient during its passage, performing like a vacuum cleaner to pump out any particles from the seafloor. Thus, this may be a reasonable explanation for the disputed results of powerful resuspension events with low bottom shear stress. Moreover, the NLIW-induced seepage could drive a significant contribution to powerful resuspension events, such as thick nepheloid layers and large subaqueous sand dune formations [24,92].

5. Conclusions and Future Outlook

In general, the seabed dynamic responses induced by NLIWs remain speculative and the guiding theories, models, and experimental data are insufficient. This is coupled with a lack of in situ observation data, especially those that measure both pressure at the seafloor and pore pressure inside the seabed. Consequently, the pressure disturbance and the pore-pressure build-up action as the NLIWs shoal and the interactions with the seabed in continental slope areas remain poorly understood. From large-scale NLIWs (wavelengths of ~1 km) to small-scale sediments (particle size of ~1 mm), the multiscale process is at fault for the difficulty in investigating the issues [2], which makes in situ observation work challenging. Furthermore, the large Reynolds numbers in the ocean cannot be reproduced in experiments, making a precise simulation experiment of these NLIW–seabed interaction processes a challenge at the laboratory scale [89]. All field-scale numerical simulations of seabed dynamic responses induced by NLIWs do not consider the influence of a permeable seabed on the NLIW evolution and the real-time interactions are not yet resolved in the current models. In actuality, the wavelengths of NLIWs are affected by the permeability of the seabed in addition to the water depth and the wave period [23]. Moreover, the waveform of an NLIW inversion weakens significantly as the seabed porosity increases [93]. Briefly, we lack the numerical method to simulate NLIWs and their interactions with the seabed in multiscale at the necessary field-relevant high resolution.

The vast majority of the NLIW shoaling events appear on gentle slopes [7,68], where the fission of the shoaling wave is the dominant process as the depression NLIW changes to an elevation type and further transforms into a small-amplitude internal wave train during shoaling [94]. However, process studies in field scales on ISW-induced seabed responses have mostly been attempted on flat bottoms [21,23]. In the published literature, the indication of the intricate dynamic response of pore pressure during the transition of NLIW polarity has received little attention. Eventually, the NLIWs arrive at the turning point, where the drawdown of the negative pore pressure and the subsequent accumulation of positive pore pressure have the potential to induce pore fluid seepage along the slope and further drive potential material exchange across sediment–water interfaces [25,95]. The effect of NLIW-induced seepage boost sediment resuspension is nonnegligible in the slope region [21], but models have still not looked into this effect. Filed observations show that the shoaling NLIWs can produce bottom nepheloid layers in horizontal length scales up to hundreds of meters [96] and further spread to the formation of intermediate nepheloid layers in horizontal length scales up to tens of kilometers [24]. In order to parameterize and quantify the potential failure and possible resuspension as a function of pore pressure and boundary slope (of which the latter governs the seabed response mechanism), seabed-related process-based research on continental slopes is necessary [35,67,71].

Our emphasis on the seabed response during the shoaling process of NLIWs is appropriate in view of the in situ observational evidence, but it is also limiting, particularly in deeper water where the pressure forcing of depression NLIWs is generally believed to vanish [43]. Considering the present numerical and experimental difficulties associated with the examination of seabed dynamic responses induced by NLIWs, there is a requirement to support in situ long-term observations on this issue. Over the past years, in situ seabed observation techniques have been developed, but technical challenges (e.g., deficiency of the sensor sensitivity and resolution) and the high cost of seafloor observations have largely limited the development of NLIW–seabed interaction studies. The limitations arise because the sensors have an inherent unpredictable drift, as well as the challenge of isolating NLIW signals from the other pressure/pore-pressure components. Short sharp signal shifts of pressure/pore pressure are easily captured, but small slow disturbances induced by NLIWs are more challenging. The precision and accuracy of sensors can likely be substantially improved in the future, for example, by using the correction factors that are usually applied when measuring surface waves with pressure sensors [97] to improvements in the continuous pressure-sensor monitoring of NLIW motions and improving the accuracy and resolution of the sensor measurements through advanced processing techniques, e.g., Kalman filtering or machine learning. We expect that the development of more advanced and lower-cost in situ seabed observation techniques will lead to more comprehensive monitoring of a wide range of NLIW–seabed interaction processes and the potential geohazards and environmental impacts that they trigger. Our understanding will be improved by these observational data, which will also help us better predict the response of the seabed dynamics to NLIW propagation on the continental slope.

Author Contributions: Conceptualization, T.C. and Y.J.; writing—original draft, T.C.; review and editing the article, Z.L., H.N., H.L. and H.S. All authors have read and agreed to the published version of the manuscript.

Funding: This work was supported by the Hainan Research Institute of China Engineering Science and Technology Development Strategy (No. 21-HN-ZD-02), the Open Research Fund of Key Laboratory of Coastal Science and Integrated Management, Ministry of Natural Resources (No. 2021COSIMQ007), the National Natural Science Foundation of China (No. 41831280), and the Fundamental Research Funds for the Central Universities (No. 202161039).

Institutional Review Board Statement: Not applicable.

Informed Consent Statement: Not applicable.

Data Availability Statement: No new data were created or analyzed in this study. Data sharing is not applicable to this article.

Acknowledgments: We are grateful to Zhihan Fan, Zhiwen Sun, and Liang Xue for their valuable comments and revisions to this review.

Conflicts of Interest: The authors declare no conflict of interest.

References

1.   Cai, S.; Xie, J.; He, J. An overview of internal solitary waves in the South China Sea. *Surv. Geophys.* **2012**, *33*, 927–943. [CrossRef]

2. Boegman, L.; Stastna, M. Sediment resuspension and transport by internal solitary waves. *Annu. Rev. Fluid Mech.* **2019**, *51*, 129–154. [CrossRef]

3. Helfrich, K.R.; Melville, W.K. Long nonlinear internal waves. *Annu. Rev. Fluid Mech.* **2006**, *38*, 395–425. [CrossRef]

4. Stastna, M.; Coutino, A.; Walter, R.K. The effect of strong shear on internal solitary-like waves. *Nonlin. Process. Geophys.* **2021**, *28*, 585–598. [CrossRef]

5. Sarkar, S.; Scotti, A. From topographic internal gravity waves to turbulence. *Annu. Rev. Fluid Mech.* **2017**, *49*, 195–220. [CrossRef]

6. Lamb, K.G. Internal wave breaking and dissipation mechanisms on the continental slope/shelf. *Annu. Rev. Fluid Mech.* **2014**, *46*, 231–254. [CrossRef]

7. Moum, J.N.; Klymak, J.M.; Nash, J.D.; Perlin, A.; Smyth, W.D. Energy Transport by Nonlinear Internal Waves. *J. Phys. Oceanogr.* **2007**, *37*, 1968–1988. [CrossRef]

8. Chen, L.; Zheng, Q.; Xiong, X.; Yuan, Y.; Xie, H.; Guo, Y.; Yu, L.; Yun, S. Dynamic and Statistical Features of Internal Solitary Waves on the Continental Slope in the Northern South China Sea Derived from Mooring Observations. *J. Geophys. Res. Oceans* **2019**, *124*, 4078–4097. [CrossRef]

9. Li, D.; Chou, W.C.; Shih, Y.Y.; Chen, G.Y.; Chang, Y.; Chow, C.H.; Lin, T.Y.; Hung, C.C. Elevated particulate organic carbon export flux induced by internal waves in the oligotrophic northern South China Sea. *Sci. Rep.* **2018**, *8*, 2042. [CrossRef]

10. Magalhaes, J.M.; da Silva, J.C.B.; Buijsman, M.C. Long lived second mode internal solitary waves in the Andaman Sea. *Sci. Rep.* **2020**, *10*, 10234. [CrossRef]

11. Thomas, J.A.; Lerczak, J.A.; Moum, J.N. Horizontal variability of high-frequency nonlinear internal waves in Massachusetts Bay detected by an array of seafloor pressure sensors. *J. Geophys. Res. Oceans* **2016**, *121*, 5587–5607. [CrossRef]

12. Stöber, U.; Moum, J.N. On the potential for automated realtime detection of nonlinear internal waves from seafloor pressure measurements. *Appl. Ocean Res.* **2011**, *33*, 275–285. [CrossRef]

13. Jackson, C. Internal wave detection using the Moderate Resolution Imaging Spectroradiometer (MODIS). *J. Geophys. Res.* **2007**, *112*, 13. [CrossRef]

14. Tian, Z.; Liu, Y.; Zhang, X.; Zhang, Y.; Zhang, M. Formation Mechanisms and Characteristics of the Marine Nepheloid Layer: A Review. *Water* **2022**, *14*, 678. [CrossRef]

15. Huang, X.; Chen, Z.; Zhao, W.; Zhang, Z.; Zhou, C.; Yang, Q.; Tian, J. An extreme internal solitary wave event observed in the northern South China Sea. *Sci. Rep.* **2016**, *6*, 30041. [CrossRef]

16. Tian, Z.; Jia, Y.; Du, Q.; Zhang, S.; Guo, X.; Tian, W.; Zhang, M.; Song, L. Shearing stress of shoaling internal solitary waves over the slope. *Ocean Eng.* **2021**, *241*, 110046. [CrossRef]

17. Deepwell, D.; Sapède, R.; Buchart, L.; Swaters, G.E.; Sutherland, B.R. Particle transport and resuspension by shoaling internal solitary waves. *Phys. Rev. Fluids* **2020**, *5*, 054303. [CrossRef]

18. Edge, W.C.; Jones, N.L.; Rayson, M.D.; Ivey, G.N. Observations of enhanced sediment transport by nonlinear internal waves. *Geophys. Res. Lett.* **2020**, *47*, e2020GL088499.

19. Ma, X.; Yan, J.; Hou, Y.; Lin, F.; Zheng, X. Footprints of obliquely incident internal solitary waves and internal tides near the shelf break in the northern South China Sea. *J. Geophys. Res. Oceans* **2016**, *121*, 8706–8719. [CrossRef]

20. Van Haren, H. High-frequency bottom-pressure and acoustic variations in a sea strait: Internal wave turbulence. *Ocean Dyn.* **2012**, *62*, 1123–1137. [CrossRef]

21. Rivera-Rosario, G.A.; Diamessis, P.J.; Jenkins, J.T. Bed failure induced by internal solitary waves. *J. Geophys. Res. Oceans* **2017**, *122*, 5468–5485. [CrossRef]

22. Olsthoorn, J.; Stastna, M.; Soontiens, N. Fluid circulation and seepage in lake sediment due to propagating and trapped internal waves. *Water Resour. Res.* **2012**, *48*, W11520. [CrossRef]

23. Chen, C.Y.; Hsu, J.R.C. Interaction between internal waves and a permeable seabed. *Ocean. Eng.* **2005**, *32*, 587–621. [CrossRef]

24. Reeder, D.B.; Ma, B.B.; Yang, Y.J. Very large subaqueous sand dunes on the upper continental slope in the South China Sea generated by episodic, shoaling deep-water internal solitary waves. *Mar. Geol.* **2011**, *279*, 12–18. [CrossRef]

25. Huettel, M.; Berg, P.; Kostka, J.E. Benthic Exchange and Biogeochemical Cycling in Permeable Sediments. *Annu. Rev. Mar. Sci.* **2014**, *6*, 23–51. [CrossRef] [PubMed]

26. Rao, A.M.F.; McCarthy, M.J.; Gardner, W.S.; Jahnke, R.A. Respiration and denitrification in permeable continental shelf deposits on the South Atlantic Bight: N-2: Ar and isotope pairing measurements in sediment column experiments. *Cont. Shelf Res.* **2008**, *28*, 602–613. [CrossRef]

27. Lien, R.-C.; Henyey, F.; Ma, B.; Yang, Y.J. Large-Amplitude Internal Solitary Waves Observed in the Northern South China Sea: Properties and Energetics. *J. Phys. Oceanogr.* **2014**, *44*, 1095–1115. [CrossRef]

28. Edge, W.C.; Jones, N.L.; Rayson, M.D.; Ivey, G.N. Calibrated suspended sediment observations beneath large amplitude non-linear internal waves. *J. Geophys. Res. Oceans* **2021**, *126*, e2021JC017538. [CrossRef]

29. Miramontes, E.; Jouet, G.; Thereau, E.; Bruno, M.; Penven, P.; Guerin, C.; Le Roy, P.; Droz, L.; Jorry, S.J.; Hernández-Molina, F.J.; et al. The impact of internal waves on upper continental slopes: Insights from the Mozambican margin (southwest Indian Ocean). *Earth Surf. Process. Landf.* **2020**, *45*, 1469–1482. [CrossRef]

30. Zhang, W.; Didenkulova, I.; Kurkina, O.; Cui, Y.; Haberkern, J.; Aepfler, R.; Santos, A.I.; Zhang, H.; Hanebuth, T.J.J. Internal solitary waves control offshore extension of mud depocenters on the NW Iberian shelf. *Mar. Geol.* **2019**, *409*, 15–30. [CrossRef]

31. Guo, X.; Liu, Z.; Zheng, J.; Luo, Q.; Liu, X. Bearing capacity factors of T-bar from surficial to stable penetration into deep-sea sediments. *Soil Dyn. Earthq. Eng.* **2023**, *165*, 107671. [CrossRef]
32. Yamamoto, T.; Koning, H.L.; Sellmeijer, H.; Hijum, E.V. On the response of a poro-elastic bed to water waves. *J. Fluid Mech.* **1978**, *87*, 193–206. [CrossRef]
33. Anderson, D.; Cox, D.; Mieras, R.; Puleo, J.A.; Hsu, T.-J. Observations of wave-induced pore pressure gradients and bed level response on a surf zone sandbar. *J. Geophys. Res. Oceans* **2017**, *122*, 5169–5193. [CrossRef]
34. Sumer, B.M. Liquefaction around Marine structures. In *Coastal Structures 2007, Proceedings of the 5th Coastal Structures International Conference, CST07, Venice, Italy, 2–4 July 2009*; World Scientific Pub. Co.: Singapore, 2009; pp. 1864–1870.
35. Tian, Z.; Chen, T.; Yu, L.; Guo, X.; Jia, Y. Penetration depth of the dynamic response of seabed induced by internal solitary waves. *Appl. Ocean Res.* **2019**, *90*, 101867. [CrossRef]
36. Guo, X.; Stoesser, T.; Zheng, D.; Luo, Q.; Liu, X.; Nian, T. A methodology to predict the run-out distance of submarine landslides. *Comput. Geotech.* **2023**, *153*, 105073. [CrossRef]
37. Guo, X.; Nian, T.; Fu, C.; Zheng, D. Numerical Investigation of the Landslide Cover Thickness Effect on the Drag Forces Acting on Submarine Pipelines. *J. Waterw. Port Coast. Ocean Eng.* **2023**, *149*, 04022032. [CrossRef]
38. Sultan, N.; Murphy, S.; Riboulot, V.; Géli, L. Creep-dilatancy development at a transform plate boundary. *Nat. Commun.* **2022**, *13*, 1913. [CrossRef]
39. Lai, J.; Richards, A.; Keller, G. In Place Measurement of Excess Pore Pressure in Gulf of Maine Clays. In *Transactions-American Geophysical Union*; American Geophysical Union: Washington, DC, USA, 1968; p. 221.
40. Richards, A.F.; Øten, K.; Keller, G.H.; Lai, J.Y. Differential piezometer probe for an in situ measurement of sea-floor. *Geotechnique* **1975**, *25*, 229–238. [CrossRef]
41. Lamb, K.G.; Warn-Varnas, A. Two-dimensional numerical simulations of shoaling internal solitary waves at the ASIAEX site in the South China Sea. *Nonlin. Process. Geophys.* **2015**, *22*, 289–312. [CrossRef]
42. Xie, J.; He, Y.; Cai, S. Bumpy Topographic Effects on the Transbasin Evolution of Large-Amplitude Internal Solitary Wave in the Northern South China Sea. *J. Geophys. Res. Oceans* **2019**, *124*, 4677–4695. [CrossRef]
43. Moum, J.N.; Nash, J.D. Seafloor Pressure Measurements of Nonlinear Internal Waves. *J. Phys. Oceanogr.* **2008**, *38*, 481–491. [CrossRef]
44. Liu, T.; Wei, G.; Kou, H.; Guo, L. Pore pressure observation: Pressure response of probe penetration and tides. *Acta Oceanol. Sin.* **2019**, *38*, 107–113. [CrossRef]
45. Higashino, M.; Clark, J.J.; Stefan, H.G. Pore water flow due to near-bed turbulence and associated solute transfer in a stream or lake sediment bed. *Water Resour. Res.* **2009**, *45*, W12414. [CrossRef]
46. Moum, J.N.; Smyth, W.D. The pressure disturbance of a nonlinear internal wave train. *J. Fluid Mech.* **2006**, *558*, 153–177. [CrossRef]
47. Williams, S.J.; Jeng, D.S. The effects of a porous-elastic seabed on interfacial wave propagation. *Ocean Eng.* **2007**, *34*, 1818–1831. [CrossRef]
48. Hsu, J.R.C.; Jeng, D.S.; Tsai, C.P. Short-crested wave-induced soil response in a porous seabed of infinite thickness. *Int. J. Numer. Anal. Methods Geomech.* **1993**, *17*, 553–576. [CrossRef]
49. Forgia, G.l.; Sciortino, G. The role of the free surface on interfacial solitary waves. *Phys. Fluids* **2019**, *31*, 106601. [CrossRef]
50. Kodaira, T.; Waseda, T.; Miyata, M.; Choi, W. Internal solitary waves in a two-fluid system with a free surface. *J. Fluid Mech.* **2016**, *804*, 201–223. [CrossRef]
51. Zhi, C.; Wang, H.; Chen, K.; You, Y. Theoretical and experimental investigation on strongly nonlinear internal solitary waves moving over slope-shelf topography. *Ocean Eng.* **2021**, *223*, 108645. [CrossRef]
52. Apel, J.R.; Ostrovsky, L.A.; Stepanyants, Y.A.; Lynch, J.F. Internal solitons in the ocean and their effect on underwater sound. *J. Acoust. Soc. Am.* **2007**, *121*, 695–722. [CrossRef]
53. Van Haren, H. Bottom-pressure observations of deep-sea internal hydrostatic and non-hydrostatic motions. *J. Fluid Mech.* **2013**, *714*, 591–611. [CrossRef]
54. Yang, Y.J.; Lien, R.-C.; Chang, M.-H.; Tang, T.Y. *Pressure Perturbations Induced by Mode-1 Depression Internal Solitary Waves*; European Geosciences Union General Assembly: Vienna, Austria, 2011.
55. Fukao, Y.; Miyama, T.; Tono, Y.; Sugioka, H.; Ito, A.; Shiobara, H.; Yamashita, M.; Varlamov, S.; Furue, R.; Miyazawa, Y. Detection of ocean internal tide source oscillations on the slope of Aogashima Island, Japan. *J. Geophys. Res. Oceans* **2019**, *124*, 4918–4933. [CrossRef]
56. Saito, T.; Tsushima, H. Synthesizing ocean bottom pressure records including seismic wave and tsunami contributions: Toward realistic tests of monitoring systems. *J. Geophys. Res. Solid Earth* **2016**, *121*, 8175–8195. [CrossRef]
57. Deng, H.; An, C.; Cai, C.; Ren, H. Theoretical Solution and Applications of Ocean Bottom Pressure Induced by Seismic Waves at High Frequencies. *Geophys. Res. Lett.* **2022**, *49*, e2021GL096952. [CrossRef]
58. Watts, D.R.; Wei, M.; Tracey, K.L.; Donohue, K.A.; He, B. Seafloor Geodetic Pressure Measurements to Detect Shallow Slow Slip Events: Methods to Remove Contributions from Ocean Water. *J. Geophys. Res. Solid Earth* **2021**, *126*, e2020JB020065. [CrossRef]
59. Van Haren, H. Internal wave–turbulence pressure above sloping sea bottoms. *J. Geophys. Res. Oceans* **2011**, *116*, C12004. [CrossRef]
60. Schultheiss, P.J. Pore pressures in marine sediments: An overview of measurement techniques and some geological and engineering applications. *Mar. Geophys. Res.* **1990**, *12*, 153–168. [CrossRef]
61. Jeng, D.S. *Porous Models for Wave-Seabed Interactions*; Springer: Berlin/Heidelberg, Germany, 2012.
62. Jeng, D.-S. *Mechanics of Wave-Seabed-Structure Interactions: Modelling, Processes and Applications*; Cambridge University Press: Cambridge, UK, 2018.
63. Biot, M.A. General Theory of Three-Dimensional Consolidation. *J. Appl. Phys.* **1941**, *12*, 155–164. [CrossRef]

64. Verruijt, A. Elastic Storage of Aquifers. In *Flow through Porous Media*; Wiest, R.J.M.D., Ed.; Academic Press: New York, NY, USA, 1969; pp. 331–376.

65. Williams, S.; Jeng, D. Viscous attenuation of interfacial waves over a porous seabed. *J. Coast. Res.* **2007**, *50*, 338–342.

66. Bear, J. *Dynamics of Fluids in Porous Media*; Courier Corporation: Chelmsford, MA, USA, 2013.

67. Qiao, L.; Guo, X.; Tian, Z.; Yu, L. Experimental analysis of pore pressure characteristics of slope sediments by shoaling internal solitary waves. *Acta Oceanol. Sin.* **2018**, *40*, 68–76.

68. Gong, Y.; Song, H.; Zhao, Z.; Guan, Y.; Zhang, K.; Kuang, Y.; Fan, W. Enhanced diapycnal mixing with polarity-reversing internal solitary waves revealed by seismic reflection data. *Nonlin. Process. Geophys.* **2021**, *28*, 445–465. [CrossRef]

69. Shroyer, E.L.; Moum, J.N.; Nash, J.D. Observations of Polarity Reversal in Shoaling Nonlinear Internal Waves. *J. Phys. Oceanogr.* **2009**, *39*, 691–701. [CrossRef]

70. Orr, M.H. Nonlinear internal waves in the South China Sea: Observation of the conversion of depression internal waves to elevation internal waves. *J. Geophys. Res.* **2003**, *108*, 16. [CrossRef]

71. Li, Y.; Liu, L.; Gao, S.; Zhang, Y.; Xiong, X. Experimental study on dynamic response characteristics of continental shelf slope to internal solitary waves. *Acta Oceanol. Sin.* **2021**, *43*, 126–134.

72. Lynch, J.F.; Ramp, S.R.; Ching-Sang, C.; Tswen Yung, T.; Yang, Y.J.; Simmen, J.A. Research highlights from the Asian Seas International Acoustics Experiment in the South China Sea. *IEEE J. Ocean. Eng.* **2004**, *29*, 1067–1074. [CrossRef]

73. Klymak, J.M.; Moum, J.N. Internal solitary waves of elevation advancing on a shoaling shelf. *Geophys. Res. Lett.* **2003**, *30*, 2045. [CrossRef]

74. Madsen, O.S. Stability of a Sand Bed under Breaking Waves. In Proceedings of the 14th International Conference on Coastal Engineering, Copenhagen, Denmark, 24–28 June 1974; pp. 776–794.

75. Liu, P.L.F.; Park, Y.S.; Lara, J.L. Long-wave-induced flows in an unsaturated permeable seabed. *J. Fluid Mech.* **2007**, *586*, 323–345. [CrossRef]

76. Torum, A. Wave-induced pore pressures—Air/gas content. *J. Waterw. Port Coast. Ocean Eng. ASCE* **2007**, *133*, 83–86. [CrossRef]

77. Cross, R.H.; Baker, V.A.; Treadwell, D.D.; Huntsman, S. Attenuation of Wave-Induced Pore Pressures in Sand. In *Civil Engineering in the Oceans IV*; ASCE: Reston, VA, USA, 1979.

78. Bennett, R.H.; Faris, J.R. Ambient and dynamic pore pressures in fine-grained submarine sediments: Mississippi Delta. *Appl. Ocean Res.* **1979**, *1*, 115–123. [CrossRef]

79. Jin, G.; Tang, H.; Gibbes, B.; Li, L.; Barry, D.A. Transport of nonsorbing solutes in a streambed with periodic bedforms. *Adv. Water Resour.* **2010**, *33*, 1402–1416. [CrossRef]

80. Elliott, A.H.; Brooks, N.H. Transfer of nonsorbing solutes to a streambed with bed forms: Theory. *Water Resour. Res.* **1997**, *33*, 123–136. [CrossRef]

81. Bear, J. *Modeling Phenomena of Flow and Transport in Porous Media*; Springer: Cham, Switzerland, 2018.

82. Huettel, M.; Ziebis, W.; Forster, S. Flow-induced uptake of particulate matter in permeable sediments. *Limnol. Oceanogr.* **1996**, *41*, 309–322. [CrossRef]

83. Boegman, L.; Ivey, G.N. Flow separation and resuspension beneath shoaling nonlinear internal waves. *J. Geophys. Res. Oceans* **2009**, *114*, C02018. [CrossRef]

84. Quaresma, L.S.; Vitorino, J.; Oliveira, A.; da Silva, J. Evidence of sediment resuspension by nonlinear internal waves on the western Portuguese mid-shelf. *Mar. Geol.* **2007**, *246*, 123–143. [CrossRef]

85. Hosegood, P.; van Haren, H. Near-bed solibores over the continental slope in the Faeroe-Shetland Channel. *Deep Sea Res. Part II Top. Stud. Oceanogr.* **2004**, *51*, 2943–2971. [CrossRef]

86. Bogucki, D.; Dickey, T.; Redekopp, L.G. Sediment Resuspension and Mixing by Resonantly Generated Internal Solitary Waves. *J. Phys. Oceanogr.* **1997**, *27*, 1181–1196. [CrossRef]

87. Johnson, D.R.; Weidemann, A.; Pegau, W.S. Internal tidal bores and bottom nepheloid layers. *Cont. Shelf Res.* **2001**, *21*, 1473–1484. [CrossRef]

88. Bluteau, C.E.; Smith, S.L.; Ivey, G.N.; Schlosser, T.L.; Jones, N.L. Assessing the relationship between bed shear stress estimates and observations of sediment resuspension in the ocean. In Proceedings of the 20th Australasian Fluid Mechanics Conference, Perth, Australia, 5–8 December 2016.

89. Aghsaee, P.; Boegman, L. Experimental investigation of sediment resuspension beneath internal solitary waves of depression. *J. Geophys. Res. Oceans* **2015**, *120*, 3301–3314. [CrossRef]

90. Richards, C.; Bourgault, D.; Galbraith, P.S.; Hay, A.; Kelley, D.E. Measurements of shoaling internal waves and turbulence in an estuary. *J. Geophys. Res. Oceans* **2013**, *118*, 273–286. [CrossRef]

91. Aghsaee, P.; Boegman, L.; Diamessis, P.J.; Lamb, K.G. Boundary-layer-separation-driven vortex shedding beneath internal solitary waves of depression. *J. Fluid Mech.* **2012**, *690*, 321–344. [CrossRef]

92. Droghei, R.; Falcini, F.; Casalbore, D.; Martorelli, E.; Mosetti, R.; Sannino, G.; Santoleri, R.; Chiocci, F.L. The role of Internal Solitary Waves on deep-water sedimentary processes: The case of up-slope migrating sediment waves off the Messina Strait. *Sci. Rep.* **2016**, *6*, 36376. [CrossRef] [PubMed]

93. Cheng, M.-H.; Hsieh, C.-M.; Hsu, J.R.C.; Hwang, R.R. Effect of porosity on an internal solitary wave propagating over a porous trapezoidal obstacle. *Ocean Eng.* **2017**, *130*, 126–141. [CrossRef]

94. Lamb, K.G.; Xiao, W. Internal solitary waves shoaling onto a shelf: Comparisons of weakly-nonlinear and fully nonlinear models for hyperbolic-tangent stratifications. *Ocean Model.* **2014**, *78*, 17–34. [CrossRef]
95. Santos, I.R.; Eyre, B.D.; Huettel, M. The driving forces of porewater and groundwater flow in permeable coastal sediments: A review. *Estuar. Coast. Shelf Sci.* **2012**, *98*, 1–15. [CrossRef]
96. Bourgault, D.; Morsilli, M.; Richards, C.; Neumeier, U.; Kelley, D.E. Sediment resuspension and nepheloid layers induced by long internal solitary waves shoaling orthogonally on uniform slopes. *Cont. Shelf Res.* **2014**, *72*, 21–33. [CrossRef]
97. Marino, M.; Rabionet, I.C.; Musumeci, R.E. Measuring free surface elevation of shoaling waves with pressure transducers. *Cont. Shelf Res.* **2022**, *245*, 104803. [CrossRef]

MDPI AG
Grosspeteranlage 5
4052 Basel
Switzerland
Tel.: +41 61 683 77 34

Journal of Marine Science and Engineering Editorial Office
E-mail: jmse@mdpi.com
www.mdpi.com/journal/jmse

9 783725 825493